Computer Modeling of Gas Lasers

OPTICAL PHYSICS AND ENGINEERING

Series Editor: **William L. Wolfe**
Optical Sciences Center, University of Arizona, Tucson, Arizona

Recent Volumes in this Series

Lucien M. Biberman and Sol Nudelman, Editors
Photoelectronic Imaging Devices
Volume 1: Physical Processes and Methods of Analysis
Volume 2: Devices and Their Evaluation

A. M. Ratner
Spectral, Spatial, and Temporal Properties of Lasers

Lucien M. Biberman, Editor
Perception of Displayed Information

W. B. Allan
Fibre Optics: Theory and Practice

Albert Rose
Vision: Human and Electronic

J. M. Lloyd
Thermal Imaging Systems

Winston E. Kock
Engineering Applications of Lasers and Holography

Shashanka S. Mitra and Bernard Bendow, Editors
Optical Properties of Highly Transparent Solids

M. S. Sodha and A. K. Ghatak
Inhomogeneous Optical Waveguides

A. K. Ghatak and K. Thyagarajan
Contemporary Optics

Kenneth Smith and R. M. Thomson
Computer Modeling of Gas Lasers

Vincent J. Corcoran
Introduction to Lasers

A Continuation Order Plan is available for this series. A continuation order will bring delivery of each new volume immediately upon publication. Volumes are billed only upon actual shipment. For further information please contact the publisher.

Computer Modeling of Gas Lasers

Kenneth Smith
and R.M. Thomson
University of Leeds
Leeds, England

Springer Science+Business Media, LLC

Library of Congress Cataloging in Publication Data

Smith, Kenneth, 1932-
Computer modeling of gas lasers.

(Optical physics and engineering)
Includes bibliographical references and index.
1. Gas lasers–Mathematical models. 2. Gas lasers–Data processing. I. Thomson, R. M., 1948- joint author. II. Title.
TA1695.S63 621.36'63'02854 77-29106
ISBN 978-1-4757-0643-7 ISBN 978-1-4757-0641-3 (eBook)
DOI 10.1007/978-1-4757-0641-3

Originally published by Plenum Press, New York in 1978

To Susan

Preface

Lasers, although invented only about eighteen years ago, are now viable devices in commercial environments. Since the laser produces light that is different in both quality and intensity from the light generated by any other source, future applications are unpredictable. Even in the few years since their invention, lasers have given a new impetus to holography (photography by wavefront reconstruction) and their potential as the carrier wave for a long-distance communication system has been recognized for some time. In manufacturing industries, lasers are mainly used for cutting and welding, while in medicine and biology the use of lasers range from eye surgery to holographic microscopy and pattern recognition to the sealing of dental enamel. Furthermore, it is now generally known that considerable investments are being made to establish the technical feasibility of laser techniques for uranium enrichment and plutonium isotope recovery as well as to establish a base of expertise for the application of laser techniques to radioactive waste processing.

Clearly, lasers are now a pivot in a major renewal of our industrial stock as numerous small companies are brought into being to supply the devices for existing and novel applications of lasers. In all this vigorous activity, the CO_2 laser occupies a position of considerable importance, and for this reason has been chosen as the standard exemplum for this book on gas lasers.

Whether or not a particular material will absorb the beam energy depends on the wavelength of the light. Most nonmetallic materials can be cut readily using the CO_2 laser, which emits infrared light ($\lambda = 10.6\ \mu m$). Using a CO_2 laser, fabric-cutting systems have been constructed which can cut garments at speeds up to several hundred feet per minute. The use of the CO_2 for laser surgery is at the beginning of serious development and has a number of potential advantages over present techniques which center around its ability to cauterize, while it cuts with a minimum amount of damage to adjacent tissue. Indeed a CO_2 laser with approximately 100 W of power, which may be delivered through a flexible optical system onto a spot diameter of approximately 1 mm, seems to be applicable to many surgical situations. The control of the power and energy distribution in time and space with the CO_2 laser seems to make it the clearly preferred choice over ruby and neodymium. One type of laser of increasing importance in communications and high-resolution spectroscopy is the waveguide CO_2 laser, a compact, high-gain, tunable CO_2 laser. Perhaps the most spectacular application of the CO_2 laser is the 100-kJ system now being designed (at a cost of $2M) at the Los Alamos Scientific Laboratory to be used in experiments connected with laser fusion.

The computer modeling of lasing systems has helped in their development to date and is expected to play an increasingly important role in the future, as the models become more realistic, as computer costs decrease, and as the number of design parameters increase, making a laboratory search for an optimal design almost impossible.

The objective of this book is to provide an extensive and in-depth treatment of gas kinetics, excitation processes, hydrodynamics, and other important aspects of gas lasers with specialization to carbon dioxide. Almost every area of modeling we consider ends with a computer program. To help the reader understand completely the processes under discussion we have presented a standardized description of each code. The FORTRAN is available from Computer Physics Communications. We hope that these features will encourage the reader to carry out his own investigations, building on the working models we present here. We do emphasize that most of the theory we discuss is readily applicable to other gas systems. For example, the whole of Chapter 3 on the Boltzmann transport equation can be taken over completely to dissociative ground state lasers, which has already been done by Rockwood and by Fournier. Chapters 5 and 7 are manifestly independent of the standard exemplum—they form the basis for understanding the importance of hydrodynamical considerations in gas-laser physics, while the theory and accompanying computer code in Chapter 6 have been designed to be as primary-species independent as possible.

It is hoped that this book will provide the reader with a passport to the current literature for both builders of lasers and graduate students. A particular purpose of the book is to describe algorithms rather than to present numbers. Once a computational physicist has his algorithm expressed as a subroutine, he can generate all the numbers he wants.

The book can be read by students who have had an introductory course in wave mechanics. It is meant to be self-contained, and most steps in the theoretical development of the formulas are given. It is particularly suitable for a two-semester course in the theory of gas lasers with Chapters 1–4 providing a homogenous course centered around kinetics, and Chapter 5–8 being based on plasmas. In Chapter 9 much of the theoretical apparatus assembled in the earlier chapters is applied to the gas dynamic laser and the waveguide laser, both of considerable current interest at several laboratories, and we conclude the book with a chapter on injection locking, which is enjoying somewhat of a revival.

The authors are most grateful to their co-worker Alan R. Davies for collaboration and considerable assistance with Chapters 2 and 3, and they owe an equal debt to their colleague Stuart A. Roberts for his lead in the work described in Chapters 4 and 6. The authors wish to acknowledge the benefit of many discussions with D. C. Tyte, Hugh Lamberton, David Hunt, Denis Hall, and Derek Day. They are also indebted to Dr. A. T. Jones for discussions on GDL lasers and to Nelson McAvoy and Gerhardt Koepf for discussions on FIR lasers. This work has been carried out with the support of the Procurement Executive, Ministry of Defense, sponsored by DCVD.

Kenneth Smith
R. M. Thomson

Contents

1

Relaxation Phenomena in Gases[(1)]

A theoretical model of a gas laser consists of a set of equations for the number densities, or energy densities, of the relevant excited states of the constituent gases of the laser, together with equations for the gas temperature and the intensity of the radiation within the laser cavity. The process by which energy is transferred between excited molecular states is called relaxation. In this chapter we shall study relaxation phenomena as a prerequisite to the formulation of laser models discussed in Chapter 2.

1.1. Definition of Relaxation Time

Consider the free fall of a sphere in a viscous liquid. The dependent variable is the velocity, which from an initial value changes monotonically to the terminal velocity. This phenomenon is known as *relaxation* and can be described by the linear differential equation

$$-dv/dt = (v - v_0)/\tau \tag{1.1}$$

where τ is a measure of the viscosity, and v_0 is the terminal velocity (an equilibrium value of v) assumed to be t-independent.

To solve this equation, let

$$y = v - v_0$$

Therefore

$$dy = dv$$

and Eq. (1.1) becomes

$$dy/y = -dt/\tau$$

Hence

$$\log y = -t/\tau + \text{const} \quad (-\log A, \text{say})$$

or

$$y = A\, e^{-t/\tau}$$

so

$$v = v_0 + A\, e^{-t/\tau} \tag{1.2}$$

where A is the arbitrary constant of integration of Eq. (1.1). To determine A we must specify a set of boundary conditions; for example, if the sphere falls from rest at $t = 0$, then Eq. (1.2) gives

$$0 = v_0 + A$$

giving

$$v = v_0(1 - e^{-t/\tau}) \tag{1.3}$$

From Eq. (1.3) we see that once v reaches its value of v_0, it does not overshoot it. The fractional change in the velocity, with respect to the equilibrium value, is

$$\frac{v_0 - v(t)}{v_0} = e^{-t/\tau} \tag{1.4}$$

The time taken for a fractional change of e^{-1}, namely, τ, is called the *relaxation time*.

1.2. Energy Exchange between Internal and External Degrees of Freedom

In Fig. 1 we consider an ideal gas compressed slowly by a piston moving inward with a speed ω. A molecule moving toward the piston with an absolute velocity component W_z normal to the piston is reflected from the moving piston head with a velocity of $W_z + \omega$ relative to the piston head. Consequently, in the laboratory frame of reference, the absolute

Fig. 1. Ideal gas compressed by a piston with (a) absolute initial velocities, and (b) molecular velocities relative to a stationary piston.

reflected velocity is $(W_z + \omega) + \omega$ and the increase in the kinetic energy will be

$$\tfrac{1}{2}m(W_z + 2\omega)^2 - \tfrac{1}{2}mW_z^2 \approx 2m\omega W_z \tag{1.5}$$

This increase in velocity of the z-component will be distributed among the other two components as well as increasing the velocity of other molecules by collision. In other words, the translational energy of the gas is increased, with an associated increase in translational temperature.

If the molecules have internal degrees of freedom, with an average internal energy of E', then this internal energy was in equilibrium with the original value of the translational energy, parameterized by the temperature T. Let $E'(T)$ be the equilibrium value of the internal energy E' when the external degrees of freedom are at T. The value of E' in equilibrium with the new, increased translational energy is higher than $E'(T)$. Therefore, energy must flow from the translational to the internal degrees of freedom. This can only occur through collisions and may take time. The time needed to decrease the maladjustment by a factor e^{-1} is called the relaxation time τ, defined by the relaxation equation

$$-dE'/dt = (1/\tau)[E' - E'(T)] \tag{1.6}$$

This relaxation equation has a single relaxation time for the whole energy E', which has one degree of freedom. Consider molecules with different energy states, with reaction rate equations governing the transition of molecules from one energy state to another. For each state there will be a rate equation—Under what conditions can these rate equations be combined to give a single equation for the internal energy as a whole?

1.3. Distorted Wave Theory of Molecular Collisions[2]

The wave equation for a system of two colliding CO_2 molecules, assuming either that each molecule has one vibrational mode, or that one molecule has two vibrational modes and the other is a structureless projectile, is

$$H\Psi = E\Psi \tag{1.7}$$

where

$$H = -\frac{\hbar^2}{2\mu}\nabla_r^2 + V(r, s_1, s_2) - \frac{\hbar^2}{2\mu_1}\nabla_{s_1}^2 + 2\pi^2\mu_1^2 s_1^2 - \frac{\hbar^2}{2\mu_2}\nabla_{s_2}^2 + 2\pi^2\mu_2^2 s_2^2 \tag{1.8}$$

where μ_1 and μ_2 are the masses of the colliding molecules and μ is the reduced mass of the collision system, given in terms of μ_1 and μ_2 by

$$\mu = \frac{\mu_1\mu_2}{\mu_1+\mu_2}$$

$V(r, s_1, s_2)$ is the potential field energy between the colliding molecules and is a function of r, the distance between the centers of mass of the two molecules, and s_1 and s_2, the normal coordinates of the vibration of the molecules. Each vibrational mode consists of a set of states (levels) and if the λ vibrational state of mode 1 and the κ vibrational state of mode 2 are excited, then the total energy of the two molecules is given by

$$E_{k\lambda\kappa} = -\frac{\hbar^2k^2}{2\mu}+\varepsilon_1^\lambda+\varepsilon_2^\kappa \tag{1.9}$$

If the vibrational states of each mode are those of a simple harmonic oscillator (SHO), then

$$\varepsilon_1^\lambda = (\lambda+\tfrac{1}{2})\varepsilon_1, \qquad \varepsilon_2^\kappa = (\kappa+\tfrac{1}{2})\varepsilon_2$$

where $\varepsilon_1 \equiv \varepsilon_1^0$ and $\varepsilon_2 \equiv \varepsilon_2^0$ are the energies of the first excited state of each mode. The overall wave function for the system can be expanded in a product orthonormal basis:

$$\Psi = \sum_{\lambda\kappa} \psi_k^{\lambda\kappa}(r)\psi_\lambda(s_1)\,\psi_\kappa(s_2) \tag{1.10}$$

and we can assume the interaction potential is separable, that is,

$$\begin{aligned} V(r_1, s_1, s_2) &= V_0V(r)V_1(s_1)V_2(s_2) \\ &= V_0\exp[-\alpha(r+A_1s_1+A_2s_2)] \end{aligned} \tag{1.11}$$

The asymptotic form of Ψ is a linear superposition of an ingoing plane wave, and outgoing elastically and inelastically scattered spherical waves:

$$\begin{aligned} \Psi \underset{r\to\infty}{\sim} & \left[\exp(ik_0z)+\frac{1}{r}g_0(\theta)\exp(ik_0r)\right]\psi_{\lambda_0}(s_1)\psi_{\kappa_0}(s_2) \\ & +\sum_{\lambda\kappa}\frac{1}{r}g_k^{\lambda\kappa}(\theta)\,e^{ikr}\psi_\lambda(s_1)\psi_\kappa(s_2) \end{aligned} \tag{1.12}$$

To simplify the analysis we consider s-wave ($l=0$) scattering only; in other words, we omit angular dependence and Eq. (1.7) becomes

$$\sum_{\lambda\kappa}\psi_\lambda(s_1)\psi_\kappa(s_2)\left(\frac{\hbar^2}{2\mu}\nabla_r^2+\frac{\hbar^2k^2}{2\mu}\right)\psi_k^{\lambda\kappa} = \sum_{\lambda\kappa} V\psi_\lambda(s_1)\psi_\kappa(s_2)\psi_k^{\lambda\kappa} \tag{1.13}$$

When we write

$$\psi_k^{\lambda\kappa} = r^{-1}f_k^{\lambda\kappa}$$

Eq. (1.13) becomes

$$\sum_{\lambda\kappa} \psi_\lambda(s_1)\psi_\kappa(s_2)\left(\frac{d^2}{dr^2}+k^2\right)f_k^{\lambda\kappa} = \sum_{\lambda\kappa} (2\mu/\hbar^2)V\psi_\lambda(s_1)\psi_\kappa(s_2)f_k^{\lambda\kappa} \quad (1.14)$$

Multiply both sides by $\psi_{\lambda'}^*(s_1)\psi_{\kappa'}^*(s_2)$ and integrate over s_1 and s_2 to obtain the system of coupled second-order ordinary differential equations:

$$\left(\frac{d^2}{dr^2}+k^2\right)f_k^{\lambda\kappa}(r) = \sum_{\lambda'\kappa'} V_0\frac{2\mu}{\hbar^2}e^{-\alpha r}f_k^{\lambda'\kappa'}(r)B_{\lambda\lambda'}C_{\kappa\kappa'} \quad (1.15)$$

where

$$B_{\lambda\lambda'} \equiv \int \psi_\lambda^*(s_1)\exp(-\alpha A_1 s_1)\psi_{\lambda'}(s_1)\,ds_1$$
$$C_{\kappa\kappa'} \equiv \int \psi_\kappa^*(s_2)\exp(-\alpha A_2 s_2)\psi_{\kappa'}(s_2)\,ds_2 \quad (1.16)$$

In the distorted wave approximation it is assumed that f_k is small compared to f_{ko}, and $B_{\kappa_0\kappa}$ is small compared to $B_{\kappa\kappa} \approx 1$, when we take only the first term in the expansion of the exponential term in Eq. (1.16). Hence, the first equation for Eq. (1.15) becomes

$$\left(\frac{d^2}{dr^2}+k_0^2-\frac{2\mu}{\hbar^2}V_0e^{-\alpha r}\right)f_{k_0}^{\lambda_0\kappa_0}(r)=0 \quad (1.17)$$

and each inelastic channel can be written as

$$\left(\frac{d^2}{dr^2}+k^2-\frac{2\mu}{\hbar^2}V_0e^{-\alpha r}\right)f_k^{\lambda\kappa}(r)=\frac{2\mu}{\hbar^2}V_0e^{-\alpha r}f_{k_0}^{\lambda_0\kappa_0}(r)B_{\lambda_0\lambda}C_{\kappa_0\kappa} \quad (1.18)$$

To solve Eqs. (1.17) and (1.18) consider the auxiliary function F_k satisfying

$$\left(\frac{d^2}{dr^2}+k^2-\frac{2\mu}{\hbar^2}V_0e^{-\alpha r}\right)F_k^{\lambda\kappa}(r)=0 \quad (1.19)$$

with boundary condition

$$F_k^{\lambda\kappa}(r=0)=0$$

and asymptotic value

$$F_k^{\lambda\kappa}(r) \underset{r\to\infty}{\sim} \sin(kr+\delta_k) \quad (1.20)$$

When this is compared with the asymptotic value of $f_{k_0}^{\lambda_0\kappa_0}(r)$ obtained from Eq. (1.12), namely,[3]

$$f_{k_0}(r) \sim \exp(i\delta_{k_0})\sin(k_0r+\delta_{k_0})k_0^{-1}$$

we have

$$f_{k_0}^{\lambda_0\kappa_0}(r) = k_0^{-1}\exp(i\delta_{k_0})F_{k_0}^{\lambda_0\kappa_0}(r) \tag{1.21}$$

By analogy with this result, we substitute

$$f_k^{\lambda\kappa}(r) = Y(r)F_k^{\lambda\kappa}(r)$$

into Eq. (1.18) and obtain

$$\frac{2dF_k^{\lambda\kappa}}{dr}\frac{dY}{dr} + F_k^{\lambda\kappa}\frac{d^2Y}{dr^2} = \frac{2\mu}{\hbar^2}e^{-\alpha r}\frac{1}{k_0}\exp(i\delta_{k_0})F_{k_0}^{\lambda_0\kappa_0}V_0B_{\lambda_0\lambda}C_{\kappa_0\kappa} \tag{1.22}$$

Problem 1. Multiply both sides of Eq. (1.22) by $F_k^{\lambda\kappa}$ and integrate with respect to r between the limits 0 and r to obtain

$$F_k^{\lambda\kappa}(r)^2\frac{dY}{dr} = \frac{2\mu}{\hbar^2}\frac{1}{k_0}\exp(i\delta_{k_0})V_0B_{\lambda_0\lambda}C_{\kappa_0\kappa}\int_0^\infty F_{k_0}^{\lambda_0\kappa_0}(r')F_k^{\lambda\kappa}(r')\exp(-\alpha r')\,dr' \tag{1.23}$$

Introduce the following into Eq. (1.23):

$$R_k \equiv \frac{2\mu}{\hbar^2}\int_0^\infty F_{k_0}^{\lambda_0\kappa_0}(r')F_k^{\lambda\kappa}(r')\exp(-\alpha r')\,dr' \tag{1.24}$$

Then Eq. (1.23) becomes

$$dY(r) = k_0^{-1}\exp(i\delta_{k_0})V_0B_{\lambda_0\lambda}C_{\kappa_0\kappa}R_k\,dr\,F_k^{\lambda\kappa}(r)^{-2}$$

When the asymptotic form of $F(r)$ is substituted into the right-hand side and the equation is integrated, we obtain

$$\begin{aligned} Y(r) &\sim k_0^{-1}\exp(i\delta_{k_0})V_0B_{\lambda_0\lambda}C_{\kappa_0\kappa}R_k\int dr\;\sin^{-2}(kr+\delta_k) \\ &\sim k_0^{-1}\exp(i\delta_{k_0})V_0B_{\lambda_0\lambda}C_{\kappa_0\kappa}R_k[-\cot(kr+\delta_k)+\beta]k^{-1} \end{aligned}$$

where β is the integration constant. When this result is substituted into the defining equation we obtain

$$f_k^{\lambda\kappa}(r) \sim (kk_0)^{-1}[-\cos(kr+\delta_k)+\beta\sin(kr+\delta_k)]\exp(i\delta_{k_0})V_0B_{\lambda_0\lambda}C_{\kappa_0\kappa}R_k \tag{1.25}$$

For $f_k^{\lambda\kappa}(r)$ to have the desired asymptotic form [see Eq. (1.12)], we must have

$$\beta = -i$$

and so

$$f_k^{\lambda\kappa}(r) \sim -(kk_0)^{-1}\exp[i(kr+\delta_k+\delta_{k_0})]V_0B_{\lambda_0\lambda}C_{\kappa_0\kappa}R_k \tag{1.26}$$

which gives the s-wave ($l = 0$) scattering amplitude:

$$g_k^{\lambda\kappa} = -(kk_0)^{-1} \exp[i(\delta_k + \delta_{k_0})] V_0 B_{\lambda_0\lambda} C_{\kappa_0\kappa} R_k \tag{1.27}$$

The total inelastic cross section is given by

$$\begin{aligned} \sigma_{\lambda_0\kappa_0,\lambda\kappa} &= 4\pi k k_0^{-1} |g_k^{\lambda\kappa}|^2 \\ &= 4\pi (k_0^3 k)^{-1} V_0^2 |R_k B_{\lambda_0\lambda} C_{\kappa_0\kappa}|^2 \end{aligned} \tag{1.28}$$

To evaluate R_k, defined by Eq. (1.24), where F satisfies Eq. (1.19), consider the change of the independent variable:

$$z = \frac{2}{\alpha}\left(\frac{2\mu}{\hbar^2} V_0\right)^{1/2} e^{-\alpha r/2} \tag{1.29}$$

Then Eq. (1.19) becomes

$$\frac{d^2F}{dz^2} + \frac{1}{z}\frac{dF}{dz} + \left(\frac{q^2}{z^2} - 1\right)F = 0 \tag{1.30}$$

where

$$q \equiv 2k/\alpha \tag{1.31}$$

The function F is Bessel's function of order iq and argument iz. The solution of Eq. (1.30) is a modified Bessel function of the second kind:

$$K_{iq}(z) = \int_0^\infty e^{-z \cosh u} \cos qu \, du$$

that is,

$$F_k = \left(\frac{q \sinh \pi q}{\pi}\right)^{1/2} K_{iq}(z)$$

Hence, we have

$$R_k = \frac{\alpha}{2\pi}(q_0 q \sinh \pi q_0 \sinh \pi q)^{1/2} \int_0^\infty K_{iq_0}(z) K_{iq}(z) z \, dz$$

Now

$$\begin{aligned} \int_0^\infty K_{iq_0}(z) K_{iq}(z) z \, dz &= \iiint_0^\infty z \, e^{-z(\cosh t + \cosh u)} \cos q_0 u \cos qt \, du \, dt \, dz \\ &= \frac{1}{2}\int_0^\infty \frac{\cos(q+q_0)T \, dT}{\cosh^2 T} \int_0^\infty \frac{\cos(q-q_0)U \, dU}{\cosh^2 U} \end{aligned}$$

where $t+u=2T$ and $t-u=2U$, and

$$\int_0^\infty \frac{\cos px\, dx}{\cosh^2 x} = \frac{\pi p}{2\sinh(\pi p/2)}$$

Therefore

$$R_k = \frac{\pi\alpha}{8}(q^2-q_0^2)\frac{(q_0 q \sinh \pi q_0 \sinh \pi q)^{1/2}}{\cosh \pi q - \cos \pi q_0} \tag{1.32}$$

1.4. Rate Coefficients

The probability that a molecule has a speed between v and $v+dv$, regardless of direction, can be given by the Maxwellian distribution,[4]

$$\chi(v)\, dv = 4\pi\left(\frac{\mu}{2\pi kT}\right)^{3/2} v^2 \exp(-\mu v^2/kT)\, dv \tag{1.33}$$

where μ is the mass of the molecule. Let us define a collision as any physical change in either of two particles, or in their momenta, as a result of their proximity. By analogy with hard-sphere collisions, we define a fictitious effective area for a target system such that a collision takes place whenever a projectile passes through this area. This fictitious effective area, the cross section, is a measure of the probability that collisions will occur among the constituents in a gas. When the collision partners are dissimilar, μ is the reduced mass of the colliding pair.

Consider a gas, possibly of several types of molecules, and let $\sigma_{if}(v)$ be the cross section for a collision between molecules (relative velocity v), where the initial states are represented by i and the final states by f. The rate that molecules with speeds between v and $v+dv$, regardless of direction, cross the effective area $\sigma_{if}(v)$ is given by

$$v\sigma_{if}(v)\chi(v)\, dv \tag{1.34}$$

The rate coefficient is obtained by integrating this expression over all relative velocities,[4] that is,

$$k_{if}(T) = 4\pi \int_0^\infty \sigma_{if}(v)\left(\frac{\mu}{2\pi kT}\right)^{3/2} v^3 \exp(-\mu v^2/2kT)\, dv \tag{1.35}$$

The dimensions of $k_{if}(T)$ are L^3T^{-1} ($\text{cm}^3\ \text{sec}^{-1}$) and $k_{if}(T)$ is dependent on the gas temperature T.

Let $N(T)$ be the number of particles per unit volume at temperature T. There are three cases to consider:

i. Gas Consisting of One Type of Particle, e.g., CO—CO. Let the number of molecules per unit volume in the state i be n_i. Collisions can only take place between like molecules, one of which will be assumed structureless, and so the number of molecules per unit volume per unit time making the transition i to f will be

$$Nn_i k_{if} \tag{1.36}$$

ii. Gas Consisting of Two Types of Particles—Structureless Projectile. Consider HeHF as an example. Two rate coefficients will now be applicable in the gas, namely,

$$k_{if}^{\mathrm{HF}} \quad \text{and} \quad k_{if}^{\mathrm{He}}$$

when the target HF molecule goes from i to f upon collision with an assumed structureless projectile, either another HF or He. Consequently, if n_i is the number density of HF molecules in the state i, then the total number of transitions from i to f per unit volume per unit time will be

$$N_{\mathrm{HF}} n_i k_{if}^{\mathrm{HF}} + N_{\mathrm{He}} n_i k_{if}^{\mathrm{He}} \tag{1.37}$$

where $N = N_{\mathrm{He}} + N_{\mathrm{HF}}$ and N_{He}/N and N_{HF}/N are the fractional particle concentrations in the gas. The general form of Eq. (1.37) is

$$N\gamma_{ij} \equiv n_i \sum_{t=1}^{C} N_t k_{if}^t \tag{1.38}$$

where the gas mixture consists of C types of particles, N_t/N are the fractional particle concentrations, and the dimensions of γ_{ij} are T^{-1} (sec^{-1}).

At low concentrations of HF,

$$N_{\mathrm{HF}}/N \sim 0, \qquad N_{\mathrm{He}}/N \sim 1$$

and at low temperatures,

$$n_0 \sim N_{\mathrm{HF}}$$

and

$$\gamma_{ij} \simeq N_{\mathrm{HF}} k_{if}^{\mathrm{He}} \tag{1.39}$$

iii. Both Colliding Particles Change Their State. Consider a gas consisting of two types of molecules with vibrational modes characterized by λ and l, respectively. Let n_{λ_i} and m_{l_i} be the number of particles per unit volume in the initial states λ_i and l_i, respectively. The number of molecules per unit volume per unit time making the transition

$$\lambda_i l_i \rightarrow \lambda_f l_f$$

will be

$$n_{\lambda_i} m_{l_i} k_{\lambda_i l_i, \lambda_f l_f} \tag{1.40}$$

We shall now examine under what conditions the rate coefficients are related to one another. In the collision between a structureless projectile and a target represented by an SHO, $C_{\kappa o \kappa} = 1$ in Eq. (1.28) and the properties of $k_{\lambda o \lambda}$ are related to the properties of the cross section $\sigma_{\lambda o \lambda}$, which in turn are related to the $B_{\lambda\lambda'}$ of Eq. (1.16), where $\lambda = \lambda_0$ and $\lambda = \lambda'$.

To evaluate this integral we recall various properties of the SHO in one dimension,[5] for which the Schrödinger equation is

$$-\frac{\hbar^2}{2m}\frac{d^2\psi}{dx^2}+\frac{ax^2}{2}\psi = E\psi \tag{1.41}$$

with eigenvalues

$$E_\kappa = (\kappa + \tfrac{1}{2})\hbar(a/m)^{1/2}, \qquad \kappa = 0, 1, 2, \ldots \tag{1.42}$$

and eigenfunctions

$$\psi_\kappa = \gamma_\kappa \exp(-y^2/2)H_\kappa(y) \tag{1.43}$$

where γ_κ is the normalization constant since

$$\int_{-\infty}^{\infty} |\psi_\kappa|^2\, dx = 1 \tag{1.44}$$

where

$$\begin{aligned} \gamma_\kappa &= \gamma_{\kappa-1}(2\kappa)^{-1/2} \\ y &= (ma/\hbar^2)^{1/4}x \end{aligned} \tag{1.45}$$

and where the Hermite polynomials are given by

$$H_\kappa(y) = (-1)^\kappa \exp(y^2)\frac{d^\kappa}{dy^\kappa}\exp(-y^2) \tag{1.46}$$

Equation (1.16) can be written as

$$B_{\kappa\lambda} = \gamma_\kappa\gamma_\lambda \int_{-\infty}^{\infty} \exp(-y^2)H_\kappa(y)H_\lambda(y)\exp(-\alpha A_1 x)\, dx \tag{1.47}$$

For the approximation

$$\exp(-\alpha A_1 x) \approx 1 - \alpha A_1 x$$

we have

$$B_{\kappa\lambda} = \delta_{\kappa\lambda} + \beta\gamma_\kappa\gamma_\lambda \int_{-\infty}^{\infty} \exp(-y^2)H_\kappa(y)H_\lambda(y)y\, dy \tag{1.48}$$

where we have the constant

$$\beta \equiv -\alpha A_1(\hbar^2/ma)^{1/2} \tag{1.49}$$

The integrand can be reexpressed as

$$B_{\kappa\lambda} = \delta_{\kappa\lambda} - \frac{\beta}{2}\gamma_\kappa\gamma_\lambda \int_{-\infty}^{\infty} H_\kappa(y)H_\lambda(y)\, d[\exp(-y^2)]$$

which can be readily integrated by parts:

$$B_{\kappa\lambda} = \delta_{\kappa\lambda} - \frac{\beta}{2}\gamma_\kappa\gamma_\lambda \Big\{ [H_\kappa(y)H_\lambda(y)\exp(-y^2)]_{-\infty}^{\infty} - \int_{-\infty}^{\infty} \exp(-y^2)\left(\frac{dH_\kappa}{dy}H_\lambda + H_\kappa\frac{dH_\lambda}{dy}\right) dy\Big\}$$

When the relation

$$dH_\kappa/dx = 2\kappa H_{\kappa-1}$$

is used,

$$B_{\kappa\lambda} = \delta_{\kappa\lambda} + \frac{\beta}{2}\gamma_\kappa\gamma_\lambda \int_{-\infty}^{\infty} \exp(-y^2)(2\kappa H_{\kappa-1}H_\lambda + H_\kappa 2\lambda H_{\lambda-1})\, dy$$

$$= \delta_{\kappa\lambda} + \frac{\beta}{2}\left[\left(\frac{\kappa}{2}\right)^{1/2}\delta_{\kappa-1\lambda} + \left(\frac{\lambda}{2}\right)^{1/2}\delta_{\kappa\lambda-1}\right] \qquad (1.50)$$

Hence the only nonzero off-diagonal matrix elements are

$$B_{\lambda\lambda+1} = \frac{\beta}{2}\left(\frac{\lambda+1}{2}\right)^{1/2} = (\lambda+1)^{1/2}B_{01} \qquad (1.51)$$

$$B_{\lambda\lambda-1} = \frac{\beta}{2}\left(\frac{\lambda}{2}\right)^{1/2} = \lambda^{1/2}B_{10} \qquad (1.52)$$

and hence symbolically the λ-dependence of the rate coefficients is given from Eqs. (1.28), (1.35), and (1.51):

$$k_{\lambda\lambda+1} = 4\pi \int_0^{\infty} 4\pi(k_0^3 k)^{-1} V_0^2 |R_k B_{\lambda\lambda+1}|^2 \left(\frac{\mu}{2\pi kT}\right)^{3/2} v^3 \exp(-\mu v^2/2kT)\, dv$$

$$= (\lambda+1)k_{01} \qquad (1.53)$$

and

$$k_{\lambda\lambda-1} = 4\pi \int_0^{\infty} \sigma_{\lambda\lambda-1}(v)\left(\frac{\mu}{2\pi kT}\right)^{3/2} v^3 \exp(-\mu v^2/2kT)\, dv$$

$$= \lambda k_{10} \qquad (1.54)$$

Since $B_{01} = B_{10}$, we have

$$k_{01} = k_{10} \qquad (1.55)$$

$$k_{\lambda\lambda+1}/k_{\lambda\lambda-1} = \lambda + 1/\lambda \qquad (1.56)$$

In collisions between a structureless projectile and a molecule with two normal modes of vibration, or between two SHOs, the cross section [see Eq. (1.28)] involves both $B_{\lambda_0\lambda}$ and $C_{\kappa_0\kappa}$. Individually, these factors satisfy the relationships derived in Eq. (1.51), but the rate coefficients are given by the equations

$$\begin{aligned}k_{\kappa\lambda,\kappa\lambda+1} &= 4\pi \int_0^\infty 4\pi (k_0^3 k)^{-1} V_0^2 |R_k B_{\kappa\kappa} C_{\lambda\lambda+1}|^2 \left(\frac{\mu}{2\pi kT}\right)^{3/2} v^3 \\ &\quad \times \exp(-\mu v^2/2kT)\, dv \\ &= (\lambda+1) \int_0^\infty |R_k B_{00} C_{01}|^2 \cdots \\ &= (\lambda+1) k_{00,01}\end{aligned} \tag{1.57}$$

$$\begin{aligned}k_{\kappa\lambda,\kappa\lambda-1} &= \lambda \int_0^\infty |R_k B_{00} C_{10}|^2 \cdots \\ &= \lambda k_{01,00}\end{aligned} \tag{1.58}$$

$$k_{\kappa\lambda,\kappa+1\lambda} = (\kappa+1) k_{00,10} \tag{1.59}$$

$$k_{\kappa\lambda,\kappa-1\lambda} = \kappa k_{10,00} \tag{1.60}$$

$$k_{\kappa\lambda,\kappa-1\lambda+1} = \kappa(\lambda+1) k_{10,01} \tag{1.61}$$

$$k_{\kappa\lambda,\kappa+1\lambda-1} = (\kappa+1)\lambda k_{01,10} \tag{1.62}$$

Problem 2. Extend the preceding results to show that

$$k_{\lambda l,\lambda+1 l-2} = (\lambda+1) l(l-1) k_{02,10}$$

$$k_{\lambda l,\lambda-1 l+2} = \lambda(l+1)(l+2) k_{10,02}$$

Problem 3. Consider a structureless projectile incident on a molecule with three modes of vibration. Show that the rate coefficients satisfy the following relations:

$$k_{l\lambda\mu,l+1\lambda-1\mu-1} = (l+1)\lambda\mu k_{011,100}$$

$$k_{l\lambda\mu,l-1\lambda+1\mu+1} = l(\lambda+1)(\mu+1) k_{100,011}$$

1.5. *Definition of Effective Vibrational Temperatures*

The energy levels of an SHO are given by

$$\varepsilon_\lambda = (\lambda + \tfrac{1}{2}) h\nu \tag{1.63}$$

and the partition function when internal and external energies are in equilibrium for a particular ambient temperature T^*, is[6]

$$z = \sum_{\lambda=0}^{\infty} \exp\left(\frac{-\varepsilon_\lambda}{kT^*}\right) = \sum_\lambda \exp\left[\frac{-(\lambda+\frac{1}{2})h\nu}{kT^*}\right]$$

$$= \exp\left(\frac{-h\nu}{2kT^*}\right) \sum_\lambda \exp\left(\frac{-\lambda h\nu}{kT^*}\right)$$

$$= \exp\left(\frac{-h\nu}{2kT^*}\right)\left[1-\exp\left(\frac{-h\nu}{kT^*}\right)\right]^{-1} \tag{1.64}$$

having used the property that the sum

$$1+x+x^2+\cdots=(1-x)^{-1} \tag{1.65}$$

If there are N oscillators per unit volume, then the number per unit volume in state λ at this equilibrium temperature, T^*, is

$$n_\lambda^* \equiv n_\lambda(T^*) = N\frac{\exp[-(\lambda+\frac{1}{2})(h\nu/kT^*)]}{z} \tag{1.66}$$

The total internal energy of vibration per unit volume at the equilibrium ambient temperature T^* will be

$$\varepsilon(T^*) = \sum_\lambda (\lambda+\tfrac{1}{2})h\nu n_\lambda(T^*)$$

$$= N\left[1-\exp\left(\frac{-h\nu}{kT^*}\right)\right]\exp\left(\frac{h\nu}{2kT^*}\right)\sum_\lambda\left(\lambda+\frac{1}{2}\right)\exp\left[\frac{-(\lambda+\frac{1}{2})h\nu}{kT^*}\right]$$

Therefore

$$\varepsilon(T^*) = \frac{1}{2}Nh\nu\left[1-\exp\left(\frac{-h\nu}{kT^*}\right)\right]\sum_\lambda\exp\left(\frac{-\lambda h\nu}{kT^*}\right)$$

$$+N\left[1-\exp\left(\frac{-h\nu}{kT^*}\right)\right]h\nu\sum_\lambda\lambda\exp\left(\frac{-\lambda h\nu}{kT^*}\right)$$

$$= \frac{1}{2}Nh\nu + Nh\nu\left[\exp\left(\frac{h\nu}{kT^*}\right)-1\right]^{-1} \tag{1.67}$$

having used Eq. (1.65) and the sum

$$1+x+2x^2+3x^3+\cdots=x(1-x)^{-2} \tag{1.68}$$

From Eq. (1.67) we can define the *internal energy* per unit volume due to the equilibrium ambient temperature as

$$E(T^*) \equiv Nh\nu[\exp(h\nu/kT^*)-1]^{-1} \tag{1.69}$$

which can be used, conversely, to define an effective vibrational temperature T_a say, in terms of *any* internal energy E_a, that is,

$$T_a \equiv \frac{h\nu/k}{\log(Nh\nu/E_a + 1)} \tag{1.70}$$

Problem 4. Given

$$\sum_{n=0}^{\infty} n^2 x^n = x(x+1)(1-x)^{-3}$$

show

$$\sum_{\lambda=0}^{\infty} \lambda^2 n_\lambda = N(1+e^{h\nu/kT})(e^{h\nu/kT}-1)^{-2}$$

1.6. Rate Equations—Structureless Projectile on an SHO

Let n_i be the number of particles per unit volume in the state i when the gas is at temperature T. Then $N_{n_i}k_{if}(T)$ will be the total number of effective collisions per unit volume per unit time that take the particles out of the i state and into the f state. The rate of change of population in the i state, that is, dn_i/dt, will be the algebraic sum of terms such as Nn_ik_{ij} (depleting the state) and $Nn_jk_{ji}(T)$ (feeding into the state).

Consider the collision of a structureless projectile with a target molecule, represented by a simple harmonic oscillator (Reference 1, pp. 86–88). It will be assumed that transitions are possible only to neighboring states, i.e., $j \to j-1$ and $j \to j+1$. The rate of change of population of the j state is given by

$$-\frac{dn_j}{N\,dt} = n_j[k_{jj+1}(T) + k_{jj-1}(T)] - n_{j-1}k_{j-1j}(T) - n_{j+1}k_{j+1j}(T) \tag{1.71}$$

where T is the ambient gas temperature. When the system is in thermal equilibrium at an ambient temperature T^*, the left-hand side equals zero, giving the equilibrium population n_j^*, that is,

$$0 = n_j^*[k_{jj+1}(T^*) + k_{jj-1}(T^*)] - n_{j-1}^*k_{j-1j}(T^*) - n_{j+1}^*k_{j+1j}(T^*) \tag{1.72}$$

Since the lowest state can only exchange with the next higher state, then

$$0 = n_0^*k_{01}(T^*) - n_1^*k_{10}(T^*)$$

which imposes pairwise balancing on Eq. (1.72); that is,

$$0 = n_j^*k_{jj+1}(T^*) - n_{j+1}^*k_{j+1j}(T^*)$$

or

$$\frac{k_{jj+1}(T^*)}{k_{j+1j}(T^*)}=\frac{n^*_{j+1}}{n^*_j}=\frac{\exp[-(j+\frac{3}{2})h\nu/kT^*]}{\exp[-(j+\frac{1}{2})(h\nu/kT^*)]}$$

having used Eq. (1.66). Hence we derive the important result,

$$\frac{k_{jj+1}(T^*)}{k_{j+1j}(T^*)}=\exp\left(\frac{-h\nu}{kT^*}\right)=\kappa(T^*) \tag{1.73}$$

which is an identity in T^* and is true for any value, T of T^*, since the gas can be in thermal equilibrium at any ambient temperature T. Hence we can replace T^* by T and write

$$\frac{k_{jj+1}(T)}{k_{j+1j}(T)}=e^{-h\nu/kT}=\kappa(T)$$

This equation is used later in the form

$$k_{01}(T)=k_{10}(T)\,e^{-h\nu/kT} \tag{1.74}$$

When Eqs. (1.73) and (1.53) are substituted into Eq. (1.71) we have

$$-\frac{dn_j(T)}{N\,dt}=n_j[\kappa(j+1)k_{10}(T)+jk_{10}(T)]-n_{j-1}\kappa jk_{10}(T)-n_{j+1}(j+1)k_{10}(T)$$

$$\frac{dn_j}{N\,dt}=k_{10}(T)[\kappa jn_{j-1}-jn_j-\kappa(j+1)n_j+(j+1)n_{j+1}] \tag{1.75}$$

When this equation is multiplied by $h\nu j$ and summed over all j we obtain

$$\frac{h\nu}{N}\frac{d}{dt}\sum_j jn_j=h\nu k_{10}(T)\sum_j[\kappa j^2n_{j-1}-j^2n_j-\kappa j(j+1)n_j+j(j+1)n_{j+1}] \tag{1.76}$$

In the last term on the right-hand side let $j'=j+1$; then

$$\sum_{j=0} j(j+1)n_{j+1}=\sum_{j'=1}(j'-1)j'n_{j'}=\sum_{j'=0}(j'-1)j'n_{j'}=\sum_{j=0}(j-1)jn_j$$

Hence the second and last terms of Eq. (1.76) give

$$\sum_{j=0}^{\infty}[-j^2+j(j-1)]n_j=-\sum_j jn_j$$

In the first term of Eq. (1.76) we have

$$\sum_{j=0} j^2n_{j-1}=\sum_{j=1} j^2n_{j-1}=\sum_{j'=0}(j'+1)^2n_{j'}=\sum_{j=0}(j+1)^2n_j$$

which, taken together with the third term, gives

$$\sum_{j=0} [(j+1)^2 - j(j+1)]n_j = \sum_{j=0} (j+1)n_j = \sum_{j=0} jn_j + N$$

When these results are substituted into Eq. (1.76) we have

$$\frac{d}{N\,dt} \sum_{j=0} jh\nu n_j = k_{10}(T)h\nu\left[\kappa\left(\sum_{j=0} jn_j + N\right) - \sum_{j=0} jn_j\right]$$

Since the total internal energy per unit volume is given by

$$E = \sum_j jh\nu n_j$$

we have

$$\begin{aligned} dE/N\,dt &= k_{10}(T)(\kappa E + \kappa h\nu N - E) \\ &= -k_{10}(T)(1-\kappa)[E - \kappa h\nu N(1-\kappa)^{-1}] \end{aligned} \tag{1.77}$$

When Eq. (1.73) is substituted for K, the term in square brackets becomes

$$\kappa h\nu N(1-\kappa)^{-1} = Nh\nu(e^{h\nu/kT} - 1) = E(T)$$

where the equilibrium value of E at temperature T is as defined in Eq. (1.69). Hence Eq. (1.77) can be written as

$$-dE/dt = [E - E(T)]/\tau(T) \tag{1.78}$$

where the relaxation time is defined by

$$\tau^{-1} = Nk_{10}(T)(1-\kappa) = Nk_{10}(T) - Nk_{01}(T) \tag{1.79}$$

when Eq. (1.73) is used.

The rates in Eq. (1.79) are given by Eq. (1.35):

$$\begin{aligned} \gamma_{10}(T) &= Nk_{10}(T) \\ &= \left(\frac{8\pi kT}{\mu}\right)^{1/2} Na_0^2 \int_0^\infty \sigma_{10}(E)\, e^{-E/kT}\left(\frac{E}{kT}\right) d\left(\frac{E}{kT}\right) \end{aligned} \tag{1.80}$$

where a_0 is the Bohr radius and $\sigma_{10}(E)$ is in units of πa_0^2. We have made the change in the integration variable:

$$E = \tfrac{1}{2}\mu v^2 \tag{1.81}$$

Since N is temperature dependent, that is, from Boyle's law

$$N = N_0 P/RT \tag{1.82}$$

where N_0 is Avogadro's number, P the pressure, and R the universal gas constant, the rate coefficient (in $\sec^{-1}$) becomes

$$\gamma_{10} = (8\pi k/\mu T)^{1/2} N_0 P a_0^2/R \int_0^\infty \sigma_{10}(xkT)\, e^{-x} x\, dx \tag{1.83}$$

having changed the integration variables:

$$x = E/kT \tag{1.84}$$

When the cgs units are substituted into the constants appearing in Eq. (1.83), we have

$$\gamma_{10} = 4.015 \times 10^{11}(M_c T)^{-1/2} \int_0^\infty \sigma_{10}(xkT)\, e^{-x} x\, dx \tag{1.85}$$

where T is in °K and M_c is the reduced mass of the collision system expressed in numbers of electron masses. To calculate τ from Eq. (1.79) we need to calculate γ_{01} as well. This quantity is related to γ_{10} through the principle of detailed balance applied to the cross sections in the form

$$k_i^2 \sigma_{if} = k_f^2 \sigma_{fi} \tag{1.86}$$

where k_n^2 are defined in Eq. (1.9) for two vibrational modes. For a structureless projectile on an SHO we have

$$k_\lambda^2 = (2\mu/\hbar^2)(E - \varepsilon_\lambda) \tag{1.87}$$

from which we see that

$$(k_\lambda a_0)^2 = (k_\eta a_0)^2 + \frac{2\mu a_0^2}{\hbar^2}(\varepsilon_\eta - \varepsilon_\lambda) \tag{1.88}$$

which is a dimensionless equation. Hence

$$\begin{aligned} \kappa &\equiv k a_0 \\ \kappa_1^2/\kappa_0^2 &= 1 - 2\mu a_0^2\, \Delta E/\hbar^2 \kappa_0^2 \\ &= 1 - \Delta E/E_c \\ &= 1 - \Delta E/(E + \Delta E) \end{aligned}$$

and so

$$\kappa_0^2/\kappa_1^2 - 1 = \Delta E/E \tag{1.89}$$

where ΔE is the vibrational energy level separation and E_c is the collision energy.

The relaxation time, τ, is obtained by using Eqs. (1.85), (1.86), and (1.89) to give[7]

$$\tau(T)^{-1} = 4.015 \times 10^{11}(M_c T)^{-1/2}\left(\frac{\Delta E}{kT}\right) \int_0^\infty \sigma_{01}(xkT)\, e^{-x}\, dx \tag{1.90}$$

in $\sec^{-1}$.

1.7. Relaxation Times in Collisions Involving Two Nonresonant Vibrational Modes[8]

Let E_a be the internal energy of N molecules of type "a" per unit volume. If a itself represents the fraction of molecules of type a, then aE_a will be the internal energy per unit volume of aN molecules. Energy is supplied to the molecules of type a by collisions amongst themselves and by collisions with molecules of type "b," where

$$a+b=1$$

The number of molecules of type a in state λ that make a transition to state $\lambda+1$ upon collision with like molecules in state λ' (which remain in λ' after the collision) per unit time per unit volume is given by

$$n_\lambda n_{\lambda'} k_{\lambda\lambda',\lambda+1\lambda'}(T)$$

The expression represents T–V collisions that lead to an increase in the internal energy. The corresponding V–T transitions leading to a decrease in the internal energy will be given by

$$n_\lambda n_{\lambda'} k_{\lambda\lambda',\lambda-1\lambda'}$$

In general, those terms involving one overall $h\nu_a$ change in the internal energy are written as

$$n_\lambda n_{\lambda'}(k_{\lambda\lambda',\lambda+\mu\lambda'-\mu+1}-k_{\lambda\lambda',\lambda-\mu\lambda'+\mu-1}) \qquad (1.91)$$

where in the first (second) term, $\mu\leqslant\lambda'+1$ ($\mu\leqslant\lambda$). We shall follow Schwartz *et al.*[8] and neglect all terms except those with $\mu=1$.

Molecules of type a will also change their internal energy during collisions with molecules of type b, whose states will be indexed by l. The number of type a molecules per unit time involved in increasing the internal energy by $h\nu_a$ upon collision with type b molecules will be given by

$$n_\lambda m_l(\bar{k}_{\lambda l,\lambda+1 l-\mu}-\bar{k}_{\lambda l,\lambda-1 l+\mu}) \qquad (1.92)$$

where we have used $\bar{k}$ to distinguish a–b collisions from a–a collisions, and where m_l/N is the fraction of type b molecules in the state l. When $\mu=0$, we have V–T transitions, and when $\mu\neq0$, the rate coefficients are calculated from the appropriate V–V cross sections. We shall adopt the approximation of Schwartz *et al.* that only $\mu=0$, 1 need be included in expressions such as (1.92).

The total rate of change of internal energy of type a molecules per unit volume will be obtained by the aggregate summing of expressions (1.91) and (1.92), that is,

$$\frac{d(aE_a)}{dt} = h\nu_a\Big[\sum_{\lambda\lambda'} n_\lambda n_{\lambda'}(k_{\lambda\lambda',\lambda+1\lambda'} - k_{\lambda\lambda',\lambda-1\lambda'})$$

$$+ \sum_{\lambda l} n_\lambda m_l(\bar{k}_{\lambda l,\lambda+1l} - \bar{k}_{\lambda l,\lambda-1l} + \bar{k}_{\lambda l,\lambda+1l-1} - \bar{k}_{\lambda l,\lambda-1l+1})\Big] \tag{1.93}$$

When Eqs. (1.57)–(1.62) for the λ-dependence of the rate coefficient in the SHO approximation are substituted into Eq. (1.93), we obtain

$$\frac{d(aE_a)}{dt} = h\nu_a\Big[\sum_{\lambda\lambda'} n_\lambda n_{\lambda'}((\lambda+1)k_{00,10} - \lambda k_{10,00})$$

$$+ \sum_{\lambda l} n_\lambda m_l((\lambda+1)\bar{k}_{00,10} - \lambda\bar{k}_{10,00} + (\lambda+1)l\bar{k}_{01,10} - \lambda(l+1)\bar{k}_{10,01})\Big] \tag{1.94}$$

1.7.1. Collisions between Unlike SHOs

The following simple summations

$$\sum_\lambda n_\lambda = aN \tag{1.95}$$

$$\sum_l m_l = bN \tag{1.96}$$

$$\sum_\lambda n_\lambda \lambda h\nu_a = aE_a \tag{1.97}$$

$$\sum_l m_l l h\nu_b = bE_b \tag{1.98}$$

can be performed and substituted into Eq. (1.94) to give

$$\frac{d(aE_a)}{N\,dt} = k_{00,10}a(aE_a + aNh\nu_a) - k_{10,00}a^2E_a + \bar{k}_{00,10}b(aE_a + aNh\nu_a)$$

$$- \bar{k}_{10,00}abE_a + \bar{k}_{01,10}\left(\frac{aE_a}{Nh\nu_a} + a\right)\frac{\nu_a bE_b}{\nu_b}$$

$$- \bar{k}_{10,01}aE_a\left(\frac{bE_b}{Nh\nu_b} + b\right) \tag{1.99}$$

Problem 5. In Eq. (1.73) we derived a relationship between k_{01} and k_{10}. Derive the following analogous relationships for the rate coefficients appearing in Eq. (1.99):

$$k_{00,10} = k_{10,00} \exp(-h\nu_a/kT)$$
$$\bar{k}_{00,10} = \bar{k}_{10,00} \exp(-h\nu_a/kT) \qquad (1.100)$$
$$\bar{k}_{01,10} = \bar{k}_{10,01} \exp(-h(\nu_a - \nu_b)/kT)$$

Problem 6. Extend the proofs of Problem 5 to verify

$$k_{110,001} = k_{001,110} \exp(-h(\nu_3 - \nu_2 - \nu_1)/kT)$$

When Eqs. (1.100) are substituted into Eq. (1.99) we obtain

$$\frac{a}{N}\frac{dE_a}{dt} = (a^2 k_{00,10} + ab\bar{k}_{10,00})\left[(Nh\nu_a + E_a)\exp\left(-\frac{h\nu_a}{kT}\right) - E_a\right]$$
$$+ \bar{k}_{10,01} ab\left[E_b\left(\frac{\nu_a}{\nu_b} + \frac{E_a}{Nh\nu_b}\right)\exp\left(\frac{-h(\nu_a - \nu_b)}{kT}\right) - E_a\left(1 + \frac{E_b}{Nh\nu_b}\right)\right] \qquad (1.101)$$

When Eq. (1.69) is used for $Nh\nu_a$,

$$(Nh\nu_a + E_a)\exp(-h\nu_a/kT) - E_a$$
$$= E_a[\exp(-h\nu_a/kT) - 1] + E_a(T)[\exp(h\nu_a/kT) - 1]\exp(-h\nu_a/kT)$$
$$= [E_a - E_a(T)][\exp(-h\nu_a/kT) - 1] \qquad (1.102)$$

Similarly, when Eq. (1.69) is used for ν_a/ν_b and $(Nh\nu_b)^{-1}$ in Eq. (1.101) we have

$$E_b\left(\frac{\nu_a Nh}{\nu_b Nh} + \frac{E_a}{Nh\nu_b}\right)\exp\left(\frac{-h(\nu_a - \nu_b)}{kT}\right) - E_a\left(1 + \frac{E_b}{Nh\nu_b}\right)$$
$$= E_b\left\{\frac{E_a(T)[\exp(h\nu_a/kT) - 1] + E_a}{E_b(T)[\exp(h\nu_b/kT) - 1]}\right\}\exp\left[\frac{-h(\nu_a - \nu_b)}{kT}\right]$$
$$- E_a\left\{\frac{E_b(T)[\exp(h\nu_b/kT) - 1] + E_b}{E_b(T)[\exp(h\nu_b/kT) - 1]}\right\}$$
$$= \frac{E_b}{E_b(T)}[E_a(T) - E_a]\frac{1 - \exp(-h\nu_a/kT)}{1 - \exp(-h\nu_b/kT)} - \frac{E_a}{E_b(T)}[E_b(T) - E_b] \qquad (1.103)$$

Equations (1.102) and (1.103) are substituted into Eq. (1.101) to give [Reference 8, Eq. (22)]

$$\frac{a}{N}\frac{dE_a}{dt} = (a^2k_{10,00} + ab\bar{k}_{10,00})[\exp(-h\nu_a/kT) - 1][E_a - E_a(T)]$$

$$+\bar{k}_{10,01}ab\left\{\frac{E_b}{E_b(T)}[E_a(T) - E_a]\frac{1-\exp(-h\nu_a/kT)}{1-\exp(-h\nu_b/kT)}\right.$$

$$\left.-\frac{E_a}{E_b(T)}[E_b(T) - E_b]\right\} \qquad (1.104)$$

where, from Eq. (1.69),

$$E_a \equiv E_a(T_a) = Nh\nu_a[\exp(h\nu_a/kT_a) - 1]^{-1} \qquad (1.105)$$

The factor a can be taken out of Eq. (1.104).

In Eq. (1.104) we can define the V–T relaxation time by

$$\tau_{(VT)}(T)^{-1} = N(ak_{10,00} + b\bar{k}_{10,00})[\exp(-h\nu_a/kT) - 1] \qquad (1.106)$$

which is the τ_a (effective) of Schwartz *et al.*[8] To derive the V–V relaxation time, consider the coefficient $-E_a$ and substitute Eq. (1.69) for E_b and $E_b(T)$, which leads to

$$\frac{\exp(h\nu_b/kT)-1}{\exp(h\nu_h/kT_b)-1}\,\frac{1-\exp(-h\nu_a/kT)}{1-\exp(-h\nu_b/kT)} + 1 - \frac{\exp(h\nu_b/kT)-1}{\exp(h\nu_b/kT_b)-1}$$

$$= \exp\left(\frac{h\nu_b}{kT}\right)\frac{[1-\exp(-h\nu_b/kT)]}{\exp(h\nu_b/kT_b)-1} + 1 - \frac{\exp(h\nu_b/kT)-1}{\exp(h\nu_b/kT_b)-1}$$

$$= \left[\exp\left(\frac{h\nu_b}{kT_b}\right) - 1\right]^{-1}\left\{\exp\left(\frac{h\nu_b}{kT}\right)\left[1 - \exp\left(\frac{-h\nu_a}{kT}\right)\right] + \exp\left(\frac{h\nu_b}{kT_b}\right) - \exp\left(\frac{h\nu_b}{kT}\right)\right\}$$

$$= \exp\left[\frac{h(\nu_b-\nu_a)}{kT}\right]\left[\exp\left(\frac{h\nu_b}{kT_b}\right) - 1\right]^{-1}\left\{\exp\left[\frac{h\nu_b}{kT_b} - \frac{h(\nu_b-\nu_a)}{kT}\right] - 1\right\}$$

$$(1.107)$$

The remaining term in Eq. (1.104) reduces to the form

$$\exp[h(\nu_b-\nu_a)/kT][\exp(h\nu_b/kT_b) - 1]^{-1}$$

and so Eq. (1.104) [Schwartz *et al.*,[8] Eq. (22)] becomes

$$\frac{dE_a}{dt} = \frac{E_a - E_a(T)}{\tau_{(VT)}} - bN\bar{k}_{10,01}\exp[h(\nu_b-\nu_a)/kT][\exp(h\nu_b/kT_b) - 1]^{-1}$$

$$\times\left\{\exp\left[\frac{h\nu_b}{kT_b} - \frac{h(\nu_b-\nu_a)}{kT}\right] - 1\right\}[E_a - E_a(T, T_b)]$$

where

$$E_a(T, T_b) = Nh\nu_a\left\{\exp\left[\frac{h\nu_b}{kT_b} - \frac{h(\nu_b - \nu_a)}{kT}\right] - 1\right\}^{-1} \tag{1.108}$$

Finally, we have

$$\frac{dE_a}{dt} = \frac{E_a - E_a(T)}{\tau_{(VT)}} - \frac{E_a - E_a(T, T_b)}{\tau_{(VV)}(T, T_b)} \tag{1.109}$$

where

$$\tau_{(VV)}(T, T_b)^{-1} = Nb\bar{k}_{10,01}\exp\left[\frac{h(\nu_b - \nu_a)}{kT}\right]\left[\exp\left(\frac{h\nu_b}{kT_b}\right) - 1\right]^{-1}$$
$$\times\left\{\exp\left[\frac{h\nu_b}{kT_b} - \frac{h(\nu_b - \nu_a)}{kT}\right] - 1\right\} \tag{1.110}$$

1.7.2. *Structureless Projectile on a Target with Two Vibrational Modes*

We shall consider a gas consisting of one type of molecule (therefore $a = 1$) which has two vibrational modes. Since the energy of each molecule is the sum of a state of each mode, e.g., $E_\lambda + E_l$, then

$$\sum_{\lambda=0}^{\infty} n_\lambda = N = \sum_{l=0}^{\infty} m_l \tag{1.111}$$

because each molecule will be in one of the λ states. If K_a is the degree of degeneracy of the a-th degree of freedom, then

$$E_a = \sum_{\lambda=0}^{\infty} n_\lambda \lambda K_a h\nu_a \tag{1.112}$$

In the analysis developed in Section 1.7, we have a redundancy for the present problem in that we consider both a–a and a–b collisions. Hence it will be sufficient to consider only the a–"b" terms of Eq. (1.94) to represent nonresonant energy change between like molecules, that is,

$$\frac{dE_a}{dt} = K_a h\nu_a \sum_{\lambda l} [(\lambda+1)k_{00,10} - \lambda k_{10,00} + (\lambda+1)lk_{01,10} - \lambda(l+1)k_{10,01}]n_\lambda m_l \tag{1.113}$$

where we have dropped the overbar on k. The λ and l modes are not degenerate. When Eqs. (1.111) and (1.112) are used to carry out the summations, we obtain

$$\frac{dE_a}{N\,dt} = k_{00,10}(E_a + K_a N h\nu_a) - k_{10,00}E_a$$
$$+ k_{01,10}(E_a + K_a N h\nu_a)E_b/NK_b h\nu_b$$
$$- k_{10,01}\left(\frac{\nu_a}{\nu_b}E_b + Nh\nu_a\right)E_a/NK_a h\nu_a \tag{1.114}$$

When Eqs. (1.100) are used for the rate coefficients, and

$$E_a(T) = K_a N h\nu_a[\exp(h\nu_a/kT) - 1]^{-1} \tag{1.115}$$

we find that

$$\frac{dE_a}{N\,dt} = k_{10,00}\left[\exp\left(\frac{-h\nu_a}{kT}\right) - 1\right][E_a - E_a(T)]$$
$$+ k_{10,01}\left\{\frac{E_b}{E_b(T)}\,\frac{1-\exp(-h\nu_a/kT)}{1-\exp(-h\nu_b/kT)}[E_a(T) - E_a]\right.$$
$$\left. - \frac{E_a}{E_b(T)}[E_b(T) - E_b]\right\} \tag{1.116}$$

We define the V–T relaxation time by

$$\tau_{(VT)}(T)^{-1} \equiv Nk_{10,00}[\exp(-h\nu_a/kT) - 1] \tag{1.117}$$

To derive $\tau_{(VV)}(T, T_b)$ we consider the coefficient of E_a, and notice that it is the same as in Eq. (1.104). Consequently, $\tau_{(VV)}(T, T_b)$ is defined as in Eq. (1.110) and Eq. (1.116) becomes

$$\frac{dE_a}{dt} = \frac{E_a - E_a(T)}{\tau_{(VT)}(T)} - \frac{E_a - E_a(T, T_b)}{\tau_{(VV)}(T, T_b)} \tag{1.118}$$

where $E_a(T, T_b)$ is defined in Eq. (1.108).

2

Vibrational Kinetics

In Chapter 1 we derived rate equations describing energy-transfer processes during collisions between vibrationally excited molecules. In this chapter we shall apply the rate-equation approach to describe the vibrational kinetics of the $CO_2:N_2:He$ gas mixture and thus obtain a theoretical model of CO_2 gas lasers.

2.1. Lasing States of the CO_2 Molecule

An N-particle system has $3N$ degrees of freedom. The motion of the center of gravity and the orientation of the axis of a linear XY_2 molecule can be specified in terms of five degrees of freedom, leaving the vibrational motion to be described in terms of $(3\times 3-5=4)$ normal coordinates. These four normal vibrations have frequencies ν_1 (symmetric mode), ν_{2a} and ν_{2b} (the doubly degenerate bending mode, vibrating in the plane of the paper and perpendicular to it, respectively), and ν_3 (asymmetric mode) as shown in Fig. 2.[9] The total vibrational energy of the CO_2 molecule can

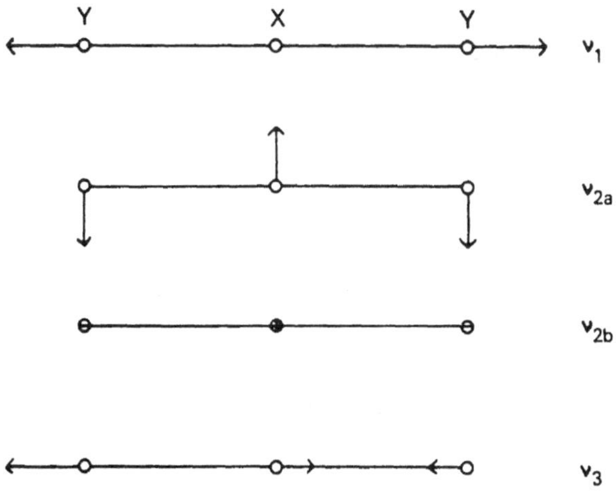

Fig. 2. Normal vibrations of the linear CO_2 molecule.

only assume the values

$$E(v_1, v_{2a}, v_{2b}, v_3) = h\nu_1(v_1 + \tfrac{1}{2}) + h\nu_{2a}(v_{2a} + \tfrac{1}{2}) + h\nu_{2b}(v_{2b} + \tfrac{1}{2}) + h\nu_3(v_3 + \tfrac{1}{2})$$

The CO_2 molecule is doubly degenerate with $\nu_{2a} = \nu_{2b} = \nu_2$, and $v_2 = v_{2a} + v_{2b}$. According to Herzberg (Reference 9, p. 80), the zero-point vibration has a unique ($v_{2a} = 0 = v_{2b}$) wave function. However, when the vibration is excited by one quantum we have either $v_{2a} = 1$, $v_{2b} = 0$, or $v_{2a} = 0$, $v_{2b} = 1$; that is, two distinct eigenfunctions at the same energy—double degeneracy. When two quanta are excited we may have $v_{2a} = 2$, $v_{2b} = 0$; $v_{2a} = 1 = v_{2b}$; or $v_{2a} = 0$, $v_{2b} = 2$; that is, there will be a triple degeneracy. In general the degree of degeneracy, when v_2 quanta of doubly degenerate vibration are excited, is equal to the number of different ways v_2 can be written as a sum of two positive integers, where the order of the integers is significant, i.e., $v_2 + 1$.

When we transform from the normal Cartesian coordinates of vibrational modes $2a$ and $2b$ to normal polar coordinates, $\xi_a = \rho_i \cos\theta_i$, $\xi_b = \rho_i \sin\theta_i$, Pauling and Wilson[10] obtained eigenfunctions of a doubly degenerate vibration to be

$$\psi_i = \exp\left(-\frac{\alpha_i \rho_i^2}{2}\right) F_{v_i}^{l_i}(\alpha_i^{1/2}\rho_i) \exp(\pm i l_i \varphi_i) \tag{2.1a}$$

where F is a polynomial of degree v_i in ρ_i and

$$l_i = v_i, v_i - 2, v_i - 4, \ldots, 1 \text{ or } 0 \tag{2.1b}$$

depending on whether v_i is odd or even. That is, for each nonzero l_i value there are a pair of eigenfunctions. The l_i values and degeneracies are presented in Fig. 3.

In the case of linear molecules, as long as only one degenerate vibration is excited, which is always the case with CO_2 since there is only one

v_2	l_2	Degeneracy (of the 2a + 2b modes)
7	7, 5, 3, 1	8
6	6, 4, 2, 0	7
5	5, 3, 1	6
4	4, 2, 0	5
3	3, 1	4
2	2,0	3
1	1	2
0	0	1

Fig. 3. Vibrational levels of a doubly degenerate vibration and their degrees of degeneracy.

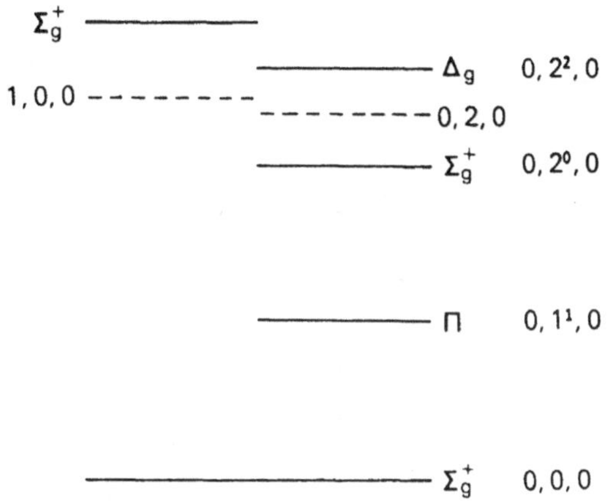

Fig. 4. Fermi resonance in CO_2. The dashed lines represent the unperturbed levels which, in first order, go over to the labeled levels.

degenerate vibration, l_2 is an exactly defined quantum number (see Herzberg,[9] p. 210) with values 0, 1, 2, 3, ... representing the vibrational angular momentum about the symmetry axis and corresponding to the species $\Sigma, \Pi, \Delta, \Phi, \ldots$. As a consequence of anharmonicity, the different levels belonging to the same v_2 are split; the higher the l_2, the higher the level.

In a polyatomic molecule, two vibrational levels belonging to different vibrational modes may have nearly the same energy, e.g., in CO_2 the level $v_1=0$, $v_2=2$, $v_3=0$ ($\nu_2=667\ \text{cm}^{-1}$) has almost the same energy as the level $v_1=1$, $v_2=0$, $v_3=0$ ($\nu_1=1337\ \text{cm}^{-1}$). That is, the levels are *accidentally degenerate*, as first recognized by Fermi, in zero-order approximation when the interactions between the modes are neglected. When the interaction between them is included as a perturbation, the wave functions are mixed and the resulting pair of mixed levels are repelled (see Fig. 4). That is, in first-order perturbation theory one of the mixed levels is pushed up and the other is pushed down, so that the level separation is greater than expected. In view of the strong mixing of the zero-approximation functions in first order, the observed levels should not, strictly speaking, be labeled in the zero-order nomenclature. That is, the observed symmetric mode is the linear superposition

$$a\psi_{1,0,0}+b\psi_{0,2,0}$$

This resonance interaction between these levels occurs between higher levels, e.g., 1, 0, 1 and 0, 2°, 1, etc.

In Fig. 5 we present a schematic of the lower vibrational levels of the electronic ground state of CO_2, showing the population distributions among the rotational sublevels of each vibrational level. In this figure, we also show the lasing transitions. These are dipole transitions that obey the selection rule

$$\Delta J = J' - J'' = +1, 0, -1$$

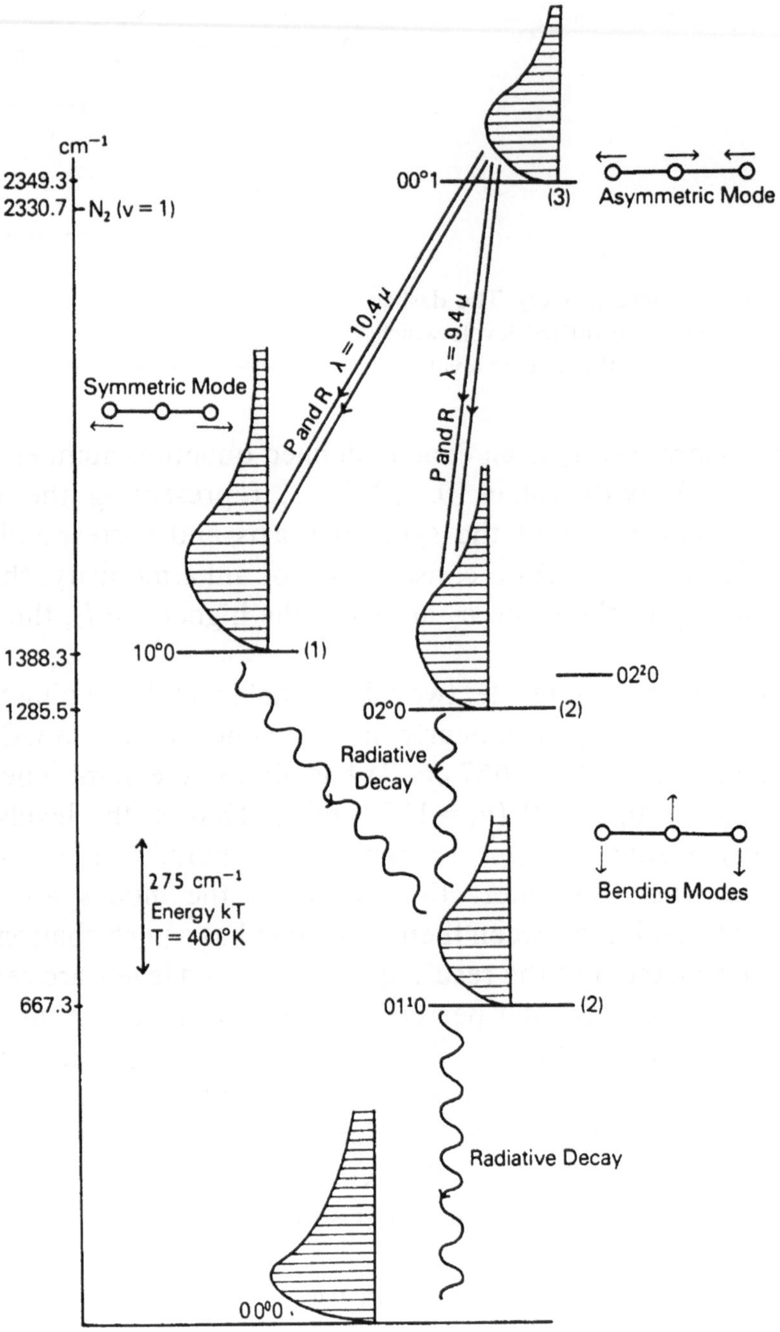

Fig. 5. Lower vibrational levels of the electronic ground state of CO_2. Vibrational levels are designated by v_1, v_2^l, v_3, where $l = v_2, v_2-2, \ldots, 0$ for even v_2, and $l = v_2, v_2-2, \ldots, 1$ for odd v_2. The populations of the rotational substates of the vibrational states are indicated by the lengths of the lines specifying the rotational states. Dipole transitions of CO_2 satisfy $\Delta J = \pm 1$, where ΔJ is the change in rotational quantum number, corresponding to the observed R and P branches of the emission spectrum. The lines are designated $P(J)$ or $R(J)$, where J is the rotational quantum number of the lower level.

corresponding to the observed R, Q, and P branches of the emission spectrum. The convention is to write $R(J)$ or $P(J)$, where J is the rotational quantum number for the lower level.

If $N_{1,0^0,0}$ is the number of CO_2 molecules in the symmetric vibrational mode, then a fraction of these molecules,

$$P(J)N_{1,0^0,0}$$

will be in the Jth rotational level, since

$$\sum_{J=0}^{\infty} P(J) = 1$$

Manes and Seguin[11] have developed a model of the TEA–CO_2 laser that considers the energy densities in the vibrational mode of N_2 (E_4) and the three modes of CO_2 (E_1, E_2, and E_3). In the following sections we shall derive the relaxation equations for these modes using the analysis of Chapter 1.

Before proceeding to develop the molecular kinetics of the lasing CO_2 system, we shall derive the number densities for the lasing states. From Eq. (1.66) we know that the equilibrium population of a level labeled v_1 (energy $h\nu_1$), with degeneracy g_{v_1}, is

$$N_{v_1} = \frac{N g_{v_1} \exp(-v_1 h\nu_1/kT)}{\sum_{v_1=0}^{\infty} g_{v_1} \exp(-v_1 h\nu_1/kT)} \tag{2.2a}$$

with three independent sets of harmonic oscillator levels:

$$N_{v_1 v_2 v_3} = \frac{N g_{v_1} g_{v_2} g_{v_3} \exp[-h(v_1\nu_1 + v_2\nu_2 + v_3\nu_3)/kT]}{\sum_{v_1}\sum_{v_2}\sum_{v_3} g_{v_1} g_{v_2} g_{v_3} \exp[-h(v_1\nu_1 + v_2\nu_2 + v_3\nu_3)/kT]} \tag{2.2b}$$

Each CO_2 molecule will have a value for v_1, v_2, and v_3. To evaluate Eq. (2.2a) for the particular case

$$g_{v_1} = 1, \qquad g_{v_2} = (v_2 + 1), \qquad g_{v_3} = 1$$

with

$$x_i = \exp(-h\nu_i/kT)$$

we have

$$N_{v_1 v_2 v_3} = \frac{N(v_2+1) x_1^{v_1} x_2^{v_2} x_3^{v_3}}{\sum_{v_1} x_1^{v_1} \sum_{v_2} (v_2+1) x_2^{v_2} \sum_{v_3} x_3^{v_3}} \tag{2.2c}$$

Since

$$\sum_{\lambda=0}^{\infty} x^{\lambda} = (1-x)^{-1}$$

and

$$\sum_{\lambda=0}^{\infty} \lambda x^{\lambda} = x(1-x)^{-2}$$

then

$$N_{v_1v_2v_3} = N \exp(-v_1 h\nu_1/kT)(v_2+1)\exp(-v_2 h\nu_2/kT)\exp(-v_3 h\nu_3/kT)$$
$$\times[1-\exp(-h\nu_1/kT)][1-\exp(-h\nu_2/kT)]^2[1-\exp(-h\nu_3/kT)] \tag{2.3a}$$

Finally, we have the number densities of the asymmetric and symmetric modes of CO_2 given, respectively, by

$$N_{001} = N_{CO_2} \exp(-h\nu_3/kT)Z \tag{2.3b}$$

and

$$N_{100} = N_{CO_2} \exp(-h\nu_1/kT)Z \tag{2.3c}$$

where

$$Z = [1-\exp(-h\nu_1/kT)][1-\exp(-h\nu_2/kT)]^2[1-\exp(-h\nu_3/kT)] \tag{2.3d}$$

2.2. V–T Energy Transfer of the CO_2 Symmetric and Bending Modes, τ_{10} and τ_{20}

We shall assume that the target CO_2 molecule can be represented by a pair of vibrational modes. In Section 1.7.2 we derived the relaxation equations for such CO_2 molecules in collision with another CO_2 molecule that was assumed to be structureless. In the laser we must consider collisions of the target CO_2 with He and N_2 as well as with other CO_2 molecules. Under these circumstances, Eq. (1.113) becomes

$$\frac{dE_1}{dt} = h\nu_1 \sum_{\lambda l} n_\lambda m_l \sum_{\iota=1}^{c} N_\iota/N_{CO_2}[(\lambda+1)k_{0,100;1,000}^{\iota\text{-}CO_2} - \lambda k_{100,000}^{\iota}$$
$$+(\lambda+1)lk_{0,100;1,000}^{\iota\text{-}CO_2} - \lambda(l+1)k_{1,000;0,100}^{\iota\text{-}CO_2}]$$

where $k_{0,100;1,000}^{N_2\text{-}CO_2}$ is the rate for the V–V process

$$N_2 + CO_2(1,0,0) \rightarrow N_2(\nu=1) + CO_2(000), \quad \text{etc.}$$

and $k_{100,000}^{He}$ is the rate for the V–V process

$$He + CO_2(1,0,0) \rightarrow He + CO_2(0,0,0), \quad \text{etc.}$$

and N_t/N are the fractional particle concentrations introduced in Eq. (1.38); Eq. (1.111) is replaced by

$$\sum_{\lambda} n_{\lambda} = N_{CO_2} = \sum_{l} m_l \tag{2.4}$$

and Eq. (1.115), the equilibrium value of E_1 at T, becomes

$$E_1(T) = N_{CO_2} h\nu_1 [\exp(h\nu_1/kT) - 1]^{-1} \tag{2.5}$$

The V–T relaxation time for the symmetric mode will be given by

$$\tau_{10}^{-1} \equiv \sum_{t=1}^{c} N_t k^t_{100;000}[\exp(-h\nu_1/kT) - 1] \tag{2.6}$$

while that for the bending mode will be given by

$$\tau_{20}^{-1} \equiv \sum_{t=1}^{c} N_t k^t_{010;000}[\exp(-h\nu_2/kT) - 1] \tag{2.7}$$

Equations (2.6) and (2.7) are in agreement with the Manes and Seguin general formula $\tau_{ij}(T)$. These authors quote temperature-independent values for

$$k^t_{ij}[\exp(-h\nu/kT) - 1]$$

The V–V relaxation formulas derived in Section 1.7.1 were for nondegenerate vibrational modes, whereas there is an accidental (Fermi) degeneracy between the symmetric and bending modes. We shall derive τ_{12} in Section 2.5.

2.3. V–V Energy Transfer between the Asymmetric Mode of CO_2 and N_2, τ_{43}

If we neglect the symmetric and bending modes in N_2–CO_2 collisions, then we have the process discussed in Section 1.7.1, that of energy transfer between two nonresonant SHOs.

In Fig. 6 we present a schematic of the energy-transfer processes involved in the collision process. The energy rate equation for N_2 is given

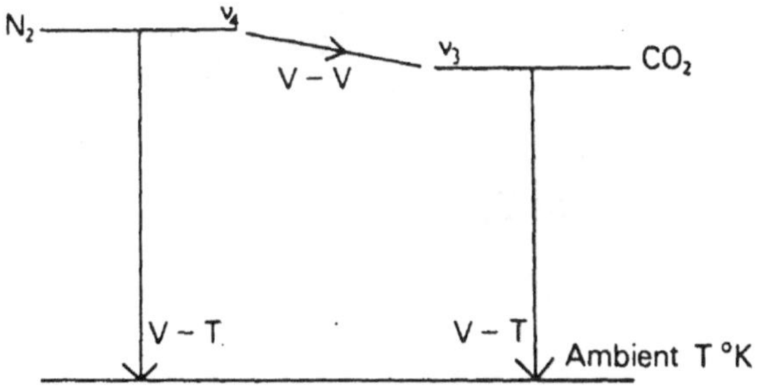

Fig. 6. Energy-transfer processes in N_2–CO_2.

in Eq. (1.104), with an analogous equation for CO_2. We shall change our notation to conform to that of Manes and Seguin,[11] that is, the subscripts change from a to 4 and b to 3; hence for N_2 we have

$$\frac{dE_4}{N\,dt}=(ak'^{N_2}_{1,000}+bk'^{N_2-CO_2}_{1,000;0,000})[\exp(-h\nu_4/kT)-1][E_4-E_4(T)]$$

$$+k^{N_2-CO_2}_{1,000;0,001}b\left\{\frac{E_3}{E_3(T)}[E_4(T)-E_4]\frac{1-\exp(-h\nu_4/kT)}{1-\exp(-h\nu_3/kT)}\right.$$

$$\left.-\frac{E_4}{E_3(T)}[E_3(T)-E_3]\right\} \qquad (2.8)$$

and for CO_2 the rate of change of energy per unit volume is given by

$$\frac{dE_3}{N\,dt}=(bk'^{CO_2}_{001;000}+ak'^{N_2-CO_2}_{0,001,0,000})[\exp(-h\nu_3/kT)-1][E_3-E_3(T)]$$

$$+k^{N_2-CO_2}_{0,001;1,000}a\left\{\frac{E_4}{E_4(T)}[E_3(T)-E_3]\frac{1-\exp(-h\nu_3/kT)}{1-\exp(-h\nu_4/kT)}\right.$$

$$\left.-\frac{E_3}{E_4(T)}[E_4(T)-E_4]\right\} \qquad (2.9)$$

where k' are the rate coefficients for V–T deexcitation of molecules upon collision with like molecules.

According to Hoffman and Vlases[12] we can approximate these equations by neglecting the V–T terms. In addition, since

$$[h(\nu_4-\nu_3)]/k=25°\text{K} \qquad (2.10)$$

we can make the further approximation that

$$\nu_4\approx\nu_3 \qquad (2.11)$$

Equation (2.8) becomes

$$\frac{dE_4}{N\,dt}=bk^{N_2-CO_2}_{1,000;0,001}\left\{\frac{E_3}{E_3(T)}[E_4(T)-E_4]-\frac{E_4}{E_3(T)}[E_3(T)-E_3]\right\}$$

$$=bk^{N_2-CO_2}_{1,000;0,001}\left[\frac{E_3E_4(T)}{E_3(T)}-E_4\right] \qquad (2.12)$$

From Eqs. (1.69) and (2.11), we have the equilibrium energy of the N_2 molecule at T, given by

$$E_4(T)=Nh\nu_4[\exp(h\nu_4/kT)-1]^{-1}\approx Nh\nu_3[\exp(h\nu_3/kT)-1]^{-1}$$

$$\approx E_3(T)$$

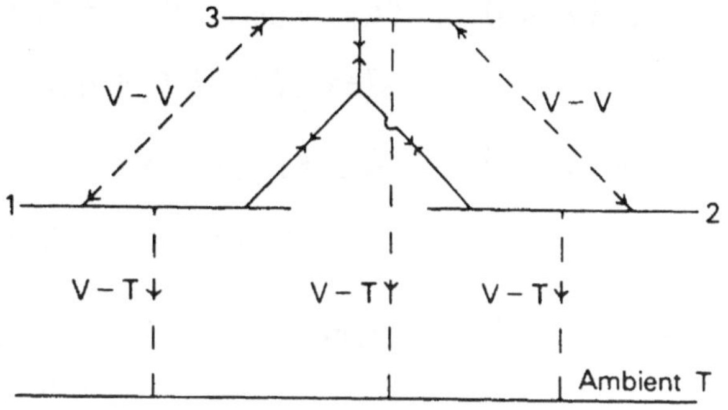

Fig. 7. Energy-transfer processes in CO_2.

while

$$E_3 \equiv E_3(T_3) = Nh\nu_3[\exp(h\nu_3/kT_3)-1]^{-1} \approx Nh\nu_4[\exp(h\nu_4/kT_3)-1]^{-1}$$
$$= E_4(T_3)$$

When this pair of results is substituted into Eq. (2.12), we have

$$\frac{dE_4}{dt} = Nbk^{N_2\text{-}CO_2}_{1,000;0,001}[E_4(T_3)-E_4] \equiv \frac{E_4(T_3)-E_4}{\tau_{43}} \tag{2.13}$$

which defines the V–V relaxation time

$$\tau_{43}^{-1}(T) = N_{CO_2}k^{N_2\text{-}CO_2}_{1,000;0,001}(T) \tag{2.14}$$

Equation (2.13) is used in Eq. (4) of Manes and Seguin.[11]

2.4. V–V Energy-Transfer Processes among the Three Vibrational Modes of CO_2, τ_3

We consider energy transfer between the asymmetric stretching mode, ν_3, the bending mode, ν_2, and the symmetric stretching mode, ν_1, as shown in Fig. 7. In this figure, the transfer processes represented by dashed lines are the three-mode generalization of the two-mode system derived in Eq. (1.118). If we label the states 1, 2, 3 as in Manes and Seguin,[11] then the generalization of Eq. (1.118) is

$$\frac{dE_3}{dt} = \frac{E_3-E_3(T)}{\tau_{VT}(T)} - \frac{E_3-E_3(T,T_2)}{\tau_{VV}(T,T_2)} - \frac{E_3-E_3(T,T_1)}{\tau_{VV}(T,T_1)} \tag{2.15}$$

where, for a gas consisting entirely of one type of molecule,

$$\tau_{VT}^{-1} = N_{CO_2}k_{100,000}[\exp(-h\nu_3/kT)-1] \tag{2.16}$$

$$\tau_{VV}(T, T_2)^{-1} = N_{CO_2}k_{100,010}\exp(+h(\nu_2-\nu_3)/kT)[\exp(h\nu_2/kT_2)-1]^{-1}$$
$$\times\{\exp[h\nu_2/kT_2 + h(\nu_3-\nu_2)/kT]-1\} \tag{2.17}$$
$$\tau_{VV}(T, T_1)^{-1} = N_{CO_2}k_{100,001}\exp(h(\nu_1-\nu_3)/kT)[\exp(h\nu_1/kT_1)-1]^{-1}$$
$$\times\{\exp[h\nu_1/kT_1 + h(\nu_3-\nu_1)/kT]-1\} \tag{2.18}$$

The equilibrium energies $E_3(T, T_2)$ and $E_3(T, T_1)$ are given by Eq. (1.108).

According to Witteman,[2] the energy-transfer processes represented by the dashed, arrowed lines of Fig. 7 are negligibly slow compared with the process where one $h\nu_3$ quantum is exchanged simultaneously with a quantum $h\nu_1$ and a quantum $h\nu_2$, as represented by the solid, arrowed lines in Fig. 7. To derive the relaxation equations for this process, we must generalize the analysis of Section 1.7 to consider the collision of several types of structureless projectiles with a molecule having three vibrational modes.

Let the energy levels associated with modes ν_1, ν_2, and ν_3 be designated by the quantum numbers μ, λ, and l, respectively. We shall assume that it is sufficient to consider single quantum transitions that result in the rate of change of energy per unit volume of mode ν_3, given by

$$\frac{dE_3}{dt} = h\nu_3\Bigg(\sum_{l=0}^{\infty}\sum_{\lambda=1}^{\infty}\sum_{\mu=1}^{\infty}\frac{m_l n_\lambda p_\mu}{N_{CO_2}^3}\sum_{t=1}^{c} N_t N_{CO_2} k^t_{\mu\lambda l,\mu-1\lambda-1l+1}$$
$$-\sum_{l=1}^{\infty}\sum_{\mu=0}^{\infty}\sum_{\lambda=0}^{\infty}\frac{m_l n_\lambda p_\mu}{N_{CO_2}^3}\sum_{t=1}^{c} N_t N_{CO_2} k^t_{\mu\lambda l,\mu+1\lambda+1l-1}\Bigg) \tag{2.19}$$

where all V–T transitions have been neglected. Since each target CO_2 molecule will be in one or another of the m_l levels, then

$$\sum_{l=0}^{\infty} m_l = N_{CO_2} \tag{2.20}$$

and similarly for n_λ and p_μ. We shall also have

$$h\nu_3\sum_{l=0}^{\infty} l m_l = E_3 \tag{2.21}$$

and similarly for ν_2 and ν_1. In Problems 2 and 5 we presented relations among the rate coefficients, which are required to simplify Eq. (2.19). With these relations, Eq. (2.19) becomes

$$\frac{dE_3}{dt} = h\nu_3\sum_{l,\lambda,\mu}\frac{n_\lambda m_l p_\mu}{N_{CO_2}^3}\sum_{t=1}^{c} N_t N_{CO_2}[(l+1)\lambda\mu k^t_{110,001} - l(\lambda+1)(\mu+1)k^t_{001,110}] \tag{2.22}$$

When Eq. (2.21) is substituted into Eq. (2.13) we have

$$\frac{dE_3}{dt}=h\nu_3\Big[\sum_t N_t N_{CO_2}k^t_{110,001}\frac{E_2}{N_{CO_2}h\nu_2}\frac{E_1}{N_{CO_2}h\nu_1}\Big(\frac{E_3}{N_{CO_2}h\nu_3}+1\Big)$$
$$-\sum_t N_t N_{CO_2}k^t_{001,110}\frac{E_3}{N_{CO_2}h\nu_3}\Big(\frac{E_2}{N_{CO_2}h\nu_2}+1\Big)\Big(\frac{E_1}{N_{CO_2}h\nu_1}+1\Big)\Big]$$

$$\frac{dE_3}{dt}=\sum_t N_t k^t_{001,110}\Big[\exp\Big(\frac{h(\nu_1+\nu_2-\nu_3)}{kT}\Big)\frac{E_2}{N_{CO_2}h\nu_2}\frac{E_1}{N_{CO_2}h\nu_1}(E_3-N_{CO_2}h\nu_3)$$
$$-E_3\Big(\frac{E_2}{N_{CO_2}h\nu_2}+1\Big)\Big(\frac{E_1}{N_{CO_2}h\nu_1}+1\Big)\Big]$$

having used the results of Problem 5. When Eq. (1.69) is substituted into this equation,

$$\frac{dE_3}{dt}=\sum_t N_t k^t_{001,110}\exp\Big[\frac{h(\nu_1+\nu_2-\nu_3)}{kT}\Big]\Big[\exp\Big(\frac{h\nu_2}{kT_2}\Big)-1\Big]^{-1}\Big[\exp\Big(\frac{h\nu_1}{kT_1}\Big)-1\Big]^{-1}$$
$$\times\Big\{E_3+N_{CO_2}h\nu_3-E_3\Big[1+\exp\Big(\frac{h\nu_2}{kT_2}\Big)-1\Big]\Big[1+\exp\Big(\frac{h\nu_1}{kT_1}\Big)-1\Big]$$
$$\times\exp\Big[-\frac{h(\nu_1+\nu_2-\nu_3)}{kT}\Big]\Big\}$$
$$=\sum_t N_t k^t_{001,110}\exp\Big[\frac{h(\nu_1+\nu_2-\nu_3)}{kT}\Big]\Big[\exp\Big(\frac{h\nu_2}{kT_2}\Big)-1\Big]^{-1}\Big[\exp\Big(\frac{h\nu_1}{kT_1}\Big)-1\Big]^{-1}$$
$$\times\Big\{\exp\Big[\frac{h\nu_2}{kT_2}+\frac{h\nu_1}{kT_1}+\frac{h(\nu_3-\nu_2-\nu_1)}{kT}\Big]-1\Big\}[E_3(T,T_1,T_2)-E_3] \quad (2.23)$$

where the equilibrium energy is defined by

$$E_3(T,T_1,T_2)=N_{CO_2}h\nu_3\Big\{\exp\Big[\frac{h\nu_1}{kT_1}+\frac{h\nu_2}{kT_2}+\frac{h(\nu_3-\nu_2-\nu_1)}{kT}\Big]-1\Big\}^{-1} \quad (2.24)$$

as in Manes and Seguin,[11] where f_i is written for our ν_i.

The V–V relaxation time for these simultaneous processes can be defined to be

$$\tau_3(T,T_1,T_2)^{-1}=\sum_t N_t k^t_{001,110}\exp\Big[\frac{h(\nu_1+\nu_2-\nu_3)}{kT}\Big]$$
$$\times\Big[\exp\Big(\frac{h\nu_2}{kT_2}\Big)-1\Big]^{-1}\Big[\exp\Big(\frac{h\nu_1}{kT_1}\Big)-1\Big]^{-1}$$
$$\times\Big\{\exp\Big[\frac{h\nu_1}{kT_1}+\frac{h\nu_2}{kT_2}+\frac{h(\nu_3-\nu_2-\nu_1)}{kT}\Big]-1\Big\} \quad (2.25)$$

and Eq. (2.22) becomes

$$\frac{dE_3}{dt}=\frac{E_3(T, T_1, T_2)-E_3}{\tau_3(T, T_1, T_2)} \tag{2.26}$$

as in Manes and Seguin,[11] Eq. (3). Their form of Eq. (2.25) replaces the factor

$$\sum_t N_t k_{001,110}^t \exp\left[\frac{h(\nu_1+\nu_2-\nu_3)}{kT}\right] \to N_{CO_2}A_3 \tag{2.27}$$

In a six-temperature model where the ambient temperature is allowed to vary, we should retain the T-dependent form of Eq. (2.25).

In deriving the rate of change of energy in the remaining pair of modes, E_1 and E_2, we note

$$h\nu_3 \approx h\nu_1 + h\nu_2$$

and so the fractional gain by dE_1/dt will be $h\nu_1/h\nu_3$; hence

$$\frac{dE_2}{dt}=-\frac{h\nu_2}{h\nu_3}\frac{dE_3}{dt} \tag{2.28}$$

and

$$\frac{dE_1}{dt}=-\frac{h\nu_1}{h\nu_3}\frac{dE_3}{dt} \tag{2.29}$$

2.5. Resonant V–V Energy Transfer between the Symmetric and Bending Modes of CO_2, τ_{12}

Consider the collision of molecules with two vibrational modes in accidental degeneracy:

$$2\nu_2 \approx \nu_1 \tag{2.30}$$

with structureless projectiles. The rate of energy change per unit volume will be

$$\frac{dE_1}{dt}=h\nu_1 \sum_{l=0}^{\infty}\sum_{\lambda=0}^{\infty}\sum_{\mu=0}^{\infty}\sum_{t=1}^{c} N_t N_{CO_2}$$

$$\times\left(\frac{n_\lambda m_{l+2}p_\mu}{N_{CO_2}^3}k_{\lambda l+2\mu,\lambda+1l\mu}^t-\frac{n_{\lambda+1}m_l p_\mu}{N_{CO_2}^3}k_{\lambda+1l\mu,\lambda l+2\mu}^t\right) \tag{2.31}$$

When the results of Problem 2 are substituted into this equation and we use detailed balance in the form

$$k_{020,100} = k_{100,020} \tag{2.32}$$

and since the energy states are in resonance, then

$$\frac{dE_1}{dt} = h\nu_1 \sum_t N_t N_{CO_2} k^t_{020,100} \sum_{l,\lambda} \left[(\lambda+1)(l+2)(l+1)\frac{n_\lambda m_{l+2}}{N^2_{CO_2}} - (\lambda+1)(l+2)(l+1)\frac{n_{\lambda+1} m_l}{N^2_{CO_2}} \right]$$

In the first (second) term on the right-hand side we can renumber the terms of the $l(\lambda)$ sum by changing $l+2(\lambda+1)$ to $l'(\lambda')$; hence, upon dropping the primes, we have

$$\begin{aligned} \frac{dE_1}{dt} &= h\nu_1 \sum_t N_t N_{CO_2} k^t_{020,100} \sum_{l,\lambda} [(\lambda+1)l(l-1) - \lambda(l+2)(l+1)] \frac{n_\lambda m_l}{N^2_{CO_2}} \\ &= h\nu_1 \sum_t N_t N_{CO_2} k^t_{020,100} \sum_{l,\lambda} (l^2 - 4l\lambda - l - 2\lambda) \frac{n_\lambda m_l}{N^2_{CO_2}} \end{aligned} \tag{2.33}$$

To perform the summations in Eq. (2.33), we note that

$$\sum_l m_l = N_{CO_2} = \sum_\lambda n_\lambda$$

and

$$\begin{aligned} h\nu_2 \sum_l l m_l &= E_2 \\ h\nu_1 \sum_\lambda \lambda n_\lambda &= E_1 \end{aligned} \tag{2.34}$$

where the concept of n_λ, and m_l, includes the counting due to degeneracy. We are now in a position to carry out the summations in Eq. (2.33) using Problem 4 for the l^2 term, and Eq. (2.34) for the linear terms, giving

$$\frac{dE_1}{dt} = 2 \sum_t N_t k^t_{020,100} \left\{ \frac{1+\exp(h\nu_2/kT_2)}{[\exp(h\nu_2/kT_2)-1]^2} N_{CO_2} h\nu_2 - \frac{2E_1E_2}{N_{CO_2} h\nu_2} - E_2 - E_1 \right\} \tag{2.35}$$

When Eq. (1.69) is substituted into the first term, we obtain

$$\begin{aligned} \frac{dE_1}{dt} &= 2 \sum_t N_t k^t_{020,100} \left\{ E_2 \left[\frac{1+\exp(h\nu_2/kT_2)}{\exp(h\nu_2/kT_2)-1} - 1 \right] - E_1 \left[1 + \frac{2}{\exp(h\nu_2/kT_2)-1} \right] \right\} \\ &= 2 \sum_t N_t k^t_{020,100} \left[E_2(T_2) \frac{2}{\exp(h\nu_2/kT_2)-1} - E_1 \frac{\exp(h\nu_2/kT_2)+1}{\exp(h\nu_2/kT_2)-1} \right] \end{aligned} \tag{2.36}$$

From Eq. (1.69) we have

$$E_1(T_2) = \frac{Nh\nu_1}{\exp(h\nu_1/kT_2) - 1} = \frac{2Nh\nu_2}{\exp(2h\nu_2/kT_2) - 1}$$

Therefore,

$$E_1(T_2) = \frac{2Nh\nu_2}{[\exp(h\nu_2/kT_2)+1][\exp(h\nu_2/kT_2)-1]} = \frac{2E_2(T_2)}{\exp(h\nu_2/kT_2)+1}$$

which, used in Eq. (2.36), gives

$$\frac{dE_1}{dt} = 2\sum_t N_t k^t_{020,100}\left[E_1(T_2)\frac{\exp(h\nu_2/kT_2)+1}{\exp(h\nu_2/kT_2)-1} - E_1\frac{\exp(h\nu_2/kT_2)+1}{\exp(h\nu_2/kT_2)-1}\right]$$
$$\equiv \tau_{12}^{-1}[E_1(T_2) - E_1] \qquad (2.37)$$

where the V–V relaxation time is given by

$$\tau_{12}(T_2)^{-1} \equiv 2\sum_t N_t k^t_{020,100}\frac{1+\exp(-h\nu_2/kT_2)}{1-\exp(-h\nu_2/kT_2)} \qquad (2.38)$$

as given in Manes and Seguin,[11] when

$$2\sum_t N_t k^t_{020,100} \to N_{CO_2}A_{12} \qquad (2.39)$$

2.6. Time Evolution of the Cavity-Field Intensity I_ν

2.6.1. Population Inversion and the $\dot{I}_\nu$ Equation

The number of transitions per unit time that a quantum mechanical system will undergo spontaneously from level n to a lower level m, emitting a quantum of radiation, is the spontaneous transition rate:

$$A_{nm} = \tau_{sp}^{-1} \qquad (\text{sec}^{-1}) \qquad (2.40)$$

Radiation emitted from a quantum mechanical system in the presence of external radiation, whose phase is the same as that of the external radiation, is called stimulated radiation. Stimulated radiation is characterized by two additional Einstein coefficients denoted B_{nm} and B_{mn}. Einstein's relations are[13]

$$\theta_n B_{nm} = \theta_m B_{mn}$$

$$A_{nm} = \frac{8\pi h\nu^3}{c^3} B_{nm} \tag{2.41}$$

where θ_i denotes the degeneracy (multiplicity) of level i, and the units of B are $E^{-1}T^{-2}L^3$.

We note that there are two forms of the B-coefficients in the literature: the density form, B, used here, and the intensity form B^i, used, for example, in Mitchell and Zemansky.[13] They are related by

$$B = \frac{c}{4\pi} B^i$$

Let $N_{0,0,1}P(J)$ $[N_{1,0,0}P(J+1)]$ be the number of CO_2 molecules per unit volume in the rotational level $J(J+1)$ of the lowest symmetric (asymmetric) mode, and let $\rho(\nu)$ denote the radiation density, the number of photons per unit volume.

The number of absorptions and stimulated emissions per unit volume per unit time will be, respectively [see Eq. (2.3)],

$$N_{1,0,0}P(J+1)B_{13}\rho(\nu) \qquad \text{and} \qquad N_{0,0,1}P(J)B_{31}\rho(\nu)$$

where

$$P(J) \equiv (2hcB/kT)\theta_J \exp\left[\frac{-hcBJ(J+1)}{kT}\right] \tag{2.42a}$$

$$N_\lambda = N_{CO_2} \exp\left(\frac{-h\nu_\lambda}{kT_\lambda}\right)\left[1-\exp\left(\frac{-h\nu_1}{kT_1}\right)\right]\left[1-\exp\left(\frac{-h\nu_2}{kT_2}\right)\right]^2\left[1-\exp\left(\frac{-h\nu_3}{kT_3}\right)\right] \tag{2.42b}$$

where λ denotes 001 or 100, B is the rotational constant, and

$$\theta_J = (2J+1) \tag{2.42c}$$

The net stimulated increase in radiation-energy density per unit time will be

$$\begin{aligned} h\nu[N_{0,0,1}P(J)B_{31} - N_{1,0,0}P(J+1)B_{13}]\rho(\nu) \\ = h\nu B_{31}[N_{0,0,1}P(J) - N_{1,0,0}P(J+1)\theta_J/\theta_{J+1}]\rho(\nu) \\ = h\nu B_{31}\,\Delta N\rho(\nu) \\ = \frac{h\nu c^3\,\Delta N\rho(\nu)}{8\pi h\nu^3 \tau_{sp}} \end{aligned} \tag{2.42d}$$

having used Eqs. (2.41), and where we have defined ΔN by

$$\Delta N \equiv N_{0,0^\circ,1}P(J)-(\theta_J/\theta_{J+1})N_{1,0^\circ,0}P(J+1) \tag{2.42e}$$

ΔN is the population inversion between the upper and lower laser levels.

Let[14] $g(\nu)\,d\nu$ be the probability that a transition between the energy levels will result in an emission, or an absorption, of a photon whose energy lies between $h\nu$ and $h(\nu+d\nu)$. The net stimulated increase in radiation density per unit time, $\mathrm{EL^{-3}T^{-1}}$, in the domain $h\nu$ to $h(\nu+d\nu)$ will be

$$\frac{h\nu c^3\,\Delta N\,\rho(\nu)g(\nu)\,d\nu}{8\pi h\nu^3\tau_{sp}} \equiv \frac{h\nu\,\Delta N\,c^2 I_\nu \delta(\nu-\nu_0)g(\nu)\,d\nu}{8\pi h\nu^3\tau_{sp}} \tag{2.43}$$

having defined the radiation flux, $\mathrm{EL^{-2}T^{-1}}$, by

$$c\rho(\nu) \equiv I_\nu \delta(\nu-\nu_0) \tag{2.44}$$

using the Dirac delta function to characterize the extremely narrow width of the laser line. When Eq. (2.43) is integrated over the frequency, we obtain

$$\frac{\Delta N\,c^2 I_{\nu 0}\,g(\nu_0)}{8\pi\nu_0^2\tau_{sp}}$$

which is expressed in units of $\mathrm{EL^{-3}T^{-1}}$. Therefore, the rate of increase of radiation flux will be

$$\left(\frac{dI_{\nu 0}}{dt}\right)_{gain} = \frac{\Delta N\,c^3 I_{\nu 0}\,g(\nu_0)}{8\pi\nu_0^2\tau_{sp}} \tag{2.45}$$

The total loss of intensity is principally due to:

(i) transmission, absorption, and scattering by the mirrors;
(ii) absorption within the amplifying medium due to the system, in fact, not being an ideal two-level;
(iii) scattering by optical inhomogeneities within the amplifying medium; and
(iv) diffraction losses by mirrors.

All these losses can be included in one parameter—the lifetime of a photon within the laser cavity, τ_c. In other words, τ_c^{-1} will be the total rate of loss of photons from the laser, per second, as a result of (i)–(iv); hence

$$\left(\frac{dI}{dt}\right)_{loss} = \frac{I_\nu}{\tau_c} \tag{2.46}$$

where an approximate form for τ_c is derived in Section 2.6.1.

For laser action to take place, the following condition must hold:

$$\left(\frac{dI}{dt}\right)_{\text{gain}} - \left(\frac{dI}{dt}\right)_{\text{loss}} \geq 0 \tag{2.47}$$

which can be written, using Eqs. (2.45) and (2.46), as

$$\frac{\Delta N c^3 g(\nu_0) I_{\nu_0}}{8\pi\nu_0^2\tau_{\text{sp}}} - \frac{I_{\nu_0}}{\tau_c} \geq 0$$

That is, the threshold population inversion required for oscillation is

$$\Delta N \geq \frac{8\pi\nu_0^2\tau_{\text{sp}}}{c^3 g(\nu_0)\tau_c} \tag{2.48}$$

For minimum threshold inversion we take the largest value of $g(\nu_0)$, i.e., at the line center ν_0. The shape of the curve $g(\nu_0)$ depends on two classes of line-broadening mechanisms called homogeneous (Lorentzian-shape collision broadening) and inhomogeneous (Gaussian-shape) (see Beesley,[14] page 41).

For a Lorentzian frequency distribution with line center ν_0, the width at half-maximum, $\Delta\nu$, is given by (Beesley,[14] p. 42):

$$g(\nu_0) = \operatorname*{Lt}_{\nu\to\nu_0} \frac{1}{2\pi} \cdot \frac{\Delta\nu}{[(\nu-\nu_0)^2 + (\Delta\nu/2)^2]} = \frac{2}{\pi\,\Delta\nu} \tag{2.49}$$

and so, neglecting spontaneous emission for the moment, the time evolution of the cavity-field intensity will be given by an equation expressed in units of $\text{EL}^{-2}\text{T}^{-2}$, that is,

$$\frac{dI_{\nu_0}}{dt} = -\frac{I_{\nu_0}}{\tau_c} + \frac{c^3\,\Delta N\,I_{\nu_0}}{4\pi^2\nu_0^2\,\Delta\nu\,\tau_{\text{sp}}}$$

$$= -\frac{I_{\nu_0}}{\tau_c} + c\nu_0\,\Delta N\left(\frac{\lambda_0^2}{4\pi^2\nu_0\,\Delta\nu\,\tau_{\text{sp}}}\right)I_{\nu_0} \tag{2.50}$$

in the notation of Manes and Seguin, provided we multiply the second term on the right-hand side by F, the ratio of the mode volume filled with gain medium to the total mode volume. F is also the "filling factor" obtained experimentally by dividing the laser gain-medium length by the optical-cavity length. The laser wavelength λ equals c/ν.

A rate equation describing the reaction of the laser medium on the photon density in the cavity has been given by Gilbert *et al.*,[15] who state that the stimulated-emission process is initiated by the spontaneous emission term, expressed in units of $\text{EL}^{-2}\text{T}^{-2}$:

$$\frac{ch\nu N_{0,0,1}P(J)}{\tau_{\text{sp}}} \times G \tag{2.51}$$

where G denotes the fraction of photons being spontaneously radiated into the small angular aperture of the resonator within the spectral width $d\nu$ of a single axial mode, near the peak of the rotational–vibrational emission line. In other words, first the small fraction of radiation retained within the resonator will be

$$\frac{\alpha}{4\pi} \tag{2.52}$$

where α is a solid angle, and second, that portion of the radiation emitted within the spectral width near ν_0 will be [see Eq. (2.49)]

$$g(\nu_0)\,d\nu = \frac{2\,d\nu}{\pi\,\Delta\nu} \tag{2.53}$$

from Eq. (2.49), where[15] the formula for $\Delta\nu$ shall be given later in Eq. (2.86); hence

$$\Delta\nu \sim 4.0\text{GHz} \quad \text{at STP} \qquad \text{and} \qquad d\nu \sim (2\pi\tau_c)^{-1} \tag{2.54}$$

Thus,

$$G = \frac{\alpha}{4\pi} \times \frac{2\,d\nu}{\pi\,\Delta\nu} \tag{2.55}$$

and the spontaneous term to be added to Eq. (2.50) becomes

$$\frac{ch\nu_0 N_{0,0,1} P(J)\alpha\,d\nu}{2\pi^2\tau_{sp}\,\Delta\nu} \tag{2.56}$$

Finally, we have the equation that describes the time evolution of the cavity-field intensity [Eqs. (2.50) and (2.56)]:

$$\frac{dI_{\nu_0}}{dt} = -\frac{I_{\nu_0}}{\tau_c} + ch\nu_0\left[\frac{\Delta N\,WI_{\nu_0}}{h} + N_{0,0,1}P(J)S\right] \tag{2.57}$$

where

$$W \equiv \frac{\lambda_0^2 g(\nu)}{8\pi\nu_0\tau_{sp}} = \frac{\lambda_0^2}{4\pi^2\nu_0\,\Delta\nu\,\tau_{sp}} \quad (\text{cm}^2\,\text{sec}) \tag{2.58}$$

excluding the "filling factor F" of Manes and Seguin, and

$$S \equiv \frac{\alpha\,d\nu}{2\pi^2\tau_{sp}\,\Delta\nu} = \frac{\lambda_0^2 2\,d\nu}{\pi A\tau_{sp}\,\Delta\nu} \tag{2.59}$$

having used the result of Gilbert *et al.*:

$$\frac{\alpha}{4\pi} = \frac{\lambda_0^2}{A} \sim 3\times 10^{-7} \tag{2.60}$$

where A is the reflecting area of the smallest cavity mirror. Gilbert *et al.* also quote

$$\frac{d\nu}{\Delta\nu} = 2 \times 10^{-3} \tag{2.61}$$

the ratio of laser to spontaneous half-widths.

The spontaneous term, Eq. (2.56), does not have the structure given by Manes and Seguin. Indeed a dimensional analysis of their Eq. (5) shows that their equation must be wrong. An alternative derivation of Eq. (2.57) by Siegman[16] leads to a result close to that of Manes and Seguin. Let n be the total number of photons, *not* per unit volume; then

$$\frac{dn}{dt} = Kn(N_3 - N_1) + KN_3 - \text{loss} \tag{2.62}$$

where it is assumed that K is the same for both spontaneous and stimulated terms. Since

$$n = \frac{IV}{ch\nu} \tag{2.63}$$

where V is the volume of gain media, then

$$\frac{ch\nu}{V}\frac{dn}{dt} = \frac{ch\nu}{V}(Kn\,\Delta N + KN_3) - \frac{I}{\tau_c}$$

$$\frac{dI}{dt} = K\left(\Delta N I + ch\nu\frac{N_3}{V}\right) - \frac{I}{\tau_c} \tag{2.64a}$$

$$= ch\nu W(\Delta N I + S) - \frac{I}{\tau_c}$$

where

$$S \equiv \frac{ch\nu N_3}{V} \quad \text{and} \quad K \equiv ch\nu W \tag{2.65}$$

The gain in photons due to the stimulating radiation can be estimated by solving Eq. (2.45), which is equivalent to neglecting the first and third terms on the right-hand side of Eq. (2.57), and assuming ΔN is independent of I_ν, the small-signal gain approximation:

$$\frac{dI_\nu}{I_\nu c\,dt} = \nu\,\Delta N\,W \equiv \alpha \tag{2.64b}$$

where ΔN is as defined in Eq. (2.42), with dimensions L^{-3}, and W [see Eq. (2.58)] has dimensions L^2T. Hence the dimensions of α are L^{-1}. The solution of this equation is

$$I_\nu = I_0\, e^{\alpha x} \tag{2.64c}$$

where α is the gain per unit length.

The effective radiative cross section is defined by

$$\sigma = \nu W \tag{2.64d}$$

and has the value $5.8 \times 10^{-20}\ \mathrm{cm}^2$ for the TEA–CO_2 medium (see Stark[17]).

2.6.2. *Laser-Cavity Lifetime and Output Power*

We consider a steady-state (CW) laser cavity of length L (cm) containing some medium that provides a constant and uniform gain per unit length α (cm^{-1}). The cavity energy flux traveling toward the output mirror is x ($\mathrm{erg\ cm^{-2}\ sec^{-1}}$), of which a proportion Rx is reflected by the output mirror and a proportion $\bar{L}x$ is lost (diffraction losses, etc.). At a distance z cm along the cavity from the output mirror, the intensity traveling away from the output mirror is increased by a factor of $e^{\alpha z}$ [Fig. 8; see also Eq. (2.64)]. After a round-trip the intensity will have increased by a factor of $e^{2\alpha L}$, and since we are considering a steady-state situation, the intensity after a round-trip must equal the original intensity. Therefore

$$e^{2\alpha L} R x = x$$

or

$$e^{2\alpha L} R = 1 \tag{2.66}$$

From this equation we obtain an expression for the gain per unit length, α, in terms of the cavity length, L, and output mirror reflectivity, R:

$$\alpha = -\frac{1}{2L} \ln R \tag{2.67}$$

From Fig. 8 we see that the output intensity is

$$I_{\mathrm{out}} = (1 - R - \bar{L})x \qquad (\mathrm{erg\ cm^{-2}\ sec^{-1}}) \tag{2.68}$$

Hence, the output power is given by

$$P_{\mathrm{out}} = A(1 - R - \bar{L})x \qquad (10^{-7}\ \mathrm{W}) \tag{2.69a}$$

where A, in cm^2, is the cross-sectional area of the output mirror.

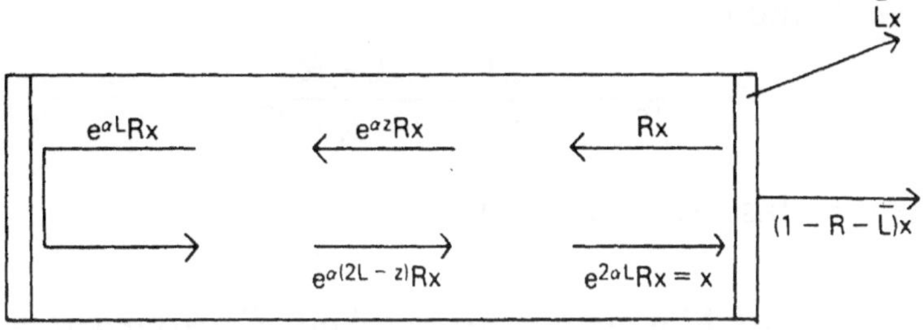

Fig. 8. Schematic of the variation of cavity-field intensity along the cavity.

The laser kinetics does not yield x, but does yield I_ν, which for a point model must be an average value for the cavity. We can define

$$I_\nu \equiv \frac{1}{L}\int_0^L I_z\, dz$$

$$= \frac{1}{L}\int_0^L (e^{\alpha z}Rx + e^{\alpha(2L-z)}Rx)\, dz$$

from Fig. 8. When Eq. (2.66) is substituted into the second term on the right-hand side, and the integration is performed, we have

$$I_\nu = \frac{x(1-R)}{\alpha L}$$

$$= \frac{2x(R-1)}{\ln R}$$

having used Eq. (2.66) again, which is substituted into Eq. (2.69a) to give

$$P_{\text{out}} = -\frac{A}{2}\ln R I_\nu(t)\frac{1-R-\bar{L}}{1-R} \qquad (10^{-7}\ \text{W}) \qquad (2.69\text{b})$$

The I/τ_c term in Eq. (2.64a) [see also Eq. (2.46)] describes $(dI/dt)_{\text{loss}}$, the rate of change of the cavity-field intensity due to the output (mirror) coupling and loss processes $\bar{L}$. We have

$$\left(\frac{dI_\nu}{dt}\right)_{\text{loss}} = (\text{output power/unit volume})\times c$$

$$= \frac{cP_{\text{out}}}{AL} = -\frac{c \ln R}{2L}\frac{1-R-\bar{L}}{1-R}I_\nu$$

$$= \frac{I_\nu}{\tau_c} \qquad (2.69\text{c})$$

where we have defined

$$\tau_c = -\frac{2L}{c \ln R}\frac{1-R}{1-R-\bar{L}} \tag{2.69d}$$

which is called the laser-cavity lifetime.

2.7. Summary of Equations of a Five-Temperature Model

2.7.1. General Considerations

E_1 represents the energy per unit volume (EL^{-3}) stored in the CO_2 symmetrical stretching mode, which relaxes via V–T transitions to the ground state, τ_{10} [see Eq. (2.6)] and via V–V transitions to the bending mode, τ_{12} [see Eq. (2.38)], and gains energy by: (i) V–V exchange among the three modes of CO_2, τ_3 [see Eq. (2.25)]; (ii) photon absorption from the stimulated emission,

$$\nu_0\,\Delta N\,WI_{\nu_0}(t) \tag{2.50a}$$

where, from Eq. (2.50), we have the form for W given in Eq. (2.58); and (iii) electronic excitation,

$$N_e(t)N_{CO_2}h\nu_1X_1(T) \tag{2.70a}$$

where $N_e(t)$ is the number of electrons per unit volume at time t, and where X_1 is the effective electron vibrational excitation rate for the symmetric stretching mode. Consequently, E_1 satisfies the equation

$$\frac{dE_1}{dt} = N_e(t)N_{CO_2}h\nu_1X_1(T) + \nu_1\,\Delta N\,WI_{\nu_0}(t) + \left(\frac{h\nu_1}{h\nu_3}\right)\frac{E_3 - E_3(T, T_1, T_2)}{\tau_3(T, T_1, T_2)} - \frac{E_1 - E_1(T)}{\tau_{10}(T)} - \frac{E_1 - E_1(T_2)}{\tau_{12}(T_2)} \tag{2.70b}$$

E_2 is the energy density stored in the doubly degenerate bending mode which relaxes via V–T transitions to the ground state τ_{20} [see Eq. (2.7)] and gains energy by: (i) electronic excitation,

$$N_e(t)N_{CO_2}h\nu_2X_2(T) \tag{2.71a}$$

(ii) V–V transitions with the symmetric mode τ_{12}; and (iii) V–V transitions among the three modes of CO_2 (τ_3). Thus, E_2 satisfies the equation

$$\frac{dE_2}{dt} = N_e(t)N_{CO_2}h\nu_2X_2(T) + \frac{E_1 - E_1(T_2)}{\tau_{12}} + \frac{h\nu_2}{h\nu_3}\frac{E_3 - E_3(T, T_1, T_2)}{\tau_3(T, T_1, T_2)} - \frac{E_2 - E_2(T)}{\tau_{20}(T)} \tag{2.71b}$$

The energy density E_3 (EL^{-3}) in the CO_2 asymmetric mode gains energy from the N_2 molecule, τ_{43} [see Eq. (2.14)] and from electronic excitation, and loses energy by the V–V process, τ_3, and by photon emission; hence

$$\frac{dE_3}{dt}=N_e(t)N_{CO_2}h\nu_3X_3(T)+\frac{E_4-E_4(T_3)}{\tau_{43}(T)} -\frac{E_3-E_3(T,T_1,T_2)}{\tau_3(T,T_1,T_2)}-\nu_3\,\Delta N\,WI_{\nu_0} \tag{2.72}$$

The energy density E_4 stored in the N_2 molecule is generated by electron excitation and is lost by pumping the CO_2 asymmetric mode, τ_{43}; hence

$$\frac{dE_4}{dt}=N_e(t)N_{N_2}h\nu_4X_4(T)-\frac{E_4-E_4(T_3)}{\tau_{43}(T)} \tag{2.73}$$

The five-temperature model consists of the preceding four equations for dE_i/dt, together with Eq. (2.57) for dI_{ν_0}/dt.

In summary, the following approximations and assumptions have been made in the Manes and Seguin model:

(a) The only collision processes to be included in the model are represented schematically in Fig. 9; additional processes are considered in Sections 2.9 and 2.10.

(b) In the V–V process between N_2 and $CO_2(0, 0, 1)$ we assumed that $\nu_3=\nu_4$; the consequences of relaxing this assumption are examined in Section 2.9.2.

(c) In the above equations, we have written the electron–molecule excitation rates, X_i, without any variable dependence, since the Manes and

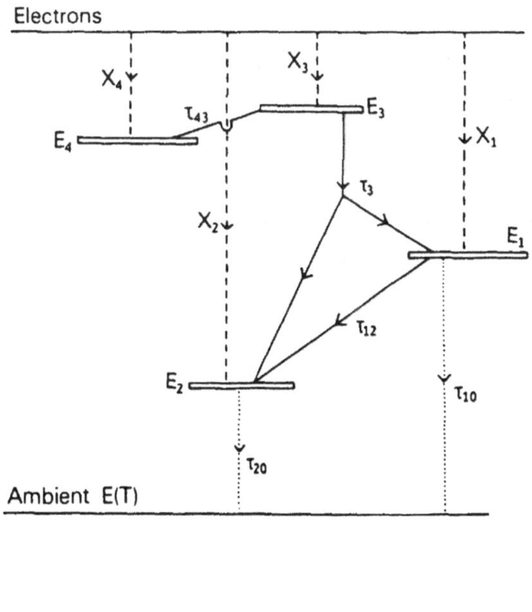

Fig. 9. Schematic energy-level diagram of the N_2–CO_2 laser, including dominant collision processes. Continuous, arrowed lines denote V–V processes, dotted, arrowed lines denote V–T processes, and dashed, arrowed lines denote electron–molecule vibrational excitation.

Seguin model treats them as constants. In Chapter 3 it shall be shown that a more realistic approximation is [see also Eq. (1.35)]

$$X_i(T, E) = \int Q^i(v) f(v, T, E) v^3 \, dv \tag{2.74}$$

where $Q^i(v) = \sum_j \alpha_j Q^i_j(v)$ is a weighted sum of the velocity-dependent electron–molecule cross sections $Q^i_j(v)$ (cm^2) for excitation of the jth vibrational level from the ground state of "species" i. The electron-velocity distribution function f, expressed in $(eV)^{-3/2}$, depends on the ambient temperature T and the externally applied electric field E. The α_j are constants, given explicitly in Eq. (3.162), and are expressed in $(eV)^{3/2}\,cm^{-3}\,sec^3$. The X_i are expressed in $cm^3\,sec^{-1}$. The problems involved in obtaining a more realistic approximation for the electron–molecule excitation rates shall be dealt with more fully in Chapter 3.

(d) In the above equations, the ambient temperature T appears implicitly in several of the quantities, yet there is no mechanism in the Manes and Seguin model for allowing for a variable ambient temperature. We examine this problem in Section 2.10 and derive an equation for the time evolution of the energy, E, stored in the kinetic motion of the molecules.

(e) A further enhancement to the Manes and Seguin model is to allow for dissociation of the CO_2 molecule into CO. Two possible schemes for including the effects of CO in the CO_2 laser are described in Section 2.9.

We recall that the term "five-temperature" refers to T_1, T_2, T_3, T_4, and T, although the latter is assumed constant in the Manes and Seguin model.

From Eq. (2.57) we see that the local gain coefficient of the medium is given by

$$g(\nu_0) = c\nu_0 \, \Delta N \, W = \frac{c\lambda_0^2}{8\pi}\left(\frac{2}{\pi \, \Delta\nu}\right)\frac{\Delta N}{\tau_{sp}} \tag{2.75}$$

which agrees with Rensch[18] apart from the factor c. In his model of the gas dynamic laser, Rensch solves Eqs. (2.71)–(2.74) in a modified form by: (a) dividing the equations by the gas-flow velocity v, which converts the integration variable from t to x; (b) dropping the X_i terms; (c) dropping τ_{10}, but including τ_{40}; and (d) replacing τ_3 by τ_{32}. He then numerically integrates Eqs. (2.69)–(2.73) along the x-axis, converting the vibrational energies so obtained into population densities that are substituted into ΔN to obtain $g(\nu_0)$. In the calculation of the relaxation times τ_j, Rensch's laser replaces helium with water.

2.7.2. Constants and Rate Data

In this section, we shall collect the values of the parameters that must be specified before Eqs. (2.57) and (2.70)–(2.73) can be solved.

In Table 1 we present the physical constants required for the CO_2 laser, while in Table 2 we present the values of various data used by Davies *et al.*[19] as the argument of subroutine called RATES.

If we let N be the total gas molecule number density (2.5×10^{19} cm^{-3}), and specify the ratio of the constituents, e.g., 1:1:8 mixture of $CO_2:N_2:He$, then N_{CO_2}, N_{N_2}, and N_{He} are all known. The applied field E equals 20 kV cm^{-1}.

Manes and Seguin give the electron-pump pulse distribution the following empirical shape:

$$N_e(t) \simeq N_0(1-e^{-t})\,e^{-2t} \tag{2.76}$$

where t is in μsec, and N_0 is chosen to give maximum $N_e(t)$ near 10^{12} electron-cm^{-3}. Equation (2.76) has a maximum at

$$t = \log \tfrac{3}{2} \tag{2.77}$$

Hence

$$N_0 = \tfrac{27}{4} N_e^{\max} \simeq 7 \times 10^{12} \text{ electron-cm}^{-3} \tag{2.78}$$

Table 1. Physical Constants and Cross Sections

Constant	Value	Units
ν_1/c	1337	cm^{-1}
ν_2/c	667	cm^{-1}
ν_3/c	2349	cm^{-1}
ν_4/c	2330	cm^{-1}
ν_5/c	2150	cm^{-1}
h	6.625×10^{-27}	erg sec
c	2.998×10^{10}	cm sec^{-1}
λ	10.6	μm
M_{CO_2}	7.3×10^{-23}	g
M_{N_2}	4.6×10^{-23}	g
M_{He}	6.7×10^{-24}	g
M_{CO}	4.6×10^{-23}	g
k	1.38044×10^{-16}	erg (°K)$^{-1}$
B	0.4	cm^{-1}
Q_{CO_2}	1.3×10^{-14}	cm^2
Q_{N_2}	1.14×10^{-14}	cm^2
Q_{He}	3.7×10^{-15}	cm^2
Q_{CO}	1.14×10^{-14}	cm^2

Table 2. Arguments of RATES *(T5, T20, T10, T43, T12, T3012, T63, T64, T612, T11) and Temperature-Dependent Equations for Corresponding Rates* k_{ij}

Argument		Description	Rate ($cm^3\ sec^{-1}$)	Reference
T5		Ambient temperature, T		
T20	τ_{20}	$k^{CO_2}_{010;000}$	$4.6\times 10^{-10}\exp(-77/T^{1/3})$	20
		$k^{N_2}_{010;000}$	$9.6\times 10^{-11}\exp(-77/T^{1/3})$	21
		$k^{He}_{010;000}$	$8.1\times 10^{-11}\exp(-45/T^{1/3})$	21
		$k^{CO}_{010;000}$	$6.82\times 10^{-8}\exp(-77/T^{1/3})$	
T10	τ_{10}		$4.5\times\tau_{20}$	
T43	τ_{43}	$k^{N_2\text{-}CO_2}_{1,000;0,001}$ largest of	$1.71\times 10^{-6}\exp(-175.3/T^{1/3})$ $6.07\times 10^{-14}\exp(15.3/T^{1/3})$	19
T12	τ_{12}	$k^{CO_2}_{020;100}$	$8.65\times 10^{-15}\times T^{3/2}$	17
		$k^{N_2}_{020;100}$	$3.68\times 10^{-16}\times T^{3/2}$	17
		$k^{He}_{020;100}$	$4.23\times 10^{-17}\times T^{3/2}$	17
		$k^{CO}_{020;100}$	$3.68\times 10^{-16}\times T^{3/2}$	
T3012	τ_3	$k^{CO_2}_{001;110}$	$9.6\times 10^{23}\times T^{-5.89}\times F(T)^a$	20
		$k^{N_2}_{001;110}$	$6.87\times 10^{23}\times T^{-5.89}\times F(T)$	20
		$k^{He}_{001;110}$	$2.43\times 10^{23}\times T^{-5.89}\times F(T)$	
		$k^{CO}_{001;110}$	$6.87\times 10^{23}\times T^{-5.89}\times F(T)$	
T63	τ_{53}	$k^{CO\text{-}CO_2}_{1,000;0,001}$	$1.56\times 10^{-11}\exp(-30.1/T^{1/3})$	22
T64	τ_{54}	$k^{CO\text{-}N_2}_{10,01}$ largest of	$1.78\times 10^{-6}\exp(-210/T^{1/3})$ $6.98\times 10^{-13}\exp(25.6/T^{1/3})$	
T612	τ_5	$k^{CO\text{-}CO_2}_{1,000;0,110}$	$5.96\times 10^{-22}\times T^{-5.86}\times F(T)$	
T11	τ_1	$k^{CO_2\text{-}CO_2}_{100,000;000,100}$	$10^{nb}\times k^{CO_2}_{020;100}$	

[a] $F(T)=\exp(-4223/T-672.7/T^{1/3}+2683/T^{2/3})$.
[b] $n\geq 0$, signifying that $k^{CO_2}_{020;100}$ is a lower limit to $k^{CO_2\text{-}CO_2}_{100,000;000,100}$.

The temperature dependence of

$$k^{CO_2}_{010;000},\quad k^{CO_2}_{001;110},\quad k^{N_2}_{001;110},\quad \text{and}\quad k^{N_2\text{-}CO_2}_{1,000;0,001}$$

was obtained by Fisher[20] by fitting a functional form of $k^{f}_{if}(T)$ to experimental values, where it is noted that Schwartz *et al.* theory[8] predicts

$$\ln k \sim T^{-1/3}$$

We have obtained the temperature dependence of $k^{N_2}_{010;000}$ and $k^{He}_{010,000}$ by fitting $\ln k \sim T^{-1/3}$ to the experimental results of Taylor and Bitterman.[21] The temperature dependence of $k^{CO_2}_{020;100}$, $k^{N_2}_{020;100}$, and $k^{He}_{020;100}$ has been obtained by noting that Schwartz *et al.* theory predicts $k \propto T^{3/2}$ for resonant (zero-energy-transfer) processes, and fitting this temperature dependence to the 400°K data of Stark.[17] The temperature dependence of $k^{\text{He}}_{001;110}$ has been taken to be the same as that given for $k^{\text{CO}_2}_{001;110}$ with a different coefficient to fit the 300°K data. The temperature dependence of $k^{\text{CO–CO}_2}_{1,000;0,110}$ has been assumed to be the same as for the $k^{t}_{001-110}$ with a coefficient chosen to give the value by Gordietz *et al.*[22] at 300°K. Fisher[20] quotes a rate of

$$k_{020,000;010,010} = 2 \times 10^{-13} T^{1/2}$$

for the zero-energy-exchange process denoted by the subscripts. We chose our resonant coefficient $k_{100,000;000,100}$ to have this same T-dependence.

The relaxation times $\tau_{\lambda 0}(T)$, for $\lambda = 1$ and 2, are given in Eqs. (2.6) and (2.7), that is,

$$\tau_{\lambda 0}(T)^{-1} \equiv \sum_{t=1}^{c} N_t k^{t}_{\lambda 0}(T)[\exp(-h\nu_\lambda/kT) - 1]$$
$$\approx \sum_{t} N_t k^{t}_{\lambda 0} \tag{2.79}$$

where

$$k^{t}_{10} \equiv k^{t}_{100,000} \qquad \text{and} \qquad k^{t}_{20} \equiv k^{t}_{010,000}$$

where we have shown the T-independent approximation used by Manes and Seguin.[11] In Table 2 we follow Manes and Seguin taking

$$\tau_{10} = 4.5\tau_{20} \tag{2.80}$$

The superscript t is used for the assumed structureless projectiles incident on CO_2 in the appropriate V–T transition.

We note that the nonapproximated form of Eq. (2.79) could be used in any T-varying calculation.

The relaxation time $\tau_{12}(T)$ is given in Eq. (2.38) to be

$$\tau_{12}(T)^{-1} \equiv 2 \sum_{t} N_t k^{t}_{020,100}(T) \tag{2.81}$$

while the relaxation time $\tau_3(T)$ is given by Eq. (2.25), namely,

$$\tau_3(T, T_1, T_2)^{-1}[\exp(h\nu_2/kT_2) - 1][\exp(h\nu_1/kT_1) - 1]$$
$$\times[\exp(h\nu_1/kT_1 + h\nu_2/kT_2 + h(\nu_3 - \nu_2 - \nu_1)/kT) - 1]$$

$$= \sum_{t} N_t k^{t}_{001,110}(T) \exp[h(\nu_1 + \nu_2 - \nu_3)/kT] \tag{2.82}$$

$$= A_3(T) \tag{2.83}$$

of Manes and Seguin. To calculate $A_3(T)$, Manes and Seguin set

$$\tau_3(T, T, T)^{-1} = \sum_t N_t k^t_{001,110}(T) \tag{2.84}$$

where the rate coefficients are given in Table 2, and substitute this value for τ_3 into Eq. (2.83) to obtain $A_3(T)$.

The final relaxation time is given by

$$\tau_{43}(T)^{-1} = N_{CO_2} k^{N_2-CO_2}_{1,000;0,001} \tag{2.85}$$

Manes and Seguin give the laser transition linewidth to be

$$\Delta\nu_L(T) = \sum_t \left[\frac{N_t Q_t}{\pi}\left(\frac{8kT}{\pi\mu_t}\right)^{1/2}\right] \tag{2.86}$$

where μ_t is the reduced mass for CO_2 in collision with the constituent heavy particles in the gas, that is,

$$\mu_t = \frac{M_{CO_2} M_t}{M_{CO_2} + M_t} \tag{2.87}$$

The laser cavity lifetime is given by [see also Eq. (2.69d)]

$$\tau_c = \frac{-2L}{c} \ln R \tag{2.88}$$

where L is the laser cavity length and R is the output mirror reflectivity.

To derive Eq. (2.86) we note that a major cause of line broadening in a gas is the collision of radiating particles with one another. As an atomic collision interrupts either the emission or the absorption of radiation, the long wavetrain, which otherwise would be present, becomes truncated. After the collision the process is restarted without memory of the phase of the radiation before the collision (Fig. 10). The resulting lineshape (spread

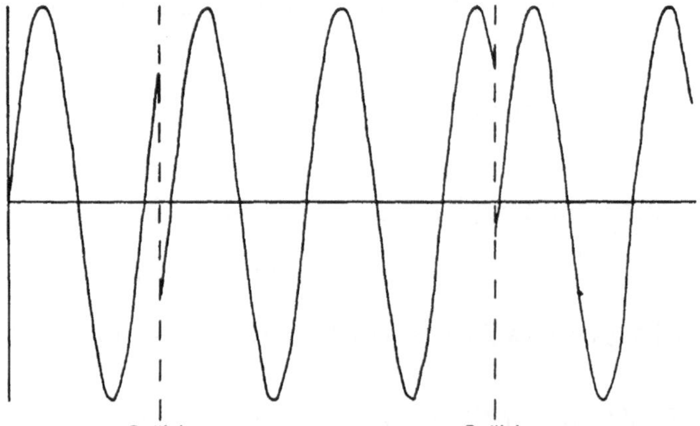

Fig. 10. Atomic collisions truncating wavetrains.

of frequencies) is Lorentzian, and the resulting linewidth $\Delta\nu_L$ is related to the average time τ between consecutive interrupting collisions by[23]

$$\Delta\nu_L = \frac{1}{\pi\tau}$$

For a gas mixture of molecules a and b of number density N_a, N_b, the number of collisions Z_{ab} per unit volume per unit time is given by Siegman (Reference 16, page 120):

$$Z_{ab} = N_a N_b Q_{ab}\left[\frac{8kT}{\pi}\left(\frac{1}{M_a}+\frac{1}{M_b}\right)\right]^{1/2}$$

where Q_{ab} (cm^2) is the collision cross section for a and b.

The average time $\tau_{a(b)}$ between a–b collisions for molecules of type a will be given by

$$\frac{1}{\tau_{a(b)}} = \frac{Z_{ab}}{N_a} = N_b Q_{ab}\left[\frac{8kT}{\pi}\left(\frac{1}{M_a}+\frac{1}{M_b}\right)\right]^{1/2}$$

In a gas mixture the collision frequency for molecules of type a due to collisions with all other types of molecules will be given by

$$\frac{1}{\tau_a} = \sum_b \frac{1}{\tau_{a(b)}}$$

i.e.,

$$\frac{1}{\tau_{CO_2}} = \frac{1}{\tau_{CO_2(CO_2)}} + \frac{1}{\tau_{CO_2(N_2)}} + \frac{1}{\tau_{CO_2(He)}}$$

Therefore

$$\Delta\nu_L = \sum_t \left\{\frac{N_t Q_t}{\pi}\left[\frac{8kT}{\pi}\left(\frac{1}{M_{CO_2}}+\frac{1}{M_t}\right)\right]^{1/2}\right\} \tag{2.86a}$$

2.8. *Electron Pulse Distribution, $N_e(t)$*

N_2, CO, and CO_2 are vibrationally excited by collisions with low-energy (about 1–10 eV) electrons. There are two techniques for producing low-energy electrons in the laser cavity: (i) transverse electrical discharge (self-sustained) lasers and (ii) electron-beam-controlled lasers. Manes and Seguin use the empirical equation (2.76) to describe the number density of electrons $N_e(t)$ in case (i). In this section we shall compare the above two mechanisms and give an equation for $N_e(t)$ applicable to case (ii).

In self-sustained lasers, discharges between overvolted cathode structures are used to produce ionization in the gas mixture. The electrons of the low-energy tail of the resulting electron energy distribution vibrationally excite the molecules. A difficulty with this technique is in optimizing the applied electric field E. A large E increases the degree of ionization but imparts a greater acceleration to the resulting electrons, leaving fewer in the low-energy tail.

In electron-beam-controlled lasers a discharge is produced in the gas mixture by injecting a high-energy electron beam to thermally ionize the gas. At the same time a field is maintained across two-electrodes to accelerate, to the required low energy, the secondary electrons (those obtained as a result of the ionization) that vibrationally excite the gas molecules. This method has the advantage over case (i) in that the separation of the electric field from the ionization mechanism allows E to be chosen to optimize the secondary electron energy.

The variation of $N_e(t)$ with time depends on the rate of secondary electron production and the loss by recombination and attachment. The loss rate due to recombination will be proportional to both the electron density and the density of ionized molecules N_i. With the approximation $N_i \simeq N_e$ the loss rate due to recombination is proportional to N_e^2. The loss rate due to attachment will be proportional to N_e only since the number of neutral molecules is taken to be a constant. The equation describing the net effect of the three processes is given by Harrach and Einwohner[24]:

$$\frac{dN_e(t)}{dt} = S - \alpha N_e^2 - \beta N_e \tag{2.89}$$

where S ($\text{cm}^{-3}\,\text{sec}^{-1}$) is the rate of secondary electron production, α ($\text{cm}^3\,\text{sec}^{-1}$) is the recombination coefficient, and β (sec^{-1}) is the attachment coefficient. Equation (2.89) neglects effects such as diffusion and dissociation.

If it is assumed that the electron beam is a square pulse of duration t_e, and that S is constant in the time interval $0 \leqslant t \leqslant t_e$ and zero when $t > t_e$, then Eq. (2.89) can be integrated to give

$$N_e(t; 0 \leqslant t \leqslant t_e) = 2S\{\beta + (\beta^2 + 4\alpha S)^{1/2} \coth[(\beta^2 + 4\alpha S)^{1/2} t/2]\}^{-1} \tag{2.90}$$

and for $t > t_e$, Eq. (2.89) with $S = 0$ is integrated to obtain

$$N_e(t; t \geqslant t_e) = N_e(t_e)\left\{\left[\frac{\alpha N_e(t_e)}{\beta} + 1\right] \exp[\beta(t - t_e)] - \frac{\alpha N_e(t_e)}{\beta}\right\}^{-1} \tag{2.91}$$

Fenstermacher *et al.*[25] assume that the attachment process is negligible compared to recombination. Thus Eq. (2.89) with $\beta = 0$ becomes, on integration:

$$N_e(t; 0 \leqslant t \leqslant t_e) = (S/\alpha)^{1/2} \tanh (\alpha S)^{1/2} t \tag{2.92}$$

$$N_e(t; t \gg t_e) = N_e(t_e)[\alpha N_e(t_e)(t - t_e) + 1]^{-1} \tag{2.93}$$

Equations (2.90) and (2.91) [or (2.92) and (2.93)] give the time development of the secondary electron number density during and after a finite and constant primary electron pulse.

2.9. Dissociation of CO_2

In Schwartz *et al.*[8] theory, E_i is the internal energy of N molecules of type "i" per unit volume. Consequently if f represents the fraction of undissociated CO_2 molecules, then fE_i will be the internal energy of fN_{CO_2} carbon dioxide molecules per unit volume and $(1-f)E_5$ will be the internal energy of $(1-f)N_{CO_2}$ carbon monoxide molecules per unit volume. In the following sections we shall make the replacement

$$fE_i \rightarrow E_i, \qquad i = 1, 2, 3$$

$$(1-f)E_5 \rightarrow E_5$$

with the understanding that E_1 is the internal energy of fN_{CO_2} carbon dioxide molecules in the 1, 0, 0 vibrational state, etc.

The simplest assumption to make is that f is time-independent, even though f will be a function of time. In other words, we would like a differential equation for $f(t)$ coupled to the functions $N_e(t)$, $E_i(t)$ and the collision processes producing dissociation.

2.9.1. Tulip's Equation for the Energy, E_5, Stored in CO ($v = 1$)

If a fraction f of the CO_2 molecules dissociates into CO, then electrons will excite the CO into the first excited state, $v = 1$. The resulting excited CO will then be available to transfer energy in V–V collisions to both N_2 and CO_2 since the quantum $h\nu_5 \simeq h\nu_4 \simeq h\nu_3$. Tulip's[26] form for the equation describing the time evolution of the energy stored in the CO ($v = 1$) vibrational state is

$$\frac{dE_5}{dt} = N_e(t)(1-f)N_{CO_2}h\nu_5 X_5 - \frac{E_5 - E_5^e(T_3)}{\tau_{53}(T)} - \frac{E_5 - E_5^e(T_4)}{\tau_{54}(T)} \tag{2.94}$$

where E_5 is the energy of the CO vibration at T_5, that is,

$$E_5 = h\nu_5(1-f)N_{CO_2}[\exp(h\nu_5/kT_5) - 1]^{-1} \tag{2.95}$$

The V–V relaxation times τ_{53} and τ_{54} will be analogous to τ_{43} derived in Section 2.3 and are given by

$$\tau_{54}(T)^{-1} = N_{N_2}\bar{\bar{k}}_{10,01}^{CO-N_2}(T) \tag{2.96}$$

and

$$\tau_{53}(T)^{-1} = fN_{CO_2}\tilde{k}_{1,000;0,001}^{CO-CO_2}(T) \tag{2.97}$$

where $\bar{\bar{k}}$ ($\tilde{k}$) are the V–V rates for CO–N_2 [CO–CO_2 (001)] energy transfer.

The presence of CO will result in Eqs. (2.72) and (2.73) being replaced by the extended pair of equations

$$\frac{dE_3}{dt} = N_e(t)fN_{CO_2}h\nu_3X_3 + \frac{E_4 - E_4^e(T_3)}{\tau_{43}(T)} + \frac{E_5 - E_5^e(T_3)}{\tau_{53}(T)} - \frac{E_3 - E_3^e(T, T_1, T_2)}{\tau_3(T, T_1, T_2)} - h\nu_3\,\Delta N\,WI_\nu \tag{2.98}$$

$$\frac{dE_4}{dt} = N_e(t)N_{N_2}h\nu_4X_4 - \frac{E_4 - E_4^e(T_3)}{\tau_{43}(T)} + \frac{E_5 - E_5^e(T_4)}{\tau_{54}(T)} \tag{2.99}$$

2.9.2. *Temperature Coupling through V–V Processes*

In the preceding section, we adopted the Manes and Seguin approximation of $\nu_4 = \nu_3$ that resulted in the term $\tau_{43}(T)^{-1}[E_4 - E_4^e(T_3)]$ coupling the E_4 and E_3 equations. If we abandon this approximation, then we have shown in Eq. (1.109) that the coupling term in the aE_4 equation is

$$\frac{-a[E_4 - E_4^e(T, T_3)]}{\tau_{VV}(T, T_3)} = -aNh\nu_a Nbk_{1,000;0,001}^{N_2-CO_2}(T)\exp\left[\frac{h(\nu_b - \nu_a)}{kT}\right] \times\left\{\exp\left[\frac{h\nu_b}{kT_b} - \frac{h(\nu_b - \nu_a)}{kT}\right] - \exp\left(\frac{-h\nu_a}{kT_a}\right)\right\} \times\left[\exp\left(\frac{h\nu_b}{kT_b}\right) - 1\right]^{-1}\left[\exp\left(\frac{h\nu_a}{kT_a}\right) - 1\right]^{-1} \tag{2.100}$$

having used Eqs. (1.105), (1.108), and (1.110). We note that a is associated with the subscript 4, and b is the fraction of the N molecules in state 3.

The analog to Eq. (1.109) for the internal energy of the asymmetric mode of CO_2 is

$$\frac{d(bE_3)}{dt} = \frac{bE_3 - bE_3^e(T)}{\tau_{VT}} - \frac{bE_3 - bE_3^e(T, T_4)}{\tau_{VV}(T, T_4)} \tag{2.101}$$

where the second coupling term is given explicitly by Eq. (2.100) when a and b are interchanged everywhere, and

$$k_{1,000;0,001}^{N_2\text{-}CO_2} \quad \text{is replaced by} \quad k_{001,0;000,1}^{CO_2\text{-}N_2}$$

That is,

$$\begin{aligned}
&\frac{-b[E_3 - E_3^e(T, T_4)]}{\tau_{VV}(T, T_4)} \\
&= -bNh\nu_b Nak_{001,0;000,1}^{CO_2\text{-}N_2}(T)\exp\left(\frac{h(\nu_a - \nu_b)}{kT}\right) \\
&\quad \times\left\{\exp\left[\frac{h\nu_a}{kT_a} - \frac{h(\nu_a - \nu_b)}{kT}\right] - \exp\left(\frac{h\nu_b}{kT_b}\right)\right\} \\
&\quad \times\left[\exp\left(\frac{h\nu_a}{kT_a}\right) - 1\right]^{-1}\left[\exp\left(\frac{h\nu_b}{kT_b}\right) - 1\right]^{-1} \\
&= -\left(\frac{h\nu_b}{h\nu_a}\right)aNh\nu_a Nbk_{1,000;0,001}^{N_2\text{-}CO_2}(T)\exp\left(\frac{h(\nu_b - \nu_a)}{kT}\right)\left[\exp\left(\frac{h\nu_a}{kT_a}\right) - 1\right]^{-1} \\
&\quad \times\left[\exp\left(\frac{h\nu_b}{kT_b}\right) - 1\right]^{-1}\left\{\exp\left(\frac{h\nu_a}{kT_a}\right) - \exp\left[\frac{h\nu_b}{kT_b} - \frac{h(\nu_b - \nu_a)}{kT}\right]\right\} \qquad (2.102)
\end{aligned}$$

having used Eq. (1.100). When the right-hand sides of Eqs. (2.100) and (2.102) are compared, we see that

$$\frac{-b[E_3 - E_3^e(T, T_4)]}{\tau_{VV}(T, T_4)} = \left(\frac{h\nu_b}{h\nu_a}\right)\frac{a[E_4 - E_4^e(T, T_3)]}{\tau_{VV}(T, T_3)} \qquad (2.103)$$

To summarize: If the coupling term of V–V processes in the aE_4 equation is written as the left-hand side of Eq. (2.100), then this same term, with opposite sign and with the factor $(h\nu_3/h\nu_4)$, appears in the coupled equation for bE_3. The residual rate of energy change

$$\left(1 - \frac{h\nu_3}{h\nu_4}\right)\frac{a[E_4 - E_4^e(T, T_3)]}{\tau_{VV}(T, T_3)} \qquad (2.104)$$

should be included in the equation describing the time evolution of the ambient temperature.

2.9.3. A Six-Temperature Model

Tulip's equation for E_5 is based on the assumption that $\nu_3 \approx \nu_5 \approx \nu_4$. If we relax this assumption, then we can use the relaxation times for collisions between unlike SHOs derived in Section 1.7.1, which gives the equilibrium energy for CO–CO_2(001) collisions to be

$$E_5^e(T, T_3) = (1 - f)N_{CO_2}h\nu_5\{\exp[h\nu_3/kT_3 - h(\nu_3 - \nu_5)/kT] - 1\}^{-1} \qquad (2.105)$$

instead of $E_5(T_3)$, and the related V–V relaxation time as

$$\tau_{53}(T, T_3)^{-1} = fN_{CO_2}k_{1,000;0,0\hat{0}1}^{CO\text{-}CO_2}(T)\exp[h(\nu_3-\nu_5)/kT]\{\exp(h\nu_3/kT_3)-1\}^{-1} \times\{\exp[h\nu_3/kT_3+h(\nu_5-\nu_3)/kT]-1\} \quad (2.106)$$

The factor $(1-f)$ is included in Eq. (2.105) because of the different definition of E_5 here, which conforms with Manes and Seguin,[11] as opposed to that of Section 1.7.1, which conforms with the Schwartz *et al.*[8] theory. Here, E_5 is the energy per unit volume stored in $(1-f)N_{CO_2}$ carbon monoxide molecules. For CO–N_2 collisions the equivalent equations are

$$E_5^e(T, T_4) = (1-f)N_{CO_2}h\nu_5\{\exp[h\nu_4/kT_4+h(\nu_5-\nu_4)/kT]-1\}^{-1} \quad (2.107)$$

and

$$\tau_{54}(T, T_4)^{-1} = N_{N_2}k_{10,01}^{CO\text{-}N_2}(T)\exp[h(\nu_4-\nu_5)/kT][\exp(h\nu_4/kT_4)-1]^{-1} \times\{\exp[h\nu_4/kT_4+h(\nu_5-\nu_4)/kT]-1\} \quad (2.108)$$

In the notation of Manes and Seguin,

$$k_{1,000;0,0\hat{0}1}^{CO\text{-}CO_2}(T)\exp[h(\nu_3-\nu_5)/kT]\rightarrow A_5 \quad (2.109)$$

Furthermore, by analogy with the τ_3 process of Fig. 8, we can expect those energy-transfer processes to be important in which one $h\nu_5$ quantum is exchanged simultaneously with one $h\nu_1$ and one $h\nu_2$ quantum. In Fig. 11, we present a schematic diagram of the set of processes that give a six-temperature model to describe the CO_2–N_2–He–CO system. These processes give rise to the same I_ν equation (2.57), but Eqs. (2.70b) and

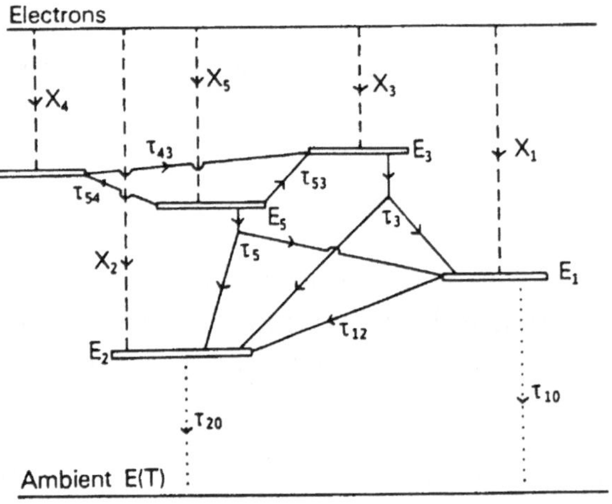

Fig. 11. Schematic energy-level diagram for the N_2–CO_2–CO system, including collision processes. Continuous, arrowed lines denote V–V transitions; dotted, arrowed lines denote V–T energy transfer processes; and dashed, arrowed lines denote electron excitation processes.

(2.71b)–(2.73) are replaced by

$$\frac{dE_1}{dt} = N_e(t)fN_{CO_2}h\nu_1X_1 - \frac{E_1 - E_1^e(T)}{\tau_{10}(T)} - \frac{E_1 - E_1^e(T_2)}{\tau_{12}(T_2)} + \left(\frac{h\nu_1}{h\nu_3}\right)\frac{E_3 - E_3^e(T, T_1, T_2)}{\tau_3(T, T_1, T_2)} + \left(\frac{h\nu_1}{h\nu_5}\right)\frac{E_5 - E_5^e(T, T_1, T_2)}{\tau_5(T, T_1, T_2)} + h\nu_1\,\Delta N\,WI_\nu \tag{2.110}$$

$$\frac{dE_2}{dt} = N_e(t)fN_{CO_2}h\nu_2X_2 + \frac{E_1 - E_1^e(T_2)}{\tau_{12}(T_2)} - \frac{E_2 - E_2^e(T)}{\tau_{20}(T)} + \left(\frac{h\nu_2}{h\nu_3}\right)\frac{E_3 - E_3^e(T, T_1, T_2)}{\tau_3(T, T_1, T_2)} + \left(\frac{h\nu_2}{h\nu_5}\right)\frac{E_5 - E_5^e(T, T_1, T_2)}{\tau_5(T, T_1, T_2)} \tag{2.111}$$

We recall that the process τ_3 involves the collisions of a structureless projectile (any molecule in the laser) with a CO_2 (001) molecule, that is,

$$M + CO_2\,(001) \rightarrow M + CO_2\,(110)$$

whereas the process τ_5 involves three vibrational modes:

$$CO\,(v = 1) + CO_2\,(000) \rightarrow CO\,(v = 0) + CO_2\,(110)$$

$$\frac{dE_3}{dt} = N_e(t)fN_{CO_2}h\nu_3X_3 - \frac{E_3 - E_3^e(T, T_1, T_2)}{\tau_3(T, T_1, T_2)} + \frac{E_4 - E_4(T_3)}{\tau_{43}(T)} + \left(\frac{h\nu_3}{h\nu_5}\right)\frac{E_5 - E_5^e(T, T_3)}{\tau_{53}(T, T_3)} - h\nu_3\,\Delta N\,WI_\nu \tag{2.112}$$

$$\frac{dE_4}{dt} = N_e(t)N_{N_2}h\nu_4X_4 - \frac{E_4 - E_4^e(T_3)}{\tau_{43}(T)} + \left(\frac{h\nu_4}{h\nu_5}\right)\frac{E_5 - E_5^e(T, T_4)}{\tau_{54}(T, T_4)} \tag{2.113}$$

$$\frac{dE_5}{dt} = N_e(t)(1-f)N_{CO_2}h\nu_5X_5 - \frac{E_5 - E_5^e(T, T_3)}{\tau_{53}(T, T_3)} - \frac{E_5 - E_5^e(T, T_1, T_2)}{\tau_5(T, T_1, T_2)} - \frac{E_5 - E_5^e(T, T_4)}{\tau_{54}(T, T_4)} \tag{2.114}$$

where

$$E_i^e(T_j) = \frac{h\nu_i f N_{CO_2}}{\exp(h\nu_i/kT_j) - 1} \qquad \text{for } i = 1, 3 \tag{2.115}$$

$$E_2^e(T_j) = \frac{2h\nu_2 f N_{CO_2}}{\exp(h\nu_2/kT_j) - 1} \tag{2.116}$$

where we recall that $h\nu_2$ is the energy of the $(0, 1^1, 0)$ excited state of CO_2:

$$E_4^e(T_j) = \frac{h\nu_4 N_{N_2}}{\exp(h\nu_4/kT_j) - 1} \tag{2.117}$$

$$E_5 = \frac{h\nu_5(1-f)N_{CO_2}}{\exp(h\nu_5/kT_5) - 1} \tag{2.118}$$

$$E_3^e(T, T_1, T_2) = \frac{fN_{CO_2} h\nu_3}{\exp[h\nu_1/kT_1 + h\nu_2/kT_2 + (h\nu_3 - h\nu_2 - h\nu_1)/kT] - 1} \tag{2.119}$$

The relaxation times $\tau_{\lambda 0}$, τ_{12}, τ_3, and τ_{43} remain as given in Section 2.7.1, while $\tau_{5\lambda}$ are given in Eqs. (2.106) and (2.108). Upon comparison with the expression for τ_3 we have

$$\begin{aligned}\tau_5(T, T_1, T_2) = &\left[\exp\left(\frac{h\nu_1}{kT_1}\right) - 1\right]\left[\exp\left(\frac{h\nu_2}{kT_2}\right) - 1\right] \\ &\times\left\{(1-f)N_{CO_2} k_{1,000;0,\bar{1}10}^{\text{CO–CO}_2}\left[\exp\left(\frac{h\nu_1}{kT_1} + \frac{h\nu_2}{kT_2}\right.\right.\right. \\ &\left.\left.\left. + \frac{h\nu_5 - h\nu_2 - h\nu_1}{kT}\right) - 1\right]\right\}^{-1}\end{aligned} \tag{2.120}$$

The quantities $\Delta\nu_L$ and τ_c are still defined by Eqs. (2.86) and (2.88) while the expressions not yet defined in this section are:

$$\Delta N = N_{0,0^\circ,1} P(J) - \left(\frac{\theta_J}{\theta_{J+1}}\right) N_{1,0^\circ,0} P(J+1) \tag{2.42e}$$

$$\begin{aligned}N_{0,0^\circ,1} = &fN_{CO_2} \exp\left(\frac{-h\nu_3}{kT_3}\right)\left[1 - \exp\left(\frac{-h\nu_1}{kT_1}\right)\right] \\ &\times\left[1 - \exp\left(\frac{-h\nu_2}{kT_2}\right)\right]^2\left[1 - \exp\left(\frac{-h\nu_3}{kT_3}\right)\right]\end{aligned} \tag{2.3e}$$

$$\begin{aligned}N_{1,0^\circ,0} = &fN_{CO_2} \exp\left(\frac{-h\nu_1}{kT_1}\right)\left[1 - \exp\left(\frac{-h\nu_1}{kT_1}\right)\right]\left[1 - \exp\left(\frac{-h\nu_2}{kT_2}\right)\right]^2 \\ &\times\left[1 - \exp\left(\frac{-h\nu_3}{kT_3}\right)\right]\end{aligned} \tag{2.3f}$$

$$P(J) = \left(\frac{2hcB}{kT}\right)\theta_J \exp\left[\frac{-hcBJ(J+1)}{kT}\right] \tag{2.42a}$$

$$W = \frac{F\lambda^2}{4\pi^2 h\nu_0 \tau_{sp} \Delta\nu_L} \tag{2.58a}$$

$$\Delta\nu_L(T)=\sum_t\left[\frac{N_tQ_t}{\pi}\left(\frac{8kT}{\pi\mu_t}\right)^{1/2}\right] \tag{2.86}$$

$$\mu_t=\frac{M_{CO_2}M_t}{M_{CO_2}+M_t} \tag{2.87}$$

2.10. Time Evolution of the Ambient Temperature T

Consider à gas consisting of molecules in their ground state at a translational temperature (ambient) T. Upon collision, some of the kinetic energy of translation will be transferred into internal vibrational energy of the colliding molecules, and the ambient temperature will decrease. This cooling of the ambient temperature will continue until there are as many molecules being deexcited, giving up energy to translation, as excited. This is the equilibrium ambient temperature. The rate at which molecules undergo transitions is given by $k_{if}(T)$ ($cm^3\,sec^{-1}$) and depends on the ambient temperature T.

In the laser the reverse process takes place. The vibrational modes are excited by collisions with electrons and the populations of the excited states relax toward their equilibrium values, giving up energy to translation and hence heating the gas.

In the Manes and Seguin model, it is assumed that the amount of energy transferred from the ambient to the internal energy is so small that the ambient temperature remains constant. In this section we shall extend the five- and six-temperature models to include the effect of a variable ambient temperature.

2.10.1. The Seventh Equation

In each of the equations for dE_i/dt, there are terms that allow for the transfer of energy from the vibrational modes to the kinetic energy (per unit volume) of the atoms and molecules of the system E, thereby changing the ambient temperature T.

The kinetic energy of an atom is related to the ambient temperature by

$$\tfrac{1}{2}Mv^2=\tfrac{3}{2}kT \tag{2.121}$$

an energy of $\frac{1}{2}kT$ being associated with each degree of translational freedom (the motion being in three dimensions). Similarly, for molecules, we must associate an energy $\frac{1}{2}kT$ with each rotational degree of freedom. N_2 and CO_2 are both linear molecules and therefore have two rotational degrees of freedom. Hence the total gas kinetic energy per unit volume will be

$$E = (\tfrac{5}{2}N_{N_2} + \tfrac{5}{2}N_{CO_2} + \tfrac{3}{2}N_{He})kT \tag{2.122}$$

From Eqs. (2.70b), (2.71b), (2.72), and (2.73) we have

$$\frac{dE}{dt} = \frac{E_1 - E_1(T)}{\tau_{10}(T)} + \frac{E_2 - E_2(T)}{\tau_{20}(T)} + \left(1 - \frac{h\nu_2}{h\nu_3} - \frac{h\nu_1}{h\nu_3}\right)\frac{E_3 - E_3(T, T_1, T_2)}{\tau_3(T, T_1, T_2)} \tag{2.123}$$

as the ambient time-dependent equation, which is consistent with the Manes and Seguin model. This same equation would also hold for the Tulip model when CO_2 dissociation to CO is taken into account.

However, for the N_2–CO_2–CO system described by Eqs. (2.110)–(2.114), the time evolution of the ambient temperature will be given by

$$\begin{aligned}\frac{dE}{dt} &= \frac{E_1 - E_1(T)}{\tau_{10}(T)} + \frac{E_2 - E_2(T)}{\tau_{20}(T)} \\ &\quad + \left(1 - \frac{h\nu_2}{h\nu_3} - \frac{h\nu_1}{h\nu_3}\right)\frac{E_3 - E_3(T, T_1, T_2)}{\tau_3(T, T_1, T_2)} + \left(1 - \frac{h\nu_4}{h\nu_5}\right)\frac{E_5 - E_5(T, T_4)}{\tau_{54}(T, T_4)} \\ &\quad + \left(1 - \frac{h\nu_2}{h\nu_5} - \frac{h\nu_1}{h\nu_5}\right)\frac{E_5 - E_5(T, T_1, T_2)}{\tau_5(T, T_1, T_2)} + \left(1 - \frac{h\nu_3}{h\nu_5}\right)\frac{E_5 - E_5(T, T_3)}{\tau_{53}(T, T_3)}\end{aligned} \tag{2.124}$$

Equations (2.123) and (2.124) represent the seventh equation when the presence of CO is taken into account in the CO_2 laser.

2.10.2. Analysis for Additional Collision Processes

In deriving Eqs. (2.72) and (2.73), Manes and Seguin[11] assumed $\nu_3 = \nu_4$ for the V–V process. If this assumption is dropped, then we have the temperature coupling term derived explicitly in Section 2.9.1. In other words, Eqs. (2.72) and (2.73) are replaced by

$$\frac{dE_3}{dt} = N_e(t)fN_{CO_2}h\nu_3X_3 - \frac{E_3 - E_3(T, T_1, T_2)}{\tau_3(T, T_1, T_2)} + \left(\frac{h\nu_3}{h\nu_4}\right)\frac{E_4 - E_4(T, T_3)}{\tau_{43}(T, T_3)} \tag{2.125}$$

$$\frac{dE_4}{dt} = N_e(t)N_{N_2}h\nu_4X_4 - \frac{E_4 - E_4(T, T_3)}{\tau_{43}(T, T_3)} \tag{2.126}$$

where, from Eqs. (2.107) and (2.108),

$$E_4(T, T_3) = N_{N_2}h\nu_4\left\{\exp\left[\frac{h\nu_3}{kT_3} + \frac{h(\nu_4 - \nu_3)}{kT}\right] - 1\right\}^{-1} \tag{2.127}$$

and

$$\tau_{43}(T, T_3)^{-1} = N_{N_2} k_{1,000;0,001}^{N_2\text{-}CO_2}(T) \exp\left(\frac{h(\nu_3 - \nu_4)}{kT}\right) \times \left[\exp\left(\frac{h\nu_3}{kT_3}\right) - 1\right]^{-1} \left\{\exp\left[\frac{h\nu_3}{kT_3} + \frac{h(\nu_4 - \nu_3)}{kT}\right] - 1\right\} \tag{2.128}$$

and Eq. (2.124) will have the following additional term on the right-hand side:

$$\left(1 - \frac{h\nu_3}{h\nu_4}\right) \frac{E_4 - E_4(T, T_3)}{\tau_{43}(T, T_3)} \tag{2.129}$$

By inspection of Fig. 11, we can see immediately that the following collision processes have been neglected: (a) V–T transfer from E_4 to the ambient temperature, τ_{40}; (b) V–T transfer from E_3 to the ambient temperature, τ_{30}; (c) V–T transfer from E_5 to the ambient temperature, τ_{50}, where $\tau_{\lambda 0}$, $\lambda = 3, 4, 5$, are defined by Eq. (2.6); (d) V–V transfer from E_4 to E_2, τ_{42}; (e) V–V transfer from E_4 to E_1, τ_{41}; (f) V–V transfer from E_3 to E_1, τ_{31}; (g) V–V transfer from E_3 to E_2, τ_{32}; (h) V–V transfer from E_5 to E_2, τ_{52}; (i) V–V transfer from E_5 to E_1, τ_{51}; (j) single quantum V–V transfer from E_1 to the ground state of E_2, $\hat{\tau}_{12}$, where $\tau_{i\lambda}(T, T_\lambda)$ for $i = 1[CO_2\,(100)]$, $3[CO_2\,(001)]$, $4[N_2]$, $5[CO]$ and $\lambda = 1[CO_2\,(100)]$, $2[CO_2\,(010)]$ are defined by

$$\tau_{i\lambda}(T, T_\lambda)^{-1} = N_\lambda k^{i-\lambda}(T) \exp\left[\frac{h(\nu_\lambda - \nu_i)}{kT}\right] \times \left[\exp\left(\frac{h\nu_\lambda}{kT_\lambda}\right) - 1\right]^{-1} \left\{\exp\left[\frac{h\nu_\lambda}{kT_\lambda} + \frac{h(\nu_i - \nu_\lambda)}{kT}\right] - 1\right\} \tag{2.128a}$$

(k) simultaneous V–V transfer from E_4 to E_2 and E_1, τ_4, where τ_4 is an equation analogous to Eq. (2.120). When all these processes are included, the six energy-coupled equations that replace either Eqs. (2.70b), (2.71b), (2.72), and (2.73) or Eqs. (2.110)–(2.114) and (2.124) are

$$\begin{aligned}
\frac{dE_1}{dt} = {} & N_e(t) f N_{CO_2} h\nu_1 X_1 - \frac{E_1 - E_1(T)}{\tau_{10}(T)} - \frac{E_1 - E_1(T_2)}{\tau_{12}(T_2)} \\
& + \left(\frac{h\nu_1}{h\nu_3}\right) \frac{E_3 - E_3(T, T_1, T_2)}{\tau_3(T, T_1, T_2)} + h\nu_1\, \Delta N\, W I_\nu \\
& - \frac{E_1 - E_1(T, T_2)}{\hat{\tau}_{12}(T, T_2)} + \left(\frac{h\nu_1}{h\nu_3}\right) \frac{E_3 - E_3(T, T_1)}{\tau_{31}(T, T_1)} \\
& + \left(\frac{h\nu_1}{h\nu_4}\right) \frac{E_4 - E_4(T, T_1)}{\tau_{41}(T, T_1)} + \left(\frac{h\nu_1}{h\nu_4}\right) \frac{E_4 - E_4(T, T_1, T_2)}{\tau_4(T, T_1, T_2)} \\
& + \left(\frac{h\nu_1}{h\nu_5}\right) \frac{E_5 - E_5(T, T_1)}{\tau_{51}(T, T_1)} + \left(\frac{h\nu_1}{h\nu_5}\right) \frac{E_5 - E_5(T, T_1, T_2)}{\tau_5(T, T_1, T_2)}
\end{aligned} \tag{2.130}$$

$$\frac{dE_2}{dt}=N_e(t)fN_{CO_2}h\nu_2X_2-\frac{E_2-E_2(T)}{\tau_{20}(T)}+\frac{E_1-E_1(T_2)}{\tau_{12}(T_2)}$$
$$+\left(\frac{h\nu_2}{h\nu_3}\right)\frac{E_3-E_3(T,T_1,T_2)}{\tau_3(T,T_1,T_2)}+\left(\frac{h\nu_2}{h\nu_4}\right)\frac{E_4-E_4(T,T_2)}{\tau_{42}(T,T_2)}$$
$$+\left(\frac{h\nu_2}{h\nu_4}\right)\frac{E_4-E_4(T,T_1,T_2)}{\tau_4(T,T_1,T_2)}+\left(\frac{h\nu_2}{h\nu_1}\right)\frac{E_1-E_1(T,T_2)}{\hat{\tau}_{12}(T,T_2)}$$
$$+\left(\frac{h\nu_2}{h\nu_3}\right)\frac{E_3-E_3(T,T_2)}{\tau_{32}(T,T_2)}+\left(\frac{h\nu_2}{h\nu_5}\right)\frac{E_5-E_5(T,T_2)}{\tau_{52}(T,T_2)}$$
$$+\left(\frac{h\nu_2}{h\nu_5}\right)\frac{E_5-E_5(T,T_1,T_2)}{\tau_5(T,T_1,T_2)} \tag{2.131}$$

$$\frac{dE_3}{dt}=N_e(t)fN_{CO_2}h\nu_3X_3-\frac{E_3-E_3(T)}{\tau_{30}(T)}-\frac{E_3-E_3(T,T_1,T_2)}{\tau_3(T,T_1,T_2)}$$
$$+\left(\frac{h\nu_3}{h\nu_4}\right)\frac{E_4-E_4(T,T_3)}{\tau_{43}(T,T_3)}+\left(\frac{h\nu_3}{h\nu_5}\right)\frac{E_5-E_5(T,T_3)}{\tau_{53}(T,T_3)}$$
$$-\sum_{\lambda=1}^{2}\frac{E_3-E_3(T,T_\lambda)}{\tau_{3\lambda}(T,T_\lambda)}-h\nu_3\,\Delta N\,WI_\nu \tag{2.132}$$

$$\frac{dE_4}{dt}=N_e(t)N_{N_2}h\nu_4X_4-\frac{E_4-E_4(T)}{\tau_{40}(T)}-\sum_{\lambda=1}^{3}\frac{E_4-E_4(T,T_\lambda)}{\tau_{4\lambda}(T,T_\lambda)}$$
$$+\left(\frac{h\nu_4}{h\nu_5}\right)\frac{E_5-E_5(T,T_4)}{\tau_{54}(T,T_4)}-\frac{E_4-E_4(T,T_1,T_2)}{\tau_4(T,T_1,T_2)} \tag{2.133}$$

$$\frac{dE_5}{dt}=N_e(t)(1-f)N_{CO_2}h\nu_5X_5-\frac{E_5-E_5(T)}{\tau_{50}(T)}$$
$$-\sum_{\lambda=1}^{4}\frac{E_5-E_5(T,T_\lambda)}{\tau_{5\lambda}(T,T_\lambda)}-\frac{E_5-E_5(T,T_1,T_2)}{\tau_5(T,T_1,T_2)} \tag{2.134}$$

$$\frac{dE}{dt}=\sum_{\lambda=1}^{5}\frac{E_\lambda-E_\lambda(T)}{\tau_{\lambda 0}(T)}+\left(1-\frac{h\nu_2}{h\nu_1}\right)\frac{E_1-E_1(T,T_2)}{\hat{\tau}_{12}(T,T_2)}$$
$$+\sum_{\lambda=1}^{2}\left(1-\frac{h\nu_\lambda}{h\nu_3}\right)\frac{E_3-E_3(T,T_\lambda)}{\tau_{31}(T,T_\lambda)}+\sum_{\lambda=1,2,3}\left(1-\frac{h\nu_\lambda}{h\nu_4}\right)\frac{E_4-E_4(T,T_\lambda)}{\tau_{4\lambda}(T,T_\lambda)}$$
$$+\sum_{\lambda=3,4,5}\left(1-\frac{h\nu_1}{h\nu_\lambda}-\frac{h\nu_2}{h\nu_\lambda}\right)\frac{E_\lambda-E_\lambda(T,T_1,T_2)}{\tau_\lambda(T,T_1,T_2)}$$
$$+\sum_{\lambda=1}^{4}\left(1-\frac{h\nu_\lambda}{h\nu_5}\right)\frac{E_5-E_5(T,T_\lambda)}{\tau_{5\lambda}(T,T_\lambda)}\,. \tag{2.135}$$

The time evolution of the radiation intensity is still given by Eq. (2.57).

2.11. Non-Boltzmann Vibrational Level Distributions

The Boltzmann population distribution of CO_2 molecules among the vibrational levels is given by Eq. (2.3a):

$$n_{ijk} = N_{CO_2} p^i s^j r^k (1-p)(1-s)^2(1-r) \tag{2.3a}$$

where

$$p = \exp\left(\frac{-h\nu_1}{kT_1}\right), \qquad s = \exp\left(\frac{-h\nu_2}{kT_2}\right), \qquad r = \exp\left(\frac{-h\nu_3}{kT_3}\right)$$

We can abandon this assumption of a Boltzmann distribution for the lower laser level, CO_2 (1, 0, 0), by constructing a rate equation for the number density $\hat{n}_{100}$ of this vibrational level, allowing for intramode and intermode vibrational relaxation with the Fermi-degenerate level, that is (dropping the 00 subscript on $\hat{n}$ and k),

$$\begin{aligned}\frac{d\hat{n}_1}{dt} &= \sum_i (-\hat{n}_1\hat{n}_i k_{1,i;0,i+1} - \hat{n}_1\hat{n}_{i+1}k_{1,i+1;2,i} \\ &\quad + \hat{n}_2\hat{n}_i k_{2,i;1,i+1} + \hat{n}_0\hat{n}_{i+1}k_{0,i+1;1,i}) \\ &\quad - \hat{n}_1 \sum_{t=1}^{c} N_t k^t_{100;020} + n_{020} \sum_{t=1}^{c} N_t k^t_{020;100} + \Delta N\, WI_\nu\end{aligned}$$

where $k^t_{i;j}$ are the rate coefficients for V–V energy transfer, between the modes i and j, of CO_2 in collision with a structureless projectile denoted by the superscript t. We note that only $\Delta E = 0$ transitions are included in the intramode terms.

When the identities (1.57) and sequence are substituted into the rate equation, we obtain

$$\begin{aligned}\frac{d\hat{n}_1}{dt} &= k_{1,0;0,1} \sum_i [-\hat{n}_1\hat{n}_i(i+1) - 2\hat{n}_1\hat{n}_{i+1}(i+1) + 2\hat{n}_2\hat{n}_i(i+1) + \hat{n}_0\hat{n}_{i+1}(i+1)] \\ &\quad - (\hat{n}_1 - n_{020}) \sum_t N_t k^t_{020;100} + \Delta N\, WI_\nu\end{aligned}$$

Since

$$\sum_i \hat{n}_i = N_{CO_2} \qquad \text{and} \qquad h\nu_1 \sum_i i\hat{n}_i \equiv E_1$$

we have

$$\begin{aligned}\frac{d\hat{n}_1}{dt} &= k_{10;01}\left[-\hat{n}_1\left(\frac{E_1}{h\nu_1} + N\right) - 2\hat{n}_1\frac{E_1}{h\nu_1} + 2\hat{n}_2\left(\frac{E_1}{h\nu_1} + N\right) + \hat{n}_0\frac{E_1}{h\nu_1}\right] \\ &\quad - \tau_{d_3}^{-1}[\hat{n}_1 - N_{CO_2}s^2(1-s)^2(1-p)(1-r)] + \Delta N\, WI_\nu\end{aligned}$$

having substituted the Boltzmann distribution for n_{020} and defining (in the notation of Douglas-Hamilton[27])

$$\tau_{d_3}^{-1} = \sum_t N_t k_{020,100}^t \tag{2.136}$$

If we assume that only the first excited level of the symmetric mode is non-Boltzmann, then we can replace $\hat{n}_0$ and $\hat{n}_2$ by n_0 and n_2, respectively, and the rate equation becomes

$$\frac{d\hat{n}_1}{dt} = -\tau_{1V}^{-1}\left\{\hat{n}_1 - \frac{2n_2[(E_1/Nh\nu_1)+1]+n_0(E_1/Nh\nu_1)}{(3E_1/Nh\nu_1)+1}\right\}$$
$$-\tau_{d_3}^{-1}[\hat{n}_1 - N_{CO_2}s^2(1-s)^2(1-p)(1-r)] + \Delta N\, WI_\nu$$

where the intramode relaxation time is defined by

$$\tau_{1V}^{-1} \equiv N_{CO_2}k_{10;01}\left(\frac{3E_1}{Nh\nu_1}+1\right) = N_{CO_2}k_{10;01}(2p+1)(1-p)^{-1} \tag{2.137}$$

having used Eq. (1.69). We define the Boltzmann equilibrium value of n_{100} by

$$n_1 \equiv \frac{2n_2[(E_1/Nh\nu_1)+1]+n_0(E_1/Nh\nu_1)}{(3E_1/Nh\nu_1)+1}$$

that is,

$$\frac{n_1}{N_{CO_2}} = \frac{[2p^2(1-p)^{-1}+p(1-p)^{-1}](1-p)(1-s)^2(1-r)}{(2p+1)(1-p)^{-1}}$$
$$= p(1-p)(1-s)^2(1-r) \tag{2.138}$$

as it should be. In conclusion, we see that the rate equation becomes[27]

$$\frac{d\hat{n}_1}{dt} = -\tau_{1V}^{-1}(\hat{n}_1 - n_1) - \tau_{d_3}^{-1}[\hat{n}_1 - N_{CO_2}s^2(1-s)^2(1-p)(1-r)] + \Delta N\, WI_\nu \tag{2.139a}$$

where

$$\Delta N \equiv N_{001}P(J) - \frac{2J+1}{2J+3}\hat{n}_1 P(J+1) \tag{2.139b}$$

where $P(J)$ are the Boltzmann factors for the Jth rotational level.

2.12. TLASER: A CO_2 Laser Kinetics Code[19,28]

The laser medium is described by eight main dependent variables, E, E_1, E_2, E_3, E_4, E_5, I_ν, N_{100}, which represent, respectively, the energy of the translational-plus-rotational motion of the gas, the energies of the three vibrational modes of CO_2, the energies of the vibrational modes of N_2 and CO, the radiation intensity within the laser cavity, and $N_{100}P(J+1)$ is the number of CO_2 molecules per unit volume in the rotational level $(J+1)$ of the lowest symmetric mode. The model consists of eight coupled ordinary differential equations, Eqs. (2.57), (2.110)–(2.114), (2.124), and (2.139), each describing the rate of change of one of the above variables with time. These equations are solved initially by the Runge–Kutta method and subsequently by a modified Hamming predictor corrector method.

The program consists of eight routines: MAIN, FXY, DFEQS2, ZNEPTS, RATES, ELEC, PRNPLT, and PLSCAL. The program structure is given in Fig. 12.

Input data is read in MAIN and all variables are initialized. MAIN calls DFEQS2 at each time step, which solves the set of differential equations. DFEQS2 calls FXY, which computes the right-hand side of the differential equations. RATES computes all the rates τ_{ij} at each step. ZNEPTS and ELEC compute either the experimental or analytical form of the electrons per unit volume. PRNPLT and PLSCAL produce a plot of the power-output profile as a function of time on the offline printer.

MAIN

The input data is read in MAIN and all constants that are independent of time are computed. All variables except the relaxation times are initialized at the initial ambient gas temperature MAIN calls RATES once to initialize the relaxation times. DFEQS2 is called at each time step, which solves the set of differential equations. MAIN prints out the solutions at each time step.

Subroutine DFEQS2 (K, NN, XX, Y, BB, EXTERN)

This routine solves a set of first-order ordinary differential equations using a modified Hamming predictor–corrector method. To start the solution initial values at four time steps are computed using the Runge–Kutta

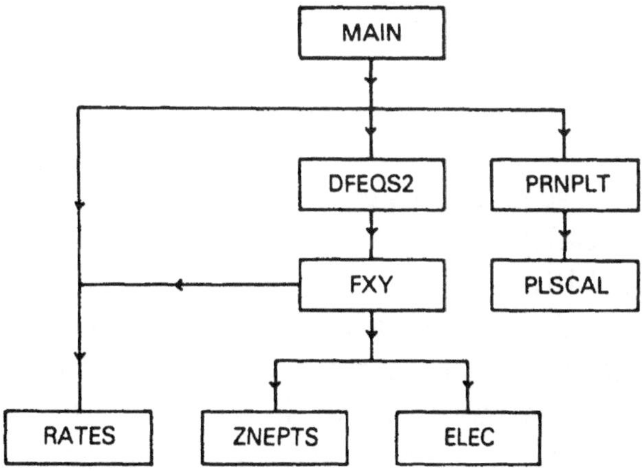

Fig. 12. Outline flow diagram for TLASER.

Table 3. Arguments of DFEQS2 (K, NN, HH, XX, Y, BB, EXTERN)

Argument	Description
K	= 1; solve by Runge–Kutta method = 2; solve by modified Hamming predictor–corrector method
NN	Number of ordinary differential equations
HH	Interval of independent variable
XX	Independent variable (time)
Y	Dependent variable array (E_i)
BB	0.0
EXTERN	Name of routine to compute right-hand side of Eqs. (2.110)–(2.114), (2.124), (2.57), (2.139)

method. The arguments of DFEQS2 are listed and described in Table 3. The Hamming method solves equations of the form

$$y' = \frac{dy}{dx} = f(x, y) \tag{2.140}$$

with $y(x_0) = y_0$. The following predictor and corrector formulas are used:

Predictor

$$p_{n+1} = y_{n-3} + \frac{4h}{3}(2y'_n - y'_{n-1} + 2y'_{n-2}) \tag{2.141}$$

Modify predictor

$$m_{n+1} = p_{n+1} + \frac{112}{121}(c_n - p_n^{(0)}) \tag{2.142}$$

$$m'_{n+1} = f(x_{n+1}, m_{n+1}) \tag{2.143}$$

Corrector

$$c_{n+1} = \frac{1}{8}(9y_n - y_{n-2}) + \frac{3h}{8}(m_{n+1} + 2y'_n - y'_{n-1}) \tag{2.144}$$

$$y_{n+1} = c_{n+1} + \frac{9}{121}(p_{n+1} - c_{n+1}) \tag{2.145}$$

This scheme uses an estimate to eliminate part of the truncation error [Eq. (2.145)]. DFEQS2 calls FXY to calculate $f(x, y)$, the right-hand side of Eq. (2.140).

Automatic halving and doubling of the interval is allowed based on an estimate of the error between the predicted and corrected value of I_ν (I_ν changes more rapidly if the ratio [(Predictor–Corrector)/Corrector] lies outside the interval 10^{-6}–10^{-5}. The interval is restrained to lie between a minimum HMIN and a maximum HMAX, which are input parameters. Due to the extremely rapid rise of the power spike, the interval is held at HMIN until the peak power is passed. Appropriate back values of the solutions are stored to be used for interval doubling. A switch ISWT is set to stop doubling at every step as the appropriate back values may not have been set. When the interval is halved the solution is restarted with the Runge–Kutta method.

Subroutine FXY (T, E, DE)

This routine computes the right-hand sides of the differential equations. FXY is the name of the routine that is equivalent to EXTERN. The name of this routine is placed in an EXTERNAL statement. FXY calls RATES for the values of the relaxation times. The arguments of FXY are listed and described in Table 4.

Subroutine RATES (T5, T20, T10, T43, T12, T3012, T63, T64, T612, T11)

This routine computes the relaxation times. The arguments of RATES, together with the temperature-dependent equations used for the rates k_{ij}, are listed in Table 2.

Function ELEC (T, S, A, B, TE)

ELEC computes $N_e(T)$, the number of electrons per unit volume at time T, from a theoretical equation used in electron-beam-controlled lasers. $N_e(T)$, the analytic pulse shape, is derived from the solution of

$$\frac{dN_e(t)}{dt} = S - \alpha N_e(t)^2 - \beta N_e(T) \tag{2.89a}$$

where $S = 0$ for $T > TE$. Solutions to this equation are given in Section 2.8.

Subroutine PRNPLT (X, Y, XMAX, XINCR, YMAX, YINCR, ISX, ISY, NPTS)

PRNPLT is a printer plot routine that plots the NPTS points given by X(I), Y(I) on a 51×101 grid, using a total of 56 lines on the printer. If ISX or ISY are nonzero, the corresponding maximum (XMAX, YMAX) and incremental step size (XINCR, YINCR) are computed. If scaling is done the corresponding new values of maximum and step size are returned.

Table 4. Arguments of FXY (T, E, DE)[a]

Argument	Description
T	Independent variable (time)
E	Array of dependent variables (E_i)
DE	Array of right-hand sides of differential equations

[a] FXY is the name of the routine that is equivalent to EXTERN.

Table 5. Program Input for TLASER

Card	Variable	Description	Units	Format
1	ZNCO2	Ratio of gas mixture: e.g., $N_{CO_2}:N_{N_2}:N_{He} = 1.0:1.0:8.0$		F5.2
	ZNN2			F5.2
	ZNHE			F5.2
	FCO2	f, fraction of undissociated CO_2		F5.2
	IDEBUG	Debug switch		I5
	NE	Switch for form of $N_e(t)$		I5
	NEQNS	If NEQNS = 8, equation for N_{100} included; if $\neq 8$ then omitted		
	ISSG	If ISSG = 1, I_ν equation omitted; small signal gain computed		I5
2	F	F, filling factor		F10.2
	ZL	L, cavity length	cm	F10.2
	R	R, output mirror reflectivity		F10.2
	A	A, area of cross section of laser volume	cm^2	F10.2
3	T5	Initial ambient gas temperature T	°K	F10.2
	P5	Initial ambient pressure	atmos	F10.2
	ZNO	N_0 or S	cm^{-3} or $sec^{-1}\ cm^{-3}$	E15.6
	TA	t_A or α. Electron pulse shape constants in Eqs. (2.146)–(2.148)	sec or 0 or $cm^3\ sec^{-1}$	E15.6
	TB	t_B or β	sec or sec^{-1}	E15.6
	TE	t_E	sec	E15.6
4		Effective electron vibrational excitation rate[a] for:		
	X1	CO_2 symmetric stretching mode	V cm^2	
	X2	CO_2 bending mode	V cm^2	
	X3	CO_2 asymmetric stretching mode	V cm^2	
	X4	N_2 vibrational mode	V cm^2	
	X6	CO vibrational mode	V cm^2	
5	TIMEND	Code runs until timend reached	sec	E10.3
	DELT	Time step	sec	E10.3
	HMIN	Minimum time step	sec	E10.3
	HMAX	Maximum time step	sec	E10.3
	PEAKPR	No printout if $I_\nu <$ PEAKPR		E10.3
	TIMEPR	Debug time intervals	sec	E10.3
	DELTPR		sec	E10.3
6	NPTS	Number of pairs of experimental points		I5
	X(I), U(I)	Pairs of numbers T, $N_e(T)$	sec cm^{-3}	2E10.3
	I = 1, NPTS	One pair to a card		

[a] These rates are strongly dependent on E/N (ratio of electric field intensity to total neutral particle density), the fractional species concentrations and the degree of fractional ionization. Curves of these values are given for example in Reference 29.

Subroutine PLSCAL (VMAX, VINCR, NPTS, DNIVIS)

This routine is a scaling program for use with PRNPLT. The full scale is adjusted to 2.5, 5.0, or 10.0 times 10**N and adjusts the maximum point VMAX to an integer multiple of 5*VINCR. Table 5 lists the input variables.

IDEBUG, the debug switch, has the value 0, 1, or 2 corresponding to a normal print out, a debug printout at every time step, or a debug printout at TIMEPR and subsequently at every DELTPR.

NE, the switch for the form of $N_e(t)$, has the values 0, 1, or 2, corresponding to the following forms for the electron pulse:

NE = 0:
$$N_e(t) = N_0(1 - e^{-t/t_A})\, e^{-t/t_B} \tag{2.146}$$

NE = 1:
$$N_e(t) = N_0 t^{t_A} e^{-t_B/t} \tag{2.147}$$

NE = 2: Form from analytic solutions of

$$\frac{dN_e(t)}{dt} = S - \alpha N_e(t)^2 - \beta N_e(t) \tag{2.89}$$

where $S = 0$ for $t > t_E$.

NE = 3: Experimental form of $N_e(t)$ as a function of t, read in as data (see Table 5).

The form for $N_e(t)$ given by Eq. (2.146) is a fit to the experimentally measured form in transverse electrical discharge (self-sustained) lasers. The form for $N_e(t)$ given by Eq. (2.147) is a fit to the experimentally measured form in some electron-beam-controlled laser experiments, and Eq. (2.89) is a theoretical equation for $N_e(t)$ in electron-beam-controlled lasers.[24] In TLASER the analytic solutions of Eq. (2.89) are coded in a function segment ELEC.

Description of Output. The input data is listed, and then the variables listed in Table 6 are output at every time step after I_ν, the cavity-field intensity, is greater than PEAKPR.

A plot of the power-output profile as a function of time is automatically plotted on the offline.

2.12.1. *Energy Balance Used in the CO_2 Laser Kinetics Code*

We define the following energy densities in units of joules/liter: E^{i}_{amb}, initial ambient (translational) energy of gas mixture; $E^{i}_{(V)}$, initial vibrational energy of gas mixture; E_{in}, discharge energy, i.e., vibrational energy gain from electrons; E^{f}_{amb}, final ambient (translational) energy of gas mixture; $E^{f}_{(V)}$, final vibrational energy of gas mixture; and E_{out}, laser output energy.

Table 6. Variables Output Every Timestep

Variable	Description	Units
T	t, time	sec
E(5)	I_ν, cavity-field intensity	erg cm^{-2} sec^{-1}
POWER	output power $= AI_\nu[(-\log R)(1-R-\bar{L})/[2(1-R)]]\times 10^{-7}$	MW
GAIN	gain $= \exp(\alpha L)$	
ALPHA	$\alpha = h_L \, \Delta NW$	cm^{-1}
DELN	ΔN population inversion per unit volume	cm^{-3}
ZN3	$N_{001}P(J)$, population density of upper laser level	cm^{-3}
ZN1	$N_{100}P(J+1)$, population density of lower laser level	cm^{-3}
T5	T, ambient gas temperature	°K
POWINT	Integrated power $\int$ POWER dt	joules
ZNEINT	Integrated electron current $\int Ne\,dt$	sec cm^{-3}

Conservation of energy implies that

$$E^{\mathrm{i}}_{\mathrm{amb}} + E^{\mathrm{i}}_{(\mathrm{V})} + E_{\mathrm{in}} = E^{\mathrm{f}}_{\mathrm{amb}} + E^{\mathrm{f}}_{(\mathrm{V})} + E_{\mathrm{out}} \tag{2.148}$$

Initial energies (denoted by superscript i) are calculated at time $t=0$ and final energies (denoted by superscript f) are calculated at time $t=t'$. The initial and final ambient energies are calculated from

$$E^{\mathrm{i,f}}_{\mathrm{amb}} = (\tfrac{5}{2}N_{\mathrm{N}_2} + \tfrac{5}{2}N_{\mathrm{CO}_2} + \tfrac{3}{2}N_{\mathrm{He}})kT^{\mathrm{i,f}} \tag{2.122a}$$

and the initial and final vibrational energies are given by the sum $\sum_{K=1}^{5} E_K$, of the solutions E_K ($K=1, 5$) of equations (2.110)–(2.114). $E^{\mathrm{f}}_{(\mathrm{V})}$ will be the total energy finally stored in the multiplicity of vibrational levels:

$$E^{\mathrm{i,f}}_{(\mathrm{V})} = \sum_{K=1}^{5} E^{\mathrm{i,f}}_{K} \tag{2.149}$$

where the superscripts i,f refer to the solutions at time $t=0$ and time $t=t'$, respectively.

From the first terms on the right-hand sides of Eqs. (2.110)–(2.114) E_{in}, the vibrational energy gain from electrons, is given by

$$E_{\mathrm{in}} = [(h\nu_1X_1 + h\nu_2X_2 + h\nu_3X_3)fN_{\mathrm{CO}_2} + h\nu_4X_4N_{\mathrm{N}_2} + h\nu_5(1-f)N_{\mathrm{CO}_2}X_5] \times \int_{t=0}^{t=t'} N_e(t)\,dt \tag{2.150}$$

The laser output energy, E_{out}, is given by the integrated power as stated in Table 6, divided by the cavity volume, that is [see also Eq. (2.69b)],

$$E_{\mathrm{out}} = \frac{1}{L}\int_{t=0}^{t=t'} I_\nu(t)\frac{-\log R}{2}\,dt\,\frac{1-R-\bar{L}}{1-R} \qquad (10^{-4}\ \mathrm{J/liter}) \tag{2.151}$$

Results

With the energy balance given by Eqs. (2.148)–(2.151) written into TLASER, the code was run with data for a laser amplifier/oscillator with the following parameters:

$$N_{CO_2}:N_{N_2}:N_{He}=1.0:1.0:3.0$$

$$F=1.0,\quad L=100.0\text{ cm},\quad R=0.3,\quad A=100.0\text{ cm}^2,\quad T_{amb}=300°\text{K},$$

$$p=1\text{ atm}$$

$$\frac{dN_e}{dt}=S-\alpha N_e^2-\beta N_e\begin{cases}s=2.34\times 10^1\ \text{cm}^{-3}\,\text{sec}^{-1}\\ \alpha=0.28\times 10^{-6}\,\text{cm}^{-3}\,\text{sec}^{-1}\\ \beta=0.0\qquad\qquad\ \ \text{sec}^{-1}\end{cases}$$

$$X_1=1.2\times 10^{-10}\,\text{cm}^3\,\text{sec}^{-1},\qquad X_2=4.8\times 10^{-9}\,\text{cm}^3\,\text{sec}^{-1}$$

$$X_3=4.8\times 10^{-9}\,\text{cm}^3\,\text{sec}^{-1},\qquad X_4=7.0\times 10^{-9}\,\text{cm}^3\,\text{sec}^{-1}$$

The energy balance part of the computer output for the above data is reproduced in Table 7. For the small-signal gain calculations the energy

Table 7. Energy Balance[a] of the Computer Output for TLASER

Small-Signal Gain (Laser Amplifier) Calculations	
Initial ambient energy	0.192778E 03
Initial vibrational energy	0.574288E 01
Vibrational energy gain from electrons	0.581005E 03
Final ambient energy	0.280219E 03
Final vibrational energy	0.499308E 03
Laser output energy	0.000000E 00
Initial energy plus energy in	0.779527E 03
Final energy plus energy out	0.779527E 03
Difference	−0.336021E−04
Laser Oscillator Calculations	
Initial ambient energy	0.192778E 03
Initial vibrational energy	0.570279E 01
Vibrational energy gain from electrons	0.587278E 03
Final ambient energy	0.479233E 03
Final vibrational energy	0.142703E 03
Laser output energy	0.162885E 03
Initial energy plus energy in	0.785759E 03
Final energy plus energy out	0.784820E 03
Difference	0.939259E 00

[a] Energy balance expressed in joules/liter.

balances to within one part in 10^7, while for the oscillator calculations the energy balances to within about one part in 10^4, which is less than 1% of the laser output energy.

2.12.2. Results of TLASER

In Fig. 13 we present the power output, calculated using Eq. (2.69b), vs. time in nanoseconds for a self-sustaining laser defined by the parameters given in Table 8, and $N_e(t)$ was calculated from the measured $i(t)$ given as the dot–dash curve in Fig. 13, according to

$$N_e(t) = i(t)/(ev_d LFA^{1/2}) \tag{2.152a}$$

where the drift velocity $v_d = 12.1 \times 10^6$ cm sec^{-1}, approximating to these conditions, was taken from Judd.[30] To prove the relationship between electron number density and current, consider

$$i = jA_e$$

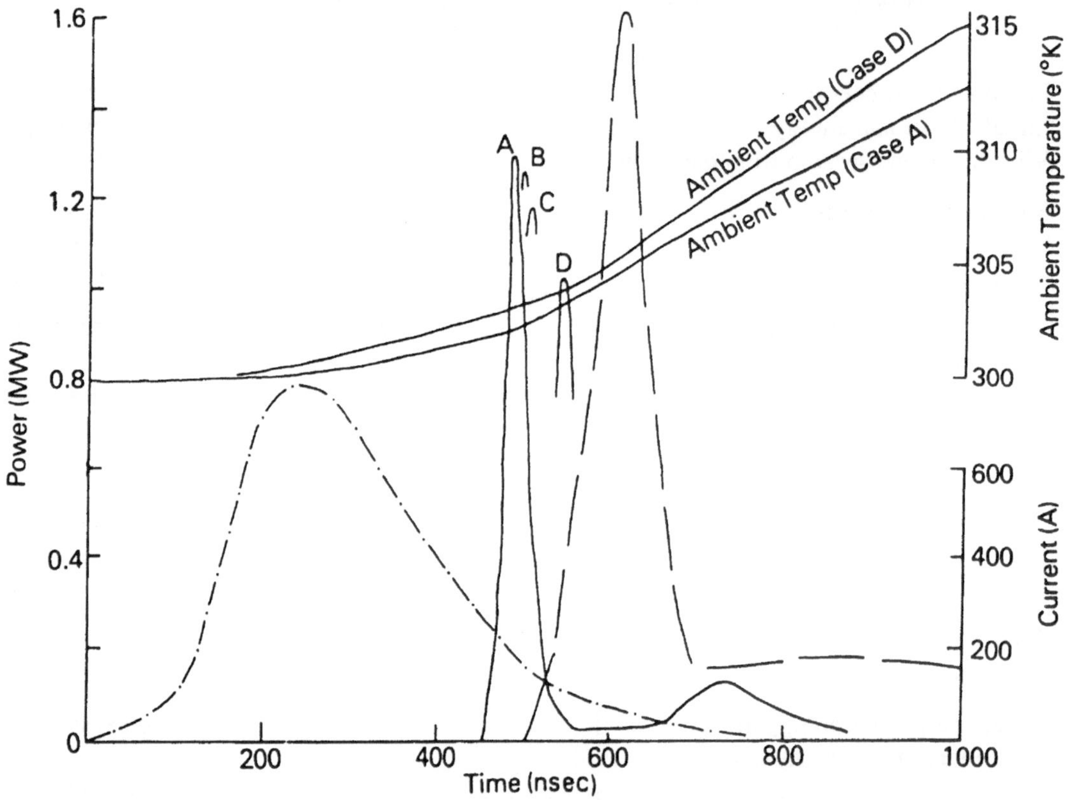

Fig. 13. Pulse shape for self-sustained laser, showing the variation of the ambient temperature. Experimental results are given by the dashed curve.

Table 8. Parameters Used in the Calculation of Figure 13

Description	Value	Units
$CO_2:N_2:He$	2:1:6	
L	28	cm
R	0.85	
F	0.71	
A	0.57	cm^2
X_1	5×10^{-10}	$cm^3\ sec^{-1}$
X_2	8×10^{-9}	$cm^3\ sec^{-1}$
X_3	5.5×10^{-9}	$cm^3\ sec^{-1}$
X_4	2.3×10^{-8}	$cm^3\ sec^{-1}$
X_6	3.0×10^{-8}	$cm^3\ sec^{-1}$
E/N	$\sim7\times10^{-16}$	$V\ cm^2$

where j is the current density, in amp cm^{-2}, and A_e is the area through which the current flows (width of electrode × length of electrode) (see Fig. 14). F, the filling factor, is the active volume divided by total volume, l/L; hence

$$l = LF$$

If the end area A is square, then

$$A_e = lA^{1/2} = LFA^{1/2}$$

while the current density is given by

$$j = N_e e v_d \tag{2.152b}$$

where N_e is the electron number density. Hence

$$i = N_e e v_d LFA^{1/2} \tag{2.152c}$$

In Fig. 13 we show the dependence of the output pulse on the percentage of CO_2 decomposition to CO, and compare our results with an experiment at the Royal Signals and Radar Establishment (Baldock). However, the uncertainties in determining $N_e(t)$ introduce a degree of uncertainty in the

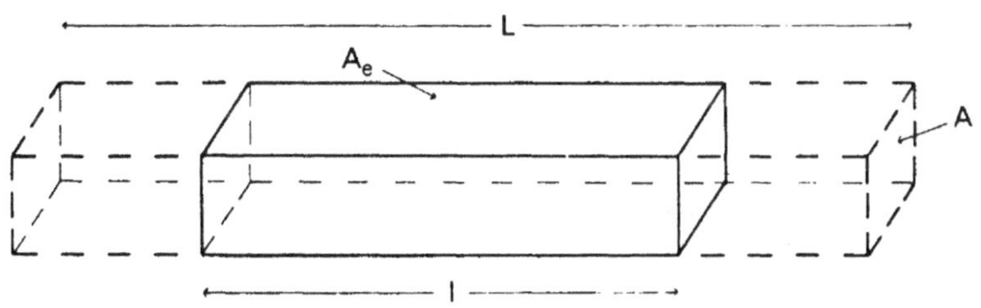

Fig. 14. Schematic of the geometry of the CO_2 laser cavity.

comparison. From the figure, we see that on the one hand, for no CO (curve A) the calculated pulse heights are in reasonable agreement with experiment, but at earlier times, while on the other hand, the calculated time of the peak approaches the experimental time with increasing CO_2 decomposition (see curves B, C, and D), but with a decreasing pulse height. We believe that the agreement between the model and experiment is quite good in view of uncertainties in $N_e(t)$. In order to eliminate the adjustable parameter f, the fractional dissociation of CO_2 to CO, it would be necessary to solve the time-dependent plasma chemistry equations (see Chapter 6) at the same time as the kinetic equation.

A number of experimental studies (see, for example, the work on self-sustaining lasers by Reid *et al.*[31,32] and on electron-beam-controlled

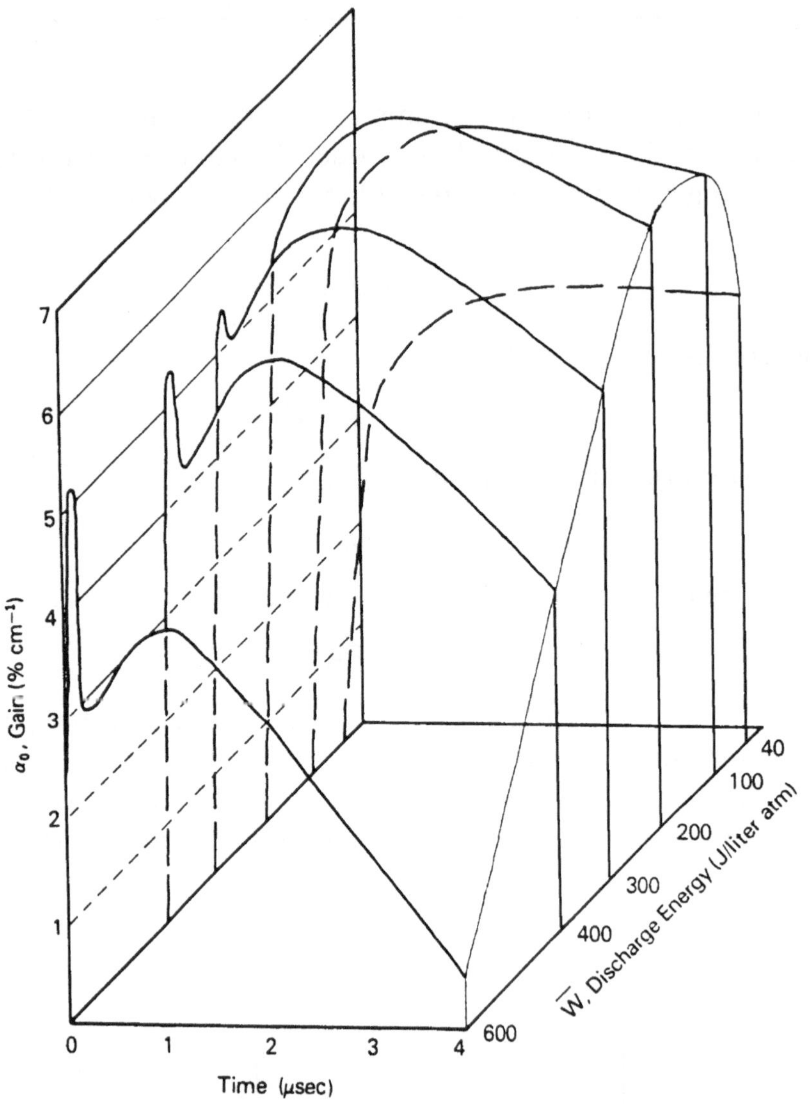

Fig. 15. Small-signal gain for differing discharge energies for an electron-beam-sustained amplifier with a primary electron-beam duration of 0.2 μ sec.

lasers by Hoffman *et al.*[33]) have shown that the small-signal gain, α_0 of Eq. (2.64b), ceases to increase with increasing discharge input energy, $\bar{W}$ (J/liter-atm), beyond a certain point. In order to investigate whether or not our model exhibits this effect, we select a value of $\bar{W}$ and then calculate S from Eq. (2.89), with

$$\beta = 0 \qquad \text{and} \qquad dN_e(t)/dt = 0$$

The equilibrium value of the electron number density, $\hat{N}_e$, to be used is given from the power density

$$\begin{aligned} P\,(\text{W m}^{-3}) &= E\,(\text{V m}^{-1}) \times j\,(\text{amp m}^{-2}) \\ &= E\,(\text{V m}^{-1}) \times e\,(\text{amp sec}) \times \hat{N}_e\,(\text{m}^{-3}) \times \nu_D\,(\text{m sec}^{-1}) \end{aligned}$$

When this equation is multiplied by t_e (sec), the duration of the electron-beam pulse, and the appropriate unit conversions are made, we have

$$\begin{aligned} \bar{W}\,(\text{J liter}^{-1}\,\text{atm}^{-1}) = e\,(\text{C}) \times E\,(\text{kV cm}^{-1}) \times \nu_D\,(\text{cm sec}^{-1}) \\ \times \hat{N}_e\,(\text{cm}^{-3}) \times t_e\,(\text{sec}) \times 10^{-1} \end{aligned} \tag{2.153}$$

In Fig. 15 we present plots of the small-signal gain vs. time for several choices of $\bar{W}$, for a 0.2 μsec electron-beam pulse, a 1 : 1 : 3 gas mixture, and $E/N = 2 \times 10^{-16}$. From this figure we see that the peak small-signal gain is achieved at 200 J liter^{-1} atm^{-1}.

2.13. CO_2 Sequence-Band Laser Calculations

The normal (regular) band of CO_2 consists of the transitions between the CO_2 (00°1) and CO_2 (10°0, 02°0) states. The terminology "band" is used to include all the transitions between a particular rotational substate of the upper CO_2 (00°1) vibrational state and a rotational substate of the lower vibrational state.

The first sequence band of CO_2 consists of the transitions between the CO_2 (00°2) and (10°1, 02°1) vibrational states. The second sequence band is the set of transitions between the CO_2 (00°3) and (10°2, 02°2) states. Lasing has been observed on the sequence bands by Reid and Siemsen.[34] Possible applications of CO_2 sequence-band lasers are concerned with CO_2 laser diagnostics and pulse propagation.

The CO_2 laser kinetics code TLASER (see Section 2.12), has been enhanced to predict the performance of CO_2 amplifiers and oscillators operating on the sequence bands. The additions to the equations of the laser model to include sequence-band operation are reviewed below.

The number density $N_{\alpha\beta\gamma}$ of molecules in the CO_2 $(\alpha\beta\gamma)$ vibrational state is given by the Boltzmann distribution

$$N_{\alpha\beta\gamma} = N \exp\left(\frac{-\alpha h\nu_1}{kT_1}\right)(\beta+1)\exp\left(\frac{-\beta h\nu_2}{kT_2}\right)\exp\left(\frac{-\gamma h\nu_3}{kT_3}\right) Z \quad (2.3a)$$

where N is the number density of CO_2 molecules, T_1, T_2, and T_3 are the effective vibrational temperatures of the symmetric, bending, and asymmetric vibrational modes, respectively, and Z is given by Eq. (2.3d).

The population inversion ΔN between the upper and lower laser levels of the normal band for P-branch transitions is given by Eq. (2.42e).

Equation (2.42e) is the equation for the population inversion used in the laser kinetics code TLASER, discussed in the previous section. Here, the equation for the population inversion for P-branch transitions on the first sequence band is

$$\Delta N = N_{002}P(J) - (\theta_J/\theta_{J+1})N_{101}P(J+1) \quad (2.154)$$

and on the second sequence band the population inversion is given by

$$\Delta N = N_{003}P(J) - (\theta_J/\theta_{J+1})N_{102}P(J+1) \quad (2.155)$$

For laser *amplifier* calculations the computer code TLASER was enhanced so that all the population inversions given by Eqs. (2.42e), (2.154), and (2.155) were calculated simultaneously by Thomson and Lamberton[35] for the various J-values. For laser oscillator calculations TLASER was modified so that the expression for the population inversion, Eq. (2.42e), could be replaced in the kinetic equations by the formulas, given by Eq. (2.154) or (2.155), for model lasing on the first or second sequence band, respectively.

Gain and output energy limitation on the normal band has been discussed in detail in Section 2.12.2. Gain and output energy limitation is expected to be observed on the sequence bands for the same reasons, and the computer experiments reported by Thompson and Lamberton give quantitative predictions concerning gain and output energy profiles on the sequence bands.

3

Electron Excitation Rates

In Eq. (2.74) the electron–molecule excitation rates X_i were expressed in terms of the relevant cross section $Q^i(v)$ and the electron velocity distribution function f (that is, for i representing a mode of CO_2 or a species):

$$X_i(T, E) = \int_0^\infty Q^i(v) f(v, T, E) v^3 \, dv \tag{3.1}$$

where f is a function of the gas temperature T, the applied field E, and the relative velocity of the electron and molecule, and is obtained by solving the relevant Boltzmann equation. This equation is derived in the following section.

3.1. Boltzmann Equation for the Electron Energy Distribution Function

3.1.1. The Boltzmann Transfer Equation

Consider an electron gas in an applied electric field $\mathbf{E}$ with a velocity distribution $f(\mathbf{x}, \mathbf{v}, t)$. The quantity $f(\mathbf{x}, \mathbf{v}, t)\, d^3x\, d^3v$ is the number of electrons that at time t are in the six-dimensional volume element $d^3x\, d^3v$ at coordinates given by $\mathbf{x}, \mathbf{v}$. The acceleration of an electron in an applied field $\mathbf{E}$ is given by $-(e\mathbf{E})/m$, so, assuming that the electrons do not interact with each other in a small time interval δt, those electrons with initial coordinates $(\mathbf{x}, \mathbf{v})$ will have coordinates $(\mathbf{x}', \mathbf{v}')$ given by

$$\mathbf{x}' = \mathbf{x} + \mathbf{v}\, \delta t, \qquad \mathbf{v}' = \mathbf{v} - \frac{e\mathbf{E}}{m} \delta t \tag{3.2}$$

Following Bond *et al.*,[36] consider two small volume elements in the six-dimensional velocity-coordinate space, at $(\mathbf{x}, \mathbf{v})$ and $(\mathbf{x}', \mathbf{v}')$ (Fig. 16). From Eq. (3.2), in a time interval δt, some of the electrons in the volume element $d^3x\, d^3v$ at $(\mathbf{x}, \mathbf{v})$ move into the volume element $d^3x'\, d^3v'$ at $(\mathbf{x}', \mathbf{v}')$.

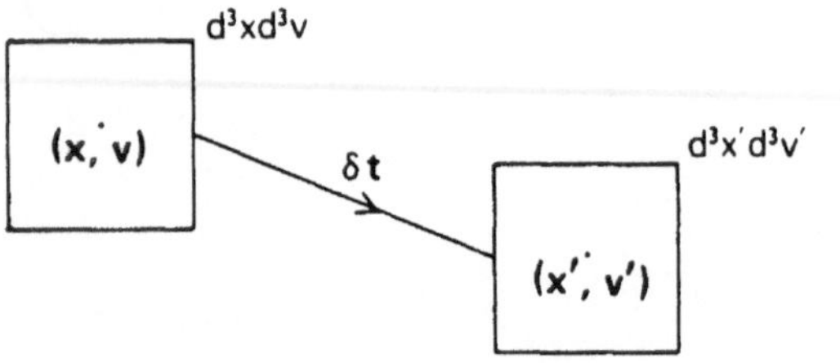

Fig. 16. Volume elements in six-dimensional velocity-coordinate space.

To first order in δt, $d^3x\, d^3v = d^3x'\, d^3v'$ and the number of molecules dN in the volume element at $(\mathbf{x}, \mathbf{v})$ is

$$dN = f(\mathbf{x}, \mathbf{v}, t)\, d^3x\, d^3v \tag{3.3}$$

while the number dN' in the volume element at $(\mathbf{x}', \mathbf{v}')$ is

$$dN' = f(\mathbf{x}', \mathbf{v}', t+\delta t)\, d^3x\, d^3v \tag{3.4}$$

Now from Eqs. (3.2)–(3.4),

$$\begin{aligned} dN' - dN &= [f(\mathbf{x}', \mathbf{v}', t+\delta t) - f(\mathbf{x}, \mathbf{v}, t)]\, d^3x\, d^3v \\ &= \left[f\left(\mathbf{x}+\mathbf{v}\,\delta t, \mathbf{v}-\frac{e\mathbf{E}}{m}\,\delta t, t+\delta t\right) - f(\mathbf{x}, \mathbf{v}, t)\right] d^3x\, d^3v \\ &= \left(\frac{\delta f}{\delta t} + \mathbf{v}\cdot\nabla f - \frac{e\mathbf{E}}{m}\cdot\nabla_v f\right)\delta t\, d^3x\, d^3v \end{aligned} \tag{3.5}$$

where ∇_v is the gradient operator in velocity space:

$$\nabla_v = \hat{\imath}\frac{\partial}{\partial v_x} + \hat{\jmath}\frac{\partial}{\partial v_y} + \hat{k}\frac{\partial}{\partial v_z} \tag{3.6}$$

$\hat{\imath}$, $\hat{\jmath}$, $\hat{k}$ being unit vectors along the v_x, v_y, and v_z axes in Cartesian velocity space.

Not all the electrons in the volume element at $(\mathbf{x}, \mathbf{v})$ at time t will be in the volume element at $(\mathbf{x}', \mathbf{v}')$ at a time δt later: Some electrons will be scattered during the time interval δt by collisions with molecules. This electron–molecule scattering will change the electron velocity distribution function $f(\mathbf{x}, \mathbf{v}, t)$. We denote by $(\delta f/\delta t)_c\, \delta t\, d^3x\, d^3v$ the net gain of electrons by the volume element $d^3x\, d^3v$ in a time δt due to this scattering. $(\delta f/\delta t)_c$ is the time rate of change of f due to electron–molecule collisions of all kinds (elastic, inelastic, superelastic, ionizing). By the above definition of $(\delta f/\delta t)_c$ we have

$$dN' - dN = +\left(\frac{\delta f}{\delta t}\right)_c \delta t\, d^3x\, d^3v \tag{3.7}$$

so that, from Eqs. (3.5) and (3.7),

$$\frac{\partial f}{\partial t}+\mathbf{v}\cdot\nabla f-\frac{e\mathbf{E}}{m}\cdot\nabla_v f=\left(\frac{\delta f}{\delta t}\right)_c \tag{3.8}$$

We now restrict our considerations to the case of a spatially uniform gas, so that f is a function of velocity only:

$$f(\mathbf{x},\mathbf{v},t)=f(\mathbf{v},t), \qquad \nabla f=0$$

and Eq. (3.8) becomes

$$\frac{\partial f}{\partial t}-\frac{e\mathbf{E}}{m}\cdot\nabla_v f=\left(\frac{\delta f}{\delta t}\right)_c \tag{3.9}$$

3.1.2. Near-Isotropic Decomposition of the Boltzmann Transfer Equation

For elastic electron–molecule collisions, the large mass difference of the colliding particles results in small energy transfers compared to the collision energy of the electrons, and also results in large electron deflections. Provided the electron mean-free path for elastic collisions is small compared to the volume occupied by the gas, it follows from the fact that the electron deflections are large, that the velocity distribution of the electrons $f(\mathbf{v})$ is nearly independent of the direction of $\mathbf{v}$. This enables $f(\mathbf{v})$ to be wiitten as

$$f(\mathbf{v})\approx f_0(v)+\frac{\mathbf{v}}{v}\cdot\mathbf{f}_1(v) \tag{3.10}$$

where f_0 and $\mathbf{f}_1$ are functions of v (the magnitude of $\mathbf{v}$ only) and $f_1 \ll f_0$. The expansion (3.10) is to be substituted into (3.9).

Since the distribution of $\mathbf{v}$ is nearly isotropic, terms depending on the direction of $\mathbf{v}$, yet quadratic in $\mathbf{v}$, will be replaced by their average over all directions of $\mathbf{v}$. Terms of the form

$$\mathbf{v}(\mathbf{v}\cdot\mathbf{A})$$

where $\mathbf{A}$ is independent of the direction of $\mathbf{v}$, will be replaced by

$$\mathbf{v}(\mathbf{v}\cdot\mathbf{A})\rightarrow\tfrac{1}{3}v^2\mathbf{A} \tag{3.11}$$

(A worse approximation would be to set equal to zero terms linear in $\mathbf{v}$, on the grounds that since $\mathbf{v}$ is nearly isotropic there is a $-\mathbf{v}$ for each $\mathbf{v}$. We shall not make this approximation since we want to evaluate to first order the anisotropic part of the distribution.)

In two dimensions, with unit vectors $\hat{\imath}$ and $\hat{\jmath}$ in the x and y directions, and an angle θ between the direction of $\mathbf{v}$ and the x-axis:

$$\mathbf{v} = v \cos\theta\hat{\imath} + v \sin\theta\hat{\jmath}$$

$$\mathbf{v}(\mathbf{v}\cdot\mathbf{A}) = \mathbf{v}(vA_x \cos\theta + vA_y \sin\theta)$$

The average over all directions of $\mathbf{v}$ is

$$[\mathbf{v}(\mathbf{v}\cdot\mathbf{A})]_{\text{av}} = \int_0^{2\pi} \mathbf{v}(\mathbf{v}\cdot\mathbf{A})\,d\theta \Big/ \int_0^{2\pi} d\theta$$

$$= \frac{1}{2\pi}\left[\hat{\imath}\int_0^{2\pi} v\cos\theta(vA_x\cos\theta + vA_y\sin\theta)\,d\theta\right.$$

$$\left. + \hat{\jmath}\int_0^{2\pi} v\sin\theta(vA_x\cos\theta + vA_y\sin\theta)\,d\theta\right]$$

$$= \frac{1}{2}v^2(A_x\hat{\imath} + A_y\hat{\jmath})$$

$$= \frac{1}{2}v^2\mathbf{A}$$

Problem 7. Repeat the above in three dimensions to prove Eq. (3.11).

When Eq. (3.10) is substituted into Eq. (3.9) we have

$$\frac{\partial f}{\partial t} - \frac{e\mathbf{E}}{m}\cdot\nabla_v\left(f_0 + \frac{\mathbf{v}}{v}\cdot\mathbf{f}_1\right) = \left(\frac{\delta f}{\delta t}\right)_c \tag{3.12}$$

Now from Eq. (3.6), consider

$$\nabla_v v_x = \left(\hat{\imath}\frac{\partial}{\partial v_x} + \hat{\jmath}\frac{\partial}{\partial v_y} + \hat{k}\frac{\partial}{\partial v_z}\right)v_x = \hat{\imath} \tag{3.13}$$

and similarly for the other components of $\mathbf{v}$. Therefore, for any two vectors $\mathbf{A}$ and $\mathbf{B}$ we have

$$\mathbf{A}\cdot(\nabla_v\mathbf{v})\cdot\mathbf{B} = \mathbf{A}\cdot(\hat{\imath}\hat{\imath} + \hat{\jmath}\hat{\jmath} + \hat{k}\hat{k})\cdot\mathbf{B}$$

$$= A_xB_x + A_yB_y + A_zB_z$$

$$= \mathbf{A}\cdot\mathbf{B} \tag{3.14}$$

where the brackets on the left-hand side of Eq. (3.14) denote that ∇_v is operating on the components of $\mathbf{v}$ only. It follows from Eq. (3.14) that

$$\mathbf{E}\cdot(\nabla_v\mathbf{v})\cdot\frac{\mathbf{f}_1}{v} = \frac{\mathbf{E}\cdot\mathbf{f}_1}{v} \tag{3.15}$$

Also, for a function $G(v)$ of the magnitude of **v** only:

$$\nabla_v G(v) = \left(\hat{\imath}\frac{\partial}{\partial v_x} + \hat{\jmath}\frac{\partial}{\partial v_y} + \hat{k}\frac{\partial}{\partial v_z}\right)G(v)$$

$$= \left(\hat{\imath}\frac{\partial v}{\partial v_x} + \hat{\jmath}\frac{\partial v}{\partial v_y} + \hat{k}\frac{\partial v}{\partial v_z}\right)\frac{\partial}{\partial v}G(v)$$

but $v = (v_x^2 + v_y^2 + v_z^2)^{1/2}$ so that

$$\frac{\partial v}{\partial v_x} = \frac{v_x}{v}, \quad \text{etc.}$$

Therefore we have

$$\nabla_v G(v) = \frac{\mathbf{v}}{v}\frac{\partial G(v)}{\partial v} \tag{3.16}$$

Using Eqs. (3.14) and (3.16) we have, for $G(v) = \mathbf{f}_1(v)/v$,

$$\mathbf{E}\cdot\nabla_v f = \frac{\mathbf{E}\cdot\mathbf{v}}{v}\frac{\partial f_0}{\partial v} + \frac{\mathbf{E}\cdot\mathbf{f}_1}{v} + \frac{(\mathbf{E}\cdot\mathbf{v})}{v}(\mathbf{v}\cdot\mathbf{f}_1)\left(-\frac{1}{v^2}\right) + \left(\frac{\mathbf{E}\cdot\mathbf{v}}{v}\right)\left(\frac{\mathbf{v}}{v}\cdot\frac{\partial \mathbf{f}_1}{\partial v}\right) \tag{3.17}$$

Since **E** is independent of **v** we apply Eq. (3.11) to the last two terms of Eq. (3.17) to obtain

$$\mathbf{E}\cdot\nabla_v f \approx \frac{\mathbf{E}\cdot\mathbf{v}}{v}\frac{\partial f_0}{\partial v} + \frac{\mathbf{E}\cdot\mathbf{f}_1}{v} - \frac{1}{3}\frac{\mathbf{E}\cdot\mathbf{f}_1}{v} + \frac{1}{3}\mathbf{E}\cdot\frac{\partial \mathbf{f}_1}{\partial v}$$

$$= \frac{\mathbf{E}\cdot\mathbf{v}}{v}\frac{\partial f_0}{\partial v} + \frac{2}{3}\frac{\mathbf{E}\cdot\mathbf{f}_1}{v} + \frac{1}{3}\mathbf{E}\cdot\frac{\partial \mathbf{f}_1}{\partial v}$$

$$= \frac{\mathbf{E}\cdot\mathbf{v}}{v}\frac{\partial f_0}{\partial v} + \frac{1}{v^2}\frac{\partial}{\partial v}\left(\frac{v^2}{3}\mathbf{E}\cdot\mathbf{f}_1\right) \tag{3.18}$$

From Eqs. (3.10), (3.12), and (3.18) the Boltzmann transfer equation (3.9) may be written

$$\frac{\partial f_0}{\partial t} + \frac{\mathbf{v}}{v}\cdot\frac{\partial \mathbf{f}_1}{\partial t} - \frac{e}{m}\left[\frac{\mathbf{E}\cdot\mathbf{v}}{v}\frac{\partial f_0}{\partial v} + \frac{1}{v^2}\frac{\partial}{\partial v}\left(\frac{v^2}{3}\mathbf{E}\cdot\mathbf{f}_1\right)\right] = \left(\frac{\delta f_0}{\delta t}\right)_c + \frac{\mathbf{v}}{v}\cdot\left(\frac{\delta \mathbf{f}_1}{\delta t}\right)_c \tag{3.19}$$

Since Eq. (3.19) must hold identically for all **v** we can equate the coefficients of zeroth and first order in **v** to obtain the two coupled equations:

$$\frac{\partial \mathbf{f}_1}{\partial t} - \frac{e\mathbf{E}}{m}\frac{\partial f_0}{\partial v} = \left(\frac{\delta \mathbf{f}_1}{\delta t}\right)_c \tag{3.20}$$

$$\frac{\partial f_0}{\partial t} - \frac{e}{3m}\frac{\mathbf{E}}{v^2}\cdot\frac{\partial}{\partial v}(v^2\mathbf{f}_1) = \left(\frac{\delta f_0}{\delta t}\right)_c \tag{3.21}$$

This is the time-dependent form of the equation for f_0 and f_1 given by Nighan[37] in his Eq. (2).

The general form of Eq. (3.10) is[38]

$$f(\mathbf{v}) = \sum_{\lambda=0}^{\infty} f_\lambda(v)P_\lambda(\cos\theta)$$

where θ is measured about the direction of the electric field. The inclusion of f_λ, $\lambda > 1$, would require a knowledge of the angular dependence of the inelastic cross sections.

3.1.3. Evaluation of $(\delta \mathbf{f}_l/\delta t)_c$

If $\lambda(v)$ (cm) is the mean distance travelled by an electron with velocity v between elastic electron–molecule collisions, then the mean frequency $\nu_e(v)$ (sec^{-1}) of elastic electron–molecule collisions for one electron will be

$$\nu_e(v) = \frac{v}{\lambda(v)} \tag{3.22}$$

Following Massey[39] we define the momentum-transfer cross section, $Q_m(v)$ (cm^2) for elastic electron–molecule collisions. If $I(\theta, v)$ is the differential cross section for the scattering of an electron with velocity v, into a range of angles $d\theta$ at an angle θ in the center-of-mass coordinate system (Reference 39, page 120), then $Q_m(v)$ is defined by

$$Q_m(v) = 2\pi \int_0^\pi (1-\cos\theta) I(\theta, v) \sin\theta \, d\theta \tag{3.23}$$

$Q_m(v)$ is related to the mean frequency $\nu_e(v)$ of elastic electron–molecule collisions by (Bond *et al.*,[36] p. 122)

$$\nu_e(v) = NQ_m(v)v \tag{3.24}$$

where N is the neutral gas density, in cm^{-3}. We assume that the electron collision frequency for elastic collisions is much larger than the collision frequency for nonelastic collisions. This is a necessary condition for the expansion (3.10) to be valid, since otherwise the electrons would gain a large component of velocity parallel to $\mathbf{E}$ between collisions, contradicting the assumption that the velocity distribution is nearly isotropic. With the above assumptions it is the elastic collisions which play the major role in reducing the asymmetry in the distribution function. Thus, following Holstein[40] we assume the rate of decrease of the asymmetric part of the velocity distribution $\mathbf{f}_1$ to be proportional to $\nu_e(v)$ times the degree of asymmetry $\mathbf{f}_1$:

$$\left(\frac{\delta \mathbf{f}_1}{\delta t}\right)_c = -\nu_e(v)\mathbf{f}_1 = -NQ_m(v)v\mathbf{f}_1(v) \tag{3.25}$$

With the assumption (3.25), Eqs. (3.20) and (3.21) become:

$$-\frac{\partial \mathbf{f}_1}{\partial t} + \frac{e\mathbf{E}}{m}\frac{\partial f_0(v)}{\partial v} = NvQ_m(v)\mathbf{f}_1(v) \tag{3.26}$$

$$\frac{\partial f_0}{\partial t} - \frac{e\mathbf{E}}{3m} \cdot \frac{\partial}{\partial v}[v^2\mathbf{f}_1(v)] = \frac{\delta f_0}{\delta t} \tag{3.27}$$

from which $\mathbf{f}_1(v)$ can be eliminated to give an equation for $f_0(v)$. We shall first evaluate the right-hand side of Eq. (3.27).

In the following sections we shall develop a numerical method for solving the coupled pair of differential equations (3.26) and (3.27), based on matrix techniques. An alternative numerical technique involving the iterative solution of an integral equation has been proposed by Sherman[(41)] and used by Long *et al.*[(42)] in their computation of electron drift velocities in molecular-gas–rare-gas mixtures.

3.1.4. The Contribution to $(\delta f_0/\delta t)_c$ from Elastic Collisions

We shall show below that the left-hand side of Eq. (3.27) is related to the gain in electron kinetic energy due to the acceleration of the electrons by the applied field $\mathbf{E}$. We shall calculate the contribution to $(\delta f_0/\delta t)_c$ from elastic collisions by correcting the mean energy gained by an electron from the applied field between collisions by an amount equal to the mean kinetic energy lost by the electrons to the molecules on each elastic collision.

The kinetic energy of an electron with velocity $\mathbf{v}$ is given by

$$\varepsilon = \tfrac{1}{2}mv^2 \tag{3.28}$$

and the rate of change of ε is given by

$$\frac{d\varepsilon}{dt} = m\dot{\mathbf{v}} \cdot \mathbf{v} = e\mathbf{E} \cdot \mathbf{v} \tag{3.29a}$$

So that the average, over all electron velocities, of the rate of change of electron energy R, due to the applied field, is given by

$$\begin{aligned} R &= \int \frac{d\varepsilon}{dt} f(\mathbf{v})\, d^3v \\ &= \int e\mathbf{E} \cdot \mathbf{v} f(\mathbf{v})\, d^3v \\ &= \int e\mathbf{E} \cdot \mathbf{v}\left[f_0(v) + \frac{\mathbf{v}}{v} \cdot \mathbf{f}_1(v)\right] d^3v \end{aligned} \tag{3.29b}$$

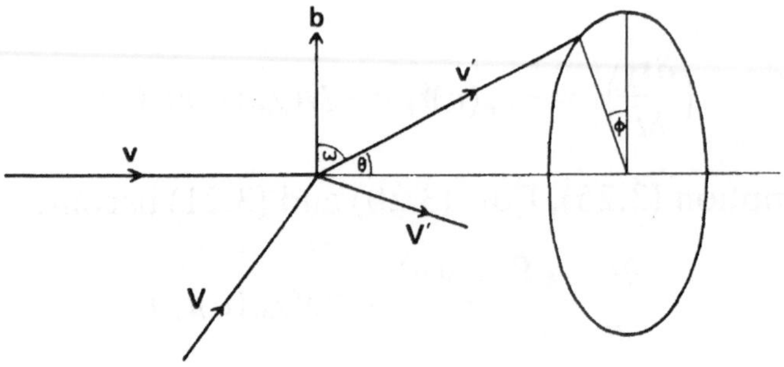

Fig. 17. Velocity vector geometry for elastic scattering.

Since $f_0(v)$ describes the isotropic part of the velocity distribution it will be an even function in v. But $\mathbf{v}$, as a function of v, is double valued, having values $\pm\mathbf{v}$ for each value of v. Therefore, the first term on the right-hand side of Eq. (3.29b) is zero, and applying the average (3.11) over all directions of $\mathbf{v}$ to the second term we obtain

$$R = \int \frac{1}{3} ev\mathbf{E} \cdot \mathbf{f}_1(v)\, d^3v \tag{3.30}$$

To evaluate the kinetic energy lost by the electrons to the molecules in elastic collisions we must consider the kinetics of a collision. For molecules of mass M and with initial electron and molecule velocities given by $\mathbf{v}$ and $\mathbf{V}$, respectively, and the corresponding final velocities given by $\mathbf{v}'$ and $\mathbf{V}'$, the energy and momentum conservation relations are:

$$\tfrac{1}{2}mv^2 + \tfrac{1}{2}MV^2 = \tfrac{1}{2}mv'^2 + \tfrac{1}{2}MV'^2 \tag{3.31}$$

$$m\mathbf{v} + M\mathbf{V} = \mathbf{v}' + M\mathbf{V}' \tag{3.32}$$

The vector $\mathbf{V}$ can be written in terms of its component along the vector $\mathbf{v}$, and a component perpendicular to $\mathbf{v}$, i.e.,

$$\mathbf{V} = a\mathbf{v} + \mathbf{b} \tag{3.33}$$

where $\mathbf{b}$ is perpendicular to $\mathbf{v}$ and lies in the plane defined by $\mathbf{v}$ and $\mathbf{V}$ (see Fig. 17). Multiplying Eq. (3.33) through by $\mathbf{v}$ we obtain, on dividing by v^2:

$$a = \frac{\mathbf{v} \cdot \mathbf{V}}{v^2} \tag{3.34}$$

On substituting Eq. (3.33) for $\mathbf{V}$ into Eqs. (3.31) and (3.32) we obtain:

$$\tfrac{1}{2}mv^2 + \tfrac{1}{2}M(a^2v^2 + b^2) = \tfrac{1}{2}mv'^2 + \tfrac{1}{2}MV'^2 \tag{3.35}$$

$$m\mathbf{v}\left(1 + \frac{M}{m}a\right) + M\mathbf{b} = m\mathbf{v}' + M\mathbf{V}' \tag{3.36}$$

The energy E' lost by the electron in the elastic collision with the molecule is, from Eq. (3.35),

$$E' = \frac{1}{2}mv^2 - \frac{1}{2}mv'^2$$

$$= \frac{1}{2}MV'^2 - \frac{1}{2}M(a^2v^2 + b^2)$$

$$= \frac{1}{2}M\left[\frac{m}{M}\mathbf{v}\left(1 + \frac{M}{m}a\right) + \mathbf{b} - \frac{m}{M}\mathbf{v}'\right]^2 - \frac{1}{2}M(a^2v^2 + b^2)$$

$$= \frac{1}{2}M\left[\frac{m^2}{M^2}v^2\left(1 + \frac{M}{m}a\right)^2 + b^2 + \frac{m^2}{M^2}v'^2 - \frac{2m^2}{M^2}\mathbf{v}\cdot\mathbf{v}'\left(1 + \frac{M}{m}a\right)\right.$$

$$\left. - \frac{2m}{M}\mathbf{b}\cdot\mathbf{v}'\right] - \frac{1}{2}M(a^2v^2 + b^2)$$

$$= \frac{1}{2}\frac{m^2}{M}(v^2 + v'^2 - 2\mathbf{v}\cdot\mathbf{v}') + ma(v^2 - \mathbf{v}\cdot\mathbf{v}') - m\mathbf{b}\cdot\mathbf{v}' \qquad (3.37)$$

where we have used Eq. (3.36) to eliminate $\mathbf{V}'$. The large mass difference between the electrons and molecules implies* that $v' \simeq v$ so that Eq. (3.37) becomes

$$E' = \left(\frac{m^2}{M}v^2 + m\mathbf{v}\cdot\mathbf{V}\right)(1 - \cos\theta) - mb\,v\cos\omega$$

where θ is the electron-scattering angle and ω is the angle between $\mathbf{b}$ and $\mathbf{v}'$.

The average rate R' of the loss of electron kinetic energy by elastic collisions with molecules will be given by E' multiplied by the differential cross section, $I(v, \theta)$, for elastic collisions, the heavy-particle number density N, the relative velocity, the electron velocity distribution $f_0(v)$, and the molecular velocity distribution $f_M(V)$, all integrated over $\mathbf{v}$ and $\mathbf{V}$:

$$R' = \int d^3v\, d^3V E'(\mathbf{v}, \mathbf{V}, \theta, \omega) I(|\mathbf{v} - \mathbf{V}|, \theta) N|\mathbf{v} - \mathbf{V}| f_0(v) f_M(V)$$

$$= \int v^2\, dv \sin\theta\, d\theta\, d\phi\, d^3V\left[\left(\frac{m^2}{M}v^2 + m\mathbf{v}\cdot\mathbf{V}\right)(1 - \cos\theta) - mbv\cos\omega\right]$$

$$\times I(|\mathbf{v} - \mathbf{V}|, \theta) N|\mathbf{v} - \mathbf{V}| f_0(v) f_M(V)$$

$$= \int v^2\, dv\, d^3V\left(\frac{m^2}{M}v^2 + m\mathbf{v}\cdot\mathbf{V}\right) Q_m(|\mathbf{v} - \mathbf{V}|) N|\mathbf{v} - \mathbf{V}| f_0(v) f_M(V) \qquad (3.38)$$

* If the initial molecule is at rest, so that $\mathbf{V} = 0$, then eliminating $\mathbf{V}'$ in Eqs. (3.31) and (3.32) gives $v^2(1 - \alpha) = v'^2(1 + \alpha) - 2\alpha\mathbf{v}\cdot\mathbf{v}'$, where $\alpha = m/M$. Since $\alpha \ll 1$, we have $v \approx v'$.

where we have used Eq. (3.23). The cos ω term in the above integral is zero since for a value of ω corresponding to some value of ϕ there is at $(\phi+\pi)$ a value of $(\pi-\omega)$ so that

$$\int_0^{2\pi} d\phi \cos\omega = 0$$

To evaluate this integral make the substitution

$$\mathbf{v}^* = \mathbf{v} - \mathbf{V} \tag{3.39}$$

Then

$$v^2 = v^{*2}\left(1+\frac{2\mathbf{v}^*\cdot\mathbf{V}}{v^{*2}}+\frac{\mathbf{V}^2}{v^{*2}}\right)$$

$$\simeq v^{*2} + 2\mathbf{v}^*\cdot\mathbf{V} \tag{3.40}$$

and

$$v \approx v^* + \frac{\mathbf{v}^*\cdot\mathbf{V}}{v^*} \tag{3.41}$$

From Eqs. (3.39)–(3.41) the factors in Eq. (3.38) become:

$$Q_m(|\mathbf{v}-\mathbf{V}|) = Q_m(v^*) \tag{3.42}$$

$$\left(v^2+\frac{M}{m}\mathbf{v}\cdot\mathbf{V}\right) \approx \left(v^{*2}+2\mathbf{v}^*\cdot\mathbf{V}+\frac{M}{m}(\mathbf{v}^*+\mathbf{V})\cdot\mathbf{V}\right)$$

$$= \left(v^{*2}+\left(2+\frac{M}{m}\right)\mathbf{v}^*\cdot\mathbf{V}+\frac{M}{m}V^2\right)$$

$$\approx \left(v^{*2}+\frac{M}{m}\mathbf{v}^*\cdot\mathbf{V}\right) \tag{3.43}$$

$$f_0(v) = f_0\left(v^*+\frac{\mathbf{v}^*\cdot\mathbf{V}}{v^*}\right)$$

$$= f_0(v^*) + \frac{\mathbf{v}^*\cdot\mathbf{V}}{v^*}\frac{\partial f_0(v^*)}{\partial v^*} + \cdots \tag{3.44}$$

keeping just the first two terms in the Taylor expression (3.44). From Eqs. (3.42)–(3.44) the expression (3.38) for R' is, upon dropping the asterisks from the v^*:

$$R' = \int d^3V \int v^2\, dv \frac{m^2}{M}\left(v^2+\frac{\mathbf{v}\cdot\mathbf{V}}{v}\frac{M}{m}\right) N Q_m(v) v \left[f_0(v)+\frac{\mathbf{v}\cdot\mathbf{V}}{v}\frac{\partial f_0}{\partial v}\right] f_M(V) \tag{3.45}$$

Since the molecular velocity distribution is isotropic, $f_M(V)$ is an even function in V, and for every $\mathbf{V}$ in Eq. (3.45) there will be a $-\mathbf{V}$, so that terms of first order in $\mathbf{V}$ disappear, giving

$$R' = \int d^3V \int v^2\, dv \left[\frac{m^2}{M} v^3 N Q_m f_0 + m \left(\frac{\mathbf{v}\cdot\mathbf{V}}{v} \right)^2 N Q_m v \frac{\partial f_0}{\partial v} \right] f_M(V) \tag{3.46}$$

With the molecular velocity distribution function normalized to unity:

$$\int f_M(V)\, d^3V = 1 \tag{3.47}$$

the average molecule kinetic energy is $\frac{3}{2}kT$, where T is the ambient gas temperature of the molecules

$$\int \frac{1}{2} M V^2 f_M(V)\, d^3V = \frac{3}{2} kT \tag{3.48}$$

Applying the averaging approximation (3.11) to the second term in the integral in Eq. (3.46) and using Eqs. (3.47) and (3.48), R' becomes

$$R' = \int v^2\, dv \left(\frac{m^2}{M} v^3 N Q_m f_0 + \frac{m}{M} kT v^2 N Q_m \frac{\partial f_0}{\partial v} \right) \tag{3.49}$$

From Eq. (3.30), the rate of change of electron energy due to the applied field, and Eq. (3.49), the rate of change of electron energy due to elastic collisions with molecules, the average rate of change of energy of electrons whose magnitude of velocity is in the range v to $v + dv$ is

$$\frac{1}{3} ev\mathbf{E}\cdot\mathbf{f}_1(v) + \frac{m^2}{M} N Q_m v^3 f_0 + \frac{mkT}{M} N Q_m v^2 \frac{\partial f_0}{\partial v} \tag{3.50}$$

which, since we are considering the electron–molecule system to be in equilibrium, will equal the rate of change of energy due to other (nonelastic) collision processes. Equation (3.50), set equal to zero, is equivalent to Eq. (19.61, 13) of Chapman and Cowling.[43]

If $\frac{1}{3}ev\mathbf{E}\cdot\mathbf{f}_1$, obtained by performing the differentiation in the second term on the left-hand side in Eq. (3.21), is replaced by the expression (3.50), then the contribution to the right-hand side of Eq. (3.21) due to elastic collisions will have been allowed for. Consequently, Eqs. (3.26) and (3.27) now become:

$$\frac{-\partial \mathbf{f}_1}{\partial t} + \frac{e\mathbf{E}}{m}\frac{\partial f_0}{\partial v} = N v Q_m \mathbf{f}_1 \tag{3.51}$$

$$\frac{\partial f_0}{\partial t} - \frac{1}{mv^2}\frac{\partial}{\partial v}\left(\frac{ev^2}{3}\mathbf{E}\cdot\mathbf{f}_1 + \frac{m^2}{M} N Q_m v^4 f_0 + \frac{mkT}{M} N Q_m v^3 \frac{\partial f_0}{\partial v} \right) = \left(\frac{\delta f_0}{\delta t} \right)_{\bar{c}} \tag{3.52}$$

where the right-hand side of Eq. (3.52) refers to inelastic collisions only.

The solution of Eq. (3.51) is, if all the quantities are assumed to be independent of time:

$$\mathbf{f}_1 = \frac{e\mathbf{E}(\partial f_0/\partial v)}{mNvQ_m} + e^{-NvQ_m t}$$

Typical values are $N = 2.45 \times 10^{19}\ \text{cm}^{-3}$ (molecular number density at atmospheric pressure) $v = 10^8\ \text{cm sec}^{-1}$ and $Q_m = 10^{-15}\ \text{cm sec}^{-1}$ so that the "relaxation time"

$$\tau = (NvQ_m)^{-1}$$

is of the order of picoseconds. Provided that **E** or N do not change in this time scale, then $\mathbf{f}_1$ is given:

$$\mathbf{f}_1 = \frac{e\mathbf{E}(\partial f_0/\partial v)}{mNvQ_m}$$

Eliminating $\mathbf{f}_1$ in Eqs. (3.51) and (3.52), and making a change in the independent variable

$$u \equiv \frac{mv^2}{2e} \quad \text{(volts)} \tag{3.53}$$

so that

$$\frac{\partial}{\partial v} = \frac{\partial u}{\partial v}\frac{\partial}{\partial u} = \frac{mv}{e}\frac{\partial}{\partial u}$$

we get

$$\frac{\partial f_0}{\partial t} - \frac{2e}{mv}\frac{\partial}{\partial u}\left[\frac{E^2 u}{3NQ_m}\frac{\partial f_0}{\partial u} + 2\frac{m}{M}NQ_m u^2 f_0 + 2\frac{mkT}{Me}NQ_m u^2\frac{\partial f_0}{\partial u}\right] = \left(\frac{\delta f_0}{\delta t}\right)_c \tag{3.54}$$

3.1.5. *The Contribution to $(\delta f_0/\delta t)_{\bar{c}}$ from Inelastic Collisions*

Following Eq. (3.53), which relates the variable u to the translational energy of an electron:

$$eu = \tfrac{1}{2}mv^2$$

we define another variable U, similarly related to the translational energy of a molecule:

$$eU = \tfrac{1}{2}MV^2$$

and eu_j is defined as the energy, E_j, of the jth internal state (vibrational, electronic, or ionized) of the molecule:

$$eu_j = E_j$$

We define $Q(u^*, U^*, 0; u, U, u_j)$ to be the inelastic cross section for the *excitation* of the molecule from the ground state to its jth internal excited state, the initial and final electron and molecule velocities being determined by u^*, u, and U^*, U, respectively.

In the notation of Section 3.1.1 (see Fig. 16), the loss of electrons of velocity v^* at time δt, due to such inelastic scattering out of the volume element $d^3x\, d^3v$ at $(\mathbf{x}, \mathbf{v})$, is

$$(dN' - dN)_{\text{loss}}$$
$$= \delta t\, d^3x\, d^3v \sum_j N_0 \int d^3V^* f(\mathbf{v}^*) f_M(\mathbf{V}^*) Q(u^*, U^*, 0; u, U, u_j) v^* \tag{3.55}$$

where N_0 represents the number of molecules in the ground state. It is assumed (Bond *et al.*,[36] p. 242) that the electrons and molecules move independently of each other prior to the collision and that the entire collision occurs within the volume element d^3x.

We now assume that Eq. (3.55) is independent of the molecular velocities (which are small compared to the electron velocities), and hence it is independent of U and U^*, and also that it does not depend on the direction of the electron velocities (the velocity distribution being nearly isotropic). With these assumptions we can write the excitation cross section Q as

$$\begin{aligned} Q(u^*, U^*, 0; u, U, u_j) &\to Q(u^*, 0; u, u_j) \\ &\equiv Q_j(u^*) \end{aligned}$$

which is in terms of the initial electron energy only, the final electron energy being determined from conservation of energy:

$$u^* = u + u_j$$

With the above assumptions, Eq. (3.55) can be written

$$\begin{aligned} (dN' - dN)_{\text{loss}} &\approx \delta t\, d^3x\, d^3v \sum_j N_0 f_0(v^*) Q_j(u^*) v^* \\ &= \delta t\, d^3x\, d^3v \sum_j N_0 f_0(u^*) Q_j(u^*) u^* \left(\frac{2e}{mv^*}\right) \\ &= \delta t\, d^3x\, d^3v \sum_j N_0 f_0(u + u_j) Q_j(u + u_j)(u + u_j) \\ &\quad \times \left[\frac{2e}{mv(1 + 2eu/mv^2)^{1/2}}\right] \\ &\approx \delta t\, d^3x\, d^3v \sum_j N_0 f_0(u + u_j) Q_j(u + u_j)(u + u_j) \frac{2e}{mv} \end{aligned} \tag{3.56}$$

Similarly, the gain in electrons for the six-volume element at time δt, due to excitation, is obtained by considering inelastic collisions with initial electron velocity $\mathbf{v}$:

$$(dN' - dN)_{\text{gain}} = \delta t\, d^3x\, d^3v \sum_j N_0 \int d^3V f(\mathbf{v}) f_M(V) \times Q(u, U, 0; u^*, U^*, u_j)|\mathbf{v} - \mathbf{V}|$$

$$\approx \delta t\, d^3x\, d^3v \sum_j N_0 f_0(v) Q_j(u) v$$

$$= \delta t\, d^3x\, d^3v \sum_j N_0 f_0(u) Q_j(u) u \left(\frac{2e}{mv}\right) \tag{3.57}$$

The net change in the number of electrons in the six-volume element at time δt is related to $(\delta f_0/\delta t)_c$ by Eq. (3.5) so that

$$(dN' - dN)_{\text{gain}} - (dN' - dN)_{\text{loss}} = -\left(\frac{\delta f_0}{\delta t}\right)_{\bar{c}} \delta t\, d^3x\, d^3v \tag{3.58}$$

So from Eqs. (3.56)–(3.58) we have (subscript e denotes excitation)

$$\left(\frac{\delta f_0}{\delta t}\right)_{e} = \sum_j N_0[f_0(u + u_j)Q_j(u + u_j)(u + u_j) - f_0(u)Q_j(u)u]\left(\frac{2e}{mv}\right) \tag{3.59}$$

The contribution to $(\delta f_0/\delta t)_{\bar{c}}$ from superelastic collisions is obtained from Eq. (3.59) by noting that, in superelastic collisions, electrons gain rather than lose energy eu_j:

$$\left(\frac{\delta f_0}{\delta t}\right)_{s} = \sum_j N_j[f_0(u - u_j)Q_{-j}(u - u_j)(u - u_j) - f_0(u)Q_{-j}(u)u]\left(\frac{2e}{mv}\right) \tag{3.60}$$

Finally, the time rate of change of f_0 due to inelastic collisions, $(\delta f_0/\delta t)_{\bar{c}}$, is given by the sum of the time rate of change of f_0 due to both excitation and superelastic collisions:

$$\left(\frac{\delta f_0}{\delta t}\right)_{\bar{c}} = \left(\frac{\delta f_0}{\delta t}\right)_{i} + \left(\frac{\delta f_0}{\delta t}\right)_{s} \tag{3.61}$$

We note that N_j is the number of molecules in the jth excited state; hence

$$N = N_0 + \sum_j N_j$$

3.1.6. The Boltzmann Equation for the Electron Energy Distribution Function

From Eqs. (3.54) and (3.59)–(3.61):

$$\left(\frac{mu}{2e}\right)^{1/2}\frac{\partial f_0}{\partial t}=\frac{E^2}{3}\frac{\partial}{\partial u}\left(\frac{u}{NQ_m}\frac{\partial f_0}{\partial u}\right)+\frac{2m}{M}\frac{\partial}{\partial u}(u^2NQ_mf_0)$$
$$+\frac{2mkT}{Me}\frac{\partial}{\partial u}\left(u^2NQ_m\frac{\partial f_0}{\partial u}\right)$$
$$+\sum_j[(u+u_j)f_0(u+u_j)N_0Q_j(u+u_j)-uf_0(u)N_0Q_j(u)]$$
$$+\sum_j[(u-u_j)f_0(u-u_j)N_jQ_{-j}(u-u_j)-uf_0(u)N_jQ_{-j}(u)] \quad (3.62)$$

Equation (3.62), with the left-hand side set equal to zero, is the Boltzmann equation for the isotropic part $f_0(v)$ of the nearly isotropic electron energy distribution, and is a second-order differential equation. The equation has been given in this form by Frost and Phelps[44] and by Sherman[41] (without the superelastic term).

Dividing the time-independent form of Eq. (3.62) by N and integrating once, we get a first-order integrodifferential equation:

$$0=\frac{1}{3}\left(\frac{E}{N}\right)^2\frac{u}{Q_m(u)}\frac{df_0}{du}+\frac{2m}{M}u^2Q_m(u)f_0+\frac{2mkT}{Me}u^2Q_m(u)\frac{df_0}{du}$$
$$+\sum_j\int_0^{u_j}du'[(u+u')Q_j(u+u')f_0(u+u')+(u-u')Q_{-j}(u-u')f_0(u-u')x_j] \quad (3.63)$$

as given by Fisher *et al.*,[45] taking $N_0\approx N$ and $x_j\equiv N_j/N$.

3.1.7. Extension of the Boltzmann Equation to a Mixture of Gases

Equations (3.62) and (3.63) apply to electrons in a gas of one type of molecule. If the gas is a mixture of different species, then let N_k be the number density of species k, whose molecules have mass M_k, and let N be the total molecule number density. Denote by Q_m^k, Q_j^k, and Q_{-j}^k the momentum transfer, inelastic, and superelastic cross sections for molecules of species k. Then instead of using Eq. (3.24), define an electron-collision frequency for elastic collisions with molecules of species k:

$$\nu_e^k(v)=N_kQ_m^k(v)v \quad (3.64)$$

Equations (3.25) and (3.26) now become

$$\left(\frac{\delta \mathbf{f}_1}{\delta t}\right) = -\sum_k \nu_e^k(v)\mathbf{f}_1 = -\sum_k N_k Q_m^k(v) v \mathbf{f}_1(v) \tag{3.65a}$$

$$-\frac{\partial \mathbf{f}_1}{\partial t} + \frac{e\mathbf{E}}{m}\frac{\partial f_0(v)}{\partial v} = \sum_k N_k v Q_m^k(v)\mathbf{f}_1(v) \tag{3.65b}$$

Consideration of the kinetics of a collision between an electron and a molecule of type k (Section 3.1.4) leads to an average rate R_k' of loss of electron kinetic energy to molecules of type k:

$$R_k' = \int dv \int dV_k \varepsilon_k' \nu_e^k(|\mathbf{v} - \mathbf{V}_k|) f_0(v) f_M^k(V_k) \tag{3.66}$$

where

$$\varepsilon_k' \equiv \frac{2m^2}{M_k}\left(v^2 + \frac{M_k}{m}\mathbf{v} \cdot \mathbf{V}_k\right) \tag{3.67}$$

in the form of Eq. (3.38). Then the total average rate R' of loss of electron kinetic energy in elastic collision with all types of molecules will be given by

$$R' = \sum_k R_k' \tag{3.68}$$

and the algebra will follow through in the same way as for Eq. (3.62) to give

$$\begin{aligned} \left(\frac{mu}{2e}\right)^{1/2}\frac{\partial f_0}{\partial t} = {} & \frac{E^2}{3}\left[\frac{\partial}{\partial u} u \left(\sum_k N_k Q_m^k\right)^{-1}\frac{\partial f_0}{\partial u}\right] + 2m\frac{\partial}{\partial u}\left[u^2\left(\sum_k \frac{N_k Q_m^k}{M_k}\right) f_0\right] \\ & + \frac{2mkT}{e}\frac{\partial}{\partial u}\left[u^2\left(\sum_k \frac{N_k Q_m^k}{M_k}\right)\frac{\partial f_0}{\partial u}\right] \\ & + \sum_j \sum_k [(u + u_{jk}) f_0(u + u_{jk}) N_k Q_j^k(u + u_{jk}) - u f_0(u) N_k Q_j^k(u)] \\ & + \sum_j \sum_k [(u - u_{jk}) f_0(u - u_{jk}) N_{kj} Q_{-j}^k(u - u_{jk}) - u f_0(u) N_{kj} Q_{-j}^k(u)] \end{aligned} \tag{3.69}$$

where N_{kj} is the number of molecules of species k, in the excited state j, and u_{jk} is the energy of the jth excited state of species k. Dividing through by N, and defining δ_k as the fractional concentration of the kth species:

$$\delta_k = \frac{N_k}{N}, \qquad \delta_{kj} = \frac{N_{kj}}{N} \tag{3.70}$$

and integrating once, gives for the steady-state case:

$$\frac{1}{3}\left(\frac{E}{N}\right)^2 u\left(\sum_k \delta_k Q_m^k\right)^{-1}\frac{df_0}{du}+2mu^2\left(\sum_k \frac{\delta_k Q_m^k}{M_k}\right)f_0\left(u,\frac{E}{N},T\right)$$
$$+\frac{2mkT}{e}u^2\left(\sum_k \frac{\delta_k Q_m^k}{M_k}\right)\frac{df_0(u,E/N,T)}{du}$$
$$+\sum_j\sum_k\int_0^{u_{jk}} du'\left[(u+u')\delta_k Q_j^k(u+u')f_0\left(u+u',\frac{E}{N},T\right)\right.$$
$$\left.+(u-u')\delta_{kj}Q_{-j}^k(u-u')f_0\left(u-u',\frac{E}{N},T\right)\right]=0 \qquad (3.71)$$

which is of the same form as the equation used by Nighan,[37] without the second and third terms. These terms may be interpreted as corrections allowing for the finite mass and the motion of the molecules, respectively.

The cross sections for superelastic collisions, Q_{-j}^k, are related to Q_j^k through detailed balance at each energy u by Mitchell and Zemansky[13] (their p. 56):

$$Q_{-j}^k(u)=\frac{u+u_{jk}}{u}Q_j^k(u+u_{jk}) \qquad (3.72)$$

Equation (3.71) is precisely the form integrated numerically in three different ways by Elliott *et al.*,[46] except for a factor specifying the fractional population of level j of the kth species.

3.2. BOLTZ: A Code Used to Calculate Electron Distributions, Transport Coefficients, and Vibrational Excitation Rates in Gases with Applied Fields[47]

BOLTZ is an important part of any program whose objective is to construct realistic mathematical models of gas lasers. Our application of the code is to provide the electron vibrational excitation rates required as input data for a CO_2 laser kinetics code (Section 2.12). Given the appropriate momentum transfer and inelastic cross sections the code will compute the electron distribution function and vibrational excitation rates, etc., for any gas mixture of up to nine gases.

The derivation of the electron Boltzmann equation for a gas mixture containing vibrating molecules, and with an applied electric field, is given in detail in Section 3.1. The computational method of solving the Boltzmann equation is the third method mentioned by Elliott *et al.*[46] and described in detail by Rockwood.[48] In common with Elliott *et al.*, we have used the cross-section data of Kieffer[49] for CO_2, N_2, and He. Kieffer lists cross-section data for other gases.

The time-dependent form of the Boltzmann equation is given in Eq. (3.69), where the independent variable u is the electron energy in volts [Eq. (3.53)]. In terms of the solution of the Boltzmann equation, $f(u, E/N, T)$, the electron–molecule excitation rate [Eq. (2.74)] is

$$X_s\left(\frac{E}{N}, T\right) = \frac{\sum_j \alpha_{js} \int_0^\infty Q_j^s f\left(u, \frac{E}{N}, T\right) 2\left(\frac{e}{m}\right)^2 u\, du}{\int_0^\infty f\left(v, \frac{E}{N}, T\right) v^2\, dv}$$

$$= \frac{\sum_j \alpha_{js} \left(\frac{2e}{m}\right)^{1/2} \int_0^\infty Q_j^s(u) f\left(u, \frac{E}{N}, T\right) u\, du}{\int_0^\infty f\left(u, \frac{E}{N}, T\right) u^{1/2}\, du} \text{ cm}^3 \text{ sec}^{-1} \tag{3.73}$$

The electron energy distribution function is normalized such that

$$\int_0^\infty f\left(u, \frac{E}{N}, T\right) u^{1/2}\, du = 1 \tag{3.74}$$

Hence

$$X_s(E/N, T) \equiv \sum_j X_{js}$$

where

$$X_{js} = \alpha_{js} \left(\frac{2e}{m}\right)^{1/2} \int_0^\infty Q_j^s(u) f\left(u, \frac{E}{N}, T\right) u\, du \tag{3.75}$$

To define the coefficients α_{js} we consider mode-energy equations such as Eq. (2.73), where we see that the X_4 term determines how much energy is pumped into the nitrogen vibrational levels in units of $h\nu_4$ quanta. If u_{js} is the energy loss of the electron (in volts) for the jth inelastic process in species s, then

$$\alpha_{js} \equiv \frac{u_{js}}{u_{1s}} \tag{3.76}$$

is the number of $h\nu_4$ quanta given to state j of N_2.

Additional quantities of physical interest are[46] the mobility

$$\mu = -\frac{1}{3N}\left(\frac{2e}{m}\right)^{1/2} \int_0^\infty \frac{u}{\sum_s \delta_s Q_m^s(u)} \frac{df}{du}\, du \tag{3.77}$$

the drift velocity

$$v_d = \mu E \tag{3.78}$$

the diffusion coefficient

$$D = \frac{1}{3N}\left(\frac{2e}{m}\right)^{1/2} \int_0^\infty \frac{uf(u, E/N, T)}{\sum_s \delta_s Q_m^s(u)}\, du \tag{3.79}$$

and the characteristic energy

$$u_k = \frac{D}{\mu} \tag{3.80}$$

and the average electron velocity $\bar{u}$, also called the electron temperature T_e, is

$$T_e = \bar{u} = \frac{2}{3}\int_0^\infty u^{3/2} f\left(u, \frac{E}{N}, T\right) du \tag{3.81}$$

where T_e is expressed in electron volts. Let Δ_j^s be the fraction of molecules of species s in the jth excited state, and related to the effective vibrational temperature, T_{js}^v, of the state by the Boltzmann distribution:

$$\begin{aligned} N_{sj} &= \Delta_j^s N_s \\ \Delta_j^s &= \exp(-u_{js}/k_B T_{js}^v) \end{aligned} \tag{3.82}$$

3.2.1. Numerical Method and Program Description

Following Rockwood,[48] the Boltzmann equation [Eq. (3.69)] is written in terms of the number density $n(u)$ of electrons with energy in the range u to $u + du$ (volts):

$$n(u) = Nu^{1/2} f(u) \tag{3.83}$$

and the electron energy, ε, is specified in electron volts using

$$eu = \varepsilon \tag{3.84}$$

Using Eqs. (3.83) and (3.84), the Boltzmann equation [Eq. (3.69)] becomes, on dividing each term by the common factor $(1/N)(e/2m)^{1/2}$:

$$\begin{aligned} \frac{\partial n}{\partial t} = &-\frac{\partial}{\partial \varepsilon}\left[\frac{2Ne^2(E/N)^2 \varepsilon}{3m(\nu/N)}\left(\frac{n}{2\varepsilon} - \frac{\partial n}{\partial \varepsilon}\right)\right] - \frac{\partial}{\partial \varepsilon}\left\{\bar{\nu}\left[n\left(\frac{kT}{2} - \varepsilon\right) - kT\varepsilon\frac{\partial n}{\partial \varepsilon}\right]\right\} \\ &+ \sum_j \sum_s N_s [R_{js}(\varepsilon + \varepsilon_{js}^*) n(\varepsilon + \varepsilon_{js}^*) - R_{js}(\varepsilon) n(\varepsilon)] \\ &+ \sum_j \sum_s N_s\, \Delta_{js} [R'_{js}(\varepsilon - \varepsilon_{js}^*) n(\varepsilon - \varepsilon_{js}^*) - R'_{js}(\varepsilon) n(\varepsilon)] \end{aligned} \tag{3.85}$$

where

$$\frac{\nu}{N} = \left(\frac{2\varepsilon}{m}\right)^{1/2} \sum_s \delta_s Q_m^s(\varepsilon) \tag{3.86}$$

$$\bar{\nu} = 2mN\left(\frac{2\varepsilon}{m}\right)^{1/2} \sum_s \left[\frac{\delta_s Q_m^s(\varepsilon)}{M_s}\right] \tag{3.87}$$

$$\varepsilon_{js}^{*} = eu_{js}$$

$$R_{js}(\varepsilon) = Q_j^s(\varepsilon)v(\varepsilon) \tag{3.88a}$$

and

$$R'_{js}(\varepsilon) = \left(\frac{\varepsilon + \varepsilon_{sj}^{*}}{\varepsilon}\right) Q_j^s(\varepsilon + \varepsilon_{sj}^{*})v(\varepsilon) = Q_{-j}^s(\varepsilon)v(\varepsilon) \tag{3.88b}$$

is the rate at which electrons at ε suffer superelastic collisions with molecules in state N_{sj} and gain energy ε_{sj}^{*}. $v(\varepsilon)$ is the electron velocity, determined by

$$v(\varepsilon) = \left(\frac{2\varepsilon}{m}\right)^{1/2} \tag{3.89}$$

The electron energy axis is partitioned into K cells of width $\Delta\varepsilon$, and n_k is defined to be the number density of electrons with energy between $(k-1)\Delta\varepsilon \equiv \varepsilon_k$ and $k\,\Delta\varepsilon \equiv \varepsilon_k^+$. Equation (3.85) is converted to a set of K-coupled ordinary differential equations[48]

$$\begin{aligned} \dot{n}_k &= a_{k-1}n_{k-1} + b_{k+1}n_{k+1} - (a_k + b_k)n_k \\ &\quad + \sum_j \sum_s N_s(R_{js,k+m_{js}}n_{k+m_{js}} - R_{js,k}n_k) \\ &\quad + \sum_j \sum_s N_s\, \Delta_{js}(R'_{js,k-m_{js}}n_{k-m_{js}} - R'_{js,k}n_k) \end{aligned} \tag{3.90}$$

where

$$a_k = \frac{2Ne^2}{3m}\left(\frac{E}{N}\right)^2 \frac{1}{\nu_k^+/N}\left(\frac{1}{\Delta\varepsilon}\right)^2\left(\varepsilon_k^+ + \frac{\Delta\varepsilon}{4}\right) + \frac{\bar{\nu}_k^+}{2\Delta\varepsilon}\left(\frac{kT}{2} - \varepsilon_k^+ + \frac{2kT}{\Delta\varepsilon}\varepsilon_k^+\right) \tag{3.91}$$

$$b_k = \frac{2Ne^2}{3m}\left(\frac{E}{N}\right)^2 \frac{1}{\nu_k^+/N}\left(\frac{1}{\Delta\varepsilon}\right)^2\left(\varepsilon_k^+ - \frac{\Delta\varepsilon}{4}\right) + \frac{\bar{\nu}_k^+}{2\Delta\varepsilon}\left(\varepsilon_k^+ - \frac{kT}{2} + \frac{2kT}{\Delta\varepsilon}\varepsilon_k^+\right) \tag{3.92}$$

$$\frac{\nu_k^+}{N} = \left(\frac{2\varepsilon_k^+}{m}\right)^{1/2} \sum_s \delta_s Q_m^s(\varepsilon_k^+) \tag{3.93}$$

$$\bar{\nu}_k^+ = 2mN\left(\frac{2\varepsilon_k^+}{m}\right)^{1/2} \sum_s [\delta_s Q_m^s(\varepsilon_k^+)/M_s] \tag{3.94}$$

$$\frac{dn}{d\varepsilon} \approx \frac{n_{k+1} - n_k}{\Delta\varepsilon} \tag{3.95}$$

$$\frac{d^2n}{d\varepsilon^2} \approx \frac{n_{k-1} - 2n_k + n_{k+1}}{(\Delta\varepsilon)^2} \tag{3.96}$$

$$R_{js,k} = R_{js}(\varepsilon_k^+) \tag{3.97}$$

and m_{js} is the nearest integer to $\varepsilon_{js}^{*}/\Delta\varepsilon$. a_k is the rate at which electrons in the kth energy cell are promoted to the $(k+1)$th cell and b_k is the rate for demotion from the kth to the $(k-1)$th cell. $R_{js,k+m_{js}}$ is a rate for electrons in the $(k+m_{js})$ cell being demoted to the kth cell by giving up some of their translational energy to molecular excitation in an inelastic collision. Similarly, $R'_{js,k-m_{js}}$ is a rate for electrons in the $(k-m_{js})$ cell being promoted to the kth cell by gaining energy from a molecule in a superelastic collision. The total number density of electrons in the energy range $K\Delta\varepsilon$ is conserved by setting $a_K=0$, and in the lowest range $\Delta\varepsilon$, by setting $b_1=0$, and by setting equal to zero rates R_{js} for which $k+m_{js}>K$ and $k-m_{js}<1$.

Equation (3.90) may be written in matrix form as

$$\dot{n}_k=\sum_l C_{kl}n_l \tag{3.98a}$$

while the steady-state electron distribution is given by the solution to the K-coupled algebraic linear homogeneous equations

$$\sum_l C_{kl}n_l=0 \tag{3.98b}$$

The matrix C is banded and sparse (nonzero elements occur only on certain diagonals). It has nonzero elements on the main diagonal and on the diagonal on each side of the main diagonal, and has a diagonal for each inelastic process on each side of, and equidistant from, the main diagonal. Figure 18 illustrates the structure of the C-matrix by writing Eq. (3.90) in its matrix form for the situation of just one gas species with one excited state, for which we take $m_{js}=4$.

BOLTZ consists of two programs, BOLTZOLD and BOLTZNEW. In BOLTZOLD the entire C-matrix of Eq. (3.98) is stored. A limit on the number of energy intervals is set by the amount of store available in the computer for the C-matrix. BOLTZOLD only provides accurate solutions for a small energy range, i.e., low values of E/N. In BOLTZNEW each diagonal of nonzero elements is stored in an array. In this way a large saving of store is made by not storing zero elements of the C-matrix. This allows a much larger number of energy intervals to be considered than is the case if the whole of the C-matrix is stored, and hence BOLTZNEW can calculate the electron distribution over a large energy range, as is required for high values of E/N. The approximate solution provided by BOLTZOLD for high values of E/N is used as the initial value for the iterative solution of the matrix equation in BOLTZNEW.

If superelastic collisions are ignored the matrix C is upper Hessenberg. Setting $n_K=1$ fixes arbitrarily the electron number-density normalization and Eq. (3.98b) is then solved by back substitution.

$$
\begin{bmatrix} \dot{n}_1 \\ \dot{n}_2 \\ \dot{n}_3 \\ \dot{n}_4 \\ \dot{n}_5 \\ \dot{n}_6 \\ \dot{n}_7 \\ \vdots \end{bmatrix}
=
\begin{bmatrix}
-a_1 - N\Delta R'_1 & b_2 & 0 & 0 & NR_5 & \cdot & \cdot & \\
a_1 & -a_2 - b_2 - N\Delta R'_2 & b_3 & 0 & 0 & NR_6 & \cdot & \\
0 & a_2 & -a_3 - b_3 - N\Delta R'_3 & b_4 & 0 & 0 & NR_7 & \ddots \\
0 & 0 & a_3 & -a_4 - b_4 - N\Delta R'_4 & b_5 & 0 & \cdot & \cdot \\
N\Delta R'_1 & 0 & 0 & a_4 & -a_5 - b_5 - NR_5 - N\Delta R'_5 & b_6 & \cdot & \\
 & N\Delta R'_2 & 0 & 0 & a_5 & -a_6 - b_6 - NR_6 - N\Delta R'_6 & b_7 & \ddots \\
 & & N\Delta R'_3 & 0 & 0 & a_6 & -a_7 - b_7 - NR_7 - N\Delta R'_7 & \\
 & & & \ddots & & & & \ddots
\end{bmatrix}
\begin{bmatrix} n_1 \\ n_2 \\ n_3 \\ n_4 \\ n_5 \\ n_6 \\ n_7 \\ \vdots \end{bmatrix}
$$

Fig. 18. Structure of the C-matrix.

With superelastic collisions included, the matrix C has as many non-zero elements below the principal diagonal as above it. Equation (3.98b) is first solved for a coarse mesh ($K = 100$) using BOLTZOLD, by supposing that the contributions from superelastic collisions are stored in a matrix D so that, if the solution of Eq. (3.98b), including superelastic collisions, is $n_k + n'_k$, then Eq. (3.98b) becomes

$$\sum_l (C_{kl} + D_{kl})(n_l + n'_l) = 0 \tag{3.99a}$$

which reduces to

$$\sum_l (C_{kl} + D_{kl})n'_l = -\sum_l D_{kl}n_l \tag{3.99b}$$

where we have used Eq. (3.98b). The matrix equation [Eq. (3.99b)] is solved for n'_l by converting the matrix $(C+D)$ to upper triangular form using column pivoting and solving by back substitution. In practice, to save store, the elements of D are stored in C plus two additional one-dimensional arrays.

The resulting solution is interpolated to the fine mesh ($K = 1000$) and is used as an approximate solution by BOLTZNEW to solve the fine mesh matrix equation by Gauss–Seidel iteration:

$$n_k^{(i+1)} = -\frac{1}{C_{kk}}\left(\sum_{l=1}^{k-1} C_{kl}n_l^{(i)} + \sum_{l=k+1}^{K} C_{kl}n_l^{(i+1)}\right) \tag{3.99c}$$

working backwards from $k=K$ to $k=1$. The matrix C is diagonally dominant. The rate of convergence is speeded by periodic use of Aitken's acceleration technique:

$$n_k^{(i+1)} = n_k^{(i-2)} - \frac{(n_k^{(i-2)} - n_k^{(i-1)})^2}{n_k^{(i-2)} - 2n_k^{(i-2)} + n_k^{(i)}} \tag{3.100a}$$

As an option, the time-dependent equation, Eq. (3.98a), may be solved using an implicit Euler method:

$$\sum_l (\delta_{kl} - hC_{kl})n_l(t+h) = n_k(t) \tag{3.100b}$$

where the time step, h, and the initial electron distribution, $n_k(0)$, must be specified.

Electron vibrational excitation rates X_{js} for excited state s of species j are calculated from

$$X_{js} = \frac{\varepsilon_{js}^*}{\varepsilon_{1s}^*}\left(\frac{2e}{m}\right)^{1/2}\frac{1}{Ne^2}\sum_k (\varepsilon_k^+)^{1/2} n_k Q_j^s(\varepsilon_k^-)\,\Delta\varepsilon \tag{3.101}$$

The X_{js} are summed over the vibrational states j' of particular modes to give effective excitation rates X_s for the modes

$$X_s = \sum_{j'} X_{j's} \tag{3.102}$$

The transport coefficients defined by Eqs. (3.77)–(3.81) are calculated from:

$$D = \frac{1}{3}\left(\frac{2}{m}\right)^{1/2}\frac{N}{n_0}\sum_k \frac{(\varepsilon_k)^{1/2} n_k}{\sum_s N_s Q_m^s(\varepsilon k)} \tag{3.103}$$

$$v_d = \sum_k \frac{(\bar{a}_k - \bar{b}_k) n_k\,\Delta\varepsilon}{E\sum_k n_k} \tag{3.104}$$

$$u_k = \frac{D}{\mu} \tag{3.105}$$

$$\mu = \frac{v_d}{E/N} \tag{3.106}$$

$$\bar{u} = \frac{2}{3e^2 N}\sum_k \varepsilon_k n_k\,\Delta\varepsilon \tag{3.107}$$

where n_0 is the total electron number density:

$$n_0 = \sum_k n_k \tag{3.108}$$

and

$$\bar{a}_k = \frac{2Ne^2}{3m}\left(\frac{E}{N}\right)^2 \frac{1}{\nu_k^+/N}\left(\frac{1}{\Delta\varepsilon}\right)^2\left(\varepsilon_k^+ + \frac{\Delta\varepsilon}{4}\right) \tag{3.109}$$

$$\bar{b}_k = \frac{2Ne^2}{3m}\left(\frac{E}{N}\right)^2 \frac{1}{\nu_k^+/N}\left(\frac{1}{\Delta\varepsilon}\right)^2\left(\varepsilon_k^+ - \frac{\Delta\varepsilon}{4}\right) \tag{3.110}$$

The code performs an energy balance check. The rate of energy gain by the electrons from the applied field is given by[46]

$$\dot{E}_g = \sum_k (\bar{a}_k - \bar{b}_k)n_k\,\Delta\varepsilon \tag{3.111}$$

The rates of electron energy losses through elastic collisions, $\dot{E}_e$, and inelastic collisions, $\dot{E}_i$, are given by:

$$\dot{E}_e = -\sum_k [(a_k - \bar{a}_k) - (b_k - \bar{b}_k)]n_k\,\Delta\varepsilon \tag{3.112}$$

$$\dot{E}_i = \sum_{ksj} [N_s(R_{js,k+m_{js}} n_{k+m_{sj}} m_{sj} - \Delta_{js} R'_{sj,k-m_{sj}} n_{k-m_{sj}}]\,\Delta\varepsilon \tag{3.113}$$

Energy conservation requires that

$$\dot{E}_g = \dot{E}_e + \dot{E}_i \tag{3.114}$$

Each program consists of 11 routines: BOLT, BLTZMN, AKBK, RSIG, SPLINE, MATZER, CONSRV, MATSOL, PRNPL2, PLSCAL, and PLSCL2. Figure 19 shows an outline flow diagram applicable to both programs. Figure 20 is a flow diagram showing the structure of a job using the two programs.

Variable input data are read in BOLT. BOLT calls BLTZMN, which in turn, calls AKBK to compute the a_k and b_k coefficients. BLTZMN then calls RSIG, which computes the $R_{sj,k}$ coefficients.

Both AKBK and RSIG read cross sections from the data deck and call SPLINE to interpolate for values required at the $\Delta\varepsilon$ energy intervals. BLTZMN calls MATZER, which solves the upper-Hessenberg matrix equation (superelastic collisions neglected), and BLTZMN then calculates the vibrational excitation rates and transport coefficients. BLTZMN then calls CONSRV for an electron-energy balance check before calling MATSOL to solve the full matrix equation (superelastic collisions included). Finally BLTZMN recalculates the vibrational excitation rates and transport coefficients with the effect of superelastic collisions included. PRNPL2, PLSCAL, and PLSCL2, produce plots of the calculated electron energy distributions on the off-line printer.

A description of these routines follows. Wherever the argument list differs between the two programs, the argument list for BOLTZOLD is given first

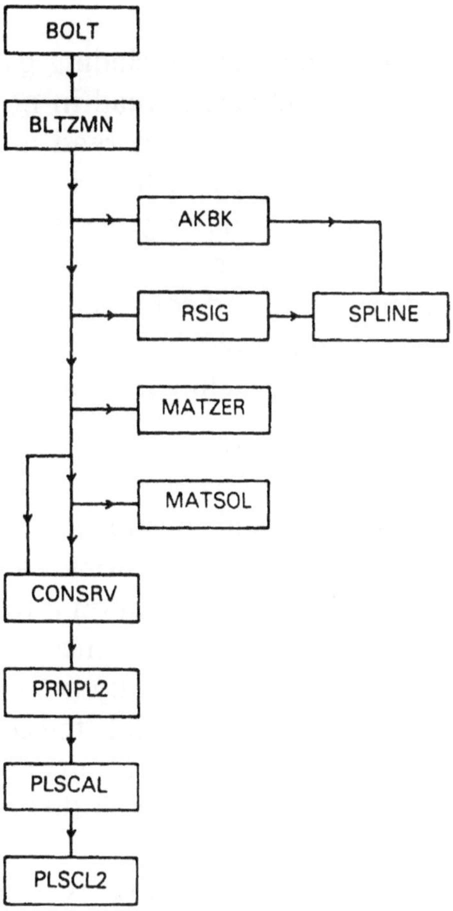

Fig. 19. Outline flow diagram for BOLTZOLD and BOLTZNEW.

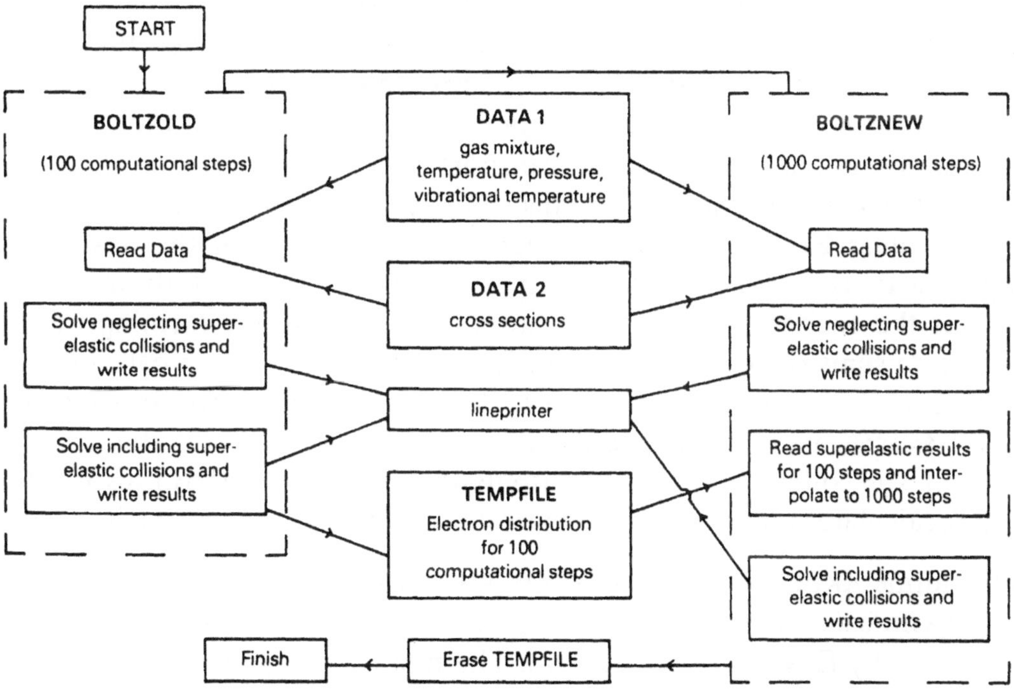

Fig. 20. Flow diagram for a job using the two programs, BOLTZOLD and BOLTZNEW.

BOLT

All input data, including gas species, mixture, and applied field, but not cross sections, is read in BOLT. T_{js}^{V}, the effective molecular excited-state temperatures, are read in and the required electron–molecule excitation rates are specified. The code can be made a subprogram of another program by removing the master segment BOLT and having a call to BLTZMN (see below) in the calling program. The calling program should contain the named common blocks BLTZ and BZRATE, and the variables in these common blocks, which are read by BOLT, should be correctly initialized in the calling routine.

Subroutine BLTZMN (ERANGE, NSUPEL)

The C-matrix is set in this routine and the electron–molecule excitation rates and the transport coefficients are calculated from the solutions to the matrix equations. ERANGE is the energy range in electron volts, over which the computation is to be performed, and KBIG is the number of discrete elements into which this energy range is partitioned. The cell width, $\Delta\varepsilon$, is then given by DELE = ERANGE/KBIG. NSUPEL takes the values 1 or 0. If NSUPEL equals 1, the computation goes on to include the effect of superelastic collisions. If superelastic collisions are to be neglected, only BOLTZNEW need be run; if they are included, then BOLTZOLD must be run first.

To set up the C-matrix, BLTZMN calls AKBK, which calculates the a_k and b_k coefficients given by Eqs. (3.91) and (3.92). BLTZMN then calls RSIG, which calculates the contribution to the C-matrix from inelastic and superelastic collisions. BLTZMN calls MATZER, which solves the upper-Hessenberg matrix equation for the electron distribution in the absence of superelastic collisions. BLTZMN outputs the electron distribution and calculates the excitation rates and transport coefficients by forming the appropriate sums over the electron distribution. BLTZMN then calls MATZER, which solves Eq. (3.98) for the electron distribution in the absence of superelastic collisions. BLTZMN outputs the electron distribution and calculates the excitation rates and transport coefficients by forming the appropriate sums over the electron distribution. BLTZMN then calls CONSRV to perform an energy balance check. If NSUPEL equals zero, BLTZMN then calls MATSOL to solve the full matrix equation to include the effect of superelastic collisions. BLTZMN outputs the resulting electron distribution and recalculates and outputs the excitation rates and transport coefficients.

Subroutine AKBK (A, B, ABAR, BBAR, KBIG, DELE)

AKBK calculates the coefficients a_k and b_k, required for the C-matrix and given by Eqs. (3.91) and (3.92). The first four arguments of AKBK are arrays of dimension KBIG: A(K) and B(K) are the coefficients a_k and b_k and ABAR(K) and BBAR(K) are the coefficients $\bar{a}_k$ and $\bar{b}_k$, required for the energy balance check performed by CONSRV and given by Eqs. (3.109) and (3.110). AKBK reads from the data deck a table of momentum-transfer cross sections $Q_{ms}(\varepsilon)$ at various energies ε for each species s. AKBK calls SPLINE to determine the values $Q_{ms}(k\ \Delta\varepsilon)$ of the momentum-transfer cross section for each cell; $k = 1, K$.

Subroutine RSIG(C, D, DL, KBIG, KBIGR, DELE. NSUPEL)
Subroutine RSIG(C, D, KST, NRJ, KBIG, KBIGR, DELE, NSUPEL)

RSIG calculates the contributions to the C-matrix due to inelastic and superelastic collisions. The first argument list refers to RSIG in BOLTZOLD. All contributions to the C-matrix are stored in the array C of dimension C(KBIG, KBIGR), where KBIGR = KBIG + 1, except for the contributions to C from superelastic collisions, which lie on the principal diagonal of C and the diagonal below the principal diagonal. Such contributions are stored in the arrays D(KBIG) and DL(KBIG), respectively. The fact that the array C, viewed as a matrix, is not square but has an extra column, is a requirement of the matrix equation solving routine MATSOL.

The second argument list refers to RSIG in BOLTZNEW. The contributions to the C-matrix from elastic and inelastic collisions are stored in the array C of dimension C(KBIG, NRJ) where NRJ is the number of diagonals in the upper-Hessenberg C-matrix. KST is an integer array of dimension KST(NRJ) and KST(I) records the distance of a diagonal from the principal diagonal of the array C(KBIG, I). The contributions to the C-matrix from superelastic collisions are stored in the array D of dimension D(KBIG, NRJ).

In each program RSIG reads lists of cross sections $Q_j^s(\varepsilon)$ for inelastic electron–molecule collisions for each of the gas species s. RSIG calls SPLINE to determine the values of the inelastic cross sections $Q_j^s(k\ \Delta\varepsilon)$ for each cell, $k = 1, K$.

Subroutine SPLINE (X, Y, T, N, M, EPS, SS, PROXIN)

SPLINE is a standard spline fitting routine, X(N) and Y(N) being the data points, where N is the number of points. M is the number of spline fit points required and these points are the points T(M) within the range of X(N). The spline fit points are given by SS(M). EPS is an accuracy criterion and PROXIN, which is not used by the calling routines, is the integral of the spline function.

Subroutine MATZER (A, M, MM, X)
Subroutine MATZER (C, KST, KBIG, NRJ, B)

MATZER solves the matrix equation

$$\sum_{j=1}^{M} A_{ij} X_j = 0 \tag{3.115}$$

by back substitution given that A_{ij} is an upper-Hessenberg matrix and $X_M = 1.0$. The dimension of A is M × MM, where MM = M + 1, and the entries

A(J, MM), J = 1, M, are ignored. In BOLTZNEW the elements of the matrix A are deduced from the elements of the arrays C(KBIG, NRJ) and KST(NRJ). The final equation in the back substitution process

$$\sum_{j=1}^{\text{KBIG}} A_{ij}X_j = 0 \tag{3.116}$$

is redundant, and this sum is calculated and is written in the output and compared with some elements of A_{ij} as a check on the self-consistency of the solution.

Subroutine MATSOL (A, M, MM, X)
Subroutine MATSOL (C, D, KST, KBIG, NRJ, N)

MATSOL in BOLTZOLD solves the matrix equation

$$\sum_j A_{ij}X_j = Y_i \tag{3.117}$$

for X_j, given Y_i and a square matrix A_{ij}. The method of solution is direct and proceeds by pivotal reduction to upper triangular form followed by back substitution. A has dimension A(M, MM), where MM = M + 1, and Y(J), J = 1, M, is stored in A(J, MM), J = 1, M. The solution is returned to the calling routine in X(M).

MATSOL in BOLTZNEW solves the homogeneous matrix equation

$$\sum_j A_{ij}N_j = 0$$

by Gauss–Seidel iteration, using Aitken's acceleration technique, Eq. (3.100a). The initial guess is given by the array N(KBIG) and the final answer is returned in the same array. The elements of the A-matrix are determined from the array C(KBIG, NRJ), D(KBIG, NRJ), specifying the nonzero diagonals, and the array KST(NRJ), specifying distance of the diagonals from the main diagonal.

Subroutine CONSRV (A, B, ABAR, BBAR, N, KBIG, DELE, VD)

This routine performs the electron energy conservation check given by Eqs. (3.111)–(3.114), and calculates the drift velocity VD, given by Eq. (3.104). The first five arguments of CONSRV are of dimension KBIG and are a_k, b_k, $\bar{a}_k$, $\bar{b}_k$, and n_k, respectively. CONSRV writes $\dot{E}_g$, $\dot{E}_e$, and $\dot{E}_i$ as well as the difference

$$E_{\text{cons}} = \dot{E}_g - (\dot{E}_e + \dot{E}_i) \tag{3.118}$$

Subroutine PRNPL2 (X, Y, Y2, XMAX, XINCR, YMAX, YINCR, ISX, ISY, NPTS)

PRNPL2 is a printer plot routine, which plots the NPTS points given by X(I), Y(I) and the NPTS points given by X(I), Y2(I) on a 51×101 grid, using a total of 56 lines on the printer. One graph of stars and one group of zeros are printed. A hash is printed at coincident points. If ISX or ISY are nonzero, the corresponding maximum (XMAX, YMAX) and incremental step size (XINCR, YINCR) are computed. If scaling is done, the corresponding new values of maximum and step size are returned.

Subroutine PLSCAL (V, VMAX, VINCR, NPTS, NDIVIS)

PLSCAL is a scaling program for scaling the X(I) values of PRNPL2. The full scale is adjusted to a digit times 10**N and the maximum point VMAX is adjusted to an integer multiple of 5*VINCR.

Subroutine PLSCL2 (V, V2, VMAX, VINCR, NPTS, NDIVIS)

PLSCL2 is a scaling program for scaling the Y(I) and Y2(I) values of PRNPL2.

3.2.2. Description of Input and Output

Table 9 lists the input variables and the form of the variables used earlier. The data deck consists of three parts, consisting of the data read by BOLTZ, AKBK, and RSIG. In column 1 of Table 9, cards are given a decimal number of the form *n.m*. Cards with $n = 1, 2, 3$ are those read by BOLTZ, AKBK, and RSIG, respectively. Only cards in the first part of the data deck need to be changed from one run to the next once the relevant cross sections are contained in the remainder of the data deck. For this reason the data deck is split into two parts. DATA1 contains the $n = 1$ cards and DATA2 contains the $n = 2$ and $n = 3$ cards.

Table 9. Program Input for BOLTZ

Card identifier	Variable	Description	Units	Format
1.1	ZNAME (NSPECS)	Names of gases, e.g., He Up to nine gases, first entry corresponds to NSPECS = 1, second to NSPECS = 2, etc.		9A8
1.2	ZQ(NSPECS)	NSPECS = 1, 9: ratio of gas mixtures, e.g., $N_{He}:N_{N_2}:N_{CO_2} = 3:3:1$ in same order as card 1.1		9F8.2
1.3	ZMASS (NSPECS)	NSPECS = 1, 9: mass of gas molecules in same order as card 1.1	g	9E8.2

continued

Table 9. (*continued*)

Card identifier	Variable	Description	Units	Format
1.4	P5	Gas pressure	atm	F5.2
	TEMP	Ambient gas temperature, T	°K	F5.2
	NS	Number of gas species		I5
	EON	Applied field/molecular number density	V cm^2	E10.3
	ERANGE	Energy range of calculation $K\ \Delta\varepsilon$	eV	F5.2
	NSUPEL	Switch to include superelastic collisions		I5
1.5	XNAME(I)	Names of up to nine electron–molecule excitation rates, e.g., $X4$		9A8
1.6	NSPEC(I)	Gas species for excitation rate I		I3
	NFRAC(I, J)	J = 1, 20 reactions, numbers J of species NSPEC(I) contributing to named electron–molecule excitation rate		20I3
More cards of type 1.6—one for each electron–molecule excitation rate				
1.7	NSPEC(I)	= −1 finish reading cards of type 1.6 specifying excitation rates		I3
1.8	NSPECS	Gas species		I5
	NREAC(J)	Excited state of molecular species NSPECS with effective vibrational temperature TV		I5
	TV	Effective vibrational temperature of excited states NREAC(J)		F10.2
More cards of type 1.8—one for each effective vibrational temperature not equal to ambient temperature				
1.9	NSPECS	= −1 finish reading cards of type 1.8 specifying vibrational temperatures		I5
1.10	H	Time interval h of Eq. (3.100b). If H is zero, time-independent equation is solved		E15.6
1.11	N(I)	I = 1, KBIG. Initial values of electron distribution for time-dependent calculation		(4E20.10)

continued

Table 9. (*continued*)

Card identifier	Variable	Description	Units	Format
2.1	E(I)	Electron energy ε	eV	F7.3
	QS(I, J)	J = 1, NS momentum-transfer cross sections; $Q_m^s(\varepsilon)$	10^{-16} cm^2	9F7.3
More cards of type 2.1 forming table of momentum-transfer cross sections				
2.2	E(I)	= −1.0 stop reading momentum-transfer cross sections		F7.3
3.1	NSPECS	Species number s		I5
	NREACJ	Reaction number j		I5
	ESTAR	Energy loss ε_{js}^*	eV	F10.1
	IONISE	= 1 if ionizing reaction; = 0 otherwise		I5
3.2	ETEMP	Electron energy ε	eV	F10.4
	SIGTEMP	Cross section for inelastic collision for reaction NREACJ of species NSPECS	10^{-16} cm^2	F10.4
More cards of type 3.2 forming table of electron energy vs. cross section				
3.3	ETEMP	= −1.0 stop reading cross sections for reaction NREACJ of species NSPECS		F10.4

The first card of the data deck, card 1.1, contains the names of the atomic and molecular species in the gas mixture. These names are read as text, to be printed in the output, and each name is a set of eight or less alphanumeric characters, e.g., N2, or NITROGEN. Card 1.2 specifies the ratio of the gas mixtures and card 1.3 the atomic and molecular masses. In card 1.4, ERANGE, expressed in electron volts, is the electron energy range $K\,\Delta\varepsilon$ over which the calculation is to be performed. The optimum value of ERANGE for any gas mixture and applied field must be found by first running the code with a judicious choice of ERANGE. An optimum value of ERANGE can be chosen by studying the spread of the calculated electron energy distribution. NSUPEL has the value 1 if superelastic collisions are to be neglected, and zero if the effect of superelastic collisions is to be included.

Card 1.5 contains the names of the required electron–molecule excitation rates. Card 1.5 is followed by a set of cards of type 1.6, one for each electron–molecule excitation rate required. NSPFC(I) is the species number, NSPECS, for excitation rate I and the NFRAC(I, J) are the reaction numbers, NREACJ, of the inelastic processes contributing to the excitation rate I. Thus NSPECS specifies the gas species s in Eq. (3.102) and the entries NFRAC(I, J) specify the reaction j' contributing to the excitation rate.

The effective vibrational temperatures T_{js}^{v} [Eq. (3.82)] are by default set equal to the ambient temperature T. For each vibrational temperature not equal to the ambient temperature, a card of type 1.7 is required, specifying the vibrational temperature TVIB (T_{js}^{v}) for species NSPECS (index "s" of the analysis) and reactions NREAC(J) (sum index "j" of the analysis).

The time interval h of Eq. (3.100b) is read from card 1.10. If h is read as zero then the time independent calculation is carried out. If h is nonzero then card 1.10 is followed by a set of cards of type 1.11 initializing the electron distribution at time zero.

The second part of the data deck is read by subroutine AKBK, and contains any number of cards forming a table of momentum-transfer cross sections at various electron energies for the different gas species.

The third part of the data deck is read by subroutine RSIG, and for each inelastic reaction j of each species, s consists of a table of inelastic cross sections at various electron energies.

The input data provided by the first part of the data deck is listed: gas mixture and E/N, etc., and nonequilibrium vibrational temperatures. The result of the consistency check [Eq. (3.116)] in subroutine MATZER is output as SUM(ZERO) and is compared to the first seven elements of the first row of the C-matrix. [SUM(ZERO) is typically less than one part in 10^4 of these elements of the C-matrix.]

The results of the electron energy conservation check performed in subroutine CONSRV are then output. In our CO_2 laser calculations we require that E_{cons} be less than one percent of $\dot{E}_g$.

The transport coefficients, and the electron–molecule excitation rates requested by the first part of the data deck (superelastic collisions ignored), are then output.

If NSUPEL equals zero, the results including the effect of superelastic collisions are then output.

A graph of the calculated electron energy distribution is then printed. Of the two curves on the graph, one (stars) includes and one (zeros) neglects the effect of superelastic collisions.

Finally, a graph of the same distribution is output, with the electron number densities plotted on a $\log_{10}$ scale.

3.2.3. *Comparisons with Analytic Forms of Electron Distribution*[50]

Following Elliott *et al.*,[46] the method of treating elastic collisions in the numerical code was checked by comparing the calculated electron distribution with the analytic forms, obtained by using the forms of the momentum-transfer cross sections appropriate to Druyvestyn and Maxwell distributions.

Druyvestyn Distribution. The Druyvestyn distribution for electrons in a single gas is obtained by ignoring the effect of inelastic collisions and gas ambient temperature, and by using a constant momentum-transfer cross section. With these conditions we have from Eq. (3.69):

$$\frac{d}{du}\left[\frac{1}{3}\left(\frac{E}{N}\right)^2 \frac{1}{Q_{\text{He}}} u \frac{df(u)}{du} + 2u^2 f(u) \frac{m}{M_{\text{He}}} Q_{\text{He}}\right] = 0 \qquad (3.119)$$

where we have chosen He to be the single-gas species. Integrating and rearranging terms we have

$$u\frac{df(u)}{du} + \frac{6mQ_{\text{He}}^2}{(E/N)^2 M_{\text{He}}} u^2 f(u) = 0 \qquad (3.120)$$

which may be rewritten in the form

$$\frac{df(u)}{f(u)} + \frac{6mQ_{\text{He}}^2}{(E/N)^2 M_{\text{He}}} u\, du = 0 \qquad (3.121)$$

The solution of Eq. (3.121) is

$$f(u) = \phi\, e^{-cu^2} \qquad (3.122)$$

where ϕ is a constant of integration and

$$c = \frac{3mQ_{\text{He}}^2}{(E/N)^2 M_{\text{He}}} \qquad (3.123)$$

Choosing $E/N = 3 \times 10^{-16}$ V cm^2 and taking $Q_{\text{He}} = 6 \times 10^{-16}$ cm^2 gives C to be 16.3×10^{-4} V^{-2}. The computed values, by BOLTZ, of

$$n(u) = Nu^{1/2} f(u)$$
$$= N\phi u^{1/2}\, e^{-cu^2} \qquad (3.124)$$

are in agreement with the analytic form of Eq. (3.124). In particular, by differentiating Eq. (3.124), it is found that $n(u)$ has a maximum at a value u_{peak} given by

$$u_{\text{peak}} = \frac{1}{2c^{1/2}} = 12.4 \quad \text{V} \qquad (3.125)$$

and the computed distribution $n(u)$ peaks at this value.

Maxwell Distribution. A Maxwell electron distribution is obtained by

$$Q_m(u) \equiv au^{-1/2} \qquad (3.126)$$

where a is constant. Equation (3.121) then becomes

$$\frac{df(u)}{f(u)}+\frac{6ma^2}{(E/N)^2 M_{He}}\,du=0 \tag{3.127}$$

and the analytic solution of Eq. (3.126) is

$$f(u)=\phi\, e^{-\bar{c}u} \tag{3.128}$$

where ϕ is a constant of integration and

$$\bar{c}=\frac{6ma^2}{(E/N)^2 M_{He}}$$

The electron distribution function computed by BOLTZ

$$\begin{aligned} n(u) &= N u^{1/2} f(u) \\ &= N\phi u^{1/2}\, e^{-\bar{c}u} \end{aligned} \tag{3.129}$$

is in agreement with the analytic form of Eq. (3.129).

3.2.4. *Results of* BOLTZ

Figure 21 shows the calculated rates for a $CO_2:N_2:He$ mixture in the proportions 3 : 3 : 4 at an ambient temperature of 300°K at one atmosphere pressure. X_T denotes the aggregate rate for the symmetric and bending

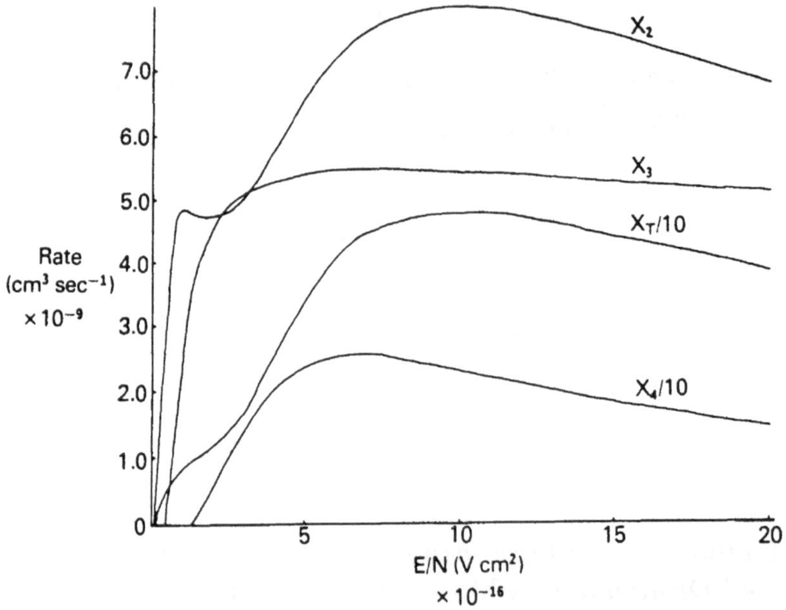

Fig. 21. Effective electron vibrational excitation rates as a function of E/N for $CO_2:N_2:He=3:3:4$, with $T=300°K$ and $p=1$ atm.

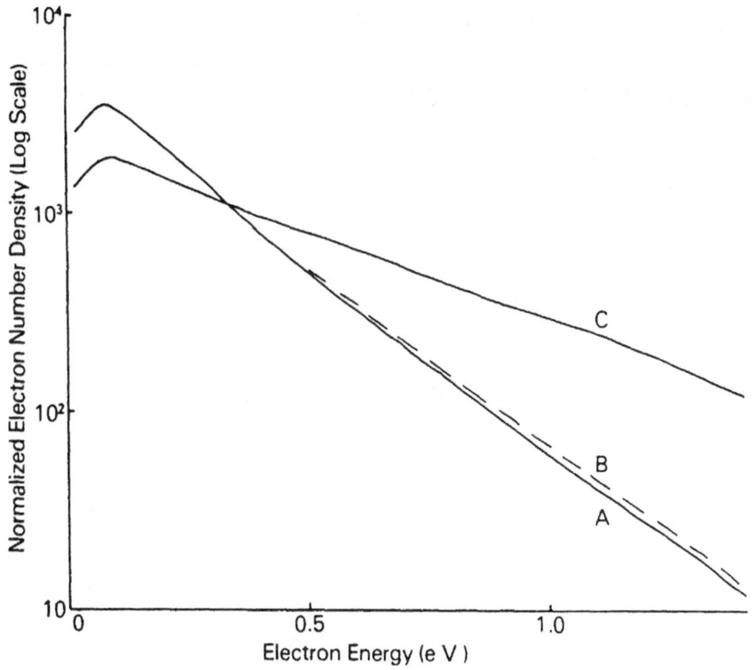

Fig. 22. Effect of superelastic collisions on electron energy distribution in $CO_2:N_2:He = 16:42:42$, with $T = 300°K$ and $p =$ 1 atm. (A) superelastic collisions neglected; (B) superelastic collisions included, $T_1 = T_2 = T_3 = T_4 = 300°K$; and (C) superelastic collisions included, $T_3 = 4000°K$, $T_1 = T_2 = T_4 =$ 300°K.

modes, neglecting the effect of superelastic collisions. Figure 22 illustrates the effect of superelastic coefficients. These results are obtained for a

$$CO_2:N_2:He = 16:42:42$$

mixture with $E/N = 0.5 \times 10^{-16}$ V cm^2. Curve A shows the electron distribution, $n(u)$, of Eq. (3.83), neglecting the effect of superelastic collisions. Curve B shows the shape of the electron distribution with superelastic collisions included, with all the excited-state populations having their equilibrium values, i.e., all effective vibrational temperatures equal to the ambient temperature, 300°K.

Curve C shows the calculated electron distribution for an excited-state population of the CO_2 asymmetric stretch mode corresponding to an effective vibrational temperature $T_3 = 4000°K$, typical of the value of T_3 for laser amplifiers at maximum gain (see Davies *et al.*[28]). The effective vibrational temperature of the other modes was kept at 300°K. The resulting electron distribution rates and transport coefficients (see Table 10) are substantially different from those of curves A and B. This result demonstrates that superelastic collisions must be taken into account when substantial excited-state populations are created in discharges.

Table 10. Rates and Transport Coefficients for Conditions of Figure 22

$CO_2:N_2:He = 16:42:42$,	$E/N = 0.5\times10^{-16}$ V cm^2,	$p = 1$ atm,	$T = 300$°K
Curve A: Neglecting superelastic collisions			
$X_1 = 2.47\times10^{-9}$,	$X_2 = 4.57\times10^{-9}$,	$X_3 = 1.02\times10^{-9}$,	$X_4 = 9.44\times10^{-12}$
$\mu = 2.24\times10^3$,	$v_d = 2.75\times10^6$,	$u_k = 0.165$,	$\bar{u} = 0.169$
Curve B: Including superelastic collisions ($T_3 = 300$°K)			
$X_1 = 2.53\times10^{-9}$,	$X_2 = 4.62\times10^{-9}$,	$X_3 = 1.06\times10^{-9}$,	$X_4 = 1.00\times10^{-11}$
$\mu = 2.23\times10^3$,	$v_d = 2.73\times10^6$,	$u_k = 0.169$,	$\bar{u} = 0.173$
Curve C: Including superelastic collisions ($T_3 = 4000$°K)			
$X_1 = 3.77\times10^{-9}$,	$X_2 = 4.94\times10^{-9}$,	$X_3 = 2.43\times10^{-9}$,	$X_4 = 1.03\times10^{-10}$
$\mu = 1.85\times10^3$,	$v_d = 2.27\times10^6$,	$u_k = 0.303$,	$\bar{u} = 0.311$

Figure 23 shows the calculated electron distribution for various values of the applied field. Figure 24 shows the results of time-dependent calculations, in which the initial electron distribution corresponds to a value of E/N of 5×10^{-16} V cm^2. The curves in Fig. 24 show the electron distributions at various times after an instantaneous change in E/N to 20×10^{-16} V cm^2. The steady-state electron distribution is attained in a few tens of picoseconds.

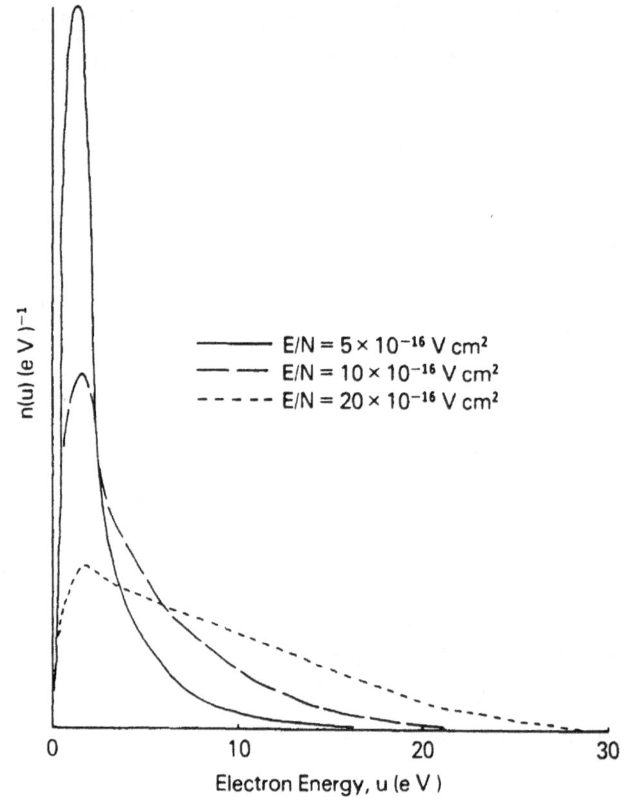

Fig. 23. Electron energy distributions for various values of (applied field)/(heavy number density) (E/N) in $CO_2:N_2:He = 1:7:12$, with $T = 300$°K and $p = 1$ atm. The effect of superelastic collisions is neglected.

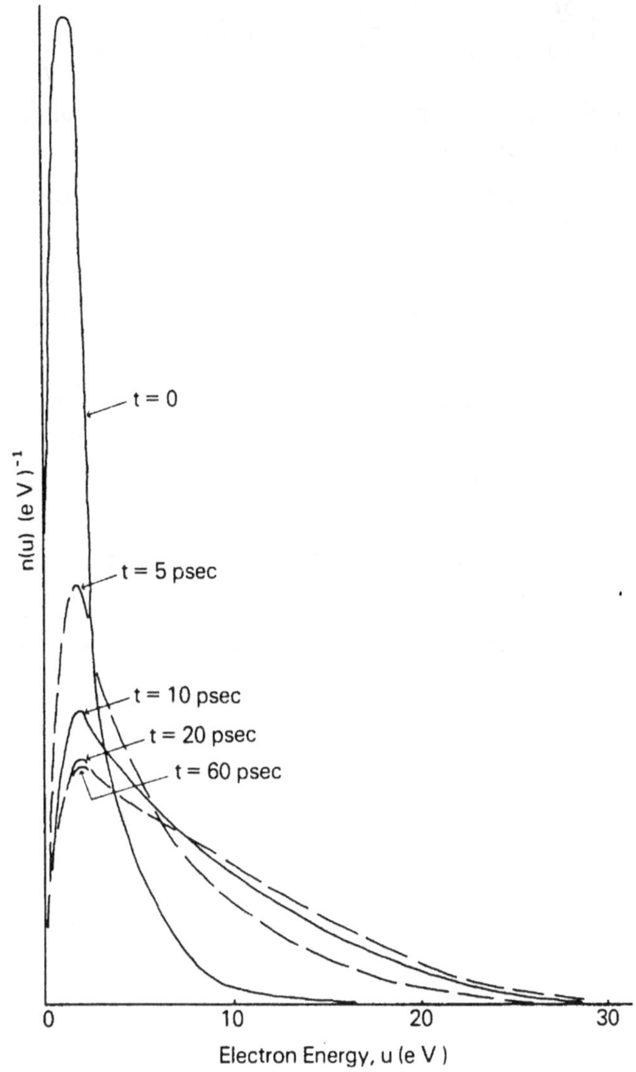

Fig. 24. Time evolution of the electron energy distribution $n(u)$, corresponding to an instantaneous change in E/N at $t = 0$ from 5×10^{-16} V cm^2 to 20×10^{-16} V cm^2. $CO_2 : N_2 : He = 1:7:12$, $T = 300°K$, $p =$ 1 atm, superelastic collisions neglected.

3.3. *The Contribution to $(\delta f_0/\delta t)_c$ from Electron–Electron Effects*

In the absence of an applied electric field, $E = 0$, and neglecting inelastic collision processes and electron–electron effects, the contribution to the time rate of change of f is given by Eqs. (3.25) and (3.54) to be

$$\left(\frac{\delta \mathbf{f}_1}{\delta t}\right)_c = -\nu_e(v)\mathbf{f}_1 \tag{3.25}$$

$$\left(\frac{\delta f_0}{\delta t}\right)_c = \frac{1}{v^2}\frac{\partial}{\partial v}\frac{m}{M}v^3\nu_e(v)\left(f_0 + \frac{kT}{e}\frac{\partial f_0}{\partial u}\right) \tag{3.130}$$

which is precisely the same form as Eqs. (7-85) of Shkarofsky *et al.*,[137] with ν_e replaced by the total collision frequency ν_g. The contribution to the collision term arising from Coulomb collisions between electrons is derived by Shkarofsky *et al.*[137]

Electron–electron collisions are not important in modeling the CO and CO_2 lasers because the pumping occurs at low electron impact energies and typical values of fractional ionization, n/N, are only 10^{-7}. However, in the electron-beam-sustained KrF laser discharge ($n/N = 3\times10^{-6}$ to 3×10^{-5}) these collisions have been shown by Long[138] to alter both the electron mobility and the rate constants for excitation and ionization. This is due to the sensitivity of the high-energy tail of the electron energy distribution toward electron–electron collisions in the KrF discharge. Because of the importance of electron–electron collisions, E/N is no longer the sole parameter characterizing the electron energy distribution in a given gas mixture; n/N must be specified too.

4

Rotational Kinetics

4.1. Introduction

The reaction kinetics associated with microsecond laser pulses through an active medium have been described in Chapters 1 and 2, assuming that rotational relaxation is instantaneous, so that the rotational sublevels within each vibrational level are populated according to a Boltzmann distribution. Experimental measurements of Cheo and Abrams[51] indicate that the relaxation time, τ_R, corresponding to the time taken for the rotational sublevels to regain a Boltzmann distribution after one sublevel has been disturbed, is of the order $\tau_R \sim 0.1$ nsec at atmospheric pressure. It is expected, therefore, that the assumption of equilibrium among the rotational levels will no longer be valid for pulses in the nanosecond regime.

The model employed to describe the coupling of rotational levels via collision processes is presented in Section 4.2. In Section 4.3 the appropriate assumptions are made for the modeling of short-pulse amplification.

It has further been demonstrated by Ballik *et al.*[52] that rotational coupling effects are important in gain-switched oscillators when the pulse length is up to a factor of 10^2 greater than the rotational relaxation time. The modification of the computer code described in Section 2.12, including these effects, is discussed in Section 4.4.

4.2. Theory of Rotational Coupling

4.2.1. The Model of Random Rotational Relaxation

Let n_J^u (cm^{-3}) be the population number density of the Jth rotational sublevel of the upper laser level, and let $k_{J,J'}^u$ be the frequency* (sec^{-1}) with

* This frequency is N, the number of particles per unit volume times an average reaction rate, denoted $\bar{k}_{if}$ in Chapter 1.

which a molecule in the J (upper) sublevel is transferred by any collision process to the J' (upper) sublevel. The rate of change of the number density, n_J^u, due to this collisional coupling between the rotational sublevels of a given vibrational level is then given by

$$\left(\frac{dn_J^u}{dt}\right)_{\text{r.c.}} = \sum_{J'} (n_{J'}^u k_{J',J}^u - n_J^u k_{J,J'}^u) \tag{4.1}$$

We include in the summation the term corresponding to $J = J'$, since $n_J^u k_{J,J}$ is added to, and subtracted from, the right-hand side of Eq. (4.1). Inclusion of this term does not affect Eq. (4.1), but it will be shown later to lead to the correct definition for the relaxation time.

Similarly, for the Jth rotational sublevel of the lower laser level:

$$\left(\frac{dn_J^l}{dt}\right)_{\text{r.c.}} = \sum_{J'} (n_{J'}^l k_{J',J}^l - n_J^l k_{J,J'}) \tag{4.2}$$

Consider now the effect of a light pulse propagating through the medium. Suppose the number density of photons is $\rho(\mathbf{r}, t, \nu_0)$ (cm^{-3}), where we suppose the pulse to be sharply peaked at frequency ν_0. The photon flux is given by $c\rho(\nu_0)$ (cm^{-2} sec^{-1}), where c is the velocity of light in the medium, and the energy flux is given by

$$I_{\nu_0}(\mathbf{r}, t) = h\nu_0 c\rho(\mathbf{r}, t, \nu_0) \qquad (\text{erg cm}^{-2}\,\text{sec}^{-1}) \tag{4.3}$$

The pulse will stimulate transitions between the J (upper) and $J+1$ or $J-1$ (lower) sublevels whenever the energy difference between the two levels is $h\nu_0$. Supposing this is the case for the J_0 (upper) and J_0+1 (lower) levels; then it can be shown [Eq. (2.57)] that the rate of change of radiation flux, I_{ν_0}, due to stimulated transitions, is given by

$$\left(\frac{dI_{\nu_0}(\mathbf{r}, t)}{dt}\right)_{\text{s.t.}} = ch\nu_0 \delta^{J_0}(\mathbf{r}, t)\, W I_{\nu_0}(\mathbf{r}, t) \tag{4.4}$$

Here W is the cross section for stimulated emission divided by $h\nu_0$, and we have replaced the equilibrium value, ΔN [Eq. (2.57)] by the nonequilibrium population difference

$$\delta^{J_0} = n_{J_0}^u - \frac{\theta_{J_0} n_{J_0+1}^l}{\theta_{J_0+1}} \tag{4.5}$$

In other words, we shall solve rate equations for $n_{J_0}^u$ and $n_{J_0+1}^l$; $\theta_{J_0} = 2J_0+1$ is the degeneracy of the J_0th rotational sublevel arising from the $2J_0+1$ possible orientations. Schappert[53] makes the approximation $\theta_{J_0}/\theta_{J_0+1} = 1$, which is correct to within 5% for $J_0 = 19$. This approximation is not necessary if the n-equations, rather than the δ-equations, are solved numerically.

From Eqs. (4.3) and (4.4),

$$\frac{d\rho(\mathbf{r}, t, \nu_0)}{dt} = \delta^{J_0} W I_{\nu_0}(\mathbf{r}, t)$$

Hence, for every net photon emission, where the number of molecules in the [J_0 (upper)/J_0+1 (lower)]-level is decreased (increased) by one, we have:

$$\left(\frac{dn_J^u}{dt}\right)_{\text{s.t.}} = -\delta^{J_0} W I_{\nu_0} \delta_{J,J_0} = -\left(\frac{dn_{J+1}^l}{dt}\right)_{\text{s.t.}} \tag{4.6}$$

where δ_{J,J_0} is the Kronecker delta, equal to unity when $J = J_0$, otherwise zero.

Our development assumes single-line operation, that is, that stimulated transitions occur between only one pair of rotational sublevels.

The population number densities are changed also through vibrational repumping and decay; we shall follow Schappert by assuming that these processes occur on a sufficiently long time scale to be uncoupled from the above equations [(4.1), (4.2), and (4.6)].

In order to express the summations in Eqs. (4.1) and (4.2) in a closed form, a relationship between the frequencies, $k_{J,J'}$, is required. We follow Hopf and Rhodes[54] by employing a model of random rotational relaxation, that is, assume there are no J selection rules. In this model, relaxation is just as likely to take place from one level J_1' as from another J_2'. The frequency $k_{J',J}$ is independent of J' and we may define

$$k_J \equiv k_{J',J} \tag{4.7}$$

With this assumption, Eq. (4.1) becomes

$$\left(\frac{dn_J^u}{dt}\right)_{\text{r.c.}} = k_J^u \sum_{J'} n_{J'}^u - n_J^u \sum_{J'} k_{J'}^u \tag{4.8}$$

At equilibrium, and when no lasing is taking place, Eq. (4.8) can be set equal to zero for each J, giving

$$k_J^u \sum_{J'} {}^e n_{J'}^u = {}^e n_J^u \sum_{J'} k_{J'}^u \tag{4.9}$$

which implies

$$\frac{k_J^u}{\sum_{J'} k_{J'}^u} = \frac{{}^e n_J^u}{\sum_{J'} {}^e n_{J'}^u} \equiv P(J) \tag{4.10}$$

where ${}^e n_J^u$ is the equilibrium value of $n_{J'}^u$ and the equation defines $P(J)$.

We now define the total population number density of the upper laser level:

$$N^u = \sum_{J'} n^u_{J'}$$

and define τ^u_R by

$$(\tau^u_R)^{-1} = \sum_{J'} k^u_{J'} \qquad (\text{sec}^{-1}) \tag{4.11}$$

where Eq. (4.8) may be written

$$\left(\frac{dn^u_J}{dt}\right)_{\text{r.c.}} = \frac{P(J)N^u - n^u_J}{\tau^u_R} \tag{4.12}$$

This equation may be interpreted physically as describing the coupling between the Jth sublevel of the upper laser level to the reservoir of N^u molecules in the upper laser level. The Jth rotational sublevel is relaxing to the "quasiequilibrium" value $N^uP(J)$, with relaxation time τ^u_R. Notice that, had we omitted the $J = J'$ term in our summations, the reservoir $\bar{N}^u$, available for transition to J, would not have included those molecules in the Jth sublevel. $\bar{N}^u$ is physically the correct picture; however, we shall see later that the corresponding definition of τ^u_R would then not be the measured relaxation time.

Adding Eqs. (4.6) and (4.12) yields the net loss of molecules in state J by rotational collisions and stimulated emission:

$$\frac{dn^u_J}{dt} = \frac{P(J)N^u - n^u_J}{\tau^u_R} - \delta^{J_0} W I_{\nu_0} \delta_{J,J_0} \tag{4.13}$$

A similar equation may be derived for the lower laser level:

$$\frac{dn^l_{J+1}}{dt} = \frac{P(J+1)N^l - n^l_{J+1}}{\tau^l_R} + \delta^{J_0} W I_{\nu_0} \delta_{J+1,J_0+1} \tag{4.14}$$

Note the Kronecker delta in Eq. (4.14) selects the correct lower laser level.

If we make the assumption that $\tau^u_R = \tau^l_R$, then Eqs. (4.13) and (4.14) may be subtracted, after multiplying Eq. (4.14) by $\theta_{J_0}/\theta_{J_0+1}$, to give for $J \neq J_0$, and therefore $J+1 \neq J_0+1$:

$$\frac{d\delta^J}{dt} = -\left\{\frac{\delta^J - [P(J)N^u - \theta_J/\theta_{J+1}P(J+1)N^l]}{\tau_R}\right\} \tag{4.15}$$

and for $J = J_0$:

$$\frac{d\delta^{J_0}}{dt} = -\delta^{J_0} W I_{\nu_0}\left(1 + \frac{\theta_{J_0}}{\theta_{J_0+1}}\right) - \left\{\frac{\delta^{J_0} - [P(J_0)N^u - \theta_{J_0}/\theta_{J_0+1}P(J_0+1)N^l]}{\tau_R}\right\} \tag{4.16}$$

The approximation

$$P(J_0) \approx \frac{\theta_{J_0}}{\theta_{J_0+1}} P(J_0+1)$$

is shown at the end of this section to be accurate to within 8% for $J_0 = 19$, assuming an ambient temperature of 300°K. Making this approximation in Eq. (4.16), we have

$$\frac{d\delta^{J_0}}{dt} = -\delta^{J_0} W I_{\nu_0}\left(1 + \frac{\theta_{J_0}}{\theta_{J_0+1}}\right) - \left[\frac{\delta^{J_0} - \Delta P(J_0)}{\tau_R}\right] \quad (4.17)$$

where

$$\Delta = N^u - N^l \quad (4.18)$$

is the population inversion density between the upper and lower laser vibrational levels. Equation (4.17), when coupled with the equations for the time rate of change of I_{ν_0} and Δ, takes into account the effects of rotational coupling. The equations for dI_{ν_0}/dt and $d\Delta/dt$, appropriate to short-pulse amplification, will be derived in Section 4.3.1, and those appropriate to the gain switched oscillator case will be discussed in Sections 4.4.1 and 4.4.2.

Derivation of P(J)

We complete this section with a derivation of $P(J)$ and show the validity of the approximation $P(J) \approx (\theta_J/\theta_{J+1}) P(J+1)$. Since $P(J)$ is defined when the rotational sublevels are in equilibrium [Eq. (4.10)], its value will be given by the Boltzmann distribution,*

$$P(J) = \frac{\theta_J \exp\{[-E(J)]/kT_R\}}{Z} \quad (4.19)$$

where $\theta_J = 2J + 1$ is the degeneracy, and $E(J)$ the energy of the Jth rotational sublevel. T_R is the effective rotational temperature. Assuming the CO_2 molecule to behave as a rigid rotator, $E(J)$ is given by

$$E(J) = J(J+1)\varepsilon \quad (4.20)$$

where $\varepsilon = hcB$, and B is the rotational constant,† inversely proportional to the moment of inertia of the CO_2 molecule, taken about an axis perpendicular to the axis of the molecule. Z is the partition function given by

$$Z = \sum_J (2J+1) \exp\left[\frac{-J(J+1)\varepsilon}{kT_R}\right] \quad (4.21)$$

* See Herzberg,[9] Vol. 1, p. 124.
† See Herzberg,[9] Vol. 1, p. 71.

The statistical weight $(2J+1)$ in Eq. (4.21) is the number of possible orientations of J in a magnetic field. However, for a linear, symmetrical molecule with nuclei of spin zero (such as the CO_2 molecule), the rotational levels, with total wave function antisymmetric under the interchange of the oxygen nuclei, are absent (see Herzberg[9]). For the 100 vibrational level, the absent rotational levels correspond to J odd, whereas for the 001 vibrational level, J even rotational levels are missing. The summation over J in Eq. (4.21) is therefore over even integers (100 level) or odd integers (001 level). The summation cannot be put into closed form, but at high temperatures we may approximate the sum to an integral, namely:

$$Z \xrightarrow[\text{high } T_R]{} \frac{1}{2}\int_0^\infty (2J+1)\exp\left[\frac{-J(J+1)\varepsilon}{kT_R}\right] dJ$$

where the factor of $\frac{1}{2}$ is due to only either odd or even terms in Eq. (4.21). Hence,

$$Z \xrightarrow[\text{high } T_R]{} \frac{1}{2}\int_0^\infty e^{-J(J+1)\varepsilon/kT_R}\, d[J(J+1)]$$

$$= \frac{kT_R}{2\varepsilon} \tag{4.22}$$

We follow Schappert[53] by assuming that the effective rotational temperature is equal to the ambient temperature T, and substitute Eq. (4.22) for Z into Eq. (4.19) to obtain

$$P(J) = \frac{2hcB}{kT}(2J+1)\exp\left[\frac{-J(J+1)hcB}{kT}\right] \tag{4.23}$$

The approximation $P(J) \approx (\theta_J/\theta_{J+1})P(J+1)$ implies

$$\exp\left[\frac{-J(J+1)hcB}{kT}\right] \approx \exp\left[\frac{-(J+2)(J+1)hcB}{kT}\right]$$

or

$$\exp\left[\frac{2(J+1)hcB}{kT}\right] \approx 1 \tag{4.24}$$

Using $B \simeq 0.4\ \text{cm}^{-1}$, the approximation may be seen to be $\sim$8% accurate for $J = 19$, at $T = 300°\text{K}$.

4.2.2. Rotational Relaxation Time τ_R

The rotational relaxation time, τ_R, has been measured for CO_2, N_2, He mixtures by Abrams and Cheo,[51] and more recently, by Jacobs *et al.*[55] The method employed was to selectively deplete the $J = 19$ rotational sublevel, and to investigate the subsequent refilling of this sublevel via measurements of small signal gain. The data were fitted to a single exponential function from which τ_R was found. The situation is described by Eq. (4.13), with $J = 19$ and I_{ν_0} set equal to zero. Since no lasing is taking place, N^u is constant, and the solution for $n^u_{J_0}$ is a single exponential function with relaxation time τ^u_R. Notice that, had we not included the term $J' = J$ in Eq. (4.1), we would have obtained instead of Eq. (4.13):

$$\frac{dn^u_J}{dt} = \frac{\bar{N}^u[P(J)^{-1} - 1]^{-1} - n^u_J}{\bar{\tau}^u_R} - \delta^{J_0} W I_{\nu_0} \delta_{J,J_0} \qquad (4.25)$$

where

$$\bar{N}^u = \sum_{J' \neq J} n^u_{J'}, \qquad (\bar{\tau}^u_R)^{-1} = \sum_{J' \neq J} k^u_{J'} \qquad (4.26)$$

Equation (4.25) is equivalent to Eq. (A-11) as given by Harrach.[56] The solution to Eq. (4.25) with $I_{\nu_0} = 0$ is a single exponential function, but not with rotational relaxation time $\bar{\tau}^u_R$ since $\bar{N}^u$ is not constant with time, unlike N^u in Eq. (4.13).

Abrams and Cheo[51] and Jacobs *et al.*[55] give values for the rate constants $k_{CO_2\text{-}CO_2}$, $k_{CO_2\text{-}He}$, and $k_{CO_2\text{-}N_2}$ (in $\sec^{-1}$ Torr^{-1}). For a mixture of gases at pressure P and in the ratio $R_{CO_2} : R_{He} : R_{N_2}$, the relaxation time is given by[51]

$$\tau_R^{-1} = P(R_{CO_2} k_{CO_2\text{-}CO_2} + R_{He} k_{CO_2\text{-}He} + R_{N_2} k_{CO_2\text{-}N_2}) \qquad (4.27)$$

4.3. *Short-Pulse Amplification*

We consider the amplification of a pulse of duration of the order of 1 nsec traveling in the x-direction through a gain medium in a cavity of length L. We shall assume that the input pulse profile, $I_{\nu_0}(x = 0, t)$, and the initial population inversion, $\delta^{J_0}(x, t = 0)$, are known. For the cases of interest, $I_{\nu_0}(x = 0, t)$ will be the output pulse from a gain-switched oscillator, but for simplicity, and in order to reproduce the results of Harrach,[56] we shall assume initially that $I_{\nu_0}(x = 0, t)$ takes a simple analytic form. If we assume the initial population inversion within the cavity to be spatially independent, then

$$\delta^{J_0}(x, t = 0) \equiv \delta^0$$

where we begin by taking values from Harrach.

Since the rotational sublevels are initially in equilibrium with one another, $\Delta(x, t = 0)$ is given by the final term in Eq. (4.17) to be

$$P(J_0)\,\Delta(x, t = 0) = \delta^0$$

4.3.1. The Two-Level Model

For short pulses of duration ~1 nsec, we may consider the kinetics of V–V and $V/R-T$ transitions to be "frozen," since these transitions occur on a 10 nsec–1 μsec time scale. Thus the population number density, N^u, of the reservoir of the upper laser vibrational level, is changed only by the overall loss due to stimulated transitions. As discussed in the derivation of $P(J)$, page 121, we sum Eq. (4.13) over all odd J to give

$$\frac{dN^u}{dt} = -WI_{\nu_0}\delta^{J_0} \tag{4.28}$$

Similarly for the lower laser level, Eq. (4.14) gives, when summed over all even J,

$$\frac{dN^l}{dt} = WI_{\nu_0}\delta^{J_0} \tag{4.29}$$

and, by the definition of Δ [Eq. (4.18)], we subtract Eq. (4.29) from Eq. (4.28) to give

$$\frac{d\Delta}{dt} = -2WI_{\nu_0}\delta^{J_0} \tag{4.30}$$

The set of coupled equations describing the pulse amplification contains Eqs. (4.1) and (4.30) and an equation governing the time rate of change of I_{ν_0}. The general form of this last equation has been derived in Eq. (2.57) to be

$$\frac{dI_{\nu_0}}{dt} = -\frac{I_{\nu_0}}{\tau_c} + ch\nu_0(\delta^{J_0}WI_{\nu_0} + n^u_{J_0}S) \tag{4.31}$$

where again we have made replacements, $\Delta N \to \delta^{J_0}$ and $P(J_0)N^u \to n^u_{J_0}$ to account for the non-Boltzmann distribution of the rotational sublevel populations.

The first term on the right-hand side of Eq. (4.31) takes into account for the case of oscillators, the dominant loss of intensity occurring at the partially reflecting mirror. The last term of Eq. (4.31), which takes account of the effects of spontaneous emission, may also be neglected, spontaneous emission being significant only when no stimulated emission is taking place. We are left with only the contribution due to stimulated transitions, as given by Eq. (4.4).

The pulse is traveling along the x-axis with velocity c within the laser cavity, so we rewrite the total derivative in Eq. (4.4) in terms of partial derivatives. Using

$$\frac{d}{dt} = \frac{\partial}{\partial t} + \frac{dx'}{dt}\frac{\partial}{\partial x'}$$

where x' is traveling with velocity c relative to the laboratory frame x, Eq. (4.4) becomes

$$\frac{\partial I_{\nu_0}}{\partial t}+c\frac{\partial I_{\nu_0}}{\partial x}=ch\nu_0\delta^{J_0}WI_{\nu_0} \tag{4.32}$$

We assume the population inversions δ^{J_0} and Δ remain stationary with respect to the laboratory frame x. Using

$$\frac{d}{dt}=\frac{\partial}{\partial t}+\frac{\partial x}{\partial t}\frac{\partial}{\partial x}=\frac{\partial}{\partial t}+0\frac{\partial}{\partial x}$$

Eqs. (4.17) and (4.30) become

$$\frac{\partial \delta^{J_0}}{\partial t}=-\delta^{J_0}WI_{\nu_0}\left(1+\frac{\theta_{J_0}}{\theta_{J_0+1}}\right)-\left[\frac{\delta^{J_0}-\Delta P(J_0)}{\tau_R}\right] \tag{4.33}$$

and

$$\frac{\partial \Delta}{\partial t}=-\delta^{J_0}WI_{\nu_0}\left(1+\frac{\theta_{J_0}}{\theta_{J_0+1}}\right) \tag{4.34}$$

If we make the approximation $\theta_{J_0}/\theta_{J_0+1}=1$, then Eqs. (4.33), (4.34), and (4.32) are Eqs. (5b), (5d), and (5a), respectively, of Schappert.[53] According to Schappert they may be solved analytically for the two limiting cases: (1) a long low-intensity pulse and (2) a very short pulse. These analytic solutions have been given also by Harrach,[56] and will be derived in the following section.

4.3.2. Analytic Solutions for Long and Short Pulse Lengths

We make the assumption $\theta_{J_0}/\theta_{J_0+1}=1$, and for ease of reference collect the relevant equations:

$$\frac{\partial \delta^{J_0}}{\partial t}=-2\delta^{J_0}WI_{\nu_0}-\left[\frac{\delta^{J_0}-P(J)\Delta}{\tau_R}\right] \tag{4.33a}$$

$$\frac{\partial \Delta}{\partial t}=-2\delta^{J_0}WI_{\nu_0} \tag{4.34a}$$

$$\frac{\partial I_{\nu_0}}{\partial t}+c\frac{\partial I_{\nu_0}}{\partial x}=ch\nu_0\delta^{J_0}WI_{\nu_0} \tag{4.35}$$

In the following, τ_p (sec) is the duration of the input pulse.

For Gaussian input pulses, that is, pulses which vary in intensity with time according to

$$I_{\nu_0}(0,t)=I_{\max}\exp[-(t-t_0)^2\alpha] \tag{4.36}$$

where $I_{\max}$, t_0, and α are constants, we define τ_p to be the full width at half-maximum measured on the time axis, that is,

$$\tau_p = 2\left(\frac{\ln 2}{\alpha}\right)^{1/2} \tag{4.37}$$

Similarly, when $I_{\nu_0}(0, t)$ is the output pulse from a gain-switched oscillator, we define τ_p to be the full width at half-maximum measured on the time axis. For rectangular input pulses, τ_p is defined by

$$I_{\nu_0}(0, t) = \begin{cases} I, & t \in (0, \tau_p) \\ 0, & \text{otherwise} \end{cases} \tag{4.38}$$

The magnitude of τ_p relative to τ_R affects the solutions to Eqs. (4.33)–(4.35), since specification of $I_{\nu_0}(0, t)$ is one of the boundary conditions.

(1) $\tau_p/\tau_R \gg 1$. Provided the input pulse is sufficiently weak,

$$[2WI_{\nu_0}(x, t=0) \ll \tau_R^{-1}]$$

Eq. (4.33) implies that δ^{J_0} will remain close to its equilibrium value, namely,

$$\delta^{J_0}(x, t) \approx P(J_0)\Delta(x, t) \tag{4.39}$$

Using this approximation, and multiplying Eq. (4.34) by $P(J_0)$, we obtain

$$\frac{\partial \delta^{J_0}(x, t)}{\partial t} = -2P(J_0)\delta^{J_0}(x, t)\, WI_{\nu_0}(x, t) \tag{4.40}$$

The pair of equations (4.35) and (4.40) have been solved analytically by Frantz and Nodvik.[57] When the change of notation,

$$\delta^{J_0} \to \Delta, \qquad h\nu_0 W \to \sigma \qquad \text{and} \qquad \frac{P(J_0)}{h\nu_0} I_{\nu_0} \to cn$$

is made, Eqs. (4.35) and (4.40) become Eqs. (7) and (6), respectively, of Frantz and Nodvik.

We may therefore write down the solutions using Eqs. (27) and (28) of Frantz and Nodvik:

$$I_{\nu_0}(x, t) = \frac{I_{\nu_0}(0, t - x/c)}{1 - \{1 - \exp[-h\nu_0 W \int_0^x \delta^{J_0}(x', 0)\, dx']\} \exp[-2WP(J_0) \int_{-\infty}^{t-x/c} I_{\nu_0}(0, t')\, dt']} \tag{4.41}$$

$$\delta^{J_0}(x, t) = \frac{\delta^{J_0}(x, 0) \exp[-h\nu_0 W \int_0^x \delta^{J_0}(x', 0)\, dx']}{\exp[2WP(J_0) \int_{-\infty}^{t-x/c} I_{\nu_0}(0, t')\, dt'] + \exp[-h\nu_0 W \int_0^x \delta^{J_0}(x', 0)\, dx'] - 1} \tag{4.42}$$

If we assume the initial population inversion $\delta^{J_0}(x, 0)$ to be spatially independent within the medium, Eqs. (4.41) and (4.42) reduce to

$$I_{\nu_0}(x, t) = \frac{I_{\nu_0}(0, t - x/c)}{1 - [1 - \exp(-g_0 x)] \exp[-2WP(J_0) \int_{-\infty}^{t-x/c} I_{\nu_0}(0, t')\, dt']} \tag{4.43}$$

and

$$\delta^{J_0}(x, t) = \frac{\delta^{J_0}(t = 0) \exp(-g_0 x)}{\exp[2WP(J_0) \int_{-\infty}^{t-x/c} I_{\nu_0}(0, t')\, dt'] + \exp(-g_0 x) - 1} \tag{4.44}$$

where

$$g_0 = \frac{h\nu_0 W}{x} \int_0^x \delta^{J_0}(t = 0)\, dx' = h\nu_0 W \delta^{J_0}(t = 0) \qquad (\mathrm{cm}^{-1}) \tag{4.45}$$

is the small signal gain per unit length.

Now $I_{\nu_0}(x, t)$ is the energy flux at the point (x, t) in space–time, expressed in ergs $\mathrm{cm}^{-2}\,\mathrm{sec}^{-1}$. The total energy input per unit cross-sectional area, E_{IN}, is therefore given by

$$E_{\mathrm{IN}} = \int_{-\infty}^{\infty} I_{\nu_0}(x = 0, t')\, dt' \tag{4.46}$$

while the output energy is given by

$$E_{\mathrm{OUT}} = \int_{\infty}^{\infty} I_{\nu_0}(x = L, t')\, dt' \tag{4.47}$$

Since we are assuming the input pulse profile $I_{\nu_0}(0, t)$ is known, E_{IN} may be calculated numerically (or analytically) from Eq. (4.46). E_{OUT} may be calculated whenever $I_{\nu_0}(x, t)$ is given by Eq. (4.43), since Eq. (4.47) may then be integrated analytically:

$$E_{\mathrm{OUT}} = \int_{-\infty}^{\infty} \frac{I_{\nu_0}(0, t - L/c)\, dt}{1 - [1 - \exp(-g_0 L)] \exp[-2WP(J_0) \int_{-\infty}^{t-L/c} I_{\nu_0}(0, t')\, dt']}$$

$$= \int_{-\infty}^{\infty} \frac{I_{\nu_0}(0, t - L/c) \exp(g_0 L) \exp[2WP(J_0) \int_{-\infty}^{t-L/c} I_{\nu_0}(0, t')\, dt']\, dt}{1 + \exp(g_0 L) \{\exp[2WP(J_0) \int_{-\infty}^{t-L/c} I_{\nu_0}(0, t')\, dt'] - 1\}}$$

Since the numerator is the differential of the denominator with respect to t, then

$$E_{\mathrm{OUT}} = \frac{1}{2WP(J_0)} \ln\left[1 + \exp(g_0 L) \left\{\exp\left[2WP(J_0) \int_{-\infty}^{\infty} I_{\nu_0}(0, t')\, dt'\right] - 1\right\}\right]$$

$$= \frac{1}{2WP(J_0)} \ln[1 + \exp(g_0 L) \{\exp[2\, WP(J_0) E_{\mathrm{IN}}] - 1\}] \tag{4.48}$$

where we have used Eq. (4.46) for E_{IN}.

For E_{IN} sufficiently small, we may expand the right-hand side of Eq. (4.48) using $\log(1+\varepsilon)\approx\varepsilon$ and, keeping only first order in E_{IN}, obtain

$$E_{\text{OUT}}=\exp(g_0L)E_{\text{IN}} \tag{4.49}$$

which is the expected linear relationship for a small signal. It is expected that this linear relationship will no longer hold wherever the population inversion is severely reduced by the passage of the pulse. We define a saturation energy, E_s, to be that input energy required to decrease the population inversion at the incident end of the cavity to $1/e$ of its original value.

From Eq. (4.44) we have, for long weak pulses:

$$\frac{\delta^{J_0}(0,\infty)}{\delta^{J_0}(t=0)}=\frac{1}{\exp[2WP(J_0)E_{\text{IN}}]}\to e^{-1} \tag{4.50}$$

when

$$E_{\text{IN}}=\frac{1}{2WP(J_0)}\equiv E_s^{\infty} \tag{4.51}$$

and the superscript ∞ is used to denote the case $\tau_p\gg\tau_{\text{R}}$. This is the saturation energy defined by Schappert[53] for long weak pulses. Equation (4.48) may now be written

$$E_{\text{OUT}}=E_s\ln\{1+\exp(g_0L)[\exp(E_{\text{IN}}/E_s)-1]\} \tag{4.52}$$

where, for long weak pulses, $E_s=E_s^{\infty}$ as defined by Eq. (4.51).

(2) $\tau_p/\tau_{\text{R}}\ll 1$. For this limiting case we may assume that no rotational relaxation takes place in time τ_p. Thus the change in δ^{J_0} is due to lasing only, and Eq. (4.33) becomes

$$\frac{d\delta^{J_0}(x,t)}{dt}=-2\delta^{J_0}(x,t)WI_{\nu_0}(x,t) \tag{4.53}$$

The equations to be solved are Eqs. (4.35) and (4.53), and since Eq. (4.53) has the same form as Eq. (4.40), but with $P(J_0)$ set equal to 1, the solution may be obtained directly from that of the previous limiting case. In particular, the output energy is now given by Eq. (4.52), but with a saturation energy E_s^0, appropriate to short pulses, defined by

$$E_s^0=\frac{1}{2W} \tag{4.54}$$

It has been observed by Schappert[53] that, for the case $E_{\text{IN}}\gg E_s$, the total available energy, E_{STORED}, will be extracted. From Eq. (4.52) we have therefore:

$$E_{\text{OUT}} - E_{\text{IN}} = E_{\text{STORED}} \approx E_s \ln[\exp(g_0 L)\exp(E_{\text{IN}}/E_s)] - E_{\text{IN}}$$
$$= E_s g_0 L \qquad (4.55)$$

so for $\tau_p \gg \tau_R$ we have

$$E_{\text{STORED}} = E_s^{\infty} g_0 L = \frac{h\nu_0 W \delta^{J_0}(t=0)L}{2WP(J_0)} = \frac{1}{2} h\nu_0 \Delta(t=0)L \qquad (4.56)$$

while for $\tau_p \ll \tau_R$,

$$E_{\text{STORED}} = E_s^0 g_0 L = \tfrac{1}{2} h\nu_0 \delta^{J_0}(t=0)L \qquad (4.57)$$

We see that the total energy available for extraction is the energy stored in all the rotational levels for the case of long weak pulses, whereas for the case of very short pulses only the energy stored in the active rotational sublevel is available.

The limiting cases of long and short pulse lengths considered above may be thought of only as mathematical limits, since the two-level model is valid only for pulse lengths long compared with the coherence time (dipole dephasing time) and short compared to V–V relaxation times. This intermediate region of validity is approximately 0.1–10 nsec.

4.3.3. Solution for Intermediate Pulse Lengths

In order to solve, *a priori*, the set of equations (4.33)–(4.35) for input pulse lengths τ_p of the same order as τ_R, numerical techniques must be employed. The question investigated by Harrach[56] is whether Eq. (4.52) might still give the correct form for the output energy if, for some intermediate value of τ_p, a corresponding effective saturation energy E_s^{eff} was used to replace E_s. If this is the case, and if $E_s^{\text{eff}}(\tau_p)$ may be defined independently of the input pulse shape, then a "once and for all" graph of E_s^{eff} vs. τ_p, together with Eq. (4.52), will be sufficient to determine the output energy density E_{OUT} for a given E_{IN}, g_0, L, and τ_p. Harrach finds it convenient to define a saturation energy

$$E_s \equiv E_s^0 \exp(-g_0 L) \qquad (4.58)$$

and a normalized extracted energy, ε, given by

$$\varepsilon \equiv \frac{E_{\text{OUT}} - E_{\text{IN}}}{E_{\text{IN}}[\exp(g_0 L) - 1]} \qquad (4.59)$$

From Eq. (4.49) we notice that $E_{\text{IN}}[\exp(g_0 L) - 1]$ is the extracted energy for $E_{\text{IN}} \approx 0$, so $\varepsilon \approx 1$ when $E_{\text{IN}} \approx 0$. Also, when $E_{\text{IN}} \gg E_s$ we see from Eq. (4.55) that $\varepsilon \approx 0$. Thus ε is a measure of the degree of saturation.

Suppose $E_{IN} = aE_s$, for $\tau_p \ll \tau_R$; then Eq. (4.52) implies

$$E_{OUT} = E_s \exp(g_0L) \ln(1 + \exp(g_0L)\{\exp[a \exp(-g_0L)] - 1\}) \tag{4.60}$$

Thus

$$E_{OUT} - E_{IN} = E_s[\exp(g_0L) \ln(1 + \exp(g_0L)\{\exp[a \exp(-g_0L)] - 1\}) - a] \tag{4.61}$$

and by the definition of ε [Eq. (4.59)]

$$\begin{aligned} \varepsilon &= \frac{\exp(g_0L) \ln(1 + \exp(g_0L)\{\exp[a \exp(-g_0L)] - 1\}) - a}{a[\exp(g_0L) - 1]} \\ &= \frac{\frac{1}{a} \exp(g_0L) \ln(1 + \exp(g_0L)\{\exp[a \exp(-g_0L)] - 1\}) - 1}{\exp(g_0L) - 1} \end{aligned} \tag{4.62}$$

which, for $\tau_p \gg \tau_R$, is the value of ε obtained for

$$E_{IN} = a\frac{E_s}{P(J_0)} \tag{4.63}$$

That is, if we plot $\ln \varepsilon$ vs. $\ln E_{IN}/E_s$ for $\tau_p \ll \tau_R$, the corresponding curve for $\tau_p \gg \tau_R$ may be obtained by translating the former curve a distance $-\ln[P(J_0)]$ along the $\ln(E_{IN}/E_s)$axis.

By numerically integrating Eqs. (4.33a)–(4.35), Harrach[56] shows that, to a good approximation, this simple translation property holds for intermediate pulse lengths with a Gaussian profile, and that an effective saturation energy may therefore be defined. He shows also that the value obtained for E_s^{eff} is changed only slightly for the case of a rectangular pulse.

It is our purpose to reproduce these results of Harrach and to investigate more fully any possible dependence of E_s^{eff} on the pulse shape. In particular we shall investigate whether the value of E_s^{eff} obtained by Harrach may be used for the case of a sharply "spiked" input pulse from a gain-switched oscillator.

4.3.4. *Outline of Numerical Technique*

We consider solving numerically Eqs. (4.33a), (4.34a), and (4.35) with boundary conditions:

$$I_{\nu_0}(x = 0, t) = I^0(t) \tag{4.64}$$

$$\Delta(x, 0) = \begin{cases} \Delta^0, & x \in [0, L] \\ 0, & x \notin [0, L] \end{cases} \tag{4.65}$$

$$\delta^{J_0}(x,0)=\begin{cases}\delta^0, & x\in[0,L]\\ 0, & x\notin[0,L]\end{cases} \tag{4.66}$$

where we are assuming the active medium fills the laser cavity.

Considering Fig. 25, the population densities associated with space–time points below the line $t=x/c$ are unaffected by the propagating pulse, so we have

$$\Delta(x,t\leq x/c)=\Delta^0, \qquad x\in[0,L] \tag{4.67}$$

and

$$\delta(x,t\leq x/c)=\delta^0, \qquad x\in[0,L] \tag{4.68}$$

A set of equations similar to Eqs. (4.33a)–(4.35) has been solved by Armandillo and Spalding.[58] We follow their treatment by using a conventional Simpson integration formula, sweeping the two-dimensional $x-t$ grid, in the strip between the parallel lines $\tau_p Z$ and OY of Fig. 25.

Suppose we define our mesh size to be $\Delta t\times c\,\Delta t$ and define

$$I(m,n)=I_{\nu_0}(mc\;\Delta t,n\;\Delta t) \tag{4.69}$$

$$\Delta(m,n)=\Delta(mc\;\Delta t,n\;\Delta t)$$

and

$$\delta(m,n)=\delta^{J_0}(mc\;\Delta t,n\;\Delta t) \tag{4.70}$$

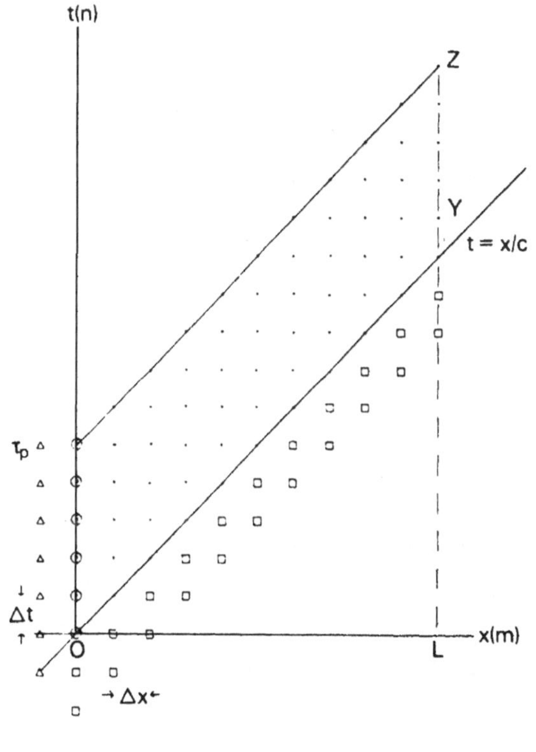

Fig. 25. Graph showing mesh points and domain over which solutions are required. The circles represent the boundary conditions given by Eq. (4.64), the squares represent those given by Eqs. (4.67) and (4.68), and the triangles represent the points at which the intensity is calculated using Eq. (4.80).

Then from Eqs. (4.33a), (4.34a), and (4.35), Simpson's rule takes the form

$$\delta(m+1, n+1) = \delta(m+1, n-1) + \frac{\Delta t}{3}[f(m+1, n+1) + 4f(m+1, n) + f(m+1, n-1)] \tag{4.71}$$

where

$$f(m, n) = -2\delta(m, n) WI(m, n) - \frac{\delta(m, n) - P(J_0)\,\Delta(m, n)}{\tau_R} \tag{4.72}$$

$$\Delta(m+1, n+1) = \Delta(m+1, n-1) + \frac{\Delta t}{3}[g(m+1, n+1) + 4g(m+1, n) + g(m+1, n-1)] \tag{4.73}$$

where

$$g(m, n) = -2\delta(m, n) WI(m, n) \tag{4.74}$$

and

$$I(m+1, n+1) = I(m-1, n-1) + \frac{\Delta t}{3}[h(m+1, n+1) + 4h(m, n) + h(m-1, n-1)] \tag{4.75}$$

where

$$h(m, n) = ch\nu_0 \delta(m, n) WI(m, n) \tag{4.76}$$

Equations (4.71), (4.73), and (4.75) are to be solved iteratively. Initial estimates of $I(m+1, n+1)$, $\delta(m+1, n+1)$, and $\Delta(m+1, n+1)$ may be obtained by linear extrapolation from the two preceding values, that is:

$$I_1(m+1, n+1) = 2I(m, n) - I(m-1, n-1) \tag{4.77}$$

$$\delta_1(m+1, n+1) = 2\delta(m+1, n) - \delta(m+1, n-1) \tag{4.78}$$

$$\Delta_1(m+1, n+1) = 2\Delta(m+1, n) - \Delta(m+1, n-1) \tag{4.79}$$

in Fig. 25.

The domain of the x–t plane, for which solutions are required, is bounded by the t-axis, the light cone $t = x/c$, and the lines t, Z, and ZY. The time at which the leading edge of the input pulse leaves the amplifier is t_1, while the time at which the trail edge leaves the amplifier is $t_1 + \tau_p$.

In Fig. 25 the mesh points designated by open squares have δ^0, Δ^0, and $I = 0$. Since $I = I^0$ along the t-axis, Eqs. (4.71) and (4.73) are linear and can be solved recursively for $\delta(0, n)$ and $\Delta(0, n)$ for $n = 0, 1, 2, \ldots$; these are the encircled points of Fig. 25. The column of mesh points $(-1, n)$, denoted by open triangles, are outside the amplifier and so

$$\delta(-1, n) = 0 = \Delta(-1, n)$$

while the intensity at these points is calculated from

$$\frac{I^0 - I(-1, n-1)}{\Delta t} = ch\nu_0 \delta(0, n) W I^0, \qquad n = 0, 1, 2, \ldots \tag{4.80}$$

4.3.5. *Calculating δ(0, n) and Δ(0, n) along the x = 0 Axis*

Since the pulse I_{ν_0} is given along the axis $x = 0$, that is, it is equal to the input pulse,

$$I_{\nu_0}(0, n) = I^0(n)$$

then Simpson's rule for δ and Δ at the point $(0, n+1)$ is given by

$$\delta(0, n+1) = \delta(0, n-1) + \frac{\Delta t}{3}\left\{ -2\delta(0, n+1) W I^0(n) - \left[\frac{\delta(0, n+1) - P(J_0)\,\Delta(0, n+1)}{\tau_R}\right] + 4f(0, n) + f(0, n-1)\right\} \tag{4.81}$$

and

$$\Delta(0, n+1) = \Delta(0, n-1) + \frac{\Delta t}{3}[-2\delta(0, n+1) W I^0(n) + 4g(0, n) + g(0, n-1)] \tag{4.82}$$

where f and g are given by Eqs. (4.72) and (4.74), respectively.

Assuming $\delta(0, m)$, $\Delta(0, m)$, $m = 0, 1, \ldots, n$, have been calculated, we may derive an analytic expression for $\delta(0, n+1)$ in terms of known quantities, as follows:

Substituting Eq. (4.82) for $\Delta(0, n+1)$ into Eq. (4.81) gives

$$\delta(0, n+1) = \delta(0, n-1) + \frac{\Delta t}{3}\left\{ -2\delta(0, n+1) W I^0(n)\left[1 + \frac{P(J_0)}{\tau_R}\frac{\Delta t}{3}\right] - \frac{\delta(0, n+1)}{\tau_R} + \frac{P(J_0)}{\tau_R}\left\{\Delta(0, n-1) + \frac{\Delta t}{3}[4g(0, n) + g(0, n-1)]\right\} + 4f(0, n) + f(0, n-1)\right\} \tag{4.83}$$

Therefore

$$\delta(0, n+1)\left\{1+\frac{2\Delta t W I^0(n)}{3}\left[1+\frac{P(J_0)}{\tau_R}\frac{\Delta t}{3}\right]+\frac{\Delta t}{3}\times\frac{1}{\tau_R}\right\}$$

$$=\delta(0, n-1)+\frac{\Delta t}{3}\left(\frac{P(J_0)}{\tau_R}\left\{\Delta(0, n-1)+\frac{\Delta t}{3}[4g(0, n)+g(0, n-1)]\right\}\right.$$

$$\left.+4f(0, n)+f(0, n-1)\right) \quad (4.84)$$

$\delta(0, n+1)$ may be calculated from Eq. (4.83) and substituted into Eq. (4.82) to give a value for $\Delta(0, n+1)$.

4.4. *The Inclusion of Rotational Coupling Effects in a Gain-Switched TEA CO_2 Laser*

It has been shown[52] necessary to include the effects of rotational coupling in order to model successfully a gain-switched TEA CO_2 laser involving pulse lengths much longer than the rotational relaxation time, τ_R. It is suggested that the importance of this coupling is its effect on the rate at which energy is removed from the lower laser level.[52]

We consider the modification of the five-temperature model, derived in Chapter 2, which includes the effects of rotational coupling. Some assumption regarding the intramode V–V transitions has to be made; in the following section we shall assume these transitions to be instantaneous, while in Section 4.4.2, the effects of a non-Boltzmann distribution within the vibrational modes of CO_2 will be considered.

4.4.1. *Neglecting the Intramode V–V Relaxation Time, $\tau_{(VV)}$*

The equations to be modified are Eqs. (2.110)–(2.114), (2.124), (2.57), and (2.139), where the population inversion, ΔN, appearing in Eqs. (2.42d), (2.110), and (2.139), was assumed to be given by Eq. (2.42e):

$$\Delta N = N_{00^01}P(J)-\left(\frac{2J+1}{2J+3}\right)N_{10^00}P(J+1) \quad (4.85)$$

In the nomenclature of Section 4.2 this becomes

$$\Delta N = {}^e n^u_{J_0} - {}^e n^u_{J_0+1} \quad (4.86)$$

the equilibrium value of δ^{J_0}, remembering that we set $(2J+1)/(2J+3)\simeq 1$.

We now relax this assumption of equilibrium and replace ΔN by the nonequilibrium value δ^{J_0}. The equation describing the time evolution of δ^{J_0}

is that derived in Eq. (4.17), namely,

$$\frac{d\delta^{J_0}}{dt} = -2\delta^{J_0} W I_{\nu_0} - \frac{\delta^{J_0} - P(J_0)\,\Delta}{\tau_R} \tag{4.87}$$

and Δ is the population inversion between the upper and lower laser vibrational levels.

If we assume, as in Eqs. (2.3), that the levels within each of the stretch modes of CO_2 are in thermal equilibrium, then N^u and N^l will be given by [see also Eq. (2.42b)]

$$N^u = N_{00^\circ 1} = fN_{CO_2} \exp(-h\nu_3/kT_3)Z \tag{4.88}$$

and

$$N^l = N_{10^\circ 0} = fN_{CO_2} \exp(-h\nu_1/kT_1)Z \tag{4.89}$$

where Z is the partition function

$$Z = [1 - \exp(-h\nu_1/kT_1)][1 - \exp(-h\nu_2/kT_2)]^2[1 - \exp(-h\nu_3/kT_3)]$$

and the vibrational temperatures T_1, T_2, and T_3 are defined by Eq. (2.5). fN_{CO_2} is the total number density of CO_2 molecules.

The population inversion Δ is given by

$$\Delta = N^u - N^l = N_{00^\circ 1} - N_{10^\circ 0} \tag{4.90}$$

Thus Eq. (4.87) becomes

$$\frac{d\delta^{J_0}}{dt} = -2\delta^{J_0} W I_{\nu_0} - \left[\frac{\delta^{J_0} - P(J_0)(N_{00^\circ 1} - N_{10^\circ 0})}{\tau_R}\right] \tag{4.91}$$

Equation (4.91), together with Eqs. (2.110)–(2.114), (2.124), (2.5), and (2.139), with ΔN replaced by δ^{J_0}, form the set of coupled equations that are to be numerically solved.

4.4.2. *Inclusion of the Effects of Intramode V–V Transitions*

The assumption that the vibrational levels within the stretch modes of CO_2 are in equilibrium will now be relaxed, and the coupling of the lower laser vibrational level (10°0) to the Fermi-degenerate bending mode will also be taken into account. Initially the populations N^u and N^l will be equal to the equilibrium values $N_{00^\circ 1}$, $N_{10^\circ 0}$, as defined by Eqs. (4.88) and (4.89); in order to derive the equation describing the subsequent time history of N^u and N^l, we make the following assumptions.

(1) The vibrational levels of the asymmetric and symmetric stretch modes are all in equilibrium within the modes, with the possible exception of the upper and lower laser vibrational levels.

(2) Nonresonant intermode V–V and V–T transitions are slow compared to the resonant intramode V–V transition and the Fermi resonant symmetric stretch–bending mode transition.

Under these assumptions the time history of the lower level population N^l is given by Eq. (2.139a):

$$\frac{dN^l}{dt} = -\frac{N^l - N_{10^00}}{\tau_{vu}^l} - \tau_{d_3}^{-1}\left[N^l - \exp\left(\frac{h\nu_1}{kT_1} - \frac{2h\nu_2}{kT_2}\right)N_{10^00}\right] + \delta^{J_0} W I_{\nu_0} \tag{4.92}$$

where

$$(\tau_{vv}^l)^{-1} = N_{CO_2} k_{100,000;000,100}(3E_1/Nh\nu_1 + 1) \tag{4.93}$$

and

$$\tau_{d_3}^{-1} = \sum_t N_t k_{020,100}^t \tag{4.94}$$

N_t is the number density of species t and N is the total particle number density.

Similarly, for the upper laser vibrational level we obtain

$$\frac{dN^u}{dt} = -\frac{(N^u - N_{00^01})}{\tau_{vv}^u} - \delta^{J_0} W I_{\nu_0} \tag{4.95}$$

where

$$(\tau_{vv}^u)^{-1} = N_{CO_2} k_{001,000;000,001}\left(\frac{3E_3}{Nh\nu_3} + 1\right) \tag{4.96}$$

The equations describing the system with intramode V–V and taking the relaxation of the lower laser vibrational level into account, are given in Section 4.4.1, with Δ given by a further rate equation:

$$\frac{d\Delta}{dt} = \frac{dN^u}{dt} - \frac{dN^l}{dt} \tag{4.97}$$

where the right-hand side is given by Eqs. (4.92) and (4.95).

4.5. *Short-Pulse Multiline–Multiband Operation*

4.5.1. *Extension of the Theory to Include Multiline–Multiband Energy Extraction*

In Sections 4.3.1 and 4.3.2 the assumption was made that stimulated transitions occur between only one pair of rotational sublevels, that is, single-line operation. The theory presented in Section 4.3.1 is here generalized to the case of multiline operation. Particular attention is paid to relaxing any unnecessary approximations previously made.

Suppose the input energy flux, $I(\mathbf{r}, t)$, contains photons of varying energies. In particular, we assume the spectrum of I is sharply peaked at the energies $h\nu_1, \ldots h\nu_M$, each resonant with a particular $J(00°1) \rightarrow J'(10°0)$ transition. We may then write the total energy flux as a finite sum over component fluxes, each of a particular frequency:

$$I(\mathbf{r}, t) = \sum_{i=1}^{M} I_{\nu_i}(\mathbf{r}, t)$$

We shall consider only P transitions, that is, such that $J' = J + 1$, and label the rotational sublevels with energy separation $h\nu_i$ by J_i and $J_i + 1$.

The rate of change of the component flux I_{ν_i} is given by the same argument that led to Eq. (4.32):

$$\frac{\partial I_{\nu_i}}{\partial t} + c\frac{\partial I_{\nu_i}}{\partial x} = ch\nu_i \delta^{J_i} W_i I_{\nu_i} \tag{4.98}$$

where δ^{J_i} is given by generalizing the form of Eq. (4.5), that is,

$$\delta^{J_i} = n^u_{J_i} - \frac{\theta_{J_i}}{\theta_{J_i+1}} n^l_{J_i+1} \tag{4.99}$$

and W_i is the cross section for spontaneous emission from J_i to $J_i + 1$ divided by $h\nu_i$, as given by Eq. (2.58). We note that τ^i_{sp} is the spontaneous-emission time given by the inverse of the Einstein coefficient (see, e.g., Reference 9, Vol. 1, p. 127)

$$(\tau^i_{sp})^{-1} = \frac{64\pi^4 |R_{12}|^2 S_{J_i} F_{J_i}}{3h\lambda_i^3 \theta_{J_i}} \tag{4.100}$$

where R_{12} and S_{J_i} are the vibrational and rotational contributions to the transition moment, respectively. F_{J_i} is the interaction factor between vibration and rotation.

Equation (4.98) is summed over all i to give

$$\frac{\partial I(\mathbf{r}, t)}{\partial t} + c\frac{\partial I(\mathbf{r}, t)}{\partial x} = ch \sum_{i=1}^{M} \nu_i \delta^{J_i} W_i I_{\nu_i} \tag{4.101}$$

Notice that we are neglecting the effects of dispersion in assuming that each component flux travels with the same velocity, c. It has been shown by Schappert and Herbert[59] that dispersion effects may be significant, but that the flux components may be kept from separating by choosing lines with nearly equal $g_0/\Delta\nu_i$ ratios.

Clearly n^u_J will be decreased (n^l_{J+1} increased) by stimulated transitions wherever $J = J_i$, $i = 1, \ldots, M$. The time rate of change of n^u_J and n^l_{J+1} due to stimulated transitions is thus given by a sum over terms similar to the right-hand side of Eq. (4.6):

$$\left(\frac{\partial n_J^u}{\partial t}\right)_{\text{s.t.}} = -\sum_{i=1}^{M} \delta^{J_i} W_i I_{\nu_i} \delta_{J,J_i} = -\left(\frac{\partial n_{J+1}^l}{\partial t}\right)_{\text{s.t.}} \tag{4.102}$$

where the Kronecker delta picks out the M active levels $J = J_i$.

The equation describing rotational coupling within the 00°1 level [Eq. (4.12)] is independent of the lasing process, and so may be combined with Eq. (4.102) to give the total time rate of change of n_J^u:

$$\frac{dn_J^u}{dt} = \frac{P(J)N^u - n_J^u}{\tau_R^u} - \sum_{i=1}^{M} \delta^{J_i} W_i I_{\nu_i} \delta_{J,J_i} \tag{4.103}$$

and similarly for the lower laser level:

$$\frac{dn_{J+1}^l}{dt} = \frac{P(J+1)N^l - n_{J+1}^l}{\tau_R^l} + \sum_{i=1}^{M} \delta^{J_i} W_i I_{\nu_i} \delta_{J+1,J_i+1} \tag{4.104}$$

Notice that we are still assuming a model of no J selection rules, expressed in the definition (4.7). We remember that this is a necessary assumption in order that the summations in Eqs. (4.1) and (4.2) may be put into a closed form. It is not necessary to make the additional assumption that the rotational relaxation times for upper and lower levels are equal; however, since the available experimental data for rotational relaxation times are measured on the upper vibrational level only,[51] this last assumption, namely,

$$\tau_R^u = \tau_R^l \equiv \tau_R \tag{4.105}$$

will be kept.

The equation for the time rate of change of δ^J is obtained by multiplying Eq. (4.104) by θ_J/θ_{J+1} and subtracting from Eq. (4.103) to give

$$\frac{d\delta^J}{dt} = -\sum_i \delta^{J_i} W_i I_{\nu_i} \delta_{J,J_i}\left(1 + \frac{\theta_J}{\theta_{J+1}}\right) + \frac{P(J)N^u - P(J+1)(\theta_J/\theta_{J+1})N^l - \delta^J}{\tau_R} \tag{4.106}$$

Upon defining the "quasi-equilibrium" population inversion, $\bar{\delta}^J$, by

$$\bar{\delta}^J \equiv P(J)N^u - P(J+1)(\theta_J/\theta_{J+1})N^l \tag{4.107}$$

Eq. (4.106) becomes

$$\frac{d\delta^J}{dt} = -\sum_i \delta^{J_i} W_i I_{\nu_i} (1 + \theta_{J_i}/\theta_{J_i+1}) \delta_{J,J_i} + \frac{\bar{\delta}^J - \delta^J}{\tau_R} \tag{4.108}$$

It remains to derive rate equations for the $\bar{\delta}^{J_i}$. Since we are considering input pulses of duration of the order of 1 nsec, we assume, as in Section 4.3.1, that dN^u/dt is given by summing Eq. (4.103) over all odd J, thus neglecting changes in N^u due to V–V and V/R–T transitions. This gives

$$\frac{dN^u}{dt}=\frac{\sum_{J=\text{odd}} P(J)N^u-\sum_{J=\text{odd}} n_J^u}{\tau_R^u}-\sum_{J=\text{odd}}\sum_i \delta^{J_i}W_iI_{\nu_i}\delta_{J,J_i}$$
$$=-\sum_i \delta^{J_i}W_iI_{\nu_i} \tag{4.109}$$

where we have used $\sum_{J=\text{odd}} P(J)=1$ and

$$\sum_{J=\text{odd}}\sum_i f(J_i)\delta_{J,J_i}=\sum_i f(J_i)$$

since the J_i are all odd.

Similarly, for the lower laser vibrational level,

$$\frac{dN^l}{dt}=\sum_i \delta^{J_i}W_iI_{\nu_i} \tag{4.110}$$

Substituting Eqs. (4.109) and (4.110) into the time rate of change of Eq. (4.107) yields

$$\frac{d\bar{\delta}^J}{dt}=-\left[P(J)+P(J+1)\left(\frac{\theta_J}{\theta_{J+1}}\right)\right]\sum_{i=1}^{M}\delta^{J_i}W_iI_{\nu_i} \tag{4.111}$$

The set of equations: Eq. (4.98) for I_{ν_i} for all $i=1,\ldots,M$, Eq. (4.108) for δ^J for all $J=J_i$, $i=1,\ldots,M$, and Eq. (4.111) for $\bar{\delta}^J$ for all $J=J_i$, $i=1,\ldots,M$, form a complete set to describe multiline operation, and may be solved using the numerical technique given in Section 4.3.4.

If we make the approximations:

$$\left.\begin{aligned}P(J_i)&\approx P(J_i+1)\\ \theta_{J_i}&\approx\theta_{J_i+1}\end{aligned}\right\}\qquad \forall i=1,\ldots,N \tag{4.112}$$

and the replacements:

$$\left.\begin{aligned}J&\to j\\ \delta^{J-1}&\to\Delta n(j,j-1)\\ W_{J-1}&\to(h\nu)^{-1}\sigma(j_l,j_l-1)\\ P(J)&\to Z_j\\ \bar{\delta}^J&\to Z_j\,\Delta N_\nu\end{aligned}\right\} \tag{4.113}$$

Then our Eqs. (4.98), (4.108), and (4.111) reduce to Eq. (11a–c) as given by Feldman.[60]

The advantage of multiline over single-line operation is that, for short input pulses, the more lines that are employed, the more energy is extracted for a given input energy. It is to further advantage if transitions are stimulated between more than one pair of vibrational levels. This latter mode of operation is termed multiband operation.

The foregoing theory is easily adapted to describe the amplification of a pulse containing transitions at 10.6μ and 9.6μ, corresponding to $J(00^\circ1)\rightarrow J+1(10^\circ0)$ and $J'(00^\circ1)\rightarrow J'+1(02^\circ0)$ transitions, respectively. A change of notation is necessary:

Let $N^{00^\circ1}$ and $n_J^{00^\circ1}$ be the population number densities of the total $00^\circ1$ vibrational level, and its Jth rotational sublevel, respectively. Similarly define $N^{10^\circ0}$, $n_J^{10^\circ0}$, $N^{02^\circ0}$ and $n_J^{02^\circ0}$.

Suppose the input pulse is of the form

$$I(\mathbf{r},t)=\sum_{i=1}^{M_1} I_{\nu_i}(\mathbf{r},t)+\sum_{i=M_1+1}^{M_2} I_{\nu_i}(\mathbf{r},t) \tag{4.114}$$

where $h\nu_1,\ldots,h\nu_{M_1}$ are resonant with transitions from $J_i(00^\circ1)\rightarrow J_i+1(10^\circ0)$, $i=1,\ldots,M_1$, and $h\nu_{M_1+1},\ldots,h\nu_{M_2}$ are resonant with transitions $J_i(00^\circ1)\rightarrow J_i+1(02^\circ0)$, $i=M_1+1,\ldots,M_2$. Furthermore we suppose that I is chosen such that $10.6\mu-$ and $9.6\mu-$ transitions do not directly interact, that is,

$$J_i\neq J_j \qquad \text{for } i=1,\ldots,M_1,\, j=M_1+1,\ldots,M_2 \tag{4.115}$$

This will usually be the case experimentally since the advantage of working with multiline–multiband energy extraction is lessened if two transitions deplete the same upper rotational sublevel.

Now define

$$\delta_1^{J_i}\equiv n_{J_i}^{00^\circ1}-(\theta_{J_i}/\theta_{J_i+1})n_{J_i+1}^{10^\circ0} \tag{4.116}$$

and

$$\delta_2^{J_j}\equiv n_{J_j}^{00^\circ1}-(\theta_{J_j}/\theta_{J_j+1})n_{J_j+1}^{02^\circ0} \tag{4.117}$$

as in Eq. (4.99), and

$$\delta^{J_i}\equiv\begin{cases}\delta_1^{J_i}, & i=1,\ldots,M_1\\ \delta_2^{J_i}, & i=M_1+1,\ldots,M_2\end{cases} \tag{4.118}$$

With this definition of δ^{J_i}, Eq. (4.99) still holds for the rate of change of I_{ν_i}. Equations (4.103) and (4.104) generalize to:

$$\frac{dn_J^{00^\circ1}}{dt}=\frac{P(J)N^{00^\circ1}-n_J^{00^\circ1}}{\tau_R}-\sum_{i=1}^{M_1}\delta_1^{J_i}W_iI_{\nu_i}\delta_{J,J_i}$$
$$-\sum_{i=M_1+1}^{M_2}\delta_2^{J_i}W_iI_{\nu_i}\delta_{J,J_i} \tag{4.119}$$

$$\frac{dn_{J+1}^{10^\circ 0}}{dt}=\frac{P(J+1)N^{10^\circ 0}-n_{J+1}^{10^\circ 0}}{\tau_R}+\sum_{i=1}^{M_1}\delta_1^{J_i}W_iI_{\nu_i}\delta_{J,J_i} \tag{4.120}$$

$$\frac{dn_{J+1}^{02^\circ 0}}{dt}=\frac{P(J+1)N^{02^\circ 0}-n_{J+1}^{02^\circ 0}}{\tau_R}+\sum_{i=M_1+1}^{M_2}\delta_2^{J_i}W_iI_{\nu_i}\delta_{J,J_i} \tag{4.121}$$

Because of the condition (4.115) it can be seen that, for a given J, at most one of the summations in Eq. (4.119) contributes; the first one if $J=J_i$, where $i=1,\ldots,M_1$, the second if $J=J_i$, where $i=M_1+1,\ldots,M_2$.

Thus, in the same way as we arrived at Eq. (4.108) we now have:

$$\frac{d\delta_1^J}{dt}=-\sum_{i=1}^{M_1}\delta_1^{J_i}W_iI_{\nu_i}(1+\theta_{J_i}/\theta_{J_i+1})\delta_{J,J_i}+\frac{\bar{\delta}_1^J-\delta_1^J}{\tau_R} \tag{4.122}$$

and

$$\frac{d\delta_2^J}{dt}=-\sum_{i=M_1+1}^{M_2}\delta_2^{J_i}W_iI_{\nu_i}(1+\theta_{J_i}/\theta_{J_i+1})\delta_{J,J_i}+\frac{\bar{\delta}_2^J-\delta_2^J}{\tau_R} \tag{4.123}$$

with

$$\bar{\delta}_1^J\equiv P(J)N^{00^\circ 1}-P(J+1)(\theta_J/\theta_{J+1})N^{10^\circ 0} \tag{4.124}$$

$$\bar{\delta}_2^J\equiv P(J)N^{00^\circ 1}-P(J+1)(\theta_J/\theta_{J+1})N^{02^\circ 0} \tag{4.125}$$

If we now sum over Eqs. (4.119)–(4.121), we obtain

$$\frac{dN^{00^\circ 1}}{dt}=-\sum_{i=1}^{M_1}\delta_1^{J_i}W_iI_{\nu_i}-\sum_{i=M_1+1}^{M_2}\delta_2^{J_i}W_iI_{\nu_i} \tag{4.126}$$

$$\frac{dN^{10^\circ 0}}{dt}=\sum_{i=1}^{M_1}\delta_1^{J_i}W_iI_{\nu_i} \tag{4.127}$$

$$\frac{dN^{02^\circ 0}}{dt}=\sum_{i=M_1+1}^{M_2}\delta_2^{J_i}W_iI_{\nu_i} \tag{4.128}$$

The complete set of coupled equations describing the multiline–multiband operation is: Eq. (4.98) for I_{ν_i}, for all $i=1,\ldots,M_2$, Eqs. (4.122) for $\dot{\delta}_1^J$, for all $J=J_i$, $i=1,\ldots,M_1$, Eq. (4.123) for $\dot{\delta}_2^J$, for all $J=J_i$, $i=M_1+1,\ldots,M_2$, and Eqs. (4.126)–(4.128).

If the approximations [Eq. (4.112)] and the replacements

$$\left.\begin{aligned} J &\to j \\ \delta_1^{J-1} &\to \Delta n_{12}(j, j-1) \\ \delta_2^{J-1} &\to \Delta n_{13}(j, j-1) \\ N^{00^\circ 1} - N^{10^\circ 0} &\to \Delta N_{12} \\ N^{00^\circ 1} - N^{02^\circ 0} &\to \Delta N_{13} \end{aligned}\right\} \tag{4.129}$$

are made then our Eqs. (4.98), (4.122), (4.123), and (4.126)–(4.128) reduce to Eqs. (17a)–(17h) given by Feldman.[60]

4.5.2. Modification of Numerical Techniques

The numerical technique outlined in Section 4.3.4 may be employed to solve the set of equations presented in the last section, describing multiline–multiband operation.

If we write

$$I_i(m, n) \equiv I_{\nu_i}(mc\ \Delta t, n\ \Delta t), \qquad i = 1, \ldots, M_2 \tag{4.130}$$

$$\left.\begin{aligned} \delta_i(m, n) &\equiv \delta_1^{J_i}(mc\ \Delta t, n\ \Delta t), \qquad i = 1, \ldots, M_1 \\ &\equiv \delta_2^{J_i}(mc\ \Delta t, n\ \Delta t), \qquad i = M_1 + 1, \ldots, M_2 \end{aligned}\right\} \tag{4.131}$$

$$N_1(m, n) \equiv N^{00^\circ 1}(mc\ \Delta t, n\ \Delta t) \tag{4.132}$$

$$N_2(m, n) \equiv N^{10^\circ 0}(mc\ \Delta t, n\ \Delta t) \tag{4.133}$$

$$N_3(m, n) \equiv N^{02^\circ 0}(mc\ \Delta t, n\ \Delta t) \tag{4.134}$$

then the boundary conditions become

$$I_i(2, n) = I_i^\circ(n), \qquad i = 1, \ldots, M_2 \tag{4.135}$$

$$\left.\begin{aligned} \delta_i(m, n < m) &= \delta_i^\circ, \qquad mc\ \Delta t \in [0, L] \\ &= 0, \qquad mc\ \Delta t \notin [0, L] \end{aligned}\right\} \quad i = 1, \ldots, M_2 \tag{4.136}$$

$$\left.\begin{aligned} N_\lambda(m, n < m) &= N_\lambda^\circ, \qquad mc\ \Delta t \in [0, L] \\ &= 0, \qquad mc\ \Delta t \notin [0, L] \end{aligned}\right\} \quad \lambda = 1, 2, 3 \tag{4.137}$$

From Eqs. (4.122) and (4.123) we see that Simpson's rule for δ_i now takes the form

$$\begin{aligned} \delta_i(m+1, n+1) &= \delta_i(m+1, n-1) \\ &\quad + \frac{\Delta t}{3}[f_i(m+1, n+1) + 4f_i(m+1, n) + f_i(m+1, n-1)] \end{aligned} \tag{4.138}$$

where

$$f_i(m,n) = -2\delta_i(m,n)W_iI_i(m,n) + \frac{\bar{\delta}_i(m,n) - \delta(m,n)}{\tau_R} \tag{4.139}$$

and

$$\begin{aligned}\bar{\delta}_i(m,n) &= P(J_i)N_1(m,n) - P(J_i+1)(\theta_{J_i}/\theta_{J_i+1})N_2(m,n), \qquad i = 1,\ldots,M_1\\ &= P(J_i)N_1(m,n) - P(J_i+1)(\theta_{J_i}/\theta_{J_i+1})N_3(m,n),\\ &\qquad i = M_1+1,\ldots,M_2\end{aligned} \tag{4.140}$$

Similarly for N_1, N_2, and N_3 we have, from Eqs. (4.126)–(4.128):

$$\begin{aligned}N_\lambda(m+1,n+1) &= N_\lambda(m+1,n-1)\\ &\quad + \frac{\Delta t}{3}[g_\lambda(m+1,n+1) + 4g_\lambda(m+1,n)\\ &\quad + g_\lambda(m+1,n-1)], \qquad \lambda = 1,2,3\end{aligned} \tag{4.141}$$

where

$$g_2(m,n) = \sum_{i=1}^{M_1} \delta_i(m,n)W_iI_i(m,n) \tag{4.142}$$

$$g_3(m,n) = \sum_{i=M_1+1}^{M_2} \delta_i(m,n)W_iI_i(m,n) \tag{4.143}$$

and

$$g_1(m,n) = -g_2(m,n) - g_3(m,n) \tag{4.144}$$

Finally Eq. (4.98) implies for I_i:

$$\begin{aligned}I_i(m+1,n+1) &= I_i(m-1,n-1)\\ &\quad + \frac{\Delta t}{3}[h_i(m+1,n+1) + 4h_i(m,n) + h_i(m-1,n-1)]\end{aligned} \tag{4.145}$$

where

$$h_i(m,n) = ch\nu_i\delta_i(m,n)W_iI_i(m,n) \tag{4.146}$$

For the general point $(m+1, n+1)$, marked as a small circle in Fig. 25, Eqs. (4.138), (4.141), and (4.145) may be solved iteratively, in the same way as discussed in Section 4.3.4.

The method described in Section 4.3.4 for calculating δ and Δ along the $x = 0$ axis cannot be generalized to the present case of multiline operation; δ_i, $i = 1, \ldots, M_2$, and N_λ, $\lambda = 1, 2, 3$, are calculated along the $x = 0$ axis by iterating Eqs. (4.138) and (4.141) and using the boundary condition (4.135) for values of I_i.

4.6. Coherence Effects in Short-Pulse Propagation

The theory of short-pulse propagation, presented in the preceding sections, neglects the effects of coherent-pulse propagation, which is of particular significance for laser-fusion applications, and includes effects of Fermi resonance coupling only for the case of longer pulse lengths. We shall now discuss the time scales for which these effects are important.

The dipole dephasing time, or coherence time, τ_2, is a measure of the time taken for a set of dipoles, initially vibrating in phase with one another, to acquire random phases relative to one another. If we consider a two-level model for the amplifying medium with initially all molecules in the *upper* level, then the propagation and amplification of a pulse through this medium will depend on the intensity and duration of the pulse. If a long ($\gg \tau_2$), weak pulse is propagated, the amplification of the pulse will be the *net* result of stimulated emission *and* absorption, the emission (or absorption) from any one molecule being independent of that from any other molecule. Consider now a very short intense pulse that saturates the medium in time less than τ_2. All molecules will emit energy within the time $< \tau_2$ and in the subsequent time interval will be able *only* to *absorb* energy from the light pulse (or remain in the lower level). Thus coherent-pulse propagation results in the pulse being initially amplified, but shortened. If the pulse is of a few τ_2 in length and sufficiently intense to saturate in time $< \tau_2$, the process described above will repeat itself, leading to pulse breakup as described by Armandillo and Spalding.[58]

The theory that leads to a description of coherent effects is presented below, from which it can be seen that the rate equations derived earlier are correct to first order in τ_2/τ_p, where τ_p is the duration of the input pulse. These new equations have been employed by Armandillo and Spalding, who give a value of 0.06 nsec for the coherence time τ_2. Thus, coherence effects will be negligible for $\tau_p \gg 0.06$ nsec, but should be included if the modeling is employed in the region $\tau_p < 0.1$ nsec.

In the analysis of short-pulse propagation presented earlier, the assumption was made that the duration of the input light pulse was shorter than the fastest time taken for V–V and V–T energy exchanges. In Section 4.4.2 we briefly discussed the inclusion of the fastest of these energy exchanges, namely, the resonant intramode V–V transitions within the symmetric and antisymmetric modes, and also Fermi resonance coupling of the lower laser level (10°0) with the bending mode (020). This discussion was in the context of modeling propagation of pulses much longer than the rotational relaxation time, but the theory is not restricted to such cases.

These fast V–V transitions will be significant for short-pulse propagation only if the pulse duration is of the same order as the rates for these transitions. Harrach[56] observes from experimental data that effects of

V–V intramode coupling are fairly negligible for pulses of duration $\tau_p \sim$ 1 nsec at atmospheric pressure. Armandillo and Spalding quote values of 1 nsec and 10 nsec for V–V intramode and Fermi resonance coupling rates, respectively. Thus we should also expect Fermi resonance to be negligible for pulses < 10 nsec in duration at atmospheric pressure. If the effect of Fermi resonance coupling is to be included then so should V–V intramode resonance coupling, for a consistent theory.

4.6.1. *The Density Operator*[61]

Let the Hamiltonian of the system be

$$H = H_0 + H_I \tag{4.147}$$

and suppose the state functions of the system, satisfying the time-dependent Schrödinger equation, are $|\psi_n\rangle$, the state being characterized by $|\psi_n\rangle$ with probability P_n. Then the expectation value of any operator θ is given by

$$\langle\theta\rangle = \sum_n P_n \langle\psi_n|\theta|\psi_n\rangle \tag{4.148}$$

If the eigenvectors of H_0 are labeled $|u_k\rangle$, then we have

$$\langle\theta\rangle = \sum_{n,k} P_n \langle\psi_n|\theta|u_k\rangle\langle u_k|\psi_n\rangle \tag{4.149}$$

since the set $\{u_k\}$ is complete. When this result is rearranged,

$$\langle\theta\rangle = \sum_{n,k} P_n \langle u_k|\psi_n\rangle\langle\psi_n|\theta|u_k\rangle \tag{4.150}$$

$$\equiv \sum_k \langle u_k|\rho\theta|u_k\rangle \tag{4.151}$$

where we are defining the (Hermitian) density operator, ρ, by

$$\rho = \sum_n P_n |\psi_n\rangle\langle\psi_n| \tag{4.152}$$

Clearly, Eq. (4.151) is the sum of the diagonal elements and so

$$\langle\theta\rangle = \mathrm{Tr}(\rho\theta) \tag{4.153}$$

If we differentiate Eq. (4.152) with respect to time we obtain

$$\dot{\rho} = \sum_n P_n \left(\frac{\partial|\psi_n\rangle}{\partial t}\langle\psi_n| + |\psi_n\rangle\frac{\partial\langle\psi_n|}{\partial t} \right) \tag{4.154}$$

Recalling that $|\psi_n\rangle$ satisfies the Schrödinger equation:

$$H|\psi_n\rangle = i\hbar\frac{d|\psi_n\rangle}{dt} \quad \text{and} \quad \langle\psi_n|H = -i\hbar\frac{d}{dt}\langle\psi_n| \tag{4.155}$$

which we substitute into Eq. (4.154):

$$\begin{aligned} i\hbar\dot{\rho} &= \sum_n P_n(H|\psi_n\rangle\langle\psi_n| - |\psi_n\rangle\langle\psi_n|H) \\ &= H\rho - \rho H \equiv [H, \rho] \end{aligned} \tag{4.156}$$

where we have introduced the commutator $[a, b]$.

If we assume that the perturbing Hamiltonian H_I contains a small time-independent part, H',

$$H_I = H^1 + H' \tag{4.157}$$

then Eq. (4.156) becomes

$$i\hbar\frac{d}{dt}\rho_{ij} = (E_i - E_j)\rho_{ij} + [H^1, \rho]_{i,j} + [H', \rho]_{i,j} \tag{4.158}$$

where we have introduced the notation

$$\rho_{ij} = \langle u_i|\rho|u_j\rangle \tag{4.159}$$

Suppose at $t = 0$ the system is characterized by $|u_1\rangle$, so that

$$\rho_{ij}(t=0) = (|u_1\rangle\langle u_1|)_{i,j} = \delta_{i1}\delta_{j1}$$

We consider the effect of H' on the system by setting H^1 to zero. H' is assumed to be small, so ρ_{11} will remain the largest matrix element of ρ, at least for some period of time.

So we write Eq. (4.158) as

$$i\hbar\frac{\partial\rho_{ij}}{\partial t} \approx (E_1 - E_j)\rho_{1j} - \rho_{11}H'_{i,j} \qquad \text{for } j \neq 1 \tag{4.160}$$

$$i\hbar\frac{\partial\rho_{11}}{\partial t} \approx \sum_k (H'_{1k}\rho_{k1} - \rho_{1k}H'_{k1}) \tag{4.161}$$

When the Laplace transforms of these equations are taken, we have

$$i\hbar\int_0^\infty e^{-st}\dot{\rho}_{1j}(t)\,dt = (E_1 - E_j)\rho_{1j}(s) - \rho_{11}(s)H'_{1,j} \tag{4.162}$$

and

$$i\hbar\int_0^\infty e^{-st}\dot{\rho}_{11}(t)\,dt = \sum_k [H'_{1,k}\rho_{k,1}(s) - \rho_{1k}(s)H'_{k,1}] \tag{4.163}$$

When the left-hand sides are integrated by parts,

$$\int_0^\infty e^{-st}\dot{\rho}_{1j}(t)\,dt = [e^{-st}\rho_{1j}(t)]_0^\infty + s\rho_{1j}(s)$$

$$= -\delta_{j,1} + s\rho_{1j}(s) \tag{4.164}$$

we obtain

$$i\hbar s\rho_{1j}(s) = (E_1 - E_j)\rho_{1j}(s) - \rho_{11}(s)H'_{1,j} \tag{4.165}$$

and

$$-i\hbar + i\hbar s\rho_{11}(s) = \sum_k [H'_{1k}\rho_{k1}(s) - \rho_{1k}(s)H'_{k1}] \tag{4.166}$$

Equation (4.165) is solved for ρ_{ij}, that is,

$$\rho_{1j}(s) = \rho_{11}(s)H'_{1,j}(E_1 - E_j - i\hbar s)^{-1} \tag{4.167}$$

Since ρ is Hermitian, then we also have

$$\rho_{j1}(s) = -\rho_{11}(s)H'_{j1}(E_j - E_1 - i\hbar s)^{-1} \tag{4.168}$$

These two results are substituted into the right-hand side of Eq. (4.166) and since H' is Hermitian, we obtain

$$\rho_{11}(s)\left\{i\hbar s - \sum_k \left[\frac{H'_{1,k}(H'_{1,k})^*}{E_1 - E_k + i\hbar s} - \frac{H'_{1,k}(H'_{1,k})^*}{E_1 - E_k - i\hbar s}\right]\right\} - i\hbar = 0 \tag{4.169}$$

When the two terms in the square brackets are combined, we have

$$\rho_{11}(s) = \frac{1}{s + \hbar^{-2}\sum_k [2s|H'_{1,k}|^2(\omega_{1k}^2 + s^2)]} \tag{4.170}$$

where

$$\omega_{1,k} = (E_1 - E_k)/\hbar \tag{4.171}$$

We now take the inverse transform of Eq. (4.170) to give

$$\rho_{11}(t) = \frac{1}{2\pi i}\int_{\varepsilon - i\infty}^{\varepsilon + i\infty} \frac{e^{st}\,ds}{s + \hbar^{-2}\sum_k [(2s|H'_{1,k}|^2)/(s^2 + \omega_{1,k}^2)]} \tag{4.172}$$

where ε is an arbitrarily small positive number. We assume that we can replace the summation over k by an integral, that is,

$$\sum_k \frac{2s|H'_{1,k}|}{s^2 + \omega_{1,k}^2} \to \int_{-\infty}^{\infty} g(\omega_{1,k})\,d\omega_{1,k}\frac{2s|H'_{1,k}|^2}{s^2 + \omega_{1,k}^2} \tag{4.173}$$

where $g(\omega_{1k})$ is the density of states in frequency space. Since we are assuming $H'_{1,k}$ to be small, the only residues of the integrand occur at

$s = \pm i\omega_{1k}$. Using Cauchy's theorem we have

$$\int_{-\infty}^{\infty} g(\omega_{1k})\, d\omega_{1k} \frac{2s|H'_{1,k}|^2}{s^2+\omega_{1k}^2} = \frac{2\pi i g(is)2s|H'_{1k}|^2}{2is}$$

$$= 2\pi g(is)|H'_{1k}|^2 \tag{4.174}$$

When these results are substituted into Eq. (4.172) we obtain

$$\rho_{11}(t) = \frac{1}{2\pi i}\int_{\varepsilon-i\infty}^{\varepsilon+i\infty} \frac{e^{st}\, ds}{s+1/\tau(s)} \tag{4.175}$$

where

$$\tau(s) = [2\pi g(is)|H'_{1k}|^2]^{-1} \tag{4.176}$$

The residue of the integrand of Eq. (4.175) occurs at s' where

$$s' = -1/\tau(s')$$

Since the density of states, $g(s)$, is bounded, we have for given ε:

$$|s'| = 2\pi g(is')|H'_{1k}|^2 < \varepsilon$$

for $|H'_{1k}|^2$ sufficiently small. That is, s' is arbitrarily close to 0 and $g(is')$ is arbitrarily close to $g(0)$ for sufficiently small $|H'_{1k}|^2$. We assume that $|H'_{1k}|^2$ is small enough for $g(is') \approx g(0)$ to be a good approximation, whence by Cauchy's theorem we have

$$\rho_{11}(t) = e^{-t}/\tau(0) \tag{4.177}$$

In the following, we shall write $\tau(0) \equiv \tau_{11}$. Equation (4.177) indicates that, for $H^1 = 0$, the effect of H^1 is to cause ρ_{11} to begin to decay exponentially with time. Clearly the proof holds for any eigenstate $|u_i\rangle$ initially characterizing the system. That is, we may assume

$$\rho_{ii}(t) = e^{-t}/\tau_{ii} \tag{4.178}$$

We further assume that H' has a similar effect on nondiagonal terms, ρ_{ij}, the assumption leading to a good description of the finite lifetimes of ρ_{ij}. Thus, for $H^1 = 0$, we assume

$$\rho_{ij}(t) = e^{-t}/\tau_{ij}$$

and from Eq. (4.158) we have

$$i\hbar\dot{\rho} = [H', \rho] \qquad \text{for } H^1 = 0$$

These last two equations imply

$$[H', \rho]_{i,j} = i\hbar\dot{\rho}_{ij} = -\frac{i\hbar\rho_{ij}}{\tau_{ij}}$$

When it is assumed that the presence of a nonzero H^1 does not interfere with the effect of the randomizing term H^r, we can use this last result in Eq. (4.160) to give

$$\frac{i\hbar\partial\rho_{ij}}{\partial t}=\hbar\omega_{ij}\rho_{ij}+[H^1,\rho]_{ij}-\frac{i\hbar\rho_{ij}}{\tau_{ij}} \tag{4.179}$$

4.6.2. *Interaction of Electromagnetic Field with Atoms*[62]

Consider now the atomic Hamiltonian

$$H=H_0+V \tag{4.180}$$

where H_0 is the free Hamiltonian with known eigenfunctions $u_k(\mathbf{r})$, and V is the perturbation due to an applied electromagnetic field of the form

$$V=-e\mathbf{E}(\mathbf{R},t)\cdot\mathbf{r} \tag{4.181}$$

Here E is the applied field evaluated at the nucleus, position $\mathbf{R}$, and we are assuming the dipole approximation (wavelength of field $\gg$ dimension of atom).

The electric field does not depend on the atomic coordinates, so we have

$$\begin{aligned} V_{n,k}&=\langle u_n|V|u_k\rangle=-e\mathbf{E}\cdot\langle u_n|\mathbf{r}|u_k\rangle \\ &=-e\mathbf{E}\cdot\mathbf{r}_{n,k} \end{aligned} \tag{4.182}$$

where $\mathbf{r}_{nn}=0$, since $|u_n|^2$ is an even function of $\mathbf{r}$, while $\mathbf{r}$ itself is an odd function. The wave function $\psi(\mathbf{r},t)$ describing the total system will satisfy the time-dependent Schrödinger equation. We consider the stationary solutions $\psi_n(\mathbf{r},t)$, where

$$\psi_n(\mathbf{r},t)=u_n(\mathbf{r})\exp(-i\omega_n t) \tag{4.183}$$

where $u_n(\mathbf{r})$ are the eigenfunctions of H_0. The general solution is a linear combination of the ψ_n, that is,

$$\psi(\mathbf{r},t)=\sum_n[C_n(t)u_n(\mathbf{r})\exp(-i\omega_n t)] \tag{4.184}$$

When Eq. (4.184) is substituted into the Schrödinger equation for the total system we obtain

$$i\hbar\sum_k[(\dot{C}_k-i\omega_k C_k)\exp(-i\omega_k t)u_k(\mathbf{r})]=\sum_k[(\hbar\omega_k+V)\exp(-i\omega_k t)u_k(\mathbf{r})C_k] \tag{4.185}$$

which upon taking the scalar product with $u_n^*(\mathbf{r})$ and integrating over the spatial coordinates we have

$$\dot{C}_n = -i/\hbar \sum_k C_k \langle n|V|k\rangle \exp[-i(\omega_k - \omega_n)t] \tag{4.186}$$

having used the orthonormality property of u_n.

Consider now an applied field $\mathbf{E}$ of the form

$$\mathbf{E} = \hat{\mathbf{x}} E_0 \cos \nu t \tag{4.187}$$

which, in a two-level model has frequency $\nu \approx (E_a - E_b)/h$. That is, the field is nearly resonant with a transition between two levels a and b. Assume further that all other transitions are way-off resonance. The probability that the atom is in level a or b will be unity and the atom can be described by the two-level wave function

$$\psi(\mathbf{r}, t) = C_a u_a(\mathbf{r}) \exp(-i\omega_a t) + C_b u_b(\mathbf{r}) \exp(-i\omega_b t) \tag{4.188}$$

The matrix elements of the interaction V are, from Eq. (4.182),

$$V_{ab} = -\gamma E_0 \cos \nu t \tag{4.189}$$

where we are defining the magnitude of the electric-dipole matrix element

$$\gamma \equiv e z_{ab} \tag{4.190}$$

which we shall show later is proportional to the off-diagonal elements of the density matrix.

Substituting this into Eq. (4.186) yields:

$$\dot{C}_a = \frac{1}{2} i \frac{\gamma E_0}{\hbar} \{\exp[i(\omega_a - \omega_b - \nu)t] + \exp[i(\omega_a - \omega_b + \nu)t]\} C_b \tag{4.191}$$

$$\dot{C}_b = \frac{1}{2} i \frac{\gamma E_0}{\hbar} \{\exp[-i(\omega_a - \omega_b - \nu)t] + \exp[-i\omega(\omega_a - \omega_b + \nu)t]\} C_a \tag{4.192}$$

If we assume $\gamma E_0 t/\hbar$ to be small, then C_a and C_b will change slowly from their initial values, and we can make a perturbation expansion. In particular, if we assume the atom is in the lower level b at $t = 0$, then we will have the boundary conditions

$$C_b(t = 0) = 1, \qquad C_a(t = 0) = 0$$

which leads to

$$\dot{C}_b \approx 0 \tag{4.193}$$

$$\dot{C}_a \approx \frac{1}{2} i \frac{\gamma E_0}{\hbar} \{\exp[i(\omega - \nu)t] + \exp[i(\omega + \nu)t]\} \tag{4.194}$$

where

$$\omega = \omega_a - \omega_b \tag{4.195}$$

Thus to first-order approximation we have

$$C_b = 1$$

$$C_a = \frac{1}{2} i \frac{\gamma E_0}{\hbar} \left[\frac{\exp i(\omega - \nu)t}{i(\omega - \nu)} + \frac{\exp i(\omega + \nu)t}{i(\omega + \nu)} \right] \tag{4.196}$$

Since the resonance condition implies $\omega - \nu \approx 0$, $\omega + \nu \gg 0$, we may neglect the second term in brackets in Eq. (4.196), an approximation called the rotating wave approximation. This approximation is equivalent to replacing Eq. (4.189) by

$$V_{ab} = -\tfrac{1}{2}\gamma E_0 \exp(i\nu t) \tag{4.197}$$

So far we have considered only the interaction of an electromagnetic field with an atom. We now consider an ensemble of atoms. Suppose the atoms are excited to state n at the rate $\lambda_n(z, t_0)$ atoms (sec^{-1} cm^{-3}), λ being a function of time and space. Let $\boldsymbol{\rho}(n, z, t_0, t)$ be the density matrix describing an atom excited to the state n at time t_0 and position z, that is, $\rho_{aa}(n, z, t_0, t)$ is the probability that the atom which was in level n at time t_0 will be in level "a" at time t; from this definition, we have

$$\rho_{ij}(a, z, t, t) = \delta_{ia}\delta_{ja} \tag{4.198}$$

Thus, if we define the population matrix (Sargent *et al.*,[62] p. 103)

$$\boldsymbol{\rho}(z, t) \equiv \sum_n \int_{-\infty}^{t} dt_0\, \lambda_n(z, t_0) \boldsymbol{\rho}(n, z, t_0, t) \tag{4.199}$$

then the diagonal element $\rho_{aa}(z, t)$ will be the total population of the level a, etc.

To understand the meaning of the off-diagonal elements of $\boldsymbol{\rho}$, we consider the macroscopic polarization $P(z, t)$, given by

$$P(z, t) = \sum_n \int_{-\infty}^{t} dt_0\, \lambda_n(z, t_0) \langle er \rangle \tag{4.200}$$

where $\langle er \rangle$ is the expectation value for the dipole moment of the atoms at z. From Eq. (4.153) we have that for a two-level system the trace has only two terms; hence

$$\begin{aligned} \langle er \rangle &= e[z\boldsymbol{\rho}(a, z, t_0, t]_{aa} + e[z\boldsymbol{\rho}(a, z, t_0, t)]_{bb} \\ &= e \sum_c (z_{ac}\delta_{cb}\rho_{ca} + z_{bc}\delta_{ca}\rho_{cb}) \\ &= \gamma\rho_{ab}(a, z, t_0, t) + \text{cc} \end{aligned} \tag{4.201}$$

Consequently the macroscopic polarization becomes

$$P(z, t) = \gamma \sum_n \int_{-\infty}^{t} dt_0\, \lambda_n(z, t_0)\rho_{ab}(n, z, t_0, t) + \text{cc} \tag{4.202}$$

The equation of motion for the population matrix [Eq. (4.199)] is now easily obtained using the equation of motion [Eq. (4.179)] for the single-atom matrix $\rho(n, z, t_0, t)$; differentiating Eq. (4.199) gives

$$\frac{d\boldsymbol{\rho}(z, t)}{dt} = \sum_n \lambda_n(z, t)\boldsymbol{\rho}(n, z, t, t) + \sum_n \int_{-\infty}^{t} dt_0\, \lambda_n(z, t_0)\dot{\boldsymbol{\rho}}(n, z, t_0, t) \tag{4.203}$$

We shall now write out this equation for the individual elements of $\dot{\boldsymbol{\rho}}$, substituting Eq. (4.198) into the first term and Eq. (4.179) in the integrand, giving

$$\begin{aligned}\frac{d\rho_{ij}}{dt} &= \sum_n \lambda_n(z, t)\delta_{in}\delta_{jn} + \sum_n \int_{-\infty}^{t} dt_0\, \lambda_n(z, t_0)\left[-i\omega_{ij}\rho_{ij} - \frac{i}{\hbar}[H^1, \rho]_{ij} - \frac{\rho_{ij}}{\tau_{ij}}\right] \\ &= \sum_n \lambda_n(z, t)\delta_{in}\delta_{jn} - \left(i\omega_{ij} + \frac{1}{\tau_{ij}}\right)\rho_{ij} - \frac{i}{\hbar}\left[H^1, \sum_n \int_{-\infty}^{t} dt_0\, \lambda_n(z, t_0)\boldsymbol{\rho}\right]_{ij} \\ &= \sum_n \lambda_n(z, t)\delta_{in}\delta_{jn} - \left(i\omega_{ij} + \frac{1}{\tau_{ij}}\right)\rho_{ij} - \frac{i}{\hbar}\left(\sum_n H^1_{in}\rho_{nj} - \sum_n \rho_{in}H^1_{nj}\right)\end{aligned}$$

where $H^1_{in} = (1 - \delta_{in})V_{in}$ for the dipole operator, and $\omega_{ij} = (1 - \delta_{ij})\omega_{ij}$ from Eq. (4.171). We now have:

$$\dot{\rho}_{ab} = -\left(i\omega_{ab} + \frac{1}{\tau_{ab}}\right)\rho_{ab} + i\hbar^{-1}V_{ab}(\rho_{aa} - \rho_{bb}) \tag{4.204}$$

$$\dot{\rho}_{aa} = \lambda_a - \frac{1}{\tau_{aa}}\rho_{aa} - (i\hbar^{-1}V_{ab}\rho_{ba} + \text{cc}) \tag{4.205}$$

$$\dot{\rho}_{bb} = \lambda_b - \frac{1}{\tau_{bb}}\rho_{bb} + (i\hbar^{-1}V_{ab}\rho_{ba} + \text{cc}) \tag{4.206}$$

We note that τ_{ab} is the coherence time, or dipole-dephasing time.

Since from Eqs. (4.202) and (4.199) we have

$$P(z, t) = \gamma\rho_{ab}(z, t) + \text{cc} \tag{4.207}$$

we see that the equation of motion for the macroscopic polarization may be obtained from Eq. (4.204).

4.6.3. *Pulse Propagation**

We may now collect the preceding results to write down the equations of motion for the variables of interest. We start by deriving the equation of motion for the electric field, which is governed by Maxwell's equations.

We assume the form for the electric field E to be

$$E(z, t) = \tfrac{1}{2}u(z, t)\exp[-i(\nu t - kz)] + \text{cc} \tag{4.208}$$

in which $u(z, t)$ is a complex field envelope that varies little in an optical wavelength, or period. Corresponding to this field, there will be an induced macroscopic polarization:

$$P(z, t) = \tfrac{1}{2}\Pi(z, t)\exp[-i(\nu t - kz)] + \text{cc} \tag{4.209}$$

where $\Pi(z, t)$ will vary little in an optical wavelength or period.

The propagation of u and Π will be governed by Maxwell's equations, which may be written in the form (Sargent *et al.*,[62] p. 98)

$$\text{div}(\varepsilon_0\mathbf{E} + \mathbf{P}) = 0, \qquad \text{curl}\,\mathbf{E} = -\frac{\partial \mathbf{B}}{\partial t} \tag{4.210}$$

$$\text{div}\,\mathbf{B} = 0, \qquad \text{curl}\,\mathbf{B} = \mu_0\left[\mathbf{J} + \frac{\partial}{\partial t}(\varepsilon_0\mathbf{E} + \mathbf{P})\right] \tag{4.211}$$

where

$$\mathbf{J} = \sigma\mathbf{E} \tag{4.212}$$

To avoid a complicated boundary value problem due to losses in diffraction and reflector transmission, we assume a medium whose conductivity σ is adjusted to account for these losses.

We substitute the derivative of the curl **B** equation into the curl of the curl **E** equation to obtain the wave equation

$$\text{curl curl}\,\mathbf{E} + \mu_0\sigma\frac{\partial \mathbf{E}}{\partial t} + \mu_0\varepsilon_0\frac{\partial^2 \mathbf{E}}{\partial t^2} = -\mu_0\frac{\partial^2 P}{\partial t^2} \tag{4.213}$$

into which we substitute Eqs. (4.208) and (4.209) to obtain

$$-2ik\frac{\partial u}{\partial z} + k^2u + \mu_0\sigma(u - i\nu u) + \mu_0\varepsilon_0(-2\dot{u}i\nu - \nu^2 u) = \mu_0(i\nu\Pi - \Pi\nu^2) \tag{4.214}$$

where we have neglected the second derivatives of Π and u. We make use of the identities

$$\nu/k = c, \qquad \mu_0\varepsilon_0 = c^{-2} \tag{4.215}$$

* See Sargent *et al.*,[62] page 198.

and further neglect $\sigma\dot{u}$ and $\dot{\Pi}$ to obtain

$$\frac{\partial u}{\partial z}+c^{-1}\frac{\partial u}{\partial t}+\frac{1}{2}\frac{\sigma}{c\varepsilon_0}u=\frac{1}{2}i\nu(c\varepsilon_0)^{-1}\Pi \tag{4.216}$$

which is the equation of propagation of u. To obtain the equation for Π we equate the right-hand sides of Eqs. (4.207) and (4.209), that is,

$$\tfrac{1}{2}\Pi(z,t)\exp[-i(\nu t-kz)]+c\cdot c=\gamma\rho_{ab}(z,t)+\text{cc} \tag{4.217}$$

whence

$$\Pi(z,t)=2\gamma\exp[i(\nu t-kz)]\rho_{ab}(z,t) \tag{4.218}$$

When we take the time derivative of this result we have

$$\begin{aligned}\dot{\Pi}(z,t)&=2\gamma\exp[i(\nu t-kz)](i\nu\rho_{ab}+\dot{\rho}_{ab})\\ &=i\nu\Pi(z,t)-\left(i\omega_{ab}+\frac{1}{\tau_{ab}}\right)\Pi(z,t)+2i\hbar^{-1}\gamma\exp[i(\nu t-kz)]\\ &\quad\times V_{ab}(\rho_{aa}-\rho_{bb})\end{aligned} \tag{4.219}$$

where we have used Eq. (4.204).

With the field E defined by Eq. (4.208), the perturbation V is given, in the rotating-wave approximation, by

$$V_{ab}=-\tfrac{1}{2}\gamma u(z,t)\exp[-i(\nu t-kz)] \tag{4.220}$$

which when substituted into Eq. (4.219) gives the equation for $\dot{\Pi}$:

$$\dot{\Pi}(z,t)=\left(i\nu-i\omega_{ab}-\frac{1}{\tau_{ab}}\right)\Pi(z,t)-i\hbar^{-1}\gamma^2u(z,t)(\rho_{aa}-\rho_{bb}) \tag{4.221}$$

We have already noticed that ρ_{aa} is just the total population of level a; thus $(\rho_{aa}-\rho_{bb})$ is the population inversion [see Eq. (4.5)]:

$$D\equiv(\rho_{aa}-\rho_{bb})\equiv\delta^{J_0} \tag{4.222}$$

The equation for the rate of change of D follows by taking the difference of Eqs. (4.205) and (4.206):

$$\dot{D}=\lambda_a-\lambda_b-\frac{D}{\tau_{aa}}+\left[\frac{i}{2\hbar}u(z,t)\Pi^*(z,t)+\text{cc}\right] \tag{4.223}$$

where we are assuming $\tau_{aa}=\tau_{bb}$, and we have used Eqs. (4.218) and (4.220).

We now collect the coupled equations (4.216), (4.221), and (4.223), valid for all ranges of the input pulse, in the laboratory frame of reference:

$$\frac{\partial u}{\partial z}+\frac{1}{c}\frac{\partial u}{\partial t}=-\kappa u+\frac{i\nu}{2c\varepsilon_0}\Pi(z,t),\qquad \kappa\equiv\sigma/2c\varepsilon_0 \tag{4.224}$$

$$\frac{\partial \Pi}{\partial t} = -\frac{\Pi(z, t)}{\tau_{ab}} - i\hbar^{-1}\gamma^2 u(z, t)D \tag{4.225}$$

$$\frac{\partial D}{\partial t} = \lambda_c - \lambda_b - \frac{D}{\tau_{aa}} + \frac{i}{2\hbar}[u(z, t)\Pi^*(z, t) - u^*(z, t)\Pi(z, t)] \tag{4.226}$$

having assumed the resonance condition $\nu = \omega_{ab}$.

The preceding formulas form the basis of the system of equations solved by Armandillo and Spalding,[58] and they reduce to Eqs. (4.33)–(4.35) under appropriate assumptions that we shall now describe.

Since Eqs. (4.33)–(4.35) are written in terms of intensity, I_ν, we must express Π in terms of u in Eq. (4.226), since

$$I_{\nu_0} \propto uu^*$$

So, we consider the formal solution to Eq. (4.221):

$$\Pi(z, t) = -i\frac{\gamma^2}{\hbar}\int_{-\infty}^{t} dt'\, u(z, t')D(z, t')\exp\left[\frac{-(t-t')}{\tau_{ab}}\right] \tag{4.227}$$

where we are assuming the resonance condition $\nu = \omega_{ab}$. Repeated integration by parts gives a series in powers of τ_{ab}:

$$\begin{aligned}\Pi(z, t) &= \{u(z, t)D(z, t)\tau_{ab} \\ &\quad - \tau_{ab}\int_{-\infty}^{t} dt' u(z, t')\dot{D}(z, t')\exp[-(t-t')/\tau_{ab}]\}\left(-\frac{i\gamma^2}{\hbar}\right) \\ &= -\frac{i\gamma^2}{\hbar}u(z, t)D(z, t)\tau_{ab} + O(\tau_{ab}^2)\end{aligned} \tag{4.228}$$

The rate-equation approximation assumes $\tau_{ab} \ll \tau_p$, where τ_p is the duration of the input pulse, and neglects all but the first term of Eq. (4.228). When this approximation for Π is substituted into Eq. (4.223) we obtain

$$\dot{D} = \lambda_a - \lambda_b - \frac{D}{\tau_{aa}} - \frac{\gamma^2}{2\hbar^2}[u(z, t)u^*(z, t) + \text{cc}]D\tau_{ab} \tag{4.229}$$

When we define the equilibrium value of the population inversion, $\bar{\delta}$:

$$\lambda_a - \lambda_b \equiv \bar{\delta}/\tau_R$$

and fix the constant of proportionality between uu^* and I_{ν_0} by

$$\tau_{ab}\frac{\gamma^2}{\hbar^2}|u(z, t)|^2 \equiv 2WI_{\nu_0} \tag{4.230}$$

then Eq. (4.229) becomes identical to Eq. (4.33).

Since the equation for I_{ν_0} [Eq. (4.35)] is independent of the normalization of I_{ν_0}, we may obtain this equation by considering the equation for $|u(z, t)|^2$.

We substitute Eq. (4.224) into the identity

$$\frac{d}{d\tau}(uu^*) \equiv u\frac{du^*}{d\tau} + u^*\frac{du}{d\tau} \tag{4.231}$$

to obtain in the laboratory reference frame:

$$\frac{\partial|u|^2}{\partial z} + \frac{1}{c}\frac{\partial|u|^2}{\partial t} = -2\kappa|u|^2 + \frac{i\nu}{2c\varepsilon_0}[\Pi u^* - \Pi^* u] \tag{4.232}$$

Finally using the approximation given by Eq. (4.228) we have

$$\frac{\partial|u|^2}{\partial z} + \frac{1}{c}\frac{\partial|u|^2}{\partial t} = -2\kappa|u|^2 - \frac{\gamma^2\nu}{c\varepsilon_0\hbar}\tau_{ab}|u|^2 D(z, t) \tag{4.233}$$

which implies, since $|u|^2 \propto I_\nu$:

$$\frac{\partial I_\nu}{\partial t} + c\frac{\partial I_\nu}{\partial z} = -2\kappa I_\nu - \frac{\gamma^2\nu}{c\varepsilon_0\hbar}\tau_{ab} I_\nu \delta^{J_0} \tag{4.234}$$

Comparison of this equation with Eq. (4.35) allows the identification

$$chW = \frac{\gamma^2\tau_{ab}}{c\varepsilon_0\hbar} \tag{4.235}$$

The coherence time τ_{ab} can be seen to be related to the band width by substituting Eq. (2.58), divided by h, for W; hence

$$\tau_{ab} = \frac{c^2h^2W\varepsilon_0}{2\pi\gamma^2} = \frac{c\hbar\lambda^3\varepsilon_0}{4\pi^2\tau_{sp}\gamma^2\Delta\nu} \tag{4.236}$$

We note that Eq. (4.234) has a loss term $-2\kappa I_\nu$, not present in Eq. (4.35). This term was dropped earlier because the most significant losses are at the oscillator mirrors, which are not present for the amplifier case.

4.7. PULSAMP: A Code Used to Predict the Amplification of Nanosecond CO_2 Laser Light Pulses

Roberts and Smith[63] have written a computer program which predicts the behavior of a light pulse of duration ~1 nsec as it propagates through a gain medium taking single-line or multiline/multiband operation into account, but neglecting the effect of coherent pulse propagation.

5

A General Plasma Model

5.1. Preliminary Definitions

We consider a plasma containing N atomic and molecular species and electrons. $f_{i\alpha}(\mathbf{v}, \mathbf{x}, t)$, $i = 1, N$, is the distribution function in position and velocity space, defined such that $f_{i\alpha}(\mathbf{v}, \mathbf{x}, t)\, d^3\mathbf{v}\, d^3\mathbf{x}$ is the number of molecules (atoms) of species i in vibrational (atomic) state α at time t in a volume element of size $d^3\mathbf{v}$ at $\mathbf{v}$ in velocity space and in volume element $d^3\mathbf{x}$ at $\mathbf{x}$ in position space. Electronically excited species, as well as distinct vibrational modes and positive and negative ions, are regarded as distinct species. In the same way we have $f_e(\mathbf{v}, \mathbf{x}, t)$, the distribution function for electrons.

The number density (in cm^{-3}) of species i at position $\mathbf{x}$ and time t is then given by

$$n_i(\mathbf{x}, t) = \sum_\alpha \int_{-\infty}^{\infty} f_{i\alpha}(\mathbf{v}, \mathbf{x}, t)\, d^3\mathbf{v} \tag{5.1}$$

and the mass density of the plasma $\rho(\mathbf{x}, t)$ (in $\mathrm{g\,cm}^{-3}$) is

$$\rho(\mathbf{x}, t) = \sum_{i=1}^{N} m_i n_i(\mathbf{x}, t) \tag{5.2}$$

where m_i is the mass of one heavy particle of species i. We define a mass-average velocity $\mathbf{v}_0(\mathbf{x}, t)$ of heavy particles to be

$$\mathbf{v}_0(\mathbf{x}, t) = \frac{1}{\rho} \sum_{i=1}^{N} \sum_\alpha m_i \int f_{i\alpha}(\mathbf{v}, \mathbf{x}, t)\mathbf{v}\, d^3\mathbf{v} \tag{5.3}$$

Following Chapman and Cowling[43] we define the variance velocity $\mathbf{V}$ of a molecule with velocity $\mathbf{v}$ as

$$\mathbf{V} \equiv \mathbf{v} - \mathbf{v}_0 \tag{5.4}$$

The average variance velocity, $\bar{\mathbf{V}}_i$, of molecules of species i is defined by

$$n_i(\mathbf{x}, t)\bar{\mathbf{V}}_i(\mathbf{x}, t) \equiv \sum_\alpha \int f_{i\alpha}(\mathbf{v}, \mathbf{x}, t)\mathbf{V}(\mathbf{x}, t)\, d^3\mathbf{v} \tag{5.5}$$

We also define the translational gas temperature $T(\mathbf{x}, t)$ by

$$\frac{3}{2}n(\mathbf{x}, t)kT(\mathbf{x}, t) \equiv \frac{1}{2}\sum_{i=1}^{N}\sum_\alpha m_i \int d^3v\, f_{i\alpha}(\mathbf{v}, \mathbf{x}, t)V^2 \tag{5.5a}$$

where $n(\mathbf{x}, t)$ is the total heavy-particle number density [see Eq. (5.17)].

5.2. Equations of Change for Heavy Particles

5.2.1. The Boltzmann Equation

The heavy particles are the molecules, atoms, and their ions, which make up the N species. The equations for the electrons shall be derived later in Section 5.4.

Our starting point is the Boltzmann equation for the distribution function* $f_{i\alpha}(\mathbf{v}, \mathbf{x}, t)$ [see Eq. (3.8)]:

$$\frac{\partial f_{i\alpha}(\mathbf{v}, \mathbf{x}, t)}{\partial t} + \mathbf{v}\cdot\nabla f_{i\alpha}(\mathbf{v}, \mathbf{x}, t) + \frac{\mathbf{X}_i(\mathbf{x}, t)}{m_i}\cdot\nabla_v f_{i\alpha}(\mathbf{v}, \mathbf{x}, t) = J_{i\alpha}(\mathbf{v}, \mathbf{x}, t) \tag{5.6}$$

$\mathbf{X}_i(\mathbf{x}, t)$ is the electrostatic force acting on charged species and will be the sum of the force due to an applied electric field $\mathbf{E}$, and the force resulting from the field $\hat{\mathbf{E}}$ due to all the other charged particles in the system, that is,

$$\mathbf{X}_i(\mathbf{x}, t) = e_i[\mathbf{E}(\mathbf{x}, t) + \hat{\mathbf{E}}(\mathbf{x}, t)] \tag{5.7}$$

where $e_i = (0, \pm e)$ is the charge on a particle of species i, and $\hat{\mathbf{E}}(\mathbf{x}, t)$, the Vlasov field,[64,65] being the gradient of the potential due to all the ions, is defined by

$$\hat{\mathbf{E}}(x, t) = -\nabla\left[\sum_{i\alpha}\int\int \frac{f_{i\alpha}(\mathbf{v}', \mathbf{x}', t)}{|\mathbf{x}-\mathbf{x}'|}\, d^3\mathbf{v}'\, d^3\mathbf{x}'\right] \tag{5.8}$$

The term on the right-hand side of the Boltzmann equation [Eq. (5.6)] represents the change in f caused by elastic, inelastic, and reactive collisions. We shall see that the contribution to $J_{i\alpha}$ from reactive collision is related to rates $k_{ab;i\gamma}$ in reactive collisions in which the initial particles are a, b and the final particles are $i\gamma$.

* We remark that in the derivation of the Boltzmann equation it is assumed that encounters with other molecules occupy a very small part of the lifetime of a molecule. This implies that only binary encounters are important.

The equations of change are continuity equations that describe the time evolution of the total density, species densities, vibrationally excited species densities, and the energy density and mass-average velocity, $\mathbf{v}_0$. These equations are derived from the Boltzmann equation.[43,50] Following Chapman and Cowling,[43] we first reexpress the Boltzmann equation [Eq. (5.6)] in terms of the variables $(\mathbf{V}, \mathbf{x}, t)$ instead of $(\mathbf{v}, \mathbf{x}, t)$. $[\partial f(\mathbf{v}, \mathbf{x}, t)]/\partial t$ is the partial differential of f with respect to t while holding $\mathbf{x}$ and $\mathbf{v}$ constant. If now $\mathbf{V}$ is taken to be the independent variable instead of $\mathbf{v}$ then

$$\frac{\partial f(\mathbf{v}, \mathbf{x}, t)}{\partial t} \rightarrow \frac{\partial f(\mathbf{V}, \mathbf{x}, t)}{\partial t} + \frac{\partial \mathbf{V}}{\partial t} \cdot \frac{\partial f(\mathbf{V}, \mathbf{x}, t)}{\partial \mathbf{V}} \tag{5.9}$$

From Eq. (5.4) we have that, if $\mathbf{v}$ and t are independent variables then

$$\left.\frac{\partial \mathbf{V}}{\partial t}\right|_{\mathbf{v}} = -\left.\frac{\partial \mathbf{v}_0}{\partial t}\right|_{\mathbf{v}} \tag{5.10}$$

so that the replacement [Eq. (5.9)] is

$$\frac{\partial f(\mathbf{v}, \mathbf{x}, t)}{\partial t} \rightarrow \frac{\partial f(\mathbf{V}, \mathbf{x}, t)}{\partial t} - \frac{\partial \mathbf{v}_0}{\partial t} \cdot \nabla_V f(\mathbf{V}, \mathbf{x}, t) \tag{5.11}$$

having substituted ∇_V for $\partial/\partial \mathbf{V}$. Similarly we have

$$\begin{aligned} \nabla f(\mathbf{v}, \mathbf{x}, t) &\rightarrow \nabla f(\mathbf{V}, \mathbf{x}, t) + \left.\frac{\partial \mathbf{V}}{\partial \mathbf{x}}\right|_{v} \frac{\partial f(\mathbf{V}, \mathbf{x}, t)}{\partial \mathbf{V}} \\ &\rightarrow \nabla f(\mathbf{V}, \mathbf{x}, t) - (\nabla \mathbf{v}_0 \cdot) \nabla_V f(\mathbf{V}, \mathbf{x}, t) \end{aligned} \tag{5.12}$$

having substituted ∇ for $\partial/\partial \mathbf{x}$. Finally, we have

$$\begin{aligned} \nabla_V f(\mathbf{v}, \mathbf{x}, t) &\rightarrow \nabla_v f(\mathbf{V}, \mathbf{x}, t) + \left.\frac{\partial V}{\partial \mathbf{v}}\right|_{x,t} \frac{\partial f(\mathbf{V}, \mathbf{x}, t)}{\partial \mathbf{V}} \\ &\rightarrow \nabla_V f(\mathbf{V}, \mathbf{x}, t) \end{aligned} \tag{5.13}$$

since from Eq. (5.4), $\partial \mathbf{V}/\partial \mathbf{v} = 1$. Substituting Eqs. (5.11)–(5.13) into the Boltzmann equation (5.6) we have

$$\begin{aligned} &\frac{\partial f_{i\alpha}(\mathbf{V}, \mathbf{x}, t)}{\partial t} - \frac{\partial \mathbf{v}_0}{\partial t} \cdot \nabla_V f_{i\alpha}(\mathbf{V}, \mathbf{x}, t) + (\mathbf{V} + \mathbf{v}_0) \cdot [\nabla f_{i\alpha}(\mathbf{V}, \mathbf{x}, t) - (\nabla \mathbf{v}_0 \cdot) \nabla_V f_{i\alpha}] \\ &+ \frac{\mathbf{X}_i}{m_i} \cdot \nabla_V f_{i\alpha}(\mathbf{V}, \mathbf{x}, t) = J_{i\alpha}(\mathbf{V}, \mathbf{x}, t) \end{aligned} \tag{5.14}$$

Defining the "time derivative following the motion" as D/Dt:

$$\frac{D}{Dt} \equiv \frac{\partial}{\partial t} + \mathbf{v}_0 \cdot \nabla \tag{5.14a}$$

then Eq. (5.14) becomes, after collecting terms:

$$\frac{Df_{i\alpha}(\mathbf{V},\mathbf{x},t)}{Dt}+\mathbf{V}\cdot\nabla f_{i\alpha}(\mathbf{V},\mathbf{x},t)+\left(\frac{\mathbf{X}_i}{m_i}-\frac{D\mathbf{v}_0}{Dt}\right)\cdot\nabla_V f_{i\alpha}(\mathbf{V},\mathbf{x},t)$$
$$-\mathbf{V}\cdot(\nabla\mathbf{v}_0)\cdot\nabla_V f_{i\alpha}(\mathbf{V},\mathbf{x},t)=J_{i\alpha}(\mathbf{V},\mathbf{x},t) \qquad (5.15)$$

In subsequent sections we shall work with this form of the Boltzmann equation for the heavy particles of the plasma.

5.2.2. The Equation of Change of $\bar{\psi}$

Let $\psi_{i\alpha}(\mathbf{V},\mathbf{x},t)$ be any time-dependent function of the variance velocity, species, and state of a molecule. We define a mean value $\bar{\psi}(\mathbf{x},t)$ as

$$\bar{\psi}(\mathbf{x},t)=\frac{1}{n(\mathbf{x},t)}\sum_i\sum_\alpha\int d^3V f_{i\alpha}(\mathbf{V},\mathbf{x},t)\psi_{i\alpha}(\mathbf{V},\mathbf{x},t) \qquad (5.16)$$

where $n(\mathbf{x},t)$ is the total heavy-particle number density:

$$n(\mathbf{x},t)=\sum_i n_i(\mathbf{x},t) \qquad (5.17)$$

We now multiply the Boltzmann equation (5.15) by $\psi_{i\alpha}(\mathbf{V},\mathbf{x},t)$, sum over i and α, and integrate over $\mathbf{V}$. We assume that the integrals are convergent and that products such as

$$\psi_{i\alpha}f_{i\alpha}\to 0$$

as $\mathbf{V}\to\infty$ in any direction. The first term of Eq. (5.15) becomes

$$\sum_{i\alpha}\int\psi_{i\alpha}\frac{Df_{i\alpha}}{Dt}d^3V=\frac{D}{Dt}\sum_{i\alpha}\int\psi_{i\alpha}f_{i\alpha}\,d^3V-\sum_{i\alpha}\int\frac{D\psi_{i\alpha}}{Dt}f_{i\alpha}\,d^3V$$
$$=\frac{D(n\bar{\psi})}{Dt}-n\left(\overline{\frac{D\psi}{Dt}}\right) \qquad (5.18)$$

In deriving Eq. (5.18), D/Dt is taken outside the integral since it does not depend on $\mathbf{V}$. The overbar on any quantity denotes its mean value defined in the same way as $\bar{\psi}$ in Eq. (5.16). The second term of Eq. (5.15) becomes

$$\sum_{i\alpha}\int\psi_{i\alpha}\mathbf{V}\cdot\nabla f_{i\alpha}\,d^3V=\nabla\cdot\sum_{i\alpha}\int\psi_{i\alpha}\mathbf{V}f_{i\alpha}\,d^3V-\sum_{i\alpha}\int(\nabla\psi_{i\alpha})\cdot\mathbf{V}f_{i\alpha}\,d^3V$$
$$=\nabla\cdot[n(\overline{\psi\mathbf{V}})]-n[\overline{\mathbf{V}\cdot\nabla\psi}] \qquad (5.19)$$

From the third term of Eq. (5.15) we have, on expanding the scalar product into Cartesian coordinates with axes labeled 1, 2, and 3:

$$\sum_{i\alpha}\int \psi_{i\alpha}\frac{\mathbf{X}_i}{m_i}\cdot\nabla_V f_{i\alpha}\,d^3V=\sum_{i\alpha}\int \frac{\psi_{i\alpha}}{m_i}\left(X_{i1}\frac{\partial}{\partial V_1}+X_{i2}\frac{\partial}{\partial V_2}+X_{i3}\frac{\partial}{\partial V_3}\right)f_{i\alpha}\,d^3V \tag{5.20}$$

We consider just the first term in the expansion (5.20) and integrate by parts:

$$\sum_{i\alpha}\int \frac{\psi_{i\alpha}}{m_i}X_{i1}\frac{\partial f_{i\alpha}}{\partial V_1}\,dV_1\,dV_2\,dV_3=\sum_{i\alpha}\int\left(\psi_{i\alpha}\frac{X_{i1}}{m_i}f_{i\alpha}\right)_{V_1=-\infty}^{V_1=+\infty}dV_2\,dV_3$$
$$-\sum_{i\alpha}\int\frac{X_{i1}}{m_i}\frac{\partial\psi_{i\alpha}}{\partial V_1}f_{i\alpha}\,d^3V \tag{5.21}$$

The first term on the right-hand side of Eq. (5.21) vanishes by the hypothesis that

$$\psi_{i\alpha}f_{i\alpha}\to 0 \qquad \text{as } V_1\to\pm\infty$$

so that, from Eqs. (5.20) and (5.21) we have

$$\sum_{i\alpha}\int \psi_{i\alpha}\frac{\mathbf{X}_i}{m_i}\cdot\nabla_V f_{i\alpha}\,d^3V=-n\overline{\mathbf{X}\left(\frac{1}{m}\cdot\nabla_V\psi\right)} \tag{5.22}$$

For the fourth term of Eq. (5.15) we have

$$-\sum_{i\alpha}\int \psi_{i\alpha}\frac{D\mathbf{v}_0}{Dt}\cdot\nabla_V f_{i\alpha}\,d^3V=-\frac{D\mathbf{v}_0}{Dt}\cdot\sum_{i\alpha}\int\psi_{i\alpha}\nabla_V f_{i\alpha}\,d^3V \tag{5.23}$$

Expanding the scalar product in Eq. (5.23) into Cartesian components and integrating by parts, we have, for the "x-component":

$$-\frac{Dv_{01}}{Dt}\sum_{i\alpha}\int\psi_{i\alpha}\frac{\partial f_{i\alpha}}{\partial V_1}\,dV_1\,dV_2\,dV_3$$
$$=-\frac{Dv_{01}}{Dt}\sum_{i\alpha}\int(\psi_{i\alpha}f_{i\alpha})_{V_1=-\infty}^{V_1=+\infty}\,dV_2\,dV_3+\frac{Dv_{01}}{Dt}\sum_{i\alpha}\int\frac{\partial\psi_{i\alpha}}{\partial V_1}f_{i\alpha}\,d^3V \tag{5.24}$$

Hence

$$-\sum_{i\alpha}\int\psi_{i\alpha}\frac{D\mathbf{v}_0}{Dt}\cdot\nabla_V f_{i\alpha}\,d^3V=n\frac{D\mathbf{v}_0}{Dt}\cdot\overline{(\nabla_V\psi)} \tag{5.25}$$

For the last term on the left-hand side of Eq. (5.15) we have, again expanding into Cartesian components:

$$-\sum_{i\alpha}\int \psi_{i\alpha}\mathbf{V}(\cdot\,\boldsymbol{\nabla}\mathbf{v}_0\,\cdot)\boldsymbol{\nabla}_V f_{i\alpha}\,d^3V$$

$$=-\sum_{i\alpha}\int \psi_{i\alpha}\mathbf{V}\cdot(\boldsymbol{\nabla}v_{01})\frac{\partial f_{i\alpha}}{\partial V_1}\,dV_1\,dV_2\,dV_3+y \text{ and } z \text{ terms}$$

$$=\sum_{i\alpha}\int f_{i\alpha}\frac{\partial}{\partial V_1}\cdot(\psi_{i\alpha}\mathbf{V}\cdot\boldsymbol{\nabla}v_{01})\,dV_1\,dV_2\,dV_3+y \text{ and } z \text{ terms}$$

$$-\sum_{i\alpha}\int (\psi_{i\alpha}f_{i\alpha}\mathbf{V}\cdot\boldsymbol{\nabla}v_{01})_{V_1=-\infty}^{V_1=\infty}\,dV_2\,dV_3$$

$$=\sum_{i\alpha}\int f_{i\alpha}\boldsymbol{\nabla}_V\cdot(\psi_{i\alpha}\mathbf{V}\cdot\boldsymbol{\nabla})\mathbf{v}_0\,d^3V$$

$$=\sum_{i\alpha}\int f_{i\alpha}\psi_{i\alpha}\boldsymbol{\nabla}\cdot\mathbf{v}_0\,d^3V+\sum_{i\alpha}\int f_{i\alpha}[(\mathbf{V}\cdot\boldsymbol{\nabla})\mathbf{v}_0]\cdot\boldsymbol{\nabla}_V\psi_{i\alpha}\,d^3V$$

$$=n\bar{\psi}\boldsymbol{\nabla}\cdot\mathbf{v}_0+n\overline{(\boldsymbol{\nabla}_V\psi)\cdot(\mathbf{V}\cdot\boldsymbol{\nabla})}\mathbf{v}_0 \tag{5.26}$$

since $\boldsymbol{\nabla}\cdot\mathbf{v}_0$ is $\mathbf{V}$ independent.

From Eqs. (5.18), (5.19), (5.22), (5.25), and (5.26) we have the Boltzmann equation (5.15) in the form

$$\frac{D(n\bar{\psi})}{Dt}+n\bar{\psi}\boldsymbol{\nabla}\cdot\mathbf{v}_0+\boldsymbol{\nabla}\cdot[n\overline{(\psi\mathbf{V})}]$$
$$-n\left[\left(\overline{\frac{D\psi}{Dt}}\right)+\overline{(\mathbf{V}\cdot\boldsymbol{\nabla}\psi)}+\mathbf{X}\overline{\left(\frac{1}{m}\cdot\boldsymbol{\nabla}_V\psi\right)}-\frac{D\mathbf{v}_0}{Dt}\cdot\overline{(\boldsymbol{\nabla}_V\psi)}-\overline{(\boldsymbol{\nabla}_V\psi)\cdot(\mathbf{V}\cdot\boldsymbol{\nabla})}\mathbf{v}_0\right]$$
$$=\sum_{i\alpha}\int \psi_{i\alpha}J_{i\alpha}\,d^3V \tag{5.27}$$

which is called the equation of change of $\bar{\psi}$ (Reference 43, page 49). We recall that $\psi_{i\alpha}(\mathbf{V}, \mathbf{x}, t)$ is any function of the variance velocity, species, and state of a molecule. We now consider particular examples of $\psi_{i\alpha}$ in order to derive the equations of change of the heavy particles from Eq. (5.27).

5.2.3. Conservation of Mass

We set

$$\psi_{i\alpha}=m_i$$

where we recall that m_i is the mass of a heavy particle of species i. From the definition of $\bar{\psi}$ [see Eq. (5.16)], we have

$$n\bar{m} = \sum_i \sum_\alpha \int d^3V f_{i\alpha} m_i$$
$$= \sum_i n_i m_i$$
$$= \rho(\mathbf{x}, t) \tag{5.28}$$

where we have used Eqs. (5.1) and (5.2). Also, using Eq. (5.16), the third term in Eq. (5.27) yields

$$n\overline{(m\mathbf{V})} = \sum_i \sum_\alpha \int d^3V f_{i\alpha} m_i \mathbf{V}$$
$$= \sum_i \sum_\alpha \int d^3v\, f_{i\alpha} m_i \mathbf{v} - \mathbf{v}_0 \sum_i \sum_\alpha \int d^3v\, f_{i\alpha} m_i$$
$$= \rho\mathbf{v}_0 - \mathbf{v}_0\rho = 0 \tag{5.29}$$

where we have transformed the variable of integration from $\mathbf{V}$ to $\mathbf{v}$ using Eq. (5.4), and have used Eqs. (5.2) and (5.3). We also have

$$\nabla m_i = 0 \qquad \text{and} \qquad \nabla_V m_i = 0$$

so that Eq. (5.27) becomes

$$\frac{D\rho}{Dt} + \rho \nabla \cdot \mathbf{v}_0 = 0 \tag{5.30}$$

where, on the right-hand side, we have put

$$\sum_{i\alpha} \int m_i J_{i\alpha}\, d^3V = 0 \tag{5.31}$$

since this term represents the change in mass caused by collisions, and mass is conserved in all collisions.

Equation (5.30) is called the *equation of conservation of mass* and expresses the fact that mass is conserved even though the total particle number density n will change due to reactive collisions.

5.2.4. Particle Rate Equations

We set

$$\psi_{i\alpha} = \delta_{ij}$$

where $\delta_{ij} = 1$ if $i = j$, and 0 otherwise. We have from Eqs. (5.1) and (5.16):

$$n\overline{\delta_{ij}} = \sum_i \sum_\alpha \int d^3V f_{i\alpha} \delta_{ij} = \sum_\alpha \int d^3V f_{j\alpha} = n_j(\mathbf{x}, t) \tag{5.32}$$

and Eq. (5.16) also gives the third term of Eq. (5.27) in the form

$$
\begin{aligned}
n\overline{(\delta_{ij}\mathbf{V})} &= \sum_i \sum_\alpha \int d^3V f_{i\alpha} \mathbf{V}\, \delta_{ij} \\
&= \sum_\alpha \int d^3V f_{j\alpha} \mathbf{V} \\
&= n_j \bar{\mathbf{V}}_j
\end{aligned}
\tag{5.33}
$$

where $\bar{\mathbf{V}}_j$ is the average variance velocity of species j [Eq. (5.5)]. Also

$$\nabla \delta_{ij} = 0, \qquad \nabla_V \delta_{ij} = 0$$

so that Eq. (5.27) becomes [Hirschfelder *et al.*,[66] Eq. (7.2-42)]

$$\frac{Dn_j}{Dt} + n_j \nabla \cdot \mathbf{v}_0 + \nabla \cdot (n_j \bar{\mathbf{V}}_j) = \sum_\alpha \int d^3V J_{j\alpha} \tag{5.34}$$

which is the rate equation for the number density n_j of species j. Only reactive collisions and external processes will contribute to the right-hand side of Eq. (5.34) so that we can write Eq. (5.34) as

$$
\begin{aligned}
\frac{Dn_j}{Dt} + n_j \nabla \cdot \mathbf{v}_0 + \nabla \cdot (n_j \cdot \bar{\mathbf{V}}_j) = S_j - \nu_j n_j + \sum_{il} n_i n_l k_{il;j\gamma} - \sum_l n_j n_l k_{jl;\gamma} \\
+ \sum_{ilq} n_i n_l n_q k_{ilq;j\gamma} - \sum_{lp} n_j n_l n_p k_{jlp;\gamma}
\end{aligned}
\tag{5.35}
$$

where S_j (in $\text{cm}^{-3}\,\text{sec}^{-1}$) denotes the rate of production of species j by external means; ν_j (in sec^{-1}) is the loss frequency of species j by means other than reactions. The loss rate of particles due to diffusion is described explicitly by the term containing $\bar{\mathbf{V}}_j$ on the left-hand side of Eq. (5.35) (see Section 5.3.5). The reaction rates k denote a loss of species j when the subscript j precedes the semicolon; γ denotes collectively all particles in the final state.

5.2.5. *Momentum Equation*

When we choose

$$\psi_{i\alpha} = m_i \mathbf{V} \tag{5.36}$$

we have, from Eq. (5.29):

$$n\overline{(mV)} = 0 \tag{5.37}$$

i.e., the first and second terms of Eq. (5.27) vanish. For the third term of Eq. (5.27) we have

$$\nabla \cdot [n(\overline{\psi \mathbf{V}})] = \nabla \cdot \left(\sum_{i\alpha} \int d^3V f_{i\alpha} m_i \mathbf{V}\mathbf{V}\right)$$
$$= \nabla \cdot \hat{P} \tag{5.38}$$

where $\hat{P}$ is the pressure tensor, evaluated later on in Section 5.3.6, defined by

$$\hat{P}_{rs} = \sum_{i\alpha} \int d^3V f_{i\alpha} m_i V_r V_s \tag{5.39}$$

Also for the fourth term in Eq. (5.27), we have

$$n\overline{\left(\frac{D\psi}{Dt}\right)} = \sum_{i\alpha} \int d^3V f_{i\alpha} \frac{D}{Dt} m_i \mathbf{V} = 0 \tag{5.40}$$

since $\mathbf{V}$ and t are independent variables. Since $\mathbf{x}$ and $\mathbf{V}$ are independent variables (see Section 5.2.1), the fifth term in Eq. (5.27) becomes

$$n\overline{(\mathbf{V} \cdot \nabla\psi)} = \sum_{i\alpha} \int d^3V f_{i\alpha} \mathbf{V} \cdot \nabla(m_i \mathbf{V}) = 0 \tag{5.41}$$

Using Eq. (5.1) we obtain the sixth term of Eq. (5.27) to be

$$n\overline{\mathbf{X} \cdot \left(\frac{1}{m} \nabla_V \psi\right)} = \sum_{i\alpha} \int d^3V f_{i\alpha} \mathbf{X}_i \frac{1}{m_i} \nabla_V m_i \mathbf{V}$$
$$= \sum_i \mathbf{X}_i n_i \tag{5.42}$$

For the seventh term of Eq. (5.27) we have

$$n \frac{D\mathbf{v}_0}{Dt} \cdot \overline{(\nabla_V \psi)} = \frac{D\mathbf{v}_0}{Dt} \cdot \sum_{i\alpha} \int d^3V f_{i\alpha} \nabla_V m_i \mathbf{V}$$
$$= \frac{D\mathbf{v}_0}{Dt} \sum_{i\alpha} \int d^3V f_{i\alpha} m_i$$
$$= \rho \frac{D\mathbf{v}_0}{Dt} \tag{5.43}$$

where $\rho(\mathbf{x}, t)$ is the mass density defined by Eq. (5.2). For the last term on the left-hand side of Eq. (5.27) we have

$$\begin{aligned} n\overline{(\nabla_V\psi)\cdot(\mathbf{V}\cdot\nabla)\mathbf{v}_0} &= \sum_{i\alpha}\int d^3V f_{i\alpha}(\nabla_V m_i\mathbf{V})\cdot(\mathbf{V}\cdot\nabla)\mathbf{v}_0 \\ &= \sum_{i\alpha}\int d^3V f_{i\alpha} m_i(\mathbf{V}\cdot\nabla)\mathbf{v}_0 \\ &= 0 \end{aligned} \tag{5.44}$$

where we have used Eq. (5.29). Since momentum is conserved in all types of collisions we set the right-hand side of Eq. (5.27) equal to zero. When the above results for $\psi_{i\alpha} = m_i\mathbf{V}$ are used, Eq. (5.27) becomes

$$\rho\frac{D\mathbf{v}_0}{Dt} + \nabla\cdot\hat{P} - \sum_i n_i\mathbf{X}_i = 0 \tag{5.45}$$

which is the equation of momentum of the gas.

5.2.6. Mode Vibrational Energy Equations

Let ε_{j1} be the energy of the first excited state of a vibrational mode of molecular species j. The total vibrational energy of the mode-species per unit volume, $E_j^{(V)}$, will be given by

$$E_j^{(V)} = \sum_\alpha \varepsilon_{j\alpha}\int d^3V f_{j\alpha}(\mathbf{V}, \mathbf{x}, t) \tag{5.46}$$

To obtain the equation of change of $E_j^{(V)}$ we set

$$\psi_{i\alpha} = \delta_{ij}\varepsilon_{j\alpha} \tag{5.47}$$

into Eq. (5.27). We have

$$\begin{aligned} n\bar{\psi} &= \sum_{i\alpha}\int d^3V f_{i\alpha}\delta_{ij}\varepsilon_{j\alpha} \\ &= \sum_\alpha\int d^3V f_{j\alpha}\varepsilon_{j\alpha} \\ &= E_j^{(V)} \end{aligned} \tag{5.48}$$

The third term of Eq. (5.27) becomes

$$n\overline{(\psi \mathbf{V})} = \sum_{i\alpha} \int d^3V f_{i\alpha}\delta_{ij}\varepsilon_{j\alpha}\mathbf{V}$$

$$= \sum_{\alpha} \int d^3V f_{j\alpha}\varepsilon_{j\alpha}\mathbf{V}$$

$$\equiv \mathbf{q}_j^{(V)} \tag{5.49}$$

where Eq. (5.49) defines the vibrational energy-flux vector (Section 5.3.8). We also have $\nabla\psi_{i\alpha} = 0$ and $\nabla_V\psi_{i\alpha} = 0$ so that Eq. (5.27) becomes

$$\frac{DE_j^{(V)}}{Dt} + E_j^{(V)}\nabla \cdot \mathbf{v}_0 + \nabla \cdot \mathbf{q}_j^{(V)} = \sum_{\alpha} \int J_{j\alpha}\varepsilon_{j\alpha}\, d^3V \tag{5.50}$$

which is a generalization of the more familiar point equations for $E_j^{(V)}$ (see Section 2.7).

To evaluate the right-hand side of Eq. (5.50), we consider three mechanisms by which mode vibrational energy can change in collisions:

(i) It can change by inelastic and superelastic collisions of electrons with an effective vibrational electron excitation rate [see Eq. (3.75)] $X_j^{(V)}(E/n, T)$, where we recall that E is the applied electric field, $n(\mathbf{x}, t)$ is the total heavy-particle number density, and $T(\mathbf{x}, t)$ the ambient gas temperature. We also define $n_e(\mathbf{x}, t)$ to be the electron number density, and then the contribution to the right-hand side of Eq. (5.50) from electron–molecule collisions is

$$\varepsilon_{j1}n_e n_j X_j^{(V)} \tag{5.51}$$

(ii) It can also change by V–T energy transfer with a relaxation time $\tau^j_{(VT)}(T)$ (see Section 2.2). The contribution to the right-hand side of Eq. (5.50) is

$$-\left[\frac{E_j^{(V)}(T_j^{(V)}) - E_j^{(V)}(T)}{\tau^j_{(VT)}(T)}\right] \tag{5.52}$$

where $T_j^{(V)}$ is the effective vibrational temperature of the mode-species, which is related to the mode vibrational energy by [see also Eq. (1.69)]

$$E_j^{(V)}(T_j^{(V)}) = \frac{n_j\varepsilon_{j1}}{e^{\varepsilon_{j1}/kT_j^{(V)}} - 1} \tag{5.53}$$

(iii) Intermode V–V energy-transfer processes will also contribute to the changes in vibrational mode energies. The form of the relaxation terms will depend on the exact details of the V–V processes for particular gas mixtures. Until we consider particular cases we shall write the contribution to Eq. (5.50) from V–V processes in the general form

$$-\left[\frac{E_j^{(V)}(T_j^{(V)})-E_k^{(V)}(T_k^{(V)})}{\tau_{(VV)}^{jk}(T, T_j^{(V)}, T_k^{(V)})}\right] \tag{5.54}$$

where the square brackets mean that there may be many terms of this form.

With Eqs. (5.51), (5.52), and (5.54) replacing the right-hand side of Eq. (5.50) we obtain the vibrational energy equation for mode-species j:

$$\frac{DE_j^{(V)}(T_j^{(V)})}{Dt}+E_j^{(V)}(T_j^{(V)})\nabla\cdot\mathbf{v}_0+\nabla\cdot\mathbf{q}_j^{(V)}$$

$$=\varepsilon_{j1}n_e n_j X_j^{(V)}-\frac{E_j^{(V)}(T_j^{(V)})-E_j^{(V)}(T)}{\tau_{(VT)}^{j}(T)}-\left[\frac{E_j^{(V)}(T_j^{(V)})-E_k^{(V)}(T_k^{(V)})}{\tau_{(VV)}^{jk}(T, T_j^{(V)}, T_k^{(V)})}\right] \tag{5.55}$$

5.2.7. Translational–Rotational Energy Equation

If we set

$$\psi_{i\alpha}=\tfrac{1}{2}m_i\mathbf{V}^2 \tag{5.56}$$

which is the kinetic energy of a molecule of species i with velocity $\mathbf{V}$, then

$$n\bar{\psi}=\sum_{i\alpha}\int d^3V f_{i\alpha}\frac{1}{2}m_i\mathbf{V}^2\equiv E^{(TR)} \tag{5.57}$$

which defines $E^{(TR)}(T, \mathbf{x}, t)$, the total translational–rotational energy density of the gas. $E^{(TR)}$ is related to the ambient gas temperature T by [see Eq. (2.122)]

$$E^{(TR)}(T)=\frac{1}{2}\left[\sum_i(3+R_i)n_i\right]kT \tag{5.58}$$

where R_i is the number of rotational degrees of freedom of a molecule of species i.

With the substitution of Eq. (5.56), the third term on the left-hand side of Eq. (5.27) is

$$\nabla\cdot[n\overline{(\psi\mathbf{V})}]=\nabla\sum_{i\alpha}\int d^3V f_{i\alpha}\frac{1}{2}m_iV^2\mathbf{V}$$

$$=\nabla\cdot\mathbf{q}^{(TR)} \tag{5.59}$$

which defines the translation–rotation energy flux $\mathbf{q}^{(TR)}$ (Section 5.3.7):

$$\mathbf{q}^{(TR)}=\sum_{i\alpha}\int d^3V f_{i\alpha}\frac{1}{2}m_iV^2\mathbf{V} \tag{5.60}$$

Since $\mathbf{V}$, t, and $\mathbf{x}$ are independent variables we have $D\psi/Dt = 0$ and $\nabla\psi = 0$. Now the sixth term of Eq. (5.27) becomes

$$\begin{aligned} n\overline{\mathbf{X}\cdot\left(\frac{1}{m}\nabla_V\psi\right)} &= \sum_{i\alpha}\int d^3V f_{i\alpha}\mathbf{X}_i\cdot\frac{1}{m_i}\nabla_V\frac{1}{2}m_iV^2 \\ &= \sum_{i\alpha}\int d^3V f_{i\alpha}\mathbf{X}_i\cdot\mathbf{V} \\ &= \sum_i \mathbf{X}_i\cdot\bar{\mathbf{V}}_i \end{aligned} \tag{5.61}$$

where we have used Eq. (5.5) and the fact that $\mathbf{X}_i$ does not depend on $\mathbf{V}$. We also have the seventh term of Eq. (5.27):

$$n\frac{D\mathbf{v}_0}{Dt}\cdot\overline{(\nabla_V\psi)} = \frac{D\mathbf{v}_0}{Dt}\sum_{i\alpha}\int d^3V f_{i\alpha}m_i\mathbf{V} = 0 \tag{5.62}$$

by Eq. (5.29), and the eighth term of Eq. (5.27) becomes

$$\begin{aligned} n\overline{(\nabla_V\psi)\cdot(\mathbf{V}\cdot\nabla)\mathbf{v}_0} &= \sum_{i\alpha}\int d^3V f_{i\alpha}m_i\mathbf{V}(\mathbf{V}\cdot\nabla)\mathbf{v}_0 \\ &= \hat{P}:(\nabla\mathbf{v}_0) \end{aligned} \tag{5.63}$$

where $\hat{P}$ is the pressure tensor defined by Eq. (5.39) and the double-dot notation means that the scalar product is taken between $\hat{P}$ and $(\nabla\mathbf{v}_0)$, the latter being regarded as a tensor.

With the above considerations we have, as the equation of change for the translational–rotational energy:

$$\begin{aligned} &\frac{DE^{(T)}}{Dt} + E^{(TR)}\nabla\cdot\mathbf{v}_0 + \nabla\cdot\mathbf{q}^{(TR)} + \hat{P}:(\nabla\mathbf{v}_0) - \sum_i \mathbf{X}_i\cdot\bar{\mathbf{V}}_i \\ &\quad = \sum_{i\alpha}\int d^3V f_{i\alpha}J_{i\alpha}\frac{1}{2}m_i\mathbf{V}^2 \end{aligned} \tag{5.64}$$

There are three types of collisions that contribute to the term on the right-hand side of Eq. (5.64) by resulting in a change of translational energy:

(i) Elastic collisions between heavy particles and electrons: Defining $\nu^{(TR)}(T_e)$ as the translation–rotation contribution to the total electron–heavy-particle collision frequency for collisions in which energy kT_e is transferred from the kinetic energy of electrons to the translational–rotational energy of the gas, then the contribution to Eq. (5.64) from such collisions is

$$n_e\nu^{(TR)}(T_e)kT_e \tag{5.65a}$$

T_e is the effective electron temperature given by [see also Eq. (3.81)]

$$T_e(E/N, T) = \frac{2}{3}\int_0^\infty u^{3/2} f(u, E/N, T)\, du \tag{5.65b}$$

(ii) The V–T energy transfer [see (ii) of Section 5.2.6]: The contribution to the right-hand side of Eq. (5.64) is, by Eq. (5.52):

$$+\sum_j \left[\frac{E_j^{(V)}(T_j^{(V)}) - E_j^{(V)}(T)}{\tau_{(VT)}^j(T)}\right] \tag{5.66}$$

(iii) The *residual energy* of the intermode V–V energy-transfer processes [see (iii) of Section 5.2.6]: Following Eq. (5.54) we denote this transfer of vibrational to translational energy by Eq. (2.104):

$$+\left\{\sum_{j,k}\left[\left(1-\frac{\varepsilon_{j1}}{\varepsilon_{k1}}\right)\frac{E_j^{(V)}(T_j^{(V)}) - E_k^{(V)}(T_k^{(V)})}{\tau_{(VV)}^{jk}(T, T_j^{(V)}, T_k^{(V)})}\right]\right\} \tag{5.67}$$

From Eqs. (5.65)–(5.67) we have, for the translational–rotational energy equation (5.64) (the superscript TR in parentheses denotes translation–rotation, the T in parentheses denotes ambient temperature):

$$\begin{aligned}
&\frac{DE^{(TR)}(T, \mathbf{x}, t)}{Dt} + E^{(TR)}(T)\nabla\cdot\mathbf{v}_0 + \nabla\cdot\mathbf{q}^{(TR)} + \hat{P}:(\nabla\mathbf{v}_0) - \sum_i \mathbf{X}_i\cdot\bar{\mathbf{V}}_i \\
&= n_e\nu^{(TR)}(T_e)kT_e + \sum_j\left[\frac{E_j^{(V)}(T_j^{(V)}) - E_j^{(V)}(T)}{\tau_{(VT)}^j(T)}\right] \\
&+\left\{\sum_{j,k}\left[\left(1-\frac{\varepsilon_{j1}}{\varepsilon_{k1}}\right)\frac{E_j^{(V)}(T_j^{(V)}) - E_j^{(V)}(T_k^{(V)})}{\tau_{(VV)}^{jk}(T, T_j^{(V)}, T_k^{(V)})}\right]\right\}
\end{aligned} \tag{5.68}$$

5.3. Perturbation Solution of the Heavy-Particle Boltzmann Equation

The mathematical description of the plasma requires the solution of the equation of change for the macroscopic quantities: ρ [Eq. (5.30)]; n_i [Eq. (5.35)], which requires $\bar{V}_i$; $\mathbf{v}_0$[Eq. (5.45)], which requires $\hat{P}$; $E_j^{(V)}$[Eq. (5.55)], which requires $\mathbf{q}_j^{(V)}$; $E^{(T)}$ [Eq. (5.68)], which requires $\bar{\mathbf{V}}_j$, $\hat{P}$, and $\mathbf{q}^{(TR)}$. In order to calculate the transport quantities $\bar{\mathbf{V}}_j$, $\hat{P}$, $\mathbf{q}_j^{(V)}$, and $\mathbf{q}^{(TR)}$, we need to know $f_{i\alpha}$, the solution to the Boltzmann equation (5.6). In the following subsections we shall describe an approximate analytical solution to Eq. (5.6), which is then used to evaluate the transport quantities.

5.3.1. Collision Integrals of the Boltzmann Equation

We consider the Boltzmann equation for heavy particles in the form given previously (where the signs of the charges e_i are included in $\mathbf{X}_i$ [see Eq. (5.7)]:

$$\frac{\partial f_{i\alpha}(\mathbf{v},\mathbf{x},t)}{\partial t}+\mathbf{v}\cdot\nabla f_{i\alpha}(\mathbf{v},\mathbf{x},t)+\frac{\mathbf{X}_i(\mathbf{x},t)}{m_i}\cdot\nabla_v f_{i\alpha}(\mathbf{v},\mathbf{x},t)=J_{i\alpha}(\mathbf{v},\mathbf{x},t) \qquad (5.6)$$

and write down an expression for $J_{i\alpha}$ in terms of collision cross sections and distribution functions under the following assumptions:

(i) Only binary collisions are significant.
(ii) There is no external field.
(iii) The Vlasov field [cf. Eq. (5.8)] does not affect the collisions.
(iv) There is no net macroscopic angular momentum.

Following Hirschfelder *et al.*[66] we rewrite $J_{i\alpha}(\mathbf{v},\mathbf{x},t)$, the collision term in the Boltzmann equation (5.6), as the algebraic sum

$$J_{i\alpha}=\sum_{j\beta}(\Gamma^{(+)}_{i\alpha,j\beta}-\Gamma^{(-)}_{i\alpha,j\beta}) \qquad (5.69)$$

where $\Gamma^{(-)}_{i\alpha,j\beta}(\mathbf{v},\mathbf{x},t)\,d^3x\,d^3v\,dt$ is the number of molecules of the ith species in vibrational state α, lost from the velocity range $\mathbf{v}$ to $\mathbf{v}+d\mathbf{v}$, in the position range $\mathbf{x}$ to $\mathbf{x}+d\mathbf{x}$, due to collisions with molecules of type j, in vibrational state β, during the time interval dt. $\Gamma^{(+)}_{i\alpha,j\beta}(\mathbf{v},\mathbf{x},t)\,d^3x\,d^3v\,dt$ is the corresponding gain.

We derive below explicit expressions for $\Gamma^{(-)}_{i\alpha,j\beta}$ and $\Gamma^{(+)}_{i\alpha,j\beta}$. Consider a molecule of type $i\alpha$ at position $\mathbf{x}$ with a velocity $\mathbf{v}_i$. We wish to find the probability that this molecule will experience a collision with a molecule of type $j\beta$, in the time interval dt, with the impact parameter b (defined in Fig. 26), in a range db about b. We assume that the intermolecular force is negligible for distances of separation greater than distances A, which is small compared to the mean-free path.

In Fig. 26 any molecule of type $j\beta$ within a sector of the cylindrical shell will undergo an encounter with molecule i during the time interval dt, characterized by an impact parameter b and an initial relative velocity $\mathbf{g}_{ij}$, given by

$$\mathbf{g}_{ij}=\mathbf{v}_j-\mathbf{v}_i \qquad (5.70)$$

The probable number of molecules of type $j\beta$ within this sector is

$$f_{j\beta}(\mathbf{x},\mathbf{v}_j,t)g_{ij}b\,db\,d\varepsilon\,dt$$

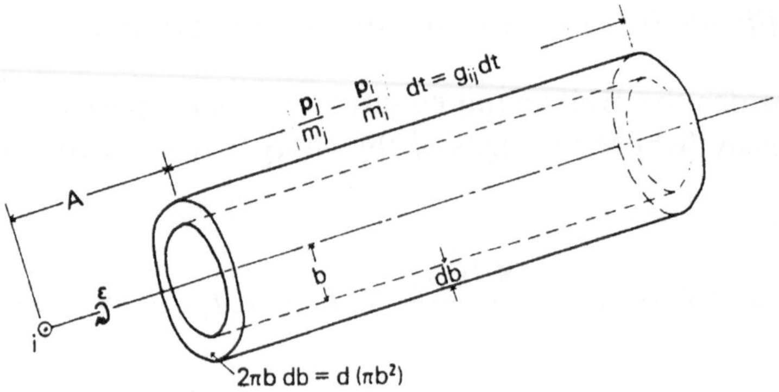

Fig. 26. Collisions of molecules of type j with one molecule of type i, in which the impact parameter is b. The distance A is essentially the intermolecular distance at which the potential begins to "take hold." Any molecule of type j which is initially located in the cylinder of base $2\pi b\, db$ and height $g_{ij}\, dt$ will undergo a collision with the molecule i during the short time interval dt. (Reproduced from Hirschfelder *et al.*,[66] with permission.)

where ε is the angle about the axis of the cylinder, and

$$g_{ij} = g_{ji} = |\mathbf{g}_{ij}| \tag{5.71}$$

The total number of collisions experienced by this molecule $i\alpha$ with molecules of type $j\beta$ is obtained by adding together the number of collisions characterized by all values of the parameters b and ε and all relative velocities g_{ji}. The result is

$$dt \iiint f_{j\beta}(\mathbf{x}, \mathbf{v}_j, t) g_{ij} b\, db\, d\varepsilon\, d\mathbf{v}_j$$

Since the probable number of molecules of type i in the volume element $d\mathbf{x}$ about $\mathbf{x}$, with velocity in the range $d\mathbf{v}_i$ about $\mathbf{v}_i$, is $f_{i\alpha}(\mathbf{x}, \mathbf{v}_i, t)\, d\mathbf{x}\, d\mathbf{v}_i$ it follows that

$$\Gamma^{(-)}_{i\alpha,j\beta}\, d\mathbf{x}\, d\mathbf{v}_i = d\mathbf{x}\, d\mathbf{v}_i\, dt \iiint f_{j\beta}(\mathbf{x}, \mathbf{v}_j, t) f_{i\alpha}(\mathbf{x}, \mathbf{v}_i, t) g_{ij} b\, db\, d\varepsilon\, d\mathbf{v}_j \tag{5.72}$$

and hence

$$\Gamma^{(-)}_{i\alpha,j\beta} = \iiint f_{j\beta} f_{i\alpha} g_{ij} b\, db\, d\varepsilon\, d\mathbf{v}_j \tag{5.73}$$

A similar argument gives $\Gamma^{(+)}_{i\alpha,j\beta}$ as

$$\Gamma^{(+)}_{i\alpha,j\beta} = \iiint f'_{i\alpha} f'_{j\beta} g_{ij} b\, db\, d\varepsilon\, d\mathbf{v}_j \tag{5.74}$$

where

$$f'_{i\alpha} = f_{i\alpha}(\mathbf{x}, \mathbf{v}'_i, t) \tag{5.75}$$

and $\mathbf{v}_i'$ is the velocity of molecule i after collision, and is related to $\mathbf{v}_i$ and $\mathbf{v}_j$ by the kinematics of the collision.

When Eqs. (5.72) and (5.74) are substituted into Eq. (5.69),

$$J_{i\alpha}=\sum_{j\beta}\iiint (f'_{i\alpha}f'_{j\beta}-f_{i\alpha}f_{j\beta})g_{ij}b\,db\,d\varepsilon\,d\mathbf{v}_j \tag{5.76}$$

or

$$J_{i\alpha}=\sum_{j\beta}J(f_{i\alpha},f_{j\beta}) \tag{5.77}$$

where

$$J(f_{i\alpha},f_{j\beta})=\iiint (f'_{i\alpha}f'_{j\beta}-f_{i\alpha}f_{j\beta})g_{ij}b\,db\,d\varepsilon\,d\mathbf{v}_j \tag{5.78}$$

The $J(f_{i\alpha}, f_{j\beta})$ are called the collision integrals of the Boltzmann equation. In the notation of Bond *et al.*,[36]

$$b\,db\,d\varepsilon\rightarrow\sigma_{i\alpha,j\beta}(\mathbf{v}_i,\mathbf{v}_j\rightarrow\mathbf{v}_i',\mathbf{v}_j')\,d\Omega \tag{5.79}$$

where $\sigma_{i\alpha,j\beta}$ is the differential cross section for the scattering of a molecule of type i, in vibrational state α, against one of type j, in vibrational state β.

In terms of differential cross sections for explicit collision processes (i.e., in which the species and internal states of the molecules are specified after the collision), Eq. (5.78) becomes

$$\begin{aligned}J(f_{i\alpha},f_{j\beta})=\sum_{k\alpha',l\beta'}\int d\mathbf{v}_j\,d\Omega[g_{kl}\sigma_{k\alpha',l\beta'}^{i\alpha,j\beta}(\mathbf{v}_k,\mathbf{v}_l\rightarrow\mathbf{v}_i,\mathbf{v}_j)f_{k\alpha'}(\mathbf{x},\mathbf{v}_k,t)f_{l\beta'}(\mathbf{x},\mathbf{v}_l,t)\\ -g_{ij}\sigma_{i\alpha,j\beta}^{k\alpha',l\beta'}(\mathbf{v}_i,\mathbf{v}_j\rightarrow\mathbf{v}_k,\mathbf{v}_l)f_{i\alpha}(\mathbf{x},\mathbf{v}_i,t)f_{j\beta}(\mathbf{x},\mathbf{v}_j,t)]\end{aligned} \tag{5.80}$$

where $\sigma_{i\alpha,j\beta}^{k\alpha',l\beta'}(\mathbf{v}_i,\mathbf{v}_j\rightarrow\mathbf{v}_k,\mathbf{v}_l)$ is the differential cross section for a molecule of species i in internal state α, colliding with a molecule of type j in internal state β, to produce molecules of species k and l in internal states α' and β', respectively, and

$$g_{ij}=|\mathbf{v}_i-\mathbf{v}_j|,\qquad g_{kl}=|\mathbf{v}_k-\mathbf{v}_l| \tag{5.81}$$

5.3.2. Quasi-Equilibrium Solution of the Boltzmann Equation

The quasi-equilibrium values, $f_{i\alpha}^{[0]}$, of the distribution functions $f_{i\alpha}$ are the values when all species have their Boltzmann distribution of kinetic and vibrational energies, and when the average velocity of each species is the same. Therefore $f_{i\alpha}^{[0]}$ will be the product of the Maxwell velocity distribution $f_i^{(\mathrm{M})}$ for species i (see Andersen in Hochstim,[50] p. 48):

$$f_i^{(\mathrm{M})}(\mathbf{V})=\left(\frac{m_i}{2\pi kT}\right)^{3/2}\exp\left(-\frac{m_iV^2}{2kT}\right) \tag{5.82}$$

and the Boltzmann distribution $n_{i\alpha}^{(B)}$ for each of the vibrational states [see Eq. (1.66)]:

$$n_{i\alpha}^{(B)} = \frac{n_i \exp(-\varepsilon_{i\alpha}/kT_i^{(V)})}{\sum_\alpha \exp(-\varepsilon_{i\alpha}/kT_i^{(V)})} \tag{5.83}$$

specified by an effective vibrational temperature $T_i^{(V)}$ for the mode-species i, where n_i is the total number density of heavy particles of mode-species i. So we have, for the quasi-equilibrium distribution functions:

$$\begin{aligned} f_{i\alpha}^{[0]}(\mathbf{V}, \mathbf{x}, t) &= n_{i\alpha}^{(B)} f_i^{(M)} \\ &= \frac{n_i}{\sum_\alpha \exp(-\varepsilon_{i\alpha}/kT_i^{(V)})}\left(\frac{m_i}{2\pi kT(\mathbf{x}, t)}\right)^{3/2} \exp\left[\frac{-\frac{1}{2}m_i V^2}{kT(\mathbf{x}, t)} - \frac{\varepsilon_{i\alpha}}{kT_i(\mathbf{x}, t)}\right] \end{aligned} \tag{5.84}$$

5.3.3. *Perturbation of the Boltzmann Equation about the Quasi-Equilibrium Solution*

Following Hirschfelder *et al.* (Reference 66, p. 467) we introduce a perturbation parameter ξ that can be arbitrarily varied from unity and has the effect of altering the frequency of collisions in the gas. This is done by rewriting the Boltzmann equation (5.6) as

$$\frac{\partial f_{i\alpha}}{\partial t} + \mathbf{v} \cdot \nabla f_{i\alpha} + \frac{\mathbf{X}_i}{m_i} \cdot \nabla_v f_{i\alpha} = \frac{1}{\xi} \sum_{j\beta} J(f_{i\alpha}, f_{j\beta}) \tag{5.85}$$

where we have introduced the perturbation parameter ξ and replaced the $J_{i\alpha}$ in Eq. (5.6) by Eq. (5.77). For small ξ ($\ll 1$) the collision term on the right-hand side of Eq. (5.85) is large, representative of a large collision frequency, so that local equilibrium will be everywhere maintained. We therefore suppose that in the limit $\xi = 0$ the distribution functions $f_{i\alpha}$ become the quasi-equilibrium solutions $f_{i\alpha}^{[0]}$ given by Eq. (5.84), and we also suppose that $f_{i\alpha}$ can be expanded in the Enskog series in ξ:

$$f_{i\alpha} = f_{i\alpha}^{[0]} + \xi f_{i\alpha}^{[1]} + \xi^2 f_{i\alpha}^{[2]} + \cdots \tag{5.86}$$

where the superscripts in square brackets indicate the degree of approximation. The expansion of Eq. (5.86) for $f_{i\alpha}$ is substituted into the modified Boltzmann equation (5.85), and multiplying through by ξ we obtain

$$\begin{aligned} &\left(\frac{\partial}{\partial t} + \mathbf{v} \cdot \nabla + \frac{\mathbf{X}_i}{m_k} \cdot \nabla_v\right)(\xi f_{i\alpha}^{[0]} + \xi^2 f_{i\alpha}^{[1]} + \xi^3 f_{i\alpha}^{[2]} + \cdots) \\ &\quad = \sum_{j\beta} J(f_{i\alpha}^{[0]} + \xi f_{i\alpha}^{[1]} + \xi^2 f_{i\alpha}^{[2]} + \cdots, f_{j\beta}^{[0]} + \xi f_{j\beta}^{[1]} + \xi^2 f_{j\beta}^{[2]} + \cdots) \\ &\qquad + J(f_{i\alpha}^{[0]} + \xi f_{i\alpha}^{[1]} + \xi^2 f_{i\alpha}^{[2]} + \cdots, f_e) \end{aligned} \tag{5.87}$$

where the summation in the first term on the right-hand side is over heavy particles only; the contribution from electrons is given explicitly by the second term. The right-hand side of Eq. (5.87) is expanded using the definition of J given in Eq. (5.78):

$$\begin{aligned}\sum_{j\beta} J(f_{i\alpha}^{[0]}, f_{j\beta}^{[0]}) + \xi \sum_{j\beta} [J(f_{i\alpha}^{[0]}, f_{j\beta}^{[1]}) + J(f_{i\alpha}^{[1]}, f_{j\beta}^{[0]})] \\ + \xi^2 \sum_{j\beta} [J(f_{i\alpha}^{[0]}, f_{j\beta}^{[2]}) + J(f_{i\alpha}^{[1]}, f_{j\beta}^{[1]}) + J(f_{i\alpha}^{[2]}, f_{j\beta}^{[0]})] \\ + J(f_{i\alpha}^{[0]}, f_e) + \xi J(f_{i\alpha}^{[1]}, f_e) + \xi^2 J(f_{i\alpha}^{[2]}, f_e) + \cdots\end{aligned} \tag{5.88}$$

Since ξ is a perturbation parameter, which we are free to vary, we can equate the coefficients of equal power of ξ on each side of Eq. (5.87). Using Eq. (5.88) we have, for the coefficients of the first three powers of ξ:

$$0 = \sum_{j\beta} J(f_{i\alpha}^{[0]}, f_{j\beta}^{[0]}) + J(f_{i\alpha}^{[0]}, f_e) \tag{5.89a}$$

$$\begin{aligned}\frac{\partial f_{i\alpha}^{[0]}}{\partial t} + \mathbf{v} \cdot \nabla f_{i\alpha}^{[0]} + \frac{\mathbf{X}_i}{m_i} \cdot \nabla_v f_{i\alpha}^{[0]} \\ = \sum_{j\beta} [J(f_{i\alpha}^{[0]}, f_{j\beta}^{[1]}) + J(f_{i\alpha}^{[1]}, f_{j\beta}^{[0]})] + J(f_{i\alpha}^{[1]}, f_e)\end{aligned} \tag{5.89b}$$

$$\begin{aligned}\frac{\partial f_{i\alpha}^{[1]}}{\partial t} + \mathbf{v} \cdot \nabla f_{i\alpha}^{[1]} + \frac{\mathbf{X}_i}{m_i} \cdot \nabla_v f_{i\alpha}^{[1]} \\ = \sum_{j\beta} [J(f_{i\alpha}^{[0]}, f_{j\beta}^{[2]}) + J(f_{i\alpha}^{[1]}, f_{j\beta}^{[1]}) + J(f_{i\alpha}^{[2]}, f_{j\beta}^{[0]})] + J(f_{i\alpha}^{[2]}, f_e)\end{aligned} \tag{5.89c}$$

Since $f_{i\alpha}^{[0]}$ is known [Eq. (5.84)], the solution of Eqs. (5.89b) will provide the first-order perturbation solution for $f_{i\alpha}^{[1]}$. We first evaluate the left-hand side of Eq. (5.89b), which we write as $Df_{i\alpha}^{[0]}$:

$$Df_{i\alpha}^{[0]} \equiv \frac{\partial f_{i\alpha}^{[0]}}{\partial t} + \mathbf{v} \cdot \nabla f_{i\alpha}^{[0]} + \frac{\mathbf{X}_i}{m_i} \cdot \nabla_v f_{i\alpha}^{[0]} \tag{5.90}$$

where $f_{i\alpha}^{[0]}$ is given by [see Eq. (5.84)]

$$\begin{aligned}f_{i\alpha}^{[0]} = \frac{n_i(\mathbf{x}, t)}{\sum_\alpha \exp[-\varepsilon_{i\alpha}/kT_i^{(V)}(\mathbf{x}, t)]} \left[\frac{m_i}{2\pi k T(\mathbf{x}, t)}\right]^{3/2} \\ \times \exp\left\{\frac{-\frac{1}{2}m_i[v - v_0(\mathbf{x}, t)]^2}{kT(\mathbf{x}, t)} - \frac{\varepsilon_{i\alpha}}{kT_i^{(V)}(\mathbf{x}, t)}\right\}\end{aligned} \tag{5.91}$$

when we substituted Eq. (5.4) for V. The independent variables in Eqs. (5.89)–(5.91) are $\mathbf{x}$, $\mathbf{v}$, and t, and from Eq. (5.91) we see that $f_{i\alpha}^{[0]}$ is an explicit function of $\mathbf{v}$ and a function of $\mathbf{x}$ and t through its dependence on $n_i(\mathbf{x}, t)$, $T(\mathbf{x}, t)$, $T_i^{(V)}(\mathbf{x}, t)$, and $v_0(\mathbf{x}, t)$. Therefore we have to reexpress the differentials occurring in Eq. (5.90), in terms of the variables n_i, T, $T_i^{(V)}$, $\mathbf{v}_0$, and $\mathbf{v}$:

$$\frac{\partial f_{i\alpha}^{[0]}}{\partial t}=\frac{\partial f_{i\alpha}^{[0]}}{\partial n_i}\frac{\partial n_i}{\partial t}+\frac{\partial f_{i\alpha}^{[0]}}{\partial T}\frac{\partial T}{\partial t}+(\nabla_{\mathbf{v}_0}f_{i\alpha}^{[0]})\cdot\frac{\partial \mathbf{v}_0}{\partial t}+\frac{\partial f_{i\alpha}^{[0]}}{\partial T_i^{(V)}}\frac{\partial T_i^{(V)}}{\partial t} \tag{5.92a}$$

$$\nabla f_{i\alpha}^{[0]}=\frac{\partial f_{i\alpha}^{[0]}}{\partial n_i}\nabla n_i+\frac{\partial f_{i\alpha}^{[0]}}{\partial T}\nabla T+(\nabla \mathbf{v}_0)\cdot\nabla_{\mathbf{v}_0}f_{i\alpha}^{[0]}+\frac{\partial f_{i\alpha}^{[0]}}{\partial T_i^{(V)}}\nabla T_i^{(V)} \tag{5.92b}$$

and $\nabla_{\mathbf{v}} f_{i\alpha}^{[0]}$ is explicit. Substituting Eqs. (5.92) into Eq. (5.90) and rearranging terms we obtain

$$\begin{aligned} Df_{i\alpha}^{[0]}&=\frac{\partial f_{i\alpha}^{[0]}}{\partial n_i}\left(\frac{\partial n_i}{\partial t}+\mathbf{v}\cdot\nabla n_i\right)+\frac{\partial f_{i\alpha}^{[0]}}{\partial T}\left(\frac{\partial T}{\partial t}+\mathbf{v}\cdot\nabla T\right)\\ &\quad+\left[\frac{\partial \mathbf{v}_0}{\partial t}+\mathbf{v}\cdot(\nabla \mathbf{v}_0)\right]\cdot\nabla_{\mathbf{v}_0}f_{i\alpha}^{[0]}+\frac{\mathbf{X}_i}{m_i}\cdot\nabla_{\mathbf{v}}f_{i\alpha}^{[0]}+\frac{\partial f_{i\alpha}^{[0]}}{\partial T_i^{(V)}}\left(\frac{\partial T_i^{(V)}}{\partial t}+\mathbf{v}\cdot\nabla T_i^{(V)}\right) \end{aligned} \tag{5.93}$$

From Eq. (5.91) we have

$$\frac{\partial f_{i\alpha}^{[0]}}{\partial n_i}=\frac{f_{i\alpha}^{[0]}}{n_i} \tag{5.94a}$$

Using the following identities

$$\begin{aligned}\frac{\partial}{\partial T_i^{(V)}}\left[\frac{1}{\sum_\alpha \exp(-\varepsilon_{i\alpha}/kT_i^{(V)})}\right]&=-\frac{1}{[\sum_\alpha \exp(-\varepsilon_{i\alpha}/kT_i^{(V)})]^2}\frac{\partial}{\partial T_i^{(V)}}\sum_\alpha \exp\left(\frac{-\varepsilon_{i\alpha}}{kT_i^{(V)}}\right)\\ &=-\frac{1}{kT_i^{(V)}}\frac{\sum_\alpha \varepsilon_{i\alpha}\exp(-\varepsilon_{i\alpha}/kT_i^{(V)})}{[\sum_\alpha \exp(-\varepsilon_{i\alpha}/kT_i^{(V)})]^2}\end{aligned}$$

$$\frac{\partial}{\partial T}\left(\frac{m_i}{2\pi kT}\right)^{3/2}=-\frac{3}{2T}\left(\frac{m_i}{2\pi kT}\right)^{3/2}$$

$$\frac{\partial}{\partial T}\exp\left(-\frac{\frac{1}{2}m_i(\mathbf{v}-\mathbf{v}_0)^2}{kT}\right)=\frac{\frac{1}{2}m_i(\mathbf{v}-\mathbf{v}_0)^2}{kT^2}\exp\left(\frac{-\frac{1}{2}m_i(\mathbf{v}-\mathbf{v}_0)^2}{kT}\right]$$

$$\frac{\partial}{\partial T_j^{(V)}}\exp\left(\frac{-\varepsilon_{j\alpha}}{kT_j^{(V)}}\right)=\frac{\varepsilon_{j\alpha}}{kT_j^{(V)2}}\exp\left(\frac{-\varepsilon_{j\alpha}}{kT_j^{(V)}}\right)$$

we have, from Eq. (5.91):

$$\frac{\partial f_{i\alpha}^{[0]}}{\partial T} = \frac{f_{i\alpha}^{[0]}}{kT^2}\left(-\frac{3}{2}kT + \frac{1}{2}m_i\mathbf{V}^2\right) \tag{5.94b}$$

$$\begin{aligned}\nabla_{\mathbf{v}_0} f_{i\alpha}^{[0]} &= \frac{m_i}{kT}(\mathbf{v}-\mathbf{v}_0)f_{i\alpha}^{[0]} \\ &= \frac{m_i}{kT}\mathbf{V}f_{i\alpha}^{[0]}\end{aligned} \tag{5.94c}$$

$$\nabla_{\mathbf{v}} f_{i\alpha}^{[0]} = -\frac{m_i}{kT}\mathbf{V}f_{i\alpha}^{[0]} \tag{5.94d}$$

$$\frac{\partial f_{i\alpha}^{[0]}}{\partial T_i^{(\mathrm{V})}} = f_{i\alpha}^{[0]}\left[-\frac{1}{kT_i^{(\mathrm{V})2}}\frac{\sum_\alpha \varepsilon_{i\alpha}\exp(-\varepsilon_{i\alpha}/kT_i^{(\mathrm{V})})}{\sum_\alpha \exp(-\varepsilon_{i\alpha}/kT_i^{(\mathrm{V})})} + \frac{\varepsilon_{i\alpha}}{kT_i^{(\mathrm{V})2}}\right] \tag{5.94e}$$

Substituting Eqs. (5.94a)–(5.94e) into Eq. (5.93) we obtain

$$\begin{aligned}Df_{i\alpha}^{[0]} = f_{i\alpha}^{[0]}\Bigg\{&\frac{1}{n_i}\left(\frac{\partial n_i}{\partial t} + \mathbf{v}\cdot\nabla n_i\right) + \left(\mathbf{W}_i^2 - \frac{3}{2}\right)\frac{1}{T}\left(\frac{\partial T}{\partial t} + \mathbf{v}\cdot\nabla T\right) \\ &+ \frac{m_i}{kT}\left[\frac{\partial \mathbf{v}_0}{\partial t} + \mathbf{v}\cdot(\nabla\mathbf{v}_0)\right]\cdot\mathbf{V} - \frac{\mathbf{X}_i\cdot\mathbf{V}}{kT} \\ &+ \left(\frac{\varepsilon_{i\alpha} - \bar{\varepsilon}_i}{kT_i^{(\mathrm{V})}}\right)\frac{1}{T_i^{(\mathrm{V})}}\left(\frac{\partial T_i^{(\mathrm{V})}}{\partial t} + \mathbf{v}\cdot\nabla T_i^{(\mathrm{V})}\right)\Bigg\}\end{aligned} \tag{5.95}$$

where we have defined the dimensionless velocity

$$\mathbf{W}_i \equiv \left(\frac{m_i}{2kT}\right)^{1/2}\mathbf{V} \tag{5.96a}$$

and introduced

$$\begin{aligned}\bar{\varepsilon}_i &\equiv \frac{\sum_\alpha \varepsilon_{i\alpha}\exp(-\varepsilon_{i\alpha}/kT_i^{(\mathrm{V})})}{\sum_\alpha \exp(-\varepsilon_{i\alpha}/kT_i^{(\mathrm{V})})} \\ &= \frac{h\nu_i}{2} + \frac{E_i(T_i^{(\mathrm{V})})}{N_i}\end{aligned} \tag{5.96b}$$

from Eq. (1.63) since for an SHO:

$$\varepsilon_{i\alpha} = (\alpha + \tfrac{1}{2})h\nu_i \tag{1.63a}$$

The quantity $\bar{\varepsilon}_i$ is the average vibrational energy of a particle of mode-species i. We now use the equations of change for n_i, $E^{(\mathrm{TR})}(T)$, $E_i^{(\mathrm{V})}(T_i^{(\mathrm{V})})$, and $\mathbf{v}_0$ to remove the time derivatives from Eq. (5.95). We will particularize the equations of change to the quasi-equilibrium condition. We recall [Section 5.3.2] that our definition of quasi-equilibrium is that all species have the Boltzmann distribution of kinetic and vibrational energies. The difference between quasi-equilibrium and full equilibrium is that in quasi-equilibrium the different effective vibrational temperatures and the ambient temperature are not equal. Therefore, in particularizing the equations of change to the quasi-equilibrium condition we shall set the collision integrals on the right-hand sides of the equations of change equal to zero (since, at quasi-equilibrium, the number of collisions per second increasing n_i, for example, will equal the number of collisions per second decreasing n_i) *except* for the collision integrals representing V–V and V–T energy-exchange collision processes (since, by our definition of quasi-equilibrium, V–V and V–T processes will still occur).

The average variance velocity $\bar{\mathbf{V}}_j$, the pressure tensor $\hat{P}$, the translation–rotation energy flux $\mathbf{q}^{(TR)}$, and the vibrational energy fluxes $\mathbf{q}_j^{(V)}$ are to be evaluated at their quasi-equilibrium values in the equations of change, which we denote by the superscript [0]. This involves replacing $f_{i\alpha}$ by $f_{i\alpha}^{[0]}$ in Eqs. (5.5), (5.39), (5.60), and (5.49) to obtain:

$$\bar{\mathbf{V}}_j^{[0]} = \frac{1}{n_j}\sum_\alpha \int d^3V f_{j\alpha}^{[0]}(\mathbf{V}, \mathbf{x}, t)\mathbf{V} = 0 \tag{5.97a}$$

$$\hat{P}^{[0]} = \sum_{i\alpha} \int d^3V f_{i\alpha}^{[0]} m_i \mathbf{V}\mathbf{V} = p\hat{1} \tag{5.97b}$$

$$\mathbf{q}^{(TR)[0]} = \sum_{i\alpha} \int d^3V f_{i\alpha}^{[0]} \frac{1}{2} m_i V^2 \mathbf{V} = 0 \tag{5.97c}$$

$$\mathbf{q}_j^{(V)[0]} = \sum_\alpha \int d^3V f_{j\alpha}^{[0]} \varepsilon_{j\alpha} \mathbf{V} = 0 \tag{5.97d}$$

$f_{i\alpha}^{[0]}$ is an even function in $\mathbf{V}$ [Eq. (5.84)], and hence the integrals for $\bar{\mathbf{V}}_j^{[0]}$, $\mathbf{q}^{(TR)[0]}$, and $\mathbf{q}_j^{(V)[0]}$ are odd functions in $\mathbf{V}$ and are zero. Similarly the nondiagonal elements of the integral in the expression for the pressure tensor are odd functions of $\mathbf{V}$ and are zero. $\hat{1}$ is the unit tensor and p is identified with the gas pressure:

$$p = nkT \tag{5.98}$$

With the above considerations we have, from Eqs. (5.34), (5.45), (5.64), and (5.50), for the quasi-equilibrium equations of change (dropping the superscript [0] on all quantities in the following equations):

$$\frac{\partial n_j}{\partial t} + \mathbf{v}_0 \cdot \nabla n_j + n_j \nabla \cdot \mathbf{v}_0 = 0 \tag{5.99a}$$

$$\frac{\partial \mathbf{v}_0}{\partial t}+\mathbf{v}_0\cdot\nabla\mathbf{v}_0+\frac{1}{\rho}\nabla p-\frac{1}{\rho}\sum_i n_i\mathbf{X}_i=0 \tag{5.99b}$$

$$\frac{\partial T}{\partial t}+\mathbf{v}_0\cdot\nabla T+T\nabla\cdot\mathbf{v}_0+\frac{2}{(3+r)nk}(p\hat{1}:\nabla\mathbf{v}_0)=\frac{2Q^{(\mathrm{T})}}{(3+r)nk} \tag{5.99c}$$

where $Q^{(\mathrm{T})}$ is given by the right-hand side of Eq. (5.68);

$$\frac{\partial E_j^{(\mathrm{V})}}{\partial T_j^{(\mathrm{V})}}\left(\frac{\partial T_j^{(\mathrm{V})}}{\partial t}+\mathbf{v}_0\cdot\nabla T_j^{(\mathrm{V})}\right)+E_j^{(\mathrm{V})}\nabla\cdot\mathbf{v}_0=-Q_j^{(\mathrm{V})}/C_{(\mathrm{V})j} \tag{5.99d}$$

where we have obtained Eq. (5.99c) from Eq. (5.68) by using Eq. (5.58), which relates the translational energy density and the gas temperature, and using Eq. (5.99a). We set

$$r=\frac{1}{n}\sum_i R_i n_i \tag{5.100a}$$

and define the vibrational specific heat of mode-species j by

$$C_{(\mathrm{V})j}\equiv\partial E_j^{(\mathrm{V})}/\partial T_j^{(\mathrm{V})} \tag{5.100b}$$

Substituting Eqs. (5.99) into Eq. (5.95) we obtain

$$\begin{aligned}Df_{i\alpha}^{[0]}=f_{i\alpha}^{[0]}\Bigg\{&\frac{1}{n_i}\mathbf{V}\cdot\nabla n_i-\nabla\cdot\mathbf{v}_0+\left(\mathbf{W}_i^2-\frac{3}{2}\right)\frac{1}{T}\\&\times\left[\mathbf{V}\cdot\nabla T-\frac{2T}{3+T}\nabla\cdot\mathbf{v}_0+\frac{2Q^{(\mathrm{T})}}{(3+r)nk}\right]\\&+\left(\frac{\varepsilon_{i\alpha}-\bar{\varepsilon}_i}{kT_i^{(\mathrm{V})}}\right)\frac{1}{T_i^{(\mathrm{V})}}\left(\frac{-Q_j^{(\mathrm{V})}}{C_{(\mathrm{V})j}}+\mathbf{V}\cdot\nabla T_i^{(\mathrm{V})}-\frac{E_i^{(\mathrm{V})}}{C_{(\mathrm{V})i}}\nabla\cdot\mathbf{v}_0\right)\\&+\frac{m_i}{kT}\left[\mathbf{V}\cdot(\nabla\mathbf{v}_0)-\frac{1}{\rho}\nabla p+\frac{1}{\rho}\sum_i n_i\mathbf{X}_i\right]\cdot\mathbf{V}-\frac{\mathbf{X}_i\cdot\mathbf{V}}{kT}\Bigg\}\end{aligned} \tag{5.101}$$

where we have used Eq. (5.98) to collect the coefficients of $\nabla\cdot\mathbf{v}_0$. Reorganizing terms in Eq. (5.101) we obtain

$$\begin{aligned}Df_{i\alpha}^{[0]}=f_{i\alpha}^{[0]}\Bigg\{&\left(W_i^2-\frac{3}{2}\right)\mathbf{V}\cdot\nabla\ln T\\&+\frac{n}{n_i}\mathbf{V}\cdot\left(\frac{1}{n}\nabla n_i-\frac{n_i m_i}{\rho}\nabla\ln p+\frac{n_i m_i}{p\rho}\sum_j n_j\mathbf{X}_j-\frac{n_i\mathbf{X}_i}{p}\right)\\&+\left(2\mathbf{W}_i\mathbf{W}_i-\hat{1}\left[1+\frac{2}{3+r}\left(\mathbf{W}_i^2-\frac{3}{2}\right)+\frac{\varepsilon_{i\alpha}-\bar{\varepsilon}_i E_i^{(\mathrm{V})}}{kT_i^{(\mathrm{V})2}C_{(\mathrm{V})i}}\right)\right]:\nabla\mathbf{v}_0\Big)\\&+\left(\frac{\varepsilon_{i\alpha}-\bar{\varepsilon}_i}{kT_i^{(\mathrm{V})}}\right)\mathbf{V}\cdot\nabla\ln T_i^{(\mathrm{V})}+g_{i\alpha}\Bigg\}\end{aligned} \tag{5.102}$$

where we have defined

$$g_{i\alpha} \equiv \left(\mathbf{W}_i^2 - \frac{3}{2}\right)\frac{2Q^{(\mathrm{T})}}{(3+r)p} - \left(\frac{\varepsilon_{i\alpha} - \bar{\varepsilon}_i}{kT_i^{(\mathrm{V})}}\right)\frac{Q_i^{(\mathrm{V})}}{T_i^{(\mathrm{V})}C_{(\mathrm{V})i}} \tag{5.102a}$$

Now

$$\begin{aligned}
\frac{1}{n}\nabla n_i &= \nabla\left(\frac{n_i}{n}\right) - n_i \nabla\left(\frac{1}{n}\right) \\
&= \nabla\left(\frac{n_i}{n}\right) - n_i \nabla\left(\frac{kT}{p}\right) \\
&= \nabla\left(\frac{n_i}{n}\right) - \frac{n_i k}{p}\nabla T + \frac{n_i kT}{p^2}\nabla p \\
&= \nabla\left(\frac{n_i}{n}\right) - \frac{n_i kT}{p}\nabla \ln T + \frac{n_i kT}{p}\nabla \ln p \\
&= \nabla\left(\frac{n_i}{n}\right) - \frac{n_i}{n}\nabla \ln T + \frac{n_i}{n}\nabla \ln p
\end{aligned} \tag{5.103}$$

When Eq. (5.103) is substituted into Eq. (5.102) we have

$$\begin{aligned}
Df_{i\alpha}^{\{0\}} = f_{i\alpha}^{\{0\}} \Bigg(& \left(W_i^2 - \frac{5}{2}\right)\mathbf{V}\cdot\nabla \ln T \\
& + \frac{n}{n_i}\mathbf{V}\cdot\left[\nabla\left(\frac{n_i}{n}\right) + \left(\frac{n_i}{n} - \frac{n_i m_i}{\rho}\right)\nabla \ln p + \frac{n_i m_i}{p\rho}\sum_j n_j \mathbf{X}_j - \frac{n_i \mathbf{X}_i}{p}\right] + g_{i\alpha}\Big\} \\
& + \left(\frac{\varepsilon_{i\alpha} - \bar{\varepsilon}_i}{kT_i^{(\mathrm{V})}}\right)\mathbf{V}\cdot\nabla \ln T_i^{(\mathrm{V})} \\
& + \left\{2\mathbf{W}_i\mathbf{W}_i - \hat{1}\left[1 + \frac{2}{3+r}\left(\mathbf{W}_i^2 - \frac{3}{2}\right) + \frac{\varepsilon_{i\alpha} - \bar{\varepsilon}_i}{kT_i^{(\mathrm{V})2}}\cdot\frac{E_i^{(\mathrm{V})}}{C_{(\mathrm{V})i}}\right] : \nabla\mathbf{v}_0\right\}\Bigg)
\end{aligned} \tag{5.104}$$

With Eq. (5.104), the integral equation [Eq. (5.89b)] for the first-order perturbation $f_{i\alpha}^{\{1\}}$ becomes

$$\begin{aligned}
f_{i\alpha}^{\{0\}} \Big\{ & \left(W_i^2 - \frac{5}{2}\right)\mathbf{V}\cdot\nabla \ln T + \left(\frac{\varepsilon_{i\alpha} - \bar{\varepsilon}_i}{kT_i^{(\mathrm{V})}}\right)\mathbf{V}\cdot\nabla \ln T_i^{(\mathrm{V})} + \frac{n}{n_i}\mathbf{V}\cdot\Big[\nabla\left(\frac{n_i}{n}\right) \\
& + \left(\frac{n_i}{n} - \frac{n_i m_i}{\rho}\right)\nabla \ln p + \frac{n_i m_i}{p\rho}\sum_j n_j \mathbf{X}_j - \frac{n_i \mathbf{X}_i}{p}\Big] + 2\left(\mathbf{W}_i\mathbf{W}_i - \frac{1}{3}W_i^2\hat{1}\right) : \nabla\mathbf{v}_0 \\
& + \left[\frac{2}{3}W_i^2 - \frac{5+r}{3+r}\left(W_i^2 - \frac{3}{2}\right) - \left(\frac{\varepsilon_{i\alpha} - \bar{\varepsilon}_i}{kT_i^{(\mathrm{V})2}}\right)\frac{E_i^{(\mathrm{V})}}{C_{(\mathrm{V})i}}\right]\nabla\cdot\mathbf{v}_0 + g_{i\alpha}\Big\} \\
= \sum_{j\beta} & \left[J(f_{i\alpha}^{[0]}, f_{j\beta}^{[1]}) + J(f_{i\alpha}^{[1]}, f_{j\beta}^{[0]})\right] + J(f_{i\alpha}^{[1]}, f_e)
\end{aligned} \tag{5.105}$$

To simplify this expression, we define (see Andersen in Hochstim,[50] p. 37)

$$\mathbf{d}_i = \nabla\left(\frac{n_i}{n}\right) + \left(\frac{n_i}{n} - \frac{n_i m_i}{\rho}\right)\nabla \ln p + \frac{n_i m_i}{p\rho}\sum_j n_j \mathbf{X}_j - \frac{n_i \mathbf{X}_i}{p} \tag{5.106a}$$

$$\hat{b}_i = 2(\mathbf{W}_i\mathbf{W}_i - \tfrac{1}{3}W_i^2\hat{1}) \tag{5.106b}$$

$$b_{i\alpha}^* = \left[\frac{2}{3}W_i^2 - \frac{5+r}{3+r}\left(W_i^2 - \frac{3}{2}\right) - \left(\frac{\varepsilon_{i\alpha} - \bar{\varepsilon}_i}{kT_i^2}\right)\frac{E_i^{(\mathrm{V})}}{C_{(\mathrm{V})i}}\right] \tag{5.106c}$$

and we express $f_{i\alpha}^{[1]}$ in terms of a perturbation function $\phi_{i\alpha}$ defined by

$$f_{i\alpha}^{[1]}(\mathbf{v}, \mathbf{x}, t) \equiv f_{i\alpha}^{[0]}(\mathbf{v}, \mathbf{x}, t)\phi_{i\alpha}(\mathbf{v}, \mathbf{x}, t) \tag{5.107}$$

From Eqs. (5.106) and (5.107) we can rewrite Eq. (5.105) as a coupled system of integral equations for determining $\phi_{j\beta}$, that is,

$$\begin{aligned} f_{i\alpha}^{[0]}\Bigg[\left(W_i^2 - \frac{5}{2}\right)&\mathbf{V}\cdot\nabla \ln T + \frac{n}{n_i}\mathbf{V}\cdot\mathbf{d}_i \\ &+ \left(\frac{\varepsilon_{i\alpha} - \bar{\varepsilon}_i}{kT_i^{(\mathrm{V})}}\right)\mathbf{V}\cdot\nabla \ln T_i^{(\mathrm{V})} + \hat{b}_i : \nabla\mathbf{v}_0 + b_{i\alpha}^*\nabla\cdot\mathbf{v}_0 + g_{i\alpha}\Bigg] \\ &= \sum_{j\beta}[J(f_{i\alpha}^{[0]}, f_{j\beta}^{[0]}\phi_{j\beta}) + J(f_{i\alpha}^{[0]}\phi_{i\alpha}, f_{j\beta}^{[0]})] + J(f_{i\alpha}^{[0]}\phi_{i\alpha}, f_e) \end{aligned} \tag{5.108}$$

having invoked Eq. (5.89a) in evaluating the right-hand side. Equation (5.108) reduces to Andersen's form of the linearized Boltzmann equation [see after his Eq. (48)], when we assume only one vibrational mode. From Eq. (5.78) we note that Eq. (5.108) is an integral equation for $\phi_{i\alpha}$, and it is linear in $\phi_{i\alpha}$. Equation (5.108) is the form of the first-order perturbation of the Boltzmann equation that we shall solve in the next section.

Equation (5.108) is the same as Eq. (7.3-26) of Hirschfelder *et al.*[66] except that they do not consider vibrational excitation of molecules and do not have the b^* term.

5.3.4. Solution of the First-Order Perturbed Form of the Boltzmann Equation

The right-hand side of Eq. (5.108) is an integral of $\phi_{i\alpha}$ over $\mathbf{V}$ weighted with a quadratic function of $f_{i\alpha}^{[0]}$. Since both $f_{i\alpha}^{[0]}$ and the left-hand side of Eq. (5.108) only depend on space and time through the quantities $n_i(\mathbf{x}, t)$, $\mathbf{v}_0(\mathbf{x}, t)$, $T(\mathbf{x}, t)$, $T_j^{(\mathrm{V})}(\mathbf{x}, t)$, $p(\mathbf{x}, t)$, $\rho(\mathbf{x}, t)$, $\mathbf{X}_i(\mathbf{x}, t)$ and their space derivatives, it follows that $\phi_{i\alpha}$ has a form dictated by the structure of the left-hand side of Eq. (5.108), that is,

$$\phi_{i\alpha} = -(\mathbf{A}_{i\alpha} \cdot \nabla \ln T) - \hat{B}_{i\alpha} : (\nabla \mathbf{v}_0) + n \sum_j \mathbf{C}^j_{i\alpha} \cdot \mathbf{d}_j - (\mathbf{A}^{(V)}_{i\alpha} \cdot \nabla \ln T^{(V)}_i) - F_{i\alpha}$$
$$+ D_{i\alpha}(W_i) \nabla \cdot \mathbf{v}_0 \tag{5.109}$$

where $\mathbf{A}_{i\alpha}$, $\hat{B}_{i\alpha}$, $\mathbf{C}^j_{i\alpha}$, $\mathbf{A}^{(V)}_{i\alpha}$, and $F_{i\alpha}$ are functions of the dimensionless velocity $\mathbf{W}_i$, the local composition $n_i(\mathbf{x}, t)$, and the local temperatures $T(\mathbf{x}, t)$ and $T^{(V)}_i(\mathbf{x}, t)$.

Now $\mathbf{A}_{i\alpha}$, $\mathbf{A}^{(V)}_{i\alpha}$, and $\mathbf{C}^j_{i\alpha}$ are vector functions of W_i, and since a vector function of a vector variable must be the vector variable itself multiplied by some scalar function of the absolute value of the vector variable, we can write

$$\mathbf{A}_{i\alpha} = \mathbf{W}_i A_{i\alpha}(|\mathbf{W}_i|) \tag{5.110a}$$

$$\mathbf{C}^j_{i\alpha} = \mathbf{W}_i C^j_{i\alpha}(|\mathbf{W}_i|) \tag{5.110b}$$

$$\mathbf{A}^{(V)}_{i\alpha} = \mathbf{W}_i A^{(V)}_{i\alpha}(|\mathbf{W}_i|) \tag{5.110c}$$

Similarly it may be shown (Hirschfelder *et al.*,[66] pages 470–471) that the only form for the tensor $\hat{B}$, which is consistent with Eqs. (5.108) and (5.109), is

$$\hat{B}_{i\alpha} = (\mathbf{W}_i \mathbf{W}_i - \tfrac{1}{3} W_i^2) B_{i\alpha}(|\mathbf{W}_i|) \tag{5.110d}$$

where $B_{i\alpha}(W_i)$ is some scalar function of $|\mathbf{W}_i|$. From Eqs. (5.109) and (5.110) we obtain an expression for $\phi_{i\alpha}$:

$$\phi_{i\alpha} = -A_{i\alpha}(W_i)\mathbf{W}_i \cdot \nabla \ln T - A^{(V)}_{i\alpha}(W_i)\mathbf{W}_i \cdot \nabla \ln T^{(V)}_i + D_{i\alpha}(W_i) \nabla \cdot \mathbf{v}_0$$
$$-B_{i\alpha}(W_i)(\mathbf{W}_i \mathbf{W}_i - \tfrac{1}{3} W_i^2) : \nabla \mathbf{v}_0 + n \sum_j C^j_{i\alpha}(W_i)\mathbf{W}_i \cdot \mathbf{d}_j - F_{i\alpha} \tag{5.111}$$

5.3.5. *The Average Variance Velocity,* $\bar{\mathbf{V}}_i$

The average variance velocity is defined by Eq. (5.5) as

$$n_i(\mathbf{x}, t)\bar{\mathbf{V}}_i \equiv \sum_\alpha \int f_{i\alpha}(\mathbf{V}, \mathbf{x}, t)\mathbf{V}(\mathbf{x}, t)\, d^3V \tag{5.112}$$

when we change integration variables using Eq. (5.4). From $f = f^{[0]} + \xi f^{[1]}$ and Eq. (5.107), we have

$$f_{i\alpha} = f^{[0]}_{i\alpha}(1 + \phi_{i\alpha}) \tag{5.86a}$$

which, when used with Eq. (5.97a), gives

$$\bar{\mathbf{V}}_i = \frac{1}{n_i} \sum_\alpha \int d^3V f^{[0]}_{i\alpha} \phi_{i\alpha} \mathbf{V} \tag{5.113}$$

We recall that $\xi=0$ gives the equilibrium solution, while $\xi=1$ gives the physical solution. When the expression for $\phi_{i\alpha}$, Eq. (5.111), is substituted into Eq. (5.113) we obtain

$$\bar{\mathbf{V}}_i=\frac{1}{n_i}\sum_\alpha\int d^3V f_{i\alpha}^{[0]}\mathbf{V}[-A_{i\alpha}(W_i)\mathbf{W}_i\cdot\nabla\ln T-A_{i\alpha}^{(V)}(W_i)\mathbf{W}_i\cdot\nabla\ln T_i^{(V)}F_{i\alpha}$$
$$\times D_{i\alpha}(W_i)\nabla\cdot\mathbf{v}_0-B_{i\alpha}(W_i)(\mathbf{W}_i\mathbf{W}_i-\tfrac{1}{3}W_i^2):\nabla\mathbf{v}_0+n\sum_j C_{i\alpha}^j(W_i)\mathbf{W}_i\cdot\mathbf{d}_j] \tag{5.114}$$

We recall that $\mathbf{W}_i$ is directly proportional to $\mathbf{V}$ [Eq. (5.96a)] and that $A_{i\alpha}$, $A_{i\alpha}^{(V)}$, $B_{i\alpha}$, $C_{i\alpha}^j$ and $F_{i\alpha}$, as functions of the scalar W_i, are even functions of $\mathbf{W}_i$. Therefore three of the last terms in the integral of Eq. (5.114) are an odd function of $\mathbf{V}$, and are zero. To evaluate the other terms of Eq. (5.114) we first prove that, for any vector $\mathbf{r}$:

$$\int F(|\mathbf{r}|)\mathbf{r}\mathbf{r}\,d^3r=\frac{1}{3}\hat{1}\int F(|\mathbf{r}|)r^2\,d^3r$$

Denoting the tensor components by the indices α, β we have

$$\left(\int F(|\mathbf{r}|)\mathbf{r}\mathbf{r}\,d^3r\right)_{\alpha\beta}=\int F(|\mathbf{r}|)r_\alpha r_\beta\,d^3r$$
$$=\begin{cases}0, & \alpha\neq\beta\\ \int F(|\mathbf{r}|)r_\alpha^2\,d^3r, & \alpha=\beta\end{cases}$$

the integral being zero for $\alpha\neq\beta$ since $F(|\mathbf{r}|)$ must be an even function of the components of $\mathbf{r}$, and hence the integral is an odd function of the components of $\mathbf{r}$. If we now define I_α by

$$I_\alpha=\int F(|\mathbf{r}|)r_\alpha^2\,d^3r$$

then we have

$$I_1=I_2=I_3$$

and

$$I_1+I_2+I_3=\int F(|\mathbf{r}|)r^2\,d^3r=3I_1=3I_2=3I_3$$

and hence

$$\left(\int F(|\mathbf{r}|)\mathbf{r}\mathbf{r}\,d^3r\right)_{\alpha\beta}=\frac{1}{3}\delta_{\alpha\beta}\int F(|\mathbf{r}|)r^2\,d^3r \tag{5.115}$$

Applying the above identity to Eq. (5.114) we obtain

$$\bar{\mathbf{V}}_i = \frac{n}{n_i}\sum_\alpha \int d^3V f_{i\alpha}^{[0]}\left(\frac{2kT}{m_i}\right)^{1/2}\mathbf{W}_i\Big[-A_{i\alpha}(W_i)\mathbf{W}_i\cdot\nabla\ln T$$

$$-A_{i\alpha}^{(V)}(W_i)\mathbf{W}_i\cdot\nabla\ln T_i + n\sum_j C_{i\alpha}^j(W_i)\mathbf{W}_i\cdot\mathbf{d}_j\Big]$$

$$=\frac{1}{n_i}\frac{1}{3}\left(\frac{2kT}{m_i}\right)^{1/2}\sum_\alpha\int d^3V f_{i\alpha}^{[0]}W_i^2\Big[-A_{i\alpha}(W_i)\nabla\ln T$$

$$-A_{i\alpha}^{(V)}(W_i)\nabla\ln T_i^{(V)} + n\sum_j C_{i\alpha}^j(W_i)\mathbf{d}_j\Big]$$

$$=-\frac{1}{3}\left(\frac{2kT}{m_i}\right)^{1/2}\frac{1}{n_i}(\nabla\ln T)\sum_\alpha\int d^3V f_{i\alpha}^{[0]}W_i^2A_{i\alpha}(W_i)$$

$$+\frac{1}{3}\left(\frac{2kT}{m_i}\right)^{1/2}\frac{n}{n_i}\sum_j\mathbf{d}_j\sum_\alpha\int d^3V f_{i\alpha}^{[0]}W_i^2C_{i\alpha}^j(W_i)$$

$$-\frac{1}{3}\left(\frac{2kT}{m_i}\right)^{1/2}\frac{1}{n_i}(\nabla\ln T_i^{(V)})\sum_\alpha\int d^3V f_{i\alpha}^{[0]}W_i^2A_{i\alpha}^{(V)}(W_i)$$

$$=-\frac{1}{m_in_i}D_i^{(T)}\nabla\ln T + \frac{n^2}{n_i\rho}\sum_j m_jD_{ij}\mathbf{d}_j - \frac{1}{m_in_i}D_i^{(V)}\nabla\ln T_i^{(V)} \tag{5.116}$$

where we have defined quantities

$$D_i^{(T)} = \frac{m_i}{3}\left(\frac{2kT}{m_i}\right)^{1/2}\sum_\alpha\int d^3V f_{i\alpha}^{[0]}W_i^2A_{i\alpha}(W_i) \tag{5.117a}$$

$$D_{ij} = \frac{\rho}{3nm_j}\left(\frac{2kT}{m_i}\right)^{1/2}\sum_\alpha\int d^3V f_{i\alpha}^{[0]}W_i^2C_{i\alpha}^j(W_i) \tag{5.117b}$$

$$D_i^{(V)} = \frac{m_i}{3}\left(\frac{2kT}{m_i}\right)^{1/2}\sum_\alpha\int d^3V f_{i\alpha}^{[0]}W_i^2A_{i\alpha}^{(V)}(W_i) \tag{5.117c}$$

which are called[66] the *multicomponent thermal diffusion coefficients*, the *multicomponent diffusion coefficients*, and the *vibrational diffusion coefficient*, respectively. These diffusion coefficients may be measured experimentally.[67] They may also be evaluated using expansions for $A_{i\alpha}$ and $C_{i\alpha}^j$ in terms of Sonine polynomials (Hirschfelder *et al.*,[66] pp. 471–491). Equations (5.116)–(5.118) are the same as the equations given by Andersen [Hochstim,[50] p. 51, Eqs. (49) et seq.] with the exception that Andersen only considers one vibrational mode, and his diffusion coefficient

D_j is related to our D_{ij} by

$$D_j \equiv \frac{m_i m_j n^2 D_{ij}}{\rho} \tag{5.118}$$

Thomson[68] has developed a numerical method for calculating these transport coefficients (see Section 5.7).

5.3.6. The Pressure Tensor, $\hat{P}$

The pressure tensor $\hat{P}(\mathbf{x}, t)$ is defined by Eq. (5.39) as

$$\hat{P}(\mathbf{x}, t) = \sum_{i\alpha} \int d^3V f_{i\alpha}(\mathbf{V}, \mathbf{x}, t) m_i \mathbf{V}\mathbf{V} \tag{5.119}$$

If we now set $f_{i\alpha} = f_{i\alpha}^{[0]}(1 + \phi_{i\alpha})$ and use Eq. (5.97b) we obtain

$$\hat{P} = p\hat{1} + \sum_{i\alpha} \int d^3V f_{i\alpha}^{[0]} \phi_{i\alpha} m_i \mathbf{V}\mathbf{V} \tag{5.120}$$

and substituting for $\phi_{i\alpha}$ as given by Eq. (5.111) we have

$$\hat{P} = p\hat{1} + \sum_{i\alpha} \int d^3V f_{i\alpha}^{[0]} m_i \mathbf{V}\mathbf{V} \Big[-A_{i\alpha}(W_i)\mathbf{W}_i \cdot \nabla \ln T - A_{i\alpha}^{(V)}(W_i)\mathbf{W}_i \cdot \nabla \ln T_i^{(V)}$$
$$- B_{i\alpha}(W_i)(\mathbf{W}_i\mathbf{W}_i - \tfrac{1}{3}W_i^2) : \nabla \mathbf{v}_0 - F_{i\alpha} + n \sum_j C_{i\alpha}^j(W_i)\mathbf{W}_i \cdot \mathbf{d}_j$$
$$+ D_{i\alpha}(W_i)\nabla \cdot \mathbf{v}_0 \Big] \tag{5.121a}$$

The first, second, and fifth terms in the above integral are odd functions of $\mathbf{W}_i$, and hence are zero. For the fourth term, we have, using Eq. (3.11):

$$-\sum_{i\alpha} \int d^3V f_{i\alpha}^{[0]} m_i \mathbf{V}\mathbf{V} F_{i\alpha} = -\frac{1}{3} \sum_{i\alpha} \int d^3V f_{i\alpha}^{[0]} m_i V^2 \hat{1} F_{i\alpha} \equiv -\zeta \hat{1}$$
$$\sum_{i\alpha} \int d^3V f_{i\alpha}^{[0]} m_i \mathbf{V}\mathbf{V} D_{i\alpha} \equiv -\eta_B \hat{1} \tag{5.121b}$$

which defines ζ, and the bulk modulus coefficient η_B.

In order to evaluate the second term we must first prove some integral identities. Consider the vector $\mathbf{V}$, with magnitude $V = |\mathbf{V}|$. Denote the Cartesian coordinates of $\mathbf{V}$ by V_x, V_y, V_z, and the polar coordinates by V, θ, and ϕ such that:

$$V_x = V \sin\theta \sin\phi$$
$$V_y = V \sin\theta \cos\phi$$
$$V_z = V \cos\theta$$

Now consider the integral $\int F(V)V_z^4\, d^3V$, where F is any function of V, the magnitude of $\mathbf{V}$, only. We have

$$\int F(V)V_z^4\, d^3V = \int F(V)V^4 \cos^4\theta V^2 \sin\theta\, dV\, d\theta\, d\phi$$
$$= \int_0^\infty F(V)V^4V^2\, dV \int_0^\pi \cos^4\theta \sin\theta\, d\theta \int_0^{2\pi} d\phi$$

Now

$$\int_{\theta=0}^{\pi} \cos^4\theta \sin\theta\, d\theta = \int_{\theta=0}^{\pi} \cos^4\theta\, d(-\cos\theta) = -\int_{X=1}^{-1} X^4\, dX = \frac{1}{5}\int_{\theta=0}^{\pi} \sin\theta\, d\theta$$

so that

$$\int F(V)V_z^4\, d^3V = \frac{1}{5}\int F(V)V^4V^2 \sin\theta\, dV\, d\theta\, d\phi = \frac{1}{5}\int F(V)V^4\, d^3V \tag{5.122}$$

and in a similar fashion it may be proved that

$$\int F(V)V_x^4\, d^3V = \int F(V)V_y^4\, d^3V = \frac{1}{5}\int F(V)V^4\, d^3V \tag{5.123}$$

$$\int F(V)V_x^2V_y^2\, d^3V = \int F(V)V_x^2V_z^2\, d^3V = \int F(V)V_y^2V_z^2\, d^3V$$
$$= \frac{1}{15}\int F(V)V^4\, d^3V \tag{5.124}$$

With Eqs. (5.122)–(5.124) we can evaluate the components P_{ab} of the pressure tensor $\hat{P}$ from Eq. (5.121a):

$$P_{ab} = (p-\zeta)\delta_{ab} - \sum_{i\alpha}\frac{m_i^2}{2kT}\int d^3V\left[V_aV_bB_{i\alpha}(W_i)\sum_{c,d=1}^{3}\left(V_cV_d - \frac{1}{3}\delta_{cd}\right)\nabla_c v_{0d}\right]$$
$$-\eta_B \nabla\cdot\mathbf{v}_0\delta_{ab} \tag{5.125}$$

having used Eq. (5.96a) to relate W_i to $\mathbf{V}$, and using the fact that the integral is zero if it is an odd function of any of the Cartesian coordinates V_a of $\mathbf{V}$, we have, for example,

$$P_{11} = p - \zeta - \sum\frac{m_i^2}{2kT}\int d^3V\Big\{V_1^2B_{i\alpha}(W_i)$$
$$\times\left[V_1^2\nabla_1 v_{01} + V_2^2\nabla_2 v_{02} + V_3^2\nabla_3 v_{03} - \frac{1}{3}V^2\nabla\cdot\mathbf{v}_0\right]\Big\} - \eta_B\nabla\cdot\mathbf{v}_0$$

$$= p - \zeta - \sum_{i\alpha}\frac{m_i^2}{2kT}\int d^3V\Big\{B_{i\alpha}(W_i)$$

$$\times\left[\frac{1}{5}\nabla_1 v_{01}+\frac{1}{15}\nabla_2 v_{02}+\frac{1}{15}\nabla_3 v_{03}-\frac{1}{3}\left(\frac{1}{5}+\frac{2}{15}\right)\boldsymbol{\nabla}\cdot\mathbf{v}_0\right]\Bigg\}-\eta_B\boldsymbol{\nabla}\cdot\mathbf{v}_0$$

$$=p-\zeta-\left(\sum_{i\alpha}\frac{m_i^2}{2kT}\int d^3V B_{i\alpha}(W_i)V^4\right)\left(\frac{2}{15}\nabla_1 v_{01}-\frac{2}{45}\boldsymbol{\nabla}\cdot\mathbf{v}_0\right)-\eta_B\boldsymbol{\nabla}\cdot\mathbf{v}_0 \tag{5.126}$$

$$P_{12}=-\sum_{i\alpha}\frac{m_i^2}{2kT}\int d^3V B_{i\alpha}(W_i)V_1^2V_2^2(\nabla_1 v_{02}+\nabla_2 v_{01})$$

$$=-\left(\sum_{i\alpha}\frac{m_i^2}{2kT}\int d^3V B_{i\alpha}(W_i)V^4\right)\frac{1}{15}(\nabla_1 v_{02}+\nabla_2 v_{01}) \tag{5.127}$$

From Eqs. (5.126) and (5.127) we may write for the components of the pressure tensor:

$$P_{ab}=(p-\zeta)\delta_{ab}-\left(\frac{2}{15}\sum_{i\alpha}\frac{m_i^2}{2kT}\int d^3V B_{i\alpha}(W_i)V^4\right)\times\left[\frac{1}{2}(\nabla_a v_{0b}+\nabla_b v_{0a})-\frac{1}{3}\delta_{ab}\boldsymbol{\nabla}\cdot\mathbf{v}_0\right] \tag{5.128}$$

or

$$\hat{P}=(p-\zeta)\hat{1}-2\eta\hat{S}-\eta_B\boldsymbol{\nabla}\cdot\mathbf{v}_0\hat{1} \tag{5.129}$$

which agrees with Hirschfelder *et al.*[66] [Eq. (7.6-29)], except for our extra tensor ζ, and where we have set

$$\eta=\frac{1}{15}\sum_{i\alpha}\frac{m_i^2}{2kT}\int d^3V B_{i\alpha}(W_i)V^4 \tag{5.130}$$

$$S_{ab}=\frac{1}{2}(\nabla_a v_{0b}+\nabla_b v_{0a})-\frac{1}{3}\delta_{ab}\boldsymbol{\nabla}\cdot\mathbf{v}_0 \tag{5.131}$$

$\hat{S}$ is known as the rate of shear tensor and η is the coefficient of shear viscosity. The coefficient of shear viscosity may be measured experimentally or calculated using expansions for $B_{i\alpha}$ in terms of Sonine polynomials (Hirschfelder *et al.*,[66] pp. 471–491).

5.3.7. *The Translation–Rotation Energy Flux,* $\mathbf{q}^{(\mathrm{TR})}$

The translation–rotation energy flux $\mathbf{q}^{(\mathrm{TR})}(\mathbf{x}, t)$ is given by Eq. (5.60) as

$$\mathbf{q}^{(\mathrm{TR})}(\mathbf{x}, t)=\sum_{i\alpha}\int d^3V f_{i\alpha}(\mathbf{V}, \mathbf{x}, t)\frac{1}{2}m_iV^2\mathbf{V} \tag{5.60a}$$

Setting $f_{i\alpha}=f_{i\alpha}^{[0]}(1+\phi_{i\alpha})$ and using Eq. (5.97c) we obtain

$$\mathbf{q}^{(\mathrm{TR})}(\mathbf{x}, t)=\sum_{i\alpha}\int d^3V f_{i\alpha}^{[0]}\phi_{i\alpha}\frac{1}{2}m_iV^2\mathbf{V} \tag{5.132}$$

Substituting for $\phi_{i\alpha}$ [Eq. (5.111)] into Eq. (5.132) we have

$$\mathbf{q}^{(\mathrm{TR})}=\sum_{i\alpha}\int d^3V f_{i\alpha}^{[0]}\frac{1}{2}m_iV^2\mathbf{V}\Big[-A_{i\alpha}(W_i)\mathbf{W}_i\cdot\nabla\ln T$$

$$-A_{i\alpha}^{(\mathrm{V})}(W_i)\mathbf{W}_i\cdot\nabla\ln T_i^{(\mathrm{V})}-B_{i\alpha}(W_i)\Big(\mathbf{W}_i\mathbf{W}_i-\frac{1}{3}W_i^2\Big):\nabla\mathbf{v}_0$$

$$+D_{i\alpha}\nabla\cdot\mathbf{v}_0-F_{i\alpha}+n\sum_j C_{i\alpha}^j(W_i)\mathbf{W}_i\cdot\mathbf{d}_j\Big]$$

$$=\sum_{i\alpha}\int d^3V f_{i\alpha}^{[0]}\frac{1}{2}m_iV^2\mathbf{V}\Big[-A_{i\alpha}(W_i)\mathbf{W}_i\cdot\nabla\ln T$$

$$-A_{i\alpha}^{(\mathrm{V})}(W_i)\mathbf{W}_i\cdot\nabla\ln T_i^{(\mathrm{V})}+n\sum_j C_{i\alpha}^j(W_i)\mathbf{W}_i\cdot\mathbf{d}_j\Big] \tag{5.133}$$

the B, D, and F terms in the integral being zero since they are odd functions of $\mathbf{V}$. We now follow Hirschfelder *et al.*[66] and decompose the integral in Eq. (5.133) into two parts by writing

$$V^2\rightarrow\left(V^2-\frac{5}{2}\frac{2kT}{m_i}\right)+\frac{5}{2}\frac{2kT}{m_i} \tag{5.134}$$

Then

$$\mathbf{q}^{(\mathrm{TR})}=\sum_{i\alpha}\int d^3V f_{i\alpha}^{[0]}\frac{5}{2}kT\mathbf{V}\Big(-A_{i\alpha}\mathbf{W}_i\cdot\nabla\ln T$$

$$-A_{i\alpha}^{(\mathrm{V})}\mathbf{W}_i\cdot\nabla\ln T_i^{(\mathrm{V})}+n\sum_j C_{i\alpha}^j\mathbf{W}_i\cdot\mathbf{d}_j\Big)$$

$$+\sum_{i\alpha}\int d^3V f_{i\alpha}^{[0]}\frac{1}{2}m_i\left(V^2-\frac{5}{2}\frac{2kT}{m_i}\right)\mathbf{V}\mathbf{V}\cdot\Big(-A_{i\alpha}\nabla\ln T$$

$$-A_{i\alpha}^{(\mathrm{V})}\nabla\ln T_i^{(\mathrm{V})}+n\sum_j C_{i\alpha}^j\mathbf{d}_j\Big)\left(\frac{m_i}{2kT}\right)^{1/2} \tag{5.135}$$

The first sum of Eq. (5.135) can be expressed in terms of the mean variance velocity $\bar{\mathbf{V}}_i$ by Eq. (5.114), and we may apply the identity (5.115) to the second sum to obtain

$$\mathbf{q}^{(\mathrm{TR})}=\frac{5}{2}kT\sum_i n_i\bar{V}_i$$

$$+\frac{1}{3}\sum_{i\alpha}\int d^3V f_{i\alpha}^{[0]}\frac{m_i}{2}\left(V^2-\frac{5}{2}\frac{2kT}{m_i}\right)V^2\Big(-A_{i\alpha}\nabla\ln T+n\sum_j C_{i\alpha}^j\mathbf{d}_j$$

$$-A_{i\alpha}^{(V)}\nabla \ln T_i^{(V)}\Big)\left(\frac{m_i}{2kT}\right)^{1/2}$$

$$=\frac{5}{2}kT\sum_i n_i\bar{\mathbf{V}}_i+\sum_{i\alpha}\frac{m_i}{6}\left(\frac{2kT}{m_i}\right)^{3/2}\int d^3V f_{i\alpha}^{[0]}\left(W_i^2-\frac{5}{2}\right)W_i^2$$

$$\times\Big(-A_{i\alpha}\nabla \ln T+n\sum_j C_{i\alpha}^j\mathbf{d}_j-A_{i\alpha}^{(V)}\nabla \ln T_i^{(V)}\Big)$$

$$=\frac{5}{2}kT\sum_i n_i\bar{\mathbf{V}}_i-\lambda'\nabla T-nkT\sum_i\frac{1}{n_im_i}D_i^{(T)*}\mathbf{d}_i-\sum_i\lambda_i^{(VT)}\nabla T_i^{(V)} \qquad (5.136)$$

where we have defined

$$\lambda'=-\frac{k}{3}\sum_{i\alpha}\left(\frac{2kT}{m_i}\right)^{1/2}\int f_{i\alpha}^{[0]}\left(\frac{5}{2}-W_i^2\right)W_i^2A_{i\alpha}(W_i)\,d^3V \qquad (5.137a)$$

$$\lambda_i^{(VT)}=-\frac{kT_i}{3T}\sum_{\alpha}\left(\frac{2kT}{m_i}\right)^{1/2}\int f_{i\alpha}^{[0]}\left(\frac{5}{2}-W_i^2\right)W_i^2A_{i\alpha}^{(V)}(W_i)\,d^3V \qquad (5.137b)$$

$$D_i^{(T)*}=-\frac{n_im_i}{3}\sum_{j\alpha}\left(\frac{2kT}{m_j}\right)^{1/2}\int f_{j\alpha}^{[0]}\left(\frac{5}{2}-W_j^2\right)W_j^2C_{j\alpha}^i\,d^3V \qquad (5.137c)$$

Hirschfelder *et al.*[66] show that

$$D_i^{(T)*}=D_i^{(T)} \qquad (5.138)$$

where $D_i^{(T)}$ is the multicomponent thermal diffusion coefficient given by Eq. (5.117a). $\lambda_i^{(VT)}$ is the coefficient of vibrational–translational conductivity for the flow of vibrational energy of the ith vibrational mode. Equations (5.136)–(5.138) are the same as those given by Andersen [see Hochstim[50]].

Equation (5.116) can be rewritten in the form

$$\mathbf{d}_j=\sum_i\mathscr{D}_{ji}^{-1}\frac{n_i\rho}{n^2}\left[\bar{\mathbf{V}}_i+\frac{1}{m_in_i}(D_i^{(T)}\nabla \ln T+D_i^{(V)}\nabla \ln T_i^{(V)})\right]$$

where the $\mathscr{D}_{ji}^{-1}$ are the elements of the matrix $\mathscr{D}^{-1}$; the elements $\mathscr{D}_{ij}$ of the matrix $\mathscr{D}$ are given by

$$\mathscr{D}_{ij}=m_jD_{ij} \qquad (5.139)$$

If the above expression for $\mathbf{d}_j$ is substituted into Eq. (5.136), then we obtain

$$\mathbf{q}^{(TR)}=\frac{5}{2}kT\sum_i[(1-\alpha_i)n_i\bar{\mathbf{V}}_i]-\lambda\nabla T-\sum_i\bar{\lambda}_i^{(VT)}\nabla T_i^{(V)} \qquad (5.136a)$$

where

$$\alpha_i = \frac{2}{5}\sum_j \frac{D_j^{(T)}\mathscr{D}_{ji}^{-1}\rho}{nn_jm_j} \tag{5.136b}$$

$$\lambda = \lambda' + \frac{\rho k}{n}\sum_j\sum_i \frac{D_j^{(T)}\mathscr{D}_{ji}^{-1}D_i^{(T)}}{n_jm_jm_i} \tag{5.136c}$$

$$\bar{\lambda}_i^{(VT)} = \lambda_i^{(VT)} + \frac{T\rho k}{nT_i^{(V)}}\sum_j \frac{D_j^{(T)}\mathscr{D}_{ji}^{-1}D_i^{(V)}}{n_jm_jm_i} \tag{5.136d}$$

Hirschfelder *et al.*[66] (pp. 483 and 491) define λ to be the usual coefficient of thermal conductivity, and state that the factor that adds to λ' to give λ is a small correction.

5.3.8. *The Vibrational Energy Flux Vector,* $\mathbf{q}_j^{(V)}$

The vibrational energy flux vector is given by Eq. (5.49) as

$$\mathbf{q}_j^{(V)} = \sum_\alpha \int d^3V f_{j\alpha}\varepsilon_{j\alpha}\mathbf{V} \tag{5.49a}$$

When we use the approximation $f_{j\alpha} = f_{j\alpha}^{[0]}(1+\phi_{j\alpha})$, we have

$$\mathbf{q}_j^{(V)} = \sum_\alpha \int d^3V f_{j\alpha}^{[0]}\phi_{j\alpha}\varepsilon_{j\alpha}\mathbf{V} \tag{5.140}$$

the term of zero order in $\phi_{j\alpha}$ vanishing because the integral is over an odd function of $\mathbf{V}$. We now substitute the expression for $\phi_{j\alpha}$ given by Eq. (5.111), and note that the term involving $B_{i\alpha}(W_i)$, $D_{i\alpha}$ and $F_{i\alpha}$ vanishes when the integral is performed, leading to

$$\mathbf{q}_j^{(V)} = \sum_\alpha \int d^3V \varepsilon_{j\alpha}\mathbf{V}\Big[-A_{j\alpha}(W_j)\mathbf{W}_j\cdot\nabla\ln T + n\sum_i C_{j\alpha}^i(W_j)\mathbf{W}_j\cdot\mathbf{d}_j$$
$$-A_{j\alpha}^{(V)}(W_j)\mathbf{W}_j\cdot\nabla\ln T_j^{(V)}\Big] \tag{5.141}$$

Decomposing this integral into two parts by writing

$$\varepsilon_{j\alpha} \to (\varepsilon_{j\alpha} - \bar{\varepsilon}_j) + \bar{\varepsilon}_j$$

where $\bar{\varepsilon}_j$ is given by Eq. (5.96b), we have

$$\mathbf{q}_j^{(V)} = \sum_\alpha \int d^3V f_{j\alpha}^{[0]}\bar{\varepsilon}_j\mathbf{V}\Big(-A_{j\alpha}\mathbf{W}_j\cdot\nabla\ln T + n\sum_i C_{j\alpha}^i\mathbf{W}_j\cdot\mathbf{d}_j$$
$$-A_{j\alpha}^{(V)}\mathbf{W}_j\cdot\nabla\ln T_j^{(V)}\Big)$$
$$+\sum_\alpha \int d^3V f_{j\alpha}^{[0]}(\varepsilon_{j\alpha} - \bar{\varepsilon}_j)\mathbf{V}\Big(-A_{j\alpha}\mathbf{W}_j\cdot\nabla\ln T + n\sum_i C_{j\alpha}^i\mathbf{W}_j\cdot\mathbf{d}_j$$

$$- A_{j\alpha}^{(V)} \mathbf{W}_j \cdot \nabla \ln T_j^{(V)} \Big)$$

$$= \bar{\varepsilon}_j n_j \bar{\mathbf{V}}_j - \frac{1}{3} \sum_\alpha \int d^3 V f_{j\alpha}^{[0]} (\varepsilon_{j\alpha} - \bar{\varepsilon}_j) W_j^2 \left(\frac{2kT}{m_i} \right)^{1/2} \Big(- A_{j\alpha} \frac{1}{T} \nabla T$$

$$- A_{j\alpha}^{(V)} \frac{1}{T_j^{(V)}} \nabla T_j^{(V)} + n \sum_i C_{j\alpha}^i \mathbf{d}_i \Big)$$

$$= \bar{\varepsilon}_j n_j \bar{\mathbf{V}}_j - \lambda_j^{(VT)} \nabla T - \sum_i D_{ij}^{(V)} \mathbf{d}_i - \lambda_j^{(V)} \nabla T_j^{(V)} \tag{5.142}$$

where we have used Eq. (5.114) for the mean variance velocity $\bar{\mathbf{V}}_j$ and defined

$$\lambda_j^{(V)} = -\frac{1}{3} k \sum_\alpha \left(\frac{2kT}{m_j} \right)^{1/2} \int d^3 V f_{j\alpha}^{[0]} \left(\frac{\bar{\varepsilon}_j - \varepsilon_{j\alpha}}{kT_j} \right) A_{j\alpha}^{(V)} W_j^2 \tag{5.143a}$$

$$\lambda_j^{(VT)} = -\frac{1}{3} k \sum_\alpha \left(\frac{2kT}{m_j} \right)^{1/2} \int d^3 V f_{j\alpha}^{[0]} \left(\frac{\bar{\varepsilon}_j}{kT} - \frac{\varepsilon_{j\alpha}}{kT} \right) A_{j\alpha} W_j^2 \tag{5.143b}$$

and

$$D_{ij}^{(V)} = -\frac{1}{3} k \sum_\alpha \left(\frac{2kT}{m_j} \right)^{1/2} nT \int d^3 V f_{j\alpha}^{[0]} \left(\frac{\bar{\varepsilon}_j}{kT} - \frac{\varepsilon_{j\alpha}}{kT} \right) C_{j\alpha}^i W_j^2 \tag{5.144}$$

$\lambda_j^{(VT)}$ and $\lambda_j^{(V)}$ are known as the *vibrational–translational* and *vibrational conductivities*, and we call $D_{ij}^{(V)}$ the *vibrational diffusion coefficient*. Equations (5.142)–(5.144) are the same as those given by Andersen.

5.4. Electron Equations

The equations of change for heavy particles contain terms corresponding to processes in which free electrons participate. These terms depend on the electron number density, n_e, and the electron temperature T_e. In this section we shall derive equations for n_e and T_e, taking as our starting point the Boltzmann transport equation for electrons. The procedure follows closely that of Section 5.2 for the heavy particles. We find that the resulting electron equations contain quantities (electron energy flux, electron pressure tensor, and electron drift velocity) that require the solution of the Boltzmann transport equation for their evaluation. The electron Boltzmann equation cannot be solved in the same way as for the heavy particles [Section 5.3] since the equilibrium electron distribution is not of the Maxwell–Boltzmann form.

5.4.1. Boltzmann Transport Equation for Electrons

We denote the electron distribution function by $f_e(\mathbf{v}, \mathbf{x}, t)$. $f_e(\mathbf{v}, \mathbf{x}, t)\, d^3v\, d^3x$ is the number of electrons in a volume element $d^3v\, d^3x$ in velocity-position space.

The electron number density n_e is given by

$$n_e(\mathbf{x}, t) = \int f_e(\mathbf{v}, \mathbf{x}, t)\, d^3v \tag{5.145}$$

The mass-average velocity $\mathbf{v}_0(\mathbf{x}, t)$ [Eq. (5.3)] should be strictly redefined to include the contribution from electrons:

$$\mathbf{v}_0(\mathbf{x}, t) = \frac{1}{\rho'}\left[\sum_{i=1}^{N}\sum_{\alpha} m_i \int f_{i\alpha}(\mathbf{v}, \mathbf{x}, t)\mathbf{v}\, d^3v + m_e \int f_e(\mathbf{v}, \mathbf{x}, t)\mathbf{v}\, d^3v\right] \tag{5.146}$$

where

$$\rho'(\mathbf{x}, t) \equiv \sum_{i=1}^{N} m_i n_i(\mathbf{x}, t) + m_e n_e(\mathbf{x}, t)$$

although, because of the small electron mass, m_e, compared to the heavy-particle masses, it will be a good approximation to use Eq. (5.3) for the mass-average velocity.

We define the electron temperature, T_e, as the translational electron temperature corresponding to a Maxwell-Boltzmann distribution (Reference 66, p. 455):

$$\frac{3}{2} n_e k T_e = \frac{1}{2} m_e \int f_e(\mathbf{v}, \mathbf{x}, t) v\, d^3v \tag{5.147}$$

The Boltzmann equation for the distribution function $f_e(\mathbf{v}, \mathbf{x}, t)$ is given by [see Eq. (3.8)]

$$\frac{\partial f_e(\mathbf{v}, \mathbf{x}, t)}{\partial t} + \mathbf{v} \cdot \nabla f_e(\mathbf{v}, \mathbf{x}, t) + \frac{\mathbf{X}_e}{m_e} \cdot \nabla_v f_e(\mathbf{v}, \mathbf{x}, t) = J_e(\mathbf{v}, \mathbf{x}, t) \tag{5.148}$$

If, as in Section 5.2.1, we transform from the variables $\mathbf{v}, \mathbf{x}, t$ to $\mathbf{V}, \mathbf{x}, t$, where $\mathbf{V}$ is the variance velocity:

$$\mathbf{V} = \mathbf{v} - \mathbf{v}_0 \tag{5.4}$$

then we obtain, in the same way as Eq. (5.15):

$$\frac{Df_e(\mathbf{V}, \mathbf{x}, t)}{Dt} + \mathbf{V} \cdot \nabla f_e(\mathbf{V}, \mathbf{x}, t) + \left(\frac{\mathbf{X}_e}{m_e} - \frac{D\mathbf{v}_0}{Dt}\right) \cdot \nabla V f_e(\mathbf{V}, \mathbf{x}, t)$$
$$- \mathbf{V} \cdot (\nabla v_0) \cdot \nabla \mathbf{v} f_e(\mathbf{V}, \mathbf{x}, t) = J_e(\mathbf{V}, \mathbf{x}, t) \tag{5.149}$$

where

$$D/Dt = \partial/\partial t + \mathbf{v}_0 \cdot \nabla \tag{5.14a}$$

5.4.2. *Equation of Change of $\bar{\psi}_e$ for Electrons*

If, following Section 5.2.2, for any function $\psi_e(\mathbf{V}, \mathbf{x}, t)$ we define a mean value $\bar{\psi}_e(x, t)$ by

$$\bar{\psi}_e(\mathbf{x}, t) = \frac{1}{n_e(\mathbf{x}, t)} \int d^3 V f_e(\mathbf{V}, \mathbf{x}, t) \psi_e(\mathbf{V}, \mathbf{x}, t) \tag{5.150}$$

then on multiplying the electron Boltzmann equation (5.149) by $\psi_e(\mathbf{V}, \mathbf{x}, t)$ and integrating over $\mathbf{V}$, we obtain, in the same way as we derived Eq. (5.27):

$$\frac{D(n_e\bar{\psi})}{Dt} + n_e\bar{\psi}_e \nabla \cdot \mathbf{v}_0 + \nabla \cdot [n_e(\overline{\psi_e \mathbf{V}_e})]$$
$$- n_e\left[\overline{\left(\frac{D\psi_e}{Dt}\right)} + \overline{(\mathbf{V} \cdot \nabla \psi_e)} + \mathbf{X}_e \cdot \overline{\left(\frac{1}{m_e} \nabla v \psi_e\right)} - \frac{D\mathbf{v}_0}{Dt} \cdot \overline{(\nabla v \psi_e)}\right.$$
$$\left. - \overline{(\nabla v \psi_e) \cdot (\mathbf{V} \cdot \nabla)\mathbf{v}_0}\right] = \int \psi_e(\mathbf{V}, \mathbf{x}, t) J_e(\mathbf{V}, \mathbf{x}, t)\, d^3 V \tag{5.151}$$

where $\bar{\mathbf{V}}_e$ is the average electron variance velocity, given by

$$n_e \bar{\mathbf{V}}_e = \int d^3 V f_e(\mathbf{V}, \mathbf{x}, t) \mathbf{V} \tag{5.152}$$

We call Eq. (5.151) the equation of change of $\bar{\psi}$ for electrons. The Boltzmann equation for the distribution function $f_e(\mathbf{v}, \mathbf{x}, t)$ is given by Eq. (5.148).

5.4.3. *Electron Number-Density Equation*

We set

$$\psi_e = 1$$

We then have

$$\nabla \psi_e = 0, \qquad \nabla_{\mathbf{V}} \psi_e = 0$$

and the equation of change, Eq. (5.151) becomes

$$\frac{Dn_e}{Dt}+n_e\nabla\cdot\mathbf{v}_0+\nabla\cdot(n_e\bar{\mathbf{V}}_e)=\int J_e(\mathbf{V},\mathbf{x},t)\,d^3V \tag{5.153}$$

which is the electron number-density equation. The right-hand side can be expressed in terms of the rates for all inelastic processes that result in a net change in the number of electrons. In the same notation as Eq. (5.35), we have

$$\frac{Dn_e}{Dt}+n_e\nabla\cdot\mathbf{v}_0+\nabla\cdot(n_e\bar{\mathbf{V}}_e)=S_e-\nu_e n_e+\sum_{ab} n_a n_b k_{ab;e\gamma}$$
$$-\sum_{c} n_e n_c k_{ec;\gamma}+\sum_{abc} n_a n_b n_c k_{abc;e\gamma}-\sum_{cd} n_e n_c n_d k_{ecd;\gamma} \tag{5.154}$$

It remains to evaluate the average electron variance velocity $\bar{\mathbf{V}}_e$, which depends on the electron distribution functions $f_e(\mathbf{V},\mathbf{x},t)$, and is defined by Eq. (5.152). Henceforth, it will be called the *electron drift velocity*.

5.4.4. Electron Energy Equation

We now set

$$\psi_e=\tfrac{1}{2}m_e\mathbf{V}^2$$

in the electron equation of change, Eq. (5.151). Then

$$n_e\bar{\psi}_e=\int d^3V f_e(\mathbf{V},\mathbf{x},t)\frac{1}{2}m_e\mathbf{V}^2$$
$$=\frac{3}{2}n_e kT_e \tag{5.155}$$

where we have used Eq. (5.147). The third term in Eq. (5.151) depends on

$$n_e\overline{(\psi_e\mathbf{V}_e)}=\int d^3V f_e(\mathbf{V},\mathbf{x},t)\frac{1}{2}m_e V^2\mathbf{V}$$
$$\equiv\mathbf{q}_e \tag{5.156}$$

defining $\mathbf{q}_e$, which we call the *electron energy flux*. We also have, since t, $\mathbf{x}$, and $\mathbf{V}$ are independent variables:

$$D\psi_e/Dt=0,\qquad \nabla\psi_e=0,\qquad \nabla_V\psi_e=m_e\mathbf{V}$$

so that

$$\overline{n_e \mathbf{X}_e \cdot \left(\frac{1}{m_e} \nabla v \psi_e\right)} = \mathbf{X}_e \cdot \bar{\mathbf{V}}_e n_e$$
$$= e\mathbf{E} \cdot \bar{\mathbf{V}}_e n_e$$
$$= -\mathbf{j}_e \cdot \mathbf{E} \tag{5.157}$$

where we have set

$$\mathbf{X}_e = -e\mathbf{E}$$

by Eq. (5.7), with the effect of the Vlasov field neglected. Also

$$\mathbf{j}_e = -en_e \bar{\mathbf{V}}_e \tag{5.158}$$

is the *electron current density*. Finally, we have

$$n_e \overline{(\nabla v \psi_e) \cdot (\mathbf{V} \cdot \nabla)} \mathbf{v}_0 = \int d^3V f_e(\mathbf{V}, \mathbf{x}, t) m_e \mathbf{V}\mathbf{V} : (\nabla \mathbf{v}_0)$$
$$= \hat{P}_e : (\nabla \mathbf{v}_0) \tag{5.159}$$

where $\hat{P}_e$, the electron pressure tensor, is defined by

$$\hat{P}_e \equiv \int d^3V f_e(\mathbf{V}, \mathbf{x}, t) m_e \mathbf{V}\mathbf{V} \tag{5.160}$$

With the above considerations Eq. (5.151) becomes

$$\frac{D}{Dt}\left(\frac{3}{2} n_e k T_e\right) + \frac{3}{2} n_e k T_e \nabla \cdot \mathbf{v}_0 + \hat{P}_e : (\nabla \mathbf{v}_0)$$
$$= \mathbf{j}_e \cdot \mathbf{E} - \nabla \cdot \mathbf{q}_e + \int d^3V J_e(\mathbf{V}, \mathbf{x}, t) \frac{1}{2} m_e \mathbf{V}^2 \tag{5.161}$$

The last term in Eq. (5.161) represents the time rate of change of total electron energy due to electron–heavy-particle collisions. We can rewrite this term in terms (Bond *et al.*,[36] p. 122) of an electron energy-exchange collision frequency, $\nu_u(T_e)$, defined to be the inverse of the mean time between collisions for an electron in which the electron loses kinetic energy kT_e. Then

$$\int d^3V J_e(\mathbf{V}, \mathbf{x}, t) \frac{1}{2} m_e \mathbf{V}^2 = -n_e \nu_u(T_e) k T_e \tag{5.162}$$

ν_u will depend on the electron heavy-particle momentum-transfer cross sections.

5.5. Evaluation of Electron Drift Velocity, Pressure Tensor, and Energy Flux

Before we can use the equations for the electron number density [Eq. (5.154)] and electron energy [Eq. (5.161)], we need to evaluate the electron drift velocity:

$$\bar{\mathbf{V}}_e(\mathbf{x}, t) = \frac{1}{n_e} \int d^3V f_e(\mathbf{V}, \mathbf{x}, t)\mathbf{V} \tag{5.152a}$$

the electron pressure tensor:

$$\hat{P}_e(\mathbf{x}, t) = \int d^3V f_e(\mathbf{V}, \mathbf{x}, t) m_e \mathbf{V}\mathbf{V} \tag{5.160a}$$

and the electron energy flux:

$$\mathbf{q}_e(\mathbf{x}, t) = \int d^3V f_e(\mathbf{V}, \mathbf{x}, t)\frac{1}{2} m_e V^2 \mathbf{V} \tag{5.156a}$$

The evaluation of the above quantities requires a knowledge of the distribution functions $f_e(\mathbf{V}, \mathbf{x}, t)$. The Boltzmann transport equation for electrons, Eq. (5.148), cannot be solved for the distribution function $f_e(\mathbf{V}, \mathbf{x}, t)$ in the same way as was done for the heavy particles in Section 5.3. In the heavy-particle case a perturbation solution of the Maxwell–Boltzmann equilibrium solution was carried out. We cannot carry out such a perturbation solution for electrons, since we know that the electron distribution function can be highly non-Maxwellian (see Chapter 3, Fig. 23).

In this section we shall express $\bar{\mathbf{V}}_e$, $\hat{P}_e$, and $\mathbf{q}_e$ in terms of quantities that depend upon the isotropic part $f_0(V, \mathbf{x}, t)$ of the electron distribution function $f_e(\mathbf{V}, \mathbf{x}, t)$. These quantities (electron mobility, electron diffusion coefficient, etc.) can either be measured experimentally or evaluated using computer solutions to the transport equation for $f_0(V, \mathbf{x}, t)$.*

5.5.1. Near Isotropic Decomposition of the Boltzmann Transfer Equation for Electrons

We consider the Boltzmann equation in the form of Eq. (5.148):

$$\frac{f_e(\mathbf{v}, \mathbf{x}, t)}{\partial t} + \mathbf{v} \cdot \nabla f_e(\mathbf{v}, \mathbf{x}, t) - \frac{e}{m_e}\mathbf{E} \cdot \nabla_v f_e(\mathbf{v}, \mathbf{x}, t) = \left(\frac{\delta f_e(\mathbf{v}, \mathbf{x}, t)}{\delta t}\right)_c \tag{5.148a}$$

* For a detailed derivation of the transport equation for the isotropic part $f_0(V, \mathbf{x}, t)$ of $f_e(\mathbf{V}, \mathbf{x}, t)$, see Chapter 3.

where we have set

$$\mathbf{X}_e = -e\mathbf{E}$$

$$J_e(\mathbf{v}, \mathbf{x}, t) = \left(\frac{\delta f_e(\mathbf{v}, \mathbf{x}, t)}{\delta t}\right)_c$$

where the subscript c denotes the rate of change of $f_e(\mathbf{v}, \mathbf{x}, t)$ due to electron–molecule collisions.

Now, for elastic electron–molecule collisions the large mass difference of the colliding particles results in small energy transfers compared to the collision energy of the electrons, and large electron deflections. Provided the electron mean-free path for elastic collisions is small compared to the volume of the gas, it follows from the fact that the electron deflections are large that the electron distribution function $f_e(\mathbf{v}, \mathbf{x}, t)$ is nearly independent of the direction of $\mathbf{v}$. This enables $f_e(\mathbf{v}, \mathbf{x}, t)$ to be written [see also Eq. (3.10)]

$$f_e(\mathbf{v}, \mathbf{x}, t) = f_0(v, \mathbf{x}, t) + \frac{\mathbf{v}}{v} \cdot \mathbf{f}_1(v, \mathbf{x}, t) \tag{5.163a}$$

where $f_0 \gg f_1$. The expansion (5.163a) is to be inserted into Eq. (5.148a).

Since the distribution of $\mathbf{v}$ is nearly isotropic, terms depending upon the direction of $\mathbf{v}$, yet quadratic in $\mathbf{v}$, will be replaced by their average over all directions of $\mathbf{v}$. Terms of the form

$$\mathbf{v}(\mathbf{v} \cdot \mathbf{A})$$

where A is independent of the direction of $\mathbf{v}$, will be replaced by

$$\mathbf{v}(\mathbf{v} \cdot \mathbf{A}) \rightarrow \tfrac{1}{3} v^2 \mathbf{A} \tag{3.11}$$

which is a form of Eq. (5.115) with $\mathbf{r}$ replaced by $\mathbf{v}$.

When Eq. (5.163a) is substituted into Eq. (5.148a) we have

$$\frac{\partial}{\partial t} f_0 + \frac{\mathbf{v}}{v} \cdot \frac{\partial}{\partial t} \mathbf{f}_1 + \mathbf{v} \cdot \nabla f_0 + \mathbf{v} \cdot \nabla\left(\frac{\mathbf{v}}{v} \cdot \mathbf{f}_1\right) - \frac{e\mathbf{E}}{m_e} \cdot \nabla_\mathbf{v}\left(f_0 + \frac{\mathbf{v}}{v} \cdot \mathbf{f}_1\right)$$
$$= \left(\frac{\delta f_0}{\delta t}\right)_c + \frac{\mathbf{v}}{v} \cdot \left(\frac{\delta \mathbf{f}_1}{\delta t}\right)_c \tag{5.148a}$$

For the fourth term on the left-hand side of Eq. (5.148b) we have

$$\mathbf{v} \cdot \nabla\left(\frac{\mathbf{v}}{v} \cdot \mathbf{f}_1\right) = \frac{\mathbf{v}\mathbf{v}}{v} : \nabla \mathbf{f}_1$$
$$= \frac{1}{3} v \nabla \cdot \mathbf{f}_1 \tag{5.163b}$$

where we have used the replacement, Eq. (3.11). Now if $\mathbf{i}, \mathbf{j}, \mathbf{k}$ are unit

Cartesian vectors in velocity space, we have

$$\nabla_v = \left(\mathbf{i}\frac{\partial}{\partial v_x} + \mathbf{j}\frac{\partial}{\partial v_y} + \mathbf{k}\frac{\partial}{\partial v_z}\right)$$

and hence that

$$\nabla_v \mathbf{v} = (\mathbf{ii} + \mathbf{jj} + \mathbf{kk})$$

Therefore, for any two vectors **A** and **B** we have

$$\mathbf{A} \cdot (\nabla_v \mathbf{v}) \cdot \mathbf{B} = \mathbf{A} \cdot \mathbf{B} \tag{5.164}$$

It follows from Eq. (5.164) that

$$\mathbf{E} \cdot (\nabla_v \mathbf{v}) \cdot \frac{\mathbf{f}_1}{v} = \frac{\mathbf{E} \cdot \mathbf{f}_1}{v} \tag{5.165}$$

Also, for a function $G(v)$ of the magnitude of $\mathbf{v}$ only:

$$\begin{aligned}\nabla_v G(v) &= \left(\mathbf{i}\frac{\partial}{\partial v_x} + \mathbf{j}\frac{\partial}{\partial v_y} + \mathbf{k}\frac{\partial}{\partial v_z}\right) G(v) \\ &= \left(\mathbf{i}\frac{\partial v}{\partial v_x} + \mathbf{j}\frac{\partial v}{\partial v_y} + \mathbf{k}\frac{\partial v}{\partial v_z}\right)\frac{\partial}{\partial v} G(v)\end{aligned}$$

but

$$v = (v_x^2 + v_y^2 + v_z^2)^{1/2}$$

so that

$$\frac{\partial v}{\partial v_x} = \frac{v_x}{v}, \quad \text{etc.}$$

and therefore we have

$$\nabla_v G(v) = \frac{\mathbf{v}}{v}\frac{\partial G(v)}{\partial v} \tag{5.166}$$

Using Eqs. (5.164) and (5.166) we have

$$\mathbf{E} \cdot \nabla_v f_0 = \frac{\mathbf{E} \cdot \mathbf{v}}{v}\frac{\partial f_0}{\partial v} \tag{5.167}$$

and the last term on the left-hand side of Eq. (5.148b) becomes

$$\mathbf{E} \cdot \nabla_v\left(\frac{\mathbf{v}}{v} \cdot \mathbf{f}_1\right) = \frac{\mathbf{E} \cdot \mathbf{f}_1}{v} + \frac{(\mathbf{E} \cdot \mathbf{v})}{v}(\mathbf{v} \cdot \mathbf{f}_1)\left(-\frac{1}{v^2}\right) + \left(\frac{\mathbf{E} \cdot \mathbf{v}}{v}\right)\left(\frac{\mathbf{v}}{v} \cdot \frac{\partial \mathbf{f}_1}{\partial v}\right) \tag{5.168}$$

Since **E** is independent of **v** we apply Eq. (3.11) to the last two terms of

Eq. (5.168) to obtain

$$\mathbf{E}\cdot\nabla_{\mathbf{v}}\left(\frac{\mathbf{v}}{v}\cdot\mathbf{f}_1\right)=\frac{\mathbf{E}\cdot\mathbf{f}_1}{v}-\frac{1}{3}\frac{\mathbf{E}\cdot\mathbf{f}_1}{v}+\frac{1}{3}\mathbf{E}\cdot\frac{\partial\mathbf{f}_1}{\partial v}$$

$$=\frac{2}{3}\frac{\mathbf{E}\cdot\mathbf{f}_1}{v}+\frac{1}{3}\mathbf{E}\cdot\frac{\partial\mathbf{f}_1}{\partial v}$$

$$=\frac{1}{v^2}\frac{\partial}{\partial v}\left(\frac{v^2}{3}\mathbf{E}\cdot\mathbf{f}_1\right) \tag{5.169}$$

From Eqs. (5.148b), (5.163b), (5.165), (5.167), and (5.169), the Boltzmann transfer equation may be written

$$\frac{\partial f_0}{\partial t}+\frac{\mathbf{v}}{v}\cdot\frac{\partial\mathbf{f}_1}{\partial t}+\mathbf{v}\cdot\nabla f_0+\frac{1}{3}v\nabla\cdot\mathbf{f}_1-\frac{e}{m_e}\left[\frac{\mathbf{E}\cdot\mathbf{v}}{v}\frac{\partial f_0}{\partial v}+\frac{1}{v^2}\frac{\partial}{\partial v}\left(\frac{v^2}{3}\mathbf{E}\cdot\mathbf{f}_1\right)\right]$$

$$=\left(\frac{\delta f_0}{\delta t}\right)_c+\frac{\mathbf{v}}{v}\cdot\left(\frac{\delta\mathbf{f}_1}{\delta t}\right)_c \tag{5.170}$$

Since Eq. (5.170) must hold identically for all $\mathbf{v}$ we can equate the coefficients of zeroth and first order in $\mathbf{v}$ to obtain the two coupled equations:

$$\frac{1}{v}\frac{\partial\mathbf{f}_1}{\partial t}+\nabla f_0-\frac{e}{m_e}\frac{\mathbf{E}}{v}\frac{\partial f_0}{\partial v}=\frac{1}{v}\left(\frac{\delta\mathbf{f}_1}{\delta t}\right)_c \tag{5.171}$$

$$\frac{\partial f_0}{\partial t}+\frac{1}{3}v\nabla\cdot\mathbf{f}_1-\frac{e}{m_e}\frac{\mathbf{E}}{v^2}\cdot\frac{\partial}{\partial v}\left(\frac{v^2}{3}\mathbf{f}_1\right)=\left(\frac{\delta f_0}{\delta t}\right)_c \tag{5.172}$$

The right-hand side of Eq. (5.171) can be expressed in terms of the electron momentum-transfer collision frequency $\nu_m(v)$, or the momentum-transfer cross section $Q^j_m(v)$ for species j, through the relation [see Section 3.1.3, Eq. (3.25)]

$$\left(\frac{\delta\mathbf{f}_1}{\delta t}\right)_c=-\nu_m(v)\mathbf{f}_1=-\sum_j n_jQ^j_m(v)v\mathbf{f}_1 \tag{5.173}$$

and Eq. (5.171) then becomes

$$\frac{1}{v}\frac{\partial\mathbf{f}_1}{\partial t}+\nabla f_0-\frac{e}{m_e}\frac{\mathbf{E}}{v}\frac{\partial f_0}{\partial v}=-\sum_j n_jQ^j_m\mathbf{f}_1 \tag{5.174}$$

or

$$\frac{\partial\mathbf{f}_1}{\partial t}+v\nabla f_0-\frac{e}{m_e}\mathbf{E}\frac{\partial f_0}{\partial v}=-\nu_m\mathbf{f}_1 \tag{5.175}$$

The term on the right-hand side of Eq. (5.175) is the collision frequency (the number of elastic electron–heavy-particle collisions per second) times $\mathbf{f}_1$. We expect this term to be very much greater than the total time rate of change of $\mathbf{f}_1$, represented by the first term on the left-hand side of Eq. (5.175). Therefore we neglect the $\partial\mathbf{f}_1/\partial t$ term in Eq. (5.174), which enables us to write Eq. (5.174) as

$$\mathbf{f}_1(v, \mathbf{x}, t) = \frac{1}{\sum_j n_j Q_m^j(v)}\left[-\nabla f_0(v, \mathbf{x}, t) + \frac{e}{m_e}\frac{\mathbf{E}}{v}\frac{\partial f_0(v, \mathbf{x}, t)}{\partial v}\right] \tag{5.176}$$

5.5.2. *Evaluation of the Electron Drift Velocity* $\bar{\mathbf{V}}_e$

The electron drift velocity, $\bar{\mathbf{V}}_e(\mathbf{x}, t)$, is given by Eq. (5.152):

$$n_e\bar{\mathbf{V}}_e = \int d^3V f_e(\mathbf{V}, \mathbf{x}, t)\mathbf{V} \tag{5.152}$$

First of all we shall calculate the quantity $\bar{\mathbf{v}}_e$, defined by

$$n_e\bar{\mathbf{v}}_e \equiv \int d^3v\, f_e(\mathbf{v}, \mathbf{x}, t)\mathbf{v} \tag{5.177}$$

which is, in fact, the drift velocity in the stationary reference frame since we have

$$\mathbf{V} = \mathbf{v} - \mathbf{v}_0 \tag{5.4}$$

Substituting for $f_e(\mathbf{v}, \mathbf{x}, t)$, given by Eq. (5.163a), into Eq. (5.177) we have

$$n_e\bar{\mathbf{v}}_e = \int d^3v\left(f_0 + \frac{\mathbf{v}}{v}\cdot\mathbf{f}_1\right)\mathbf{v} \tag{5.178}$$

Now $f_0(v, \mathbf{x}, t)$ as a function of v, the argument of $\mathbf{v}$, is an even function of $\mathbf{v}$. Therefore $[f_0(v, \mathbf{x}, t)]\,\mathbf{v}$ is an odd function in $\mathbf{v}$ and hence

$$\int d^3v\, f_0\mathbf{v} = 0 \tag{5.179}$$

Thus Eq. (5.178) becomes

$$n_e\bar{\mathbf{v}}_e = \int d^3v\frac{\mathbf{v}}{v}\cdot\mathbf{f}_1\mathbf{v} = \frac{1}{3}\int d^3v\, v\mathbf{f}_1 \tag{5.180}$$

where we have used Eq. (3.11). We now substitute for $\mathbf{f}_1$, as given by Eq. (5.176), into Eq. (5.180) to obtain

$$\begin{aligned} n_e\bar{\mathbf{v}}_e &= \frac{1}{3}\int d^3v\frac{1}{\sum_j n_j Q_m^j(v)}\left[-v\nabla f_0(v, \mathbf{x}, t) + \frac{e}{m_e}\mathbf{E}\frac{\partial f_0(v, \mathbf{x}, t)}{\partial v}\right] \\ &= \frac{1}{3n}\int d^3v\frac{1}{\sum_j \delta_j Q_m^j(v)}\left[-v\nabla f_0(v, \mathbf{x}, t) + \frac{e}{m_e}\mathbf{E}\frac{\partial f_0(v, \mathbf{x}, t)}{\partial v}\right] \end{aligned} \tag{5.181}$$

where we have defined $\delta_j(\mathbf{x}, t)$ to be the fractional population of heavy-particle species j, n being the total heavy-particle number density:

$$\delta_j(\mathbf{x}, t) = \frac{n_j(\mathbf{x}, t)}{n(\mathbf{x}, t)} \tag{5.182}$$

$$n(\mathbf{x}, t) = \sum_j n_j(\mathbf{x}, t) \tag{5.17a}$$

If we now assume that $\nabla(1/\delta_j)$ is negligible compared with ∇f_0, then we can rewrite Eq. (5.181) in the form

$$n_e\bar{\mathbf{v}}_e(\mathbf{x}, t) = -\frac{1}{n}\nabla\left[\frac{1}{3}\int d^3v \frac{vf_0(v, \mathbf{x}, t)}{\sum_j \delta_j Q_m^j(v)}\right] + \mathbf{E}\left[\frac{e}{3nm_e}\int d^3v \frac{\partial f_0/\partial v}{\sum_j Q_m^j(v)\delta_j}\right] \tag{5.183}$$

If we now define [see also Eqs. (3.77) and (3.79)]

$$D_e(\mathbf{x}, t) \equiv \frac{1}{n_e n}\frac{1}{3}\int d^3v \frac{vf_0(v, \mathbf{x}, t, E/N)}{\sum_j \delta_j Q_m^j(v)} \tag{5.184}$$

and

$$\mu_e(\mathbf{x}, t) \equiv -\frac{1}{n_e}\frac{e}{3nm_e}\int d^3v \frac{[\partial f_0(v, \mathbf{x}, t, E/N)]/\partial v}{\sum_j \delta_j Q_m^j(v)} \tag{5.185}$$

then we can rewrite Eq. (5.183) as

$$\begin{aligned} n_e\bar{\mathbf{v}}_e &= -\frac{1}{n}\nabla(n_e n D_e) - \mathbf{E}n_e\mu_e \\ &= -\nabla(n_e D_e) - n_e D_e \nabla(\ln n) - n_e\mu_e\mathbf{E} \end{aligned} \tag{5.186}$$

$D_e(\mathbf{x}, t)$ and $\mu_e(\mathbf{x}, t)$ are known as the electron diffusion coefficient and the electron mobility, respectively. We shall show in the next section that their definitions, Eqs. (5.184) and (5.185), are equivalent to those given by other authors.

In terms of the electron momentum-transfer collision frequency, $\nu_m(v)$, we see from Eq. (5.173) that we can write Eqs. (5.184) and (5.185) as

$$D_e(\mathbf{x}, t) = \frac{1}{3n_e}\int d^3v \frac{v^2 f_0(v, \mathbf{x}, t)}{\nu_m(v)} \tag{5.187}$$

$$\mu_e(\mathbf{x}, t) = -\frac{e}{3n_e m_e}\int d^3v \frac{v}{\nu_m(v)}\frac{\partial f_0(v, \mathbf{x}, t)}{\partial v} \tag{5.188}$$

We now suppose that Eqs. (5.187) and (5.188) may be written in terms of an *effective* electron momentum-transfer collision[69] (see also Frost and

Phelps[44]) frequency, ν'_m, which depends on the gas mixture but not on v; we then have

$$D_e(\mathbf{x}, t) = \frac{1}{3n_e\nu'_m} \int d^3v\, v\, v^2 f_0(v, \mathbf{x}, t) \tag{5.189}$$

$$\mu_e(\mathbf{x}, t) = -\frac{e}{3n_e m_e \nu'_m} \int d^3v\, v \frac{\partial f_0(v, \mathbf{x}, t)}{\partial v} \tag{5.190}$$

From Eq. (5.147) we have

$$\frac{3}{2} n_e(\mathbf{x}, t) k T_e(\mathbf{x}, t) = \frac{1}{2} m_e \int \left[f_0(v, \mathbf{x}, t) + \frac{\mathbf{v}}{v} \cdot \mathbf{f}_1(v, \mathbf{x}, t) \right] v^2\, d^3v$$

$$= \frac{1}{2} m_e \int f_0(v, \mathbf{x}, t) v^2\, d^3v \tag{5.191}$$

the other term in the integrand being an odd function of **v**. From Eqs. (5.189) and (5.191) we obtain

$$D_e(\mathbf{x}, t) = \frac{kT_e(\mathbf{x}, t)}{m_e \nu'_m} \tag{5.192}$$

To express $\mu_e(\mathbf{x}, t)$ in a similar form we integrate Eq. (5.190) by parts:

$$\mu_e(\mathbf{x}, t) = -\frac{e}{3n_e m_e \nu'_m} \int d^3v\, v \frac{\partial f_0(v, \mathbf{x}, t)}{\partial v}$$

$$= -\frac{e4\pi}{3n_e m_e \nu'_m} \int v^3\, d[f_0(v, \mathbf{x}, t)]$$

$$= -\frac{e4\pi}{3n_e m_e \nu'_m} [v^3 f_0(v, \mathbf{x}, t)]_{v=0}^{v=\infty} + \frac{e4\pi}{n_e m_e \nu'_m} \int f_0(v, \mathbf{x}, t) v^2\, dv$$

$$= \frac{e}{m_e \nu'_m} \tag{5.193}$$

where we have used

$$n_e(\mathbf{x}, t) = 4\pi \int f_0(v, \mathbf{x}, t) v^2\, dv$$

which is obtained from Eqs. (5.145) and (5.163a), and we have assumed that $f_0(v, \mathbf{x}, t) \to 0$ as $v \to \infty$ such that $v^3 f_0 \to 0$ as $v \to \infty$. Equations (5.192) and (5.193) are in agreement with the expressions given by Haas.[70]

We recall that we have evaluated $\bar{\mathbf{v}}_e(\mathbf{x}, t)$ given by Eq. (5.177), whereas the electron drift velocity $\tilde{\mathbf{V}}_e(\mathbf{x}, t)$ is given by Eq. (5.152). Using

Eqs. (5.152), (5.163a), (5.177), and (5.4) we have

$$n_e \bar{\mathbf{V}}_e(\mathbf{x}, t) = \int d^3V f_e(\mathbf{V}, \mathbf{x}, t)\mathbf{V}$$

$$= \int d^3v\, f_e(\mathbf{v}, \mathbf{x}, t)(\mathbf{v} - \mathbf{v}_0)$$

$$= \int d^3v \left[f_0(v, \mathbf{x}, t) + \frac{\mathbf{v}}{v} \cdot \mathbf{f}_1(v, \mathbf{x}, t) \right] (\mathbf{v} - \mathbf{v}_0)$$

$$= n_e \bar{\mathbf{v}}_e - n_e \mathbf{v}_0 \tag{5.194}$$

From Eqs. (5.186) and (5.194) we have

$$n_e \bar{\mathbf{V}}_e = -\nabla(n_e D_e) - n_e D_e \nabla(\ln n) - n_e \mu_e \mathbf{E} - n_e \mathbf{v}_0 \tag{5.195}$$

which is in agreement with the expression given by Haas [Reference 70, Eq. (11)] except for the presence of the last term on the right-hand side of Eq. (5.195).

5.5.3. Electron Mobility and Diffusion Coefficient in Terms of Energy Normalization

In this section we shall see that the equations (5.184) and (5.185) for the diffusion coefficient $D_e(\mathbf{x}, t)$ and electron mobility $\mu_e(\mathbf{x}, t)$ are equivalent to those given by Elliott *et al.*[46] and other authors in terms of a different normalization.

Our expressions for $D_e(x, t)$ and $\mu_e(x, t)$ are

$$D_e(\mathbf{x}, t) = \frac{1}{n_e n} \frac{1}{3} \int d^3v \frac{v f_0(v, \mathbf{x}, t)}{\sum_j \delta_j Q_m^j(v)} \tag{5.184}$$

$$\mu_e(\mathbf{x}, t) = -\frac{1}{n_e} \frac{e}{3nm_e} \int d^3v \frac{\partial f_0(v, \mathbf{x}, t)/\partial v}{\sum_j \delta_j Q_m^j(v)} \tag{5.185}$$

and the normalization we have used is given by Eq. (5.145) as

$$n_e(\mathbf{x}, t) = \int f_e(\mathbf{v}, \mathbf{x}, t)\, d^3v \tag{5.145}$$

which using Eq. (5.163a), becomes

$$n_e(\mathbf{x}, t) = \int f_0(v, \mathbf{x}, t)\, d^3v \tag{5.196}$$

A common alternative (called here the "energy") normalization is

$$\int u^{1/2} f_0(u, \mathbf{x}, t)\, du = 1 \tag{5.197}$$

where eu is the electron kinetic energy, expressed in terms of v as

$$u = m_e v^2/2e \tag{3.53}$$

In terms of u our normalization, Eq. (5.196) may be written, using Eq. (3.53):

$$n_e(\mathbf{x}, t) = 4\pi\left(\frac{e}{m_e}\right)\int f_0(u, \mathbf{x}, t)\left(\frac{2e}{m_e}\right)^{1/2} u^{1/2}\, du \tag{5.198}$$

or

$$\frac{4\pi}{n_e(\mathbf{x}, t)}\left(\frac{2e}{m_e}\right)^{1/2}\left(\frac{e}{m_e}\right)\int f_0(u, \mathbf{x}, t) u^{1/2}\, du = 1 \tag{5.199}$$

Comparing Eqs. (5.197) and (5.199) we see that, to express $D_e(\mathbf{x}, t)$ and $\mu_e(\mathbf{x}, t)$ in terms of the energy normalization, we should replace $f_0(u, \mathbf{x}, t)$ in Eqs. (5.184) and (5.185) by

$$f_0(u, \mathbf{x}, t) \rightarrow \frac{1}{4\pi} n_e(\mathbf{x}, t)\left(\frac{2e}{m_e}\right)^{-1/2}\left(\frac{m_e}{e}\right) f_0(u, \mathbf{x}, t) \tag{5.200}$$

In terms of u, Eqs. (5.184) and (5.185) are given by

$$D_e(\mathbf{x}, t) = \frac{1}{n_e(\mathbf{x}, t) n(\mathbf{x}, t)} \frac{8\pi}{3}\left(\frac{e}{m_e}\right)^2 \int du \frac{f_0(u, \mathbf{x}, t) u}{\sum_j \delta_j Q_m^j(u)} \tag{5.201}$$

$$\mu_e(\mathbf{x}, t) = \frac{-e}{3 n_e(\mathbf{x}, t) n(\mathbf{x}, t) m_e} \int du \frac{\partial f_0(u, \mathbf{x}, t)/\partial u}{\sum_j \delta_j Q_m^j(u)} \tag{5.202}$$

By making the replacement given by Eq. (5.200) in Eqs. (5.201) and (5.202), we obtain expressions for $D_e(\mathbf{x}, t)$ and $\mu_e(\mathbf{x}, t)$ in terms of the normalization of Eq. (5.197):

$$D_e(\mathbf{x}, t) = \frac{1}{n(\mathbf{x}, t)} \frac{1}{3}\left(\frac{2e}{m_e}\right)^{1/2} \int du \frac{u f_0(u, \mathbf{x}, t)}{\sum_j \delta_j Q_m^j(u)} \tag{5.203}$$

$$\mu_e(\mathbf{x}, t) = -\frac{1}{n(\mathbf{x}, t)} \frac{1}{3}\left(\frac{2e}{m_e}\right)^{1/2} \int du \frac{u\, \partial f_0(u, \mathbf{x}, t)/\partial u}{\sum_j \delta_j Q_m^j(u)} \tag{5.204}$$

Equations (5.203) and (5.204) are in agreement with the expressions for $D_e(\mathbf{x}, t)$ and $\mu_e(\mathbf{x}, t)$ given by Elliott *et al.* [Reference 46, Eqs. (1) and (3)].

5.5.4. *Evaluation of the Electron Pressure Tensor* $\hat{P}_e(\mathbf{x}, t)$

The electron pressure tensor $\hat{P}_e(\mathbf{x}, t)$ is given by Eq. (5.160) to be

$$\hat{P}_e(\mathbf{x}, t) = \int d^3V f_e(\mathbf{V}, \mathbf{x}, t) m_e \mathbf{V}\mathbf{V} \tag{5.160}$$

Using Eq. (5.4) for the variance velocity $\mathbf{V}$ we have

$$\hat{P}_e(\mathbf{x}, t) = \int d^3v\, f_e(\mathbf{v}, \mathbf{x}, t) m_e(\mathbf{v} - \mathbf{v}_0)(\mathbf{v} - \mathbf{v}_0) \tag{5.205}$$

Making the substitution of Eq. (5.163a) for $f_e(\mathbf{v}, \mathbf{x}, t)$ we obtain

$$\begin{aligned}\hat{P}_e(\mathbf{x}, t) &= \int d^3v \left[f_0(v, \mathbf{x}, t) + \frac{\mathbf{v}}{v} \cdot \mathbf{f}_1(v, \mathbf{x}, t) \right] m_e(\mathbf{vv} - 2\mathbf{vv}_0 + \mathbf{v}_0\mathbf{v}_0) \\ &= \int d^3v\, m_e \left[f_0(v, \mathbf{x}, t)(\mathbf{vv} + \mathbf{v}_0\mathbf{v}_0) - 2\frac{\mathbf{v}}{v} \cdot \mathbf{f}_1(v, \mathbf{x}, t)\mathbf{vv}_0 \right]\end{aligned} \tag{5.206}$$

where, to obtain Eq. (5.206), we have used the fact that the integral of odd functions of $\mathbf{v}$ is zero. We can now use Eq. (3.11) to write Eq. (5.206) as

$$\hat{P}_e(\mathbf{x}, t) = \int d^3v\, m_e \left[f_0(v, \mathbf{x}, t)(\mathbf{vv} + \mathbf{v}_0\mathbf{v}_0) - \frac{2}{3} v \mathbf{f}_1(v, \mathbf{x}, t)\mathbf{v}_0 \right] \tag{5.207}$$

Since we have assumed that $f_1 \ll f_0$ we shall neglect the second term in Eq. (5.207). We now use the fact that the off-diagonal elements of the tensor $\mathbf{vv}$ are odd functions of the components of $\mathbf{v}$. Therefore the off-diagonal elements of the integral will be zero, and from Eq. (5.207) we obtain

$$\begin{aligned}\hat{P}_e(\mathbf{x}, t) &= \int d^3v\, m_e \frac{v^2}{3} f_0(v, \mathbf{x}, t)\hat{1} + \mathbf{v}_0\mathbf{v}_0 \int d^3v\, m_e f_0(v, \mathbf{x}, t) \\ &= n_e(\mathbf{x}, t)kT_e(\mathbf{x}, t)\hat{1} + n_e(\mathbf{x}, t)m_e\mathbf{v}_0\mathbf{v}_0\end{aligned} \tag{5.208}$$

where we have used Eqs. (5.145), (3.11), and (5.191).

Since $\mathbf{v}_0$ is a mass-average velocity over both heavy particles and electrons it follows that, provided the electron temperature T_e is comparable to or greater than the heavy-particle temperature T, the mean electron velocity $\bar{\mathbf{v}}$:

$$\tfrac{1}{2}m_e\bar{v}^2 = \tfrac{3}{2}kT_e$$

is very much greater than the mass-average heavy-particle velocity $\mathbf{v}_0$:

$$\tfrac{1}{2}Mv_0^2 \approx \tfrac{3}{2}kT$$

where M is the averaged heavy-particle mass. Therefore we can neglect the second term in Eq. (5.208) and write

$$\hat{P}_e(\mathbf{x}, t) = p_e(\mathbf{x}, t)\hat{1} \tag{5.209}$$

with

$$p_e(\mathbf{x}, t) = n_e(\mathbf{x}, t)kT_e(\mathbf{x}, t) \tag{5.210}$$

5.5.5. Evaluation of the Electron Energy Flux $\mathbf{q}_e(\mathbf{x}, t)$

The electron energy flux $\mathbf{q}_e(\mathbf{x}, t)$ is given by Eq. (5.156):

$$\mathbf{q}_e(\mathbf{x}, t) = \int d^3V f_e(\mathbf{V}, \mathbf{x}, t) \tfrac{1}{2} m_e V^2 \mathbf{V} \tag{5.156a}$$

Provided the electron drift velocity $\bar{\mathbf{V}}_e$ [Eq. (5.152)] is large compared to the mass-average velocity $\mathbf{v}_0$ [Eq. (5.146)], then by Eq. (5.4) we can replace $\mathbf{V}$ by $\mathbf{v}$ in Eq. (5.156) and write

$$\mathbf{q}_e(\mathbf{x}, t) = \int d^3v\, f_e(\mathbf{v}, \mathbf{x}, t) \frac{1}{2} m_e v^2 \mathbf{v} \tag{5.211}$$

We now substitute for $f_e(\mathbf{v}, \mathbf{x}, t)$ using Eq. (5.163a):

$$\begin{aligned} \mathbf{q}_e(\mathbf{x}, t) &= \int d^3v \left[f_0(v, \mathbf{x}, t) + \frac{\mathbf{v}}{v} \cdot \mathbf{f}_1(v, \mathbf{x}, t) \right] \frac{1}{2} m_e v^2 \mathbf{v} \\ &= \frac{1}{3} \int d^3v \frac{1}{2} m_e v^3 \mathbf{f}_1(v, \mathbf{x}, t) \end{aligned} \tag{5.212}$$

where the term containing $f_0(v, \mathbf{x}, t)$ in the integral vanishes since it is an odd function of $\mathbf{v}$, and we have used Eq. (5.163b) to simplify the other term. We now use Eq. (5.176) to rewrite Eq. (5.212) as

$$\mathbf{q}_e(\mathbf{x}, t) = \frac{1}{6} m_e \int d^3v \frac{v^3}{\sum_j n_j Q_m^j(v)} \left[-\nabla f_0(v, \mathbf{x}, t) - \frac{e}{m_e} \frac{\mathbf{E}}{v} \frac{\partial f_0(v, \mathbf{x}, t)}{\partial v} \right] \tag{5.213}$$

Equation (5.213) for the electron energy flux cannot be written in terms of the electron mobility μ_e and diffusion D_e unless assumptions are made concerning the form of the electron distribution $f_0(v, \mathbf{x}, t)$ and the momentum-transfer cross sections $Q_m^j(v)$. As an example, the energy flux is expressed below in terms of D_e and μ_e for a Maxwell electron distribution:

$$f_0(v) = n_e \left(\frac{m_e}{2\pi k T_e} \right)^{3/2} \exp\left(-\frac{m_e v^2}{2kT_e} \right)$$

with the momentum-transfer cross sections being given by [see Section 3.2.3, Eq. (3.126)]

$$Q_m^j(v) = c^j v^{-1}$$

From Eq. (5.184),

$$D_e = \frac{1}{3n_e n}\int d^3v \frac{f_0(v)}{\sum_j \delta_j Q^j_m(v)}$$

$$= \frac{4\pi}{3n_e n}\int dv\, v^3 n_e\left(\frac{m_e}{2\pi k T_e}\right)^{3/2} \exp\left(-\frac{m_e v^2}{2kT_e}\right)\left(\sum_j \delta_j c^j\right)^{-1}$$

$$= \frac{kT_e}{nm_e \sum_j \delta_j c^j} \tag{5.214}$$

where we have put $v^2 = t$ and used the integral identity

$$\int_0^\infty t^{3/2} e^{-\alpha t}\, dt = \frac{3}{4}\frac{\pi^{1/2}}{\alpha^{5/2}} \tag{5.215}$$

Similarly, from Eq. (5.185):

$$\mu_e = -\frac{e}{3nn_e m_e}\int d^3v \frac{\partial f_0(v)/\partial v}{\sum_j \delta_j Q^j_m(v)}$$

$$= \frac{e}{nm_e \sum_j \delta_j c^j} \tag{5.216}$$

where we have again used the identity (5.215).

Now using the integral identity

$$\int_0^\infty t^{5/2} e^{-\alpha t}\, dt = \frac{15}{8}\frac{\pi^{1/2}}{\alpha^{7/2}} \tag{5.217}$$

Eq. (5.213) may be evaluated to give

$$\mathbf{q}_e(\mathbf{x}, t) = -\frac{5}{2n}\nabla(n_e k T_e n D_e) - \frac{5}{2} n_e k T_e \mathbf{E} \mu_e \tag{5.218}$$

Now from Eq. (5.158)

$$\mathbf{j}_e = -en_e \bar{\mathbf{V}}_e \tag{5.158}$$

and Eq. (5.186) together with Eq. (5.4)

$$n_e \bar{\mathbf{V}}_e = -\nabla(n_e D_e) - n_e D_e \nabla(\ln n) - n_e \mu_e \mathbf{E} - n\mathbf{v}_0 \tag{5.186a}$$

we have, on neglecting the term containing $\mathbf{v}_0$:

$$n_e \mu_e \mathbf{E} = -\frac{1}{n}\nabla(nn_e D_e) + \frac{\mathbf{j}_e}{e} \tag{5.219}$$

From Eqs. (5.218) and (5.219) we obtain our final expression for $\mathbf{q}_e(\mathbf{x}, t)$:

$$\mathbf{q}_e(\mathbf{x}, t) = -\frac{5}{2} n_e k D_e \nabla T_e - \frac{5}{2}\frac{kT_e}{e}\mathbf{j}_e \tag{5.220}$$

or

$$\mathbf{q}_e(\mathbf{x}, t) = -\lambda_e \nabla T_e - \frac{5}{2}\frac{kT_e}{e}\mathbf{j}_e \tag{5.221}$$

where

$$\lambda_e(\mathbf{x}, t) = \tfrac{5}{2} k n_e D_e \tag{5.222}$$

is called the effective electron thermal conductivity.

Equation (5.220) agrees with the expression given by Haas [Reference 70, Eq. (13)].

5.6. *Equations for the Electromagnetic Field*

Throughout the development of the preceding sections we have implicitly assumed that the effects of magnetic fields can be neglected. We expect this to be the case in laser discharges where there is an applied electric field but not applied magnetic field. In the absence of a magnetic field, Maxwell's equations are

$$\nabla \cdot \mathbf{E} = \rho_c/\varepsilon \tag{5.223}$$

$$\nabla \times \mathbf{E} = 0 \tag{5.224}$$

$$0 = \mu_e \mathbf{j}_e + \mu_- \mathbf{j}_- + \mu_+ \mathbf{j}_+ + \varepsilon \frac{\partial \mathbf{E}}{\partial t} \tag{5.225}$$

where ρ_c is the charge density, which will be given by

$$\rho_c = -e\left(n_e - \sum_j n_j^{(+)} + \sum_j n_j^{(-)}\right) \tag{5.226}$$

where $n_j^{(+)}$ ($n_j^{(-)}$) is the number density of heavy-particle species j which are positive (negative) ions. μ_e is the electron mobility [see Eqs. (5.185) and (5.193)] and $\mathbf{j}_e$ is the electron current density, given by Eqs. (5.158) and (5.195). $\mathbf{j}_-$ and $\mathbf{j}_+$ are the negative-ion and positive-ion current densities, given by analogy with Eq. (5.158) to be

$$\begin{aligned} \mathbf{j}_- &= -e \sum_j n_j^{(-)} \bar{\mathbf{V}}_j \\ \mathbf{j}_+ &= e \sum_j n_j^{(+)} \bar{\mathbf{V}}_j \end{aligned} \tag{5.227}$$

where $\bar{\mathbf{V}}_j$ is given by Eq. (5.5) and is the average variance velocity of a heavy particle of species j. μ_- and μ_+ are the negative- and positive-ion mobilities and ε is the capacitance of the laser medium.

The equations for the electric field given in this section refer to the field in the plasma due to an applied external field and the presence of charged particles. They do not take into account the high-frequency electromagnetic field due to spontaneous and stimulated emission. The determination of the spatial dependence of this field is a separate research program.

5.7. HPBE: A Code Used to Solve the Heavy-Particle Boltzmann Equation and Calculate Transport Coefficients of Molecular Species in Laser Plasmas

The transport coefficients of molecular species [the coefficients of diffusion and conductivity defined by Eqs. (5.117a, b, c), (5.137a, b, c), (5.143a, b), and (5.144)] are required for spatially dependent laser calculations; for example, for a stability analysis of lasing media (Chapter 7), and for modeling CO_2 waveguide lasers (Section 9.2). These coefficients can only be calculated if the molecular velocity distribution functions [Eq. (5.86a)] are known.

The heavy-particle Boltzmann equations are reduced to integral equations [for the unknown coefficients $A_{i\alpha}(W_i)$, etc., of Eq. (5.111)], obtained by substituting Eq. (5.111) for $\phi_{i\alpha}$ into the collision integrals of the right-hand side of Eq. (5.108) and equating to zero the coefficients of $\nabla \ln T, \nabla \ln T_i^{(V)}, \nabla \mathbf{v}_0, \nabla \cdot \mathbf{v}_0$, and $\mathbf{d}_j$. The unknown coefficients $A_{i\alpha}(W_i)$, etc., also satisfy certain auxiliary conditions (see Chapman and Cowling,[43] page 112). Thomson[68] has evaluated the collision integrals under the approximation that the dominant collision processes which give rise to non-Boltzmann molecular distribution functions are electron–molecule vibrational excitations and heavy-particle–heavy-particle elastic collisions.

The integral equations for $A_{i\alpha}(W_i)$, etc., are transformed into coupled sets of infinite sets of linear algebraic equations for the coefficients introduced upon expanding $A_{i\alpha}(W_i)$, etc., in terms of Sonine polynomials (see Chapman and Cowling,[43] p. 127). Thomson obtained approximate solutions to the algebraic equations while keeping only two terms in the Sonine expansion.

6

Plasma Chemistry Models of Gas Lasers

6.1. Low-Energy Reactions

6.1.1. Discharge Chemistry

In a laser cavity containing primary species CO_2, N_2, He, H_2, and CO, low-energy electrons are used to excite the vibrational modes of N_2 and CO_2. These same electrons also react with the primary species to form secondary species. The various types of reaction, with an example of each, are given below.

Dissociation

$$e + CO_2 \rightarrow CO + O + e \tag{6.1}$$

Ionization

$$e + CO_2 \rightarrow CO_2^+ + 2e \tag{6.2}$$

Dissociative attachment

$$e + CO_2 \rightarrow CO + O^- \tag{6.3}$$

The minority species so formed react chemically to form further minority species, e.g.,

$$O + N \rightarrow NO + h\nu \tag{6.4}$$

as well as themselves suffering electron impact dissociation, ionization, and dissociative attachment. Furthermore, positive ions participate in *dissociative recombination*, e.g.,

$$e + CO_2^+ \rightarrow CO + O \tag{6.5}$$

negative ions in *associative detachment*, e.g.,

$$O^- + CO \rightarrow CO_2 + e \tag{6.6}$$

and *negative-ion clustering*, e.g.,

$$O^- + CO_2 + M \rightarrow CO_3^- + M \tag{6.7}$$

Positive and negative ions participate in *ionic recombination*, e.g.,

$$NO_2^- + NO^+ \rightarrow NO_2 + NO \tag{6.8}$$

As a result of these various types of reactions, the CO_2 laser discharge will contain many secondary species. Laser performance is very sensitive to the relative number densities of these secondary species. Wiegand and Nighan[71,72] have shown theoretically that negative-ion concentration may become comparable to the electron density at low pressures, leading to plasma instability. Elsewhere[73] it has been shown that negative-ion concentration is likely to be at least an order of magnitude smaller than electron concentration at atmospheric pressure, but that this may nevertheless be sufficient to disturb the discharge stability.

At atmospheric pressure the dominant negative ion is likely to be[73] CO_3^-, formed mainly by reaction (6.7); O_2^- is also likely to be important. NO_2^- and NO_3^- have been shown both experimentally[74] and theoretically[73] to be dominant at low pressures, their concentrations depending largely on the amounts of NO_2 and NO present. Bletzinger *et al.*[75] have shown that laser gain is markedly reduced by the presence of 0.1% NO_2 or N_2O.

Another important factor influencing laser performance is the rate of dissociation and recombination of CO_2. The loss of CO_2 and the buildup of excess O_2 is of particular concern in sealed-off devices in which the CO fractional concentration has time to accumulate to greater than one part in ten, the dominant dissociation process of CO_2 being reaction (6.1). Stark *et al.*[76] have shown that the addition of H_2 and CO to the initial gas mixture can lead to a stable equilibrium mixture. According to Smith and Browne[77] the dominant effect of adding H_2, which leads to recombination of CO_2, is described by the reaction

$$OH + CO \rightarrow CO_2 + H \tag{6.9}$$

It is the purpose of this plasma chemistry model to investigate the temporal evolution of all the species within the plasma. A list of the species to be considered is given in Table 11.

6.1.2. *Rate Equations and Rate Coefficients*

It is a good approximation for atmospheric lasers[73] to treat the number densities of N_2 and He as constant. For particular initial ratios of the primary species CO_2, N_2, He, H_2, and CO, the time rate of change of

Table 11. List of Species within the Plasma

Species number	Species	Species number	Species
1	*e*	31	O_2^+
2	CO_2	32	N_2^+
3	N_2	33	NO^+
4	He	34	O^+
5	H_2	35	N^+
6	CO	36	CO_2^+
7	O	37	He^+
8	O_2	38	CO^+
9	NO_2	39	C^+
10	O_3	40	OH^+
11	N_2O	41	N_2O^+
12	NO	42	H_2^+
13	N	43	H_3^+
14	H	44	HO_2^+
15	H_2O	45	He_2^+
16	NO_3	46	N_2OH^+
17	HO_2	47	H_2O^+
18	OH	48	H_3O^+
19	$O_2(^1\Delta_g)$	49	$O_2^+\cdot N_2$
20	O^-	50	$NO^+\cdot NO$
21	O_2^-	51	$NO^+\cdot H_2O$
22	NO_2^-	52	$O_2^+\cdot N_2O$
23	O_3^-	53	$O_2^+\cdot H_2O$
24	H^-	54	$H_3O^+\cdot OH$
25	NO^-	55	N_3^+
26	N_2O^-	56	NO_2^+
27	OH^-	57	O_4^+
28	NO_3^-	58	$NO_2^-\cdot H_2O$
29	CO_3^-	59	$CO_3^-\cdot H_2O$
30	CO_4^-	60	$O_2^-\cdot H_2O$

the number density, n_i, of the species i depends on:

(1) the rate of production, S_i, of species i from external means

(2) the frequency, ν_i, of loss of the species i by processes other than reactions

(3) the rates of production and loss of the species i by reactions of the types outlined in Section 6.1

(4) reactions involving collisions between molecules and the cavity walls.

According to Smith (Reference 73, Appendix 1), at atmospheric pressure the diffusion rate to the tube walls or electrodes will be too slow for heterogeneous processes to compete with homogeneous processes*; we therefore shall ignore reactions involving collisions between molecules and the cavity walls. The space-independent rate equation for the species i may then be written in the general form given by Niles[78]:

$$dn_i/dt = S_i - \nu_i n_i + \sum_{a,b} n_a n_b k_{ab;i\gamma}$$

$$-\sum_c n_i n_c k_{ic;\gamma} + \sum_{abc} n_a n_b n_c k_{abc;i\gamma} - \sum_{c,d} n_c n_d n_i k_{cdi;\gamma} \tag{6.10}$$

The rate coefficients, $k_{ab;i\gamma}$, for two initial particle reactions are expressed in $cm^3\ sec^{-1}$. $k_{abc;i\gamma}$, expressed in $cm^6\ sec^{-1}$, are the rates for three initial particle reactions. The number of final particles depends on the particular reaction; the final particles are represented collectively by γ (or $i\gamma$). In general the rate coefficients are dependent on the ambient gas temperature T. For reactions in which electrons are initial particles, the corresponding rate coefficients are also dependent on an effective electron temperature T_e, which is related to the mean electron energy of the non-Maxwellian electron-energy distribution function f by

$$T_e\left(\frac{E}{N}, T\right) = \frac{2}{3}\int_0^\infty u^{3/2} f\left(u, \frac{E}{N}, T\right) du \tag{3.81a}$$

For these latter reactions the rate coefficients are given by

$$k_{\alpha,\beta}\left(\frac{E}{N}, T\right) = \left(\frac{2e}{m}\right)^{1/2} \int_0^\infty u f\left(u, \frac{E}{N}, T\right) Q_{\alpha,\beta}(u)\, du \tag{3.75a}$$

where E is the applied electric field used to accelerate the low-energy electrons, N is the total atom–molecule number density, and $Q_{\alpha,\beta}(u)$ is the cross section for the reaction $\alpha \to \beta$.

For a sealed-off laser the source term S_i in Eq. (6.10) will be zero for all primary species, whereas for a gas-flow laser, S_i is given by

$$S_i = R_i(dN/dt)_{\mathrm{IN}}$$

where $(dN/dt)_{\mathrm{IN}}$ is the inflow number density rate (in $cm^{-3}\ sec^{-1}$) and R_i is the fraction of primary species.

For self-sustained lasers no further source terms are included.

For electron-beam-controlled lasers a source term S_e for secondary electrons is required. S_e is usually taken to be a top-hat function, constant

* We take this to imply that reactions or losses, which are dependent on the presence of the cavity walls, have negligible effect compared to particle–particle interactions occurring within the cavity.

during the primary electron pulse. The corresponding source terms S_{i+} for the positive ions of primary species i are proportional to the rate of ionization of species i and the number density of i, normalized such that

$$\sum_i S_{i+} = S_e$$

For gas-flow lasers the loss frequency ν_i will include the term

$$N^{-1}(dN/dT)_{\mathrm{OUT}}$$

accounting for loss due to outflow of all species, both primary and secondary.

We assume all other losses e.g., diffusion (i.e., other than gas outflow and reactions), to be zero. This assumption has been made by Nighan and Wiegand[72] and is good in the high-pressure limit.[78]

6.1.2.1. *Rate Coefficients*

The rate coefficients $k_{ab;\gamma}$, $k_{abc;\gamma}$ are for the most part available from the literature; values are presented in Tables 12–17. Where it is known, the dependence on ambient temperature has been given.

For reactions with an electron as an initial particle, the rate coefficient is given by Eq. (3.75). The computer code (BOLTZ) (see Section 3.2) is used to calculate the electron-energy distribution function $f(u, E/N, T)$. The relevant cross sections $Q_{\alpha,\beta}(u)$ are mostly obtainable from the literature, according to the following: ionization cross sections for H_2, N_2, O_2, CO, NO, He, and CO_2 (see Browne[79]); cross sections for dissociative ionization of N_2, CO, H_2, NO, O_2, N_2O, and CO_2 (see Browne[79]) cross sections for dissociation and dissociative attachment of CO_2 (see Kieffer[49]); and cross sections for dissociative attachment of CO, NO, O_2, and N_2O (see Rapp and Briglia[80]).

Since some of the reactions are exothermic and some endothermic, and since there may also be V–T transitions taking place, the ambient temperature is expected to vary in time. However, the ambient temperature dependence of only a few of the rate coefficients is known, and investigations have been performed assuming that the ambient temperature is constant at 300°K.

Wiegand and Nighan[71] demonstrate theoretically that the electron temperature T_e is coupled to the electron density. Since the rates given by Eq. (3.75) are more sensitive to changes in T_e than to changes in T, an equation describing the time evolution of T_e is required to be coupled with Eq. (6.10). Such an equation has been given by Haas[70] and shall be

Table 12. Reactions Involving Electrons as Initial Particles

Reaction	k^a ($cm^3\ sec^{-1}$)	Reference
$O + e \rightarrow O^- + h\nu$	1.3(−15)[b]	78
$O_2 + e \rightarrow O_2^- + h\nu$	2.0(−19)	78
$NO_2 + e \rightarrow NO_2^- + h\nu$	1.0(−17)	78
$O_3 + e \rightarrow O_3^- + h\nu$	1.0(−17)	78
$O_3 + e \rightarrow O^- + O_2$	3.0(−12)	78
$O_3 + e \rightarrow O_2^- + O$	3.8(−22)	78
$O + e + O_2 \rightarrow O^- + O_2$	1.0(−31)	78
$O + e + N_2 \rightarrow O^- + N_2$	1.0(−31)	78
$O_2 + e + O \rightarrow O_2^- + O$	1.89(−30)	78
$NO_2 + e + O_2 \rightarrow NO_2^- + O_2$	3.0(−28)	78
$NO_2 + e + N_2 \rightarrow NO_2^- + N_2$	8.0(−28)	78
$O_2 + e + N_2 \rightarrow O_2^- + N_2$	1.0(−31)	78
$O_2^+ + e \rightarrow O + O$	2.2(−7)	78
$N_2^+ + e \rightarrow N + N$	2.8(−7)	78
$NO^+ + e \rightarrow N + O$	5.0(−7)	78
$O^+ + e + M \rightarrow O + M$	1.0(−26)	78
$O_2^+ + e + M \rightarrow O_2 + M$	1.0(−26)	78
$N_2^+ + e + M \rightarrow N_2 + M$	1.0(−26)	78
$NO^+ + e + M \rightarrow N + O + M$	1.0(−27)	78
$NO^+ + e + M \rightarrow NO + M$	1.0(−26)	78
$O^+ + e \rightarrow O + h\nu$	3.5(−12)	78
$O_2^+ + e \rightarrow O_2 + h\nu$	1.0(−12)	78
$N_2^+ + e \rightarrow N_2 + h\nu$	1.0(−12)	78
$NO^+ + e \rightarrow NO + h\nu$	1.0(−12)	78
$N^+ + e + M \rightarrow N + M$	1.0(−26)	78
$N^+ + e \rightarrow N + h\nu$	3.5(−12)	78
$CO_2^+ + e \rightarrow CO + O$	6.0(−8)	72
$NO^+ \cdot NO + e \rightarrow NO + NO$	1.7(−6)	78
$NO^+ \cdot H_2O + e \rightarrow NO + H_2O$	1.0(−6)	78
$NO_2 + He + e \rightarrow NO_2^- + He$	2.0(−11)[c]	73
$CO_2 + e \rightarrow CO + O^-$	5.0(−13)	73
$CO + e \rightarrow C + O^-$	3.0(−14)	73
$O_2 + e \rightarrow O + O^-$	3.0(−12)	73
$N_2O + e \rightarrow N_2 + O^-$	2.0(−10)	73
$NO + e \rightarrow N + O^-$	1.0(−12)	73
$NO_2 + e \rightarrow NO + O^-$	1.0(−11)	73

[a] k is expressed in $cm^6\ sec^{-1}$ for three initial particle reactions.
[b] Numbers in parentheses denote the power of 10.
[c] This is a saturated three-body process, expressed in $cm^3\ sec^{-1}$.

Table 13. Binary Negative-Ion Molecule Reactions

Reaction	ΔE (eV)	k ($cm^3\ sec^{-1}$)	Reference
$H^- + H \rightarrow H_2 + e$	3.8	1.3(−9)[a]	81
$H^- + NO_2 \rightarrow NO_2^- + H$		2.9(−9)	81
$O^- + NO_2 \rightarrow NO_2^- + O$		1.1(−9)	81
$O_2^- + NO_2 \rightarrow NO_2^- + O_2$		1.9(−9)	78, 81
$O^- + NO_2 \rightarrow O_2^- + NO$		1.0(−10)	73, 81
$O^- + O_2 \rightarrow O_3^- + h\nu$		1.0(−17)	78
$O^- + O \rightarrow O_2 + e$	3.6	1.9(−10)	78, 81
$O^- + O_2 \rightarrow O_3 + e$		5.0(−15)	78
$O^- + N \rightarrow NO + e$	5.1	2.6(−10)	81
$O^- + H_2 \rightarrow H_2O + e$	3.6	6.0(−10)	78, 81
$O^- + NO \rightarrow NO_2 + e$	1.4	2.6(−10)	81
$O^- + CO \rightarrow CO_2 + e$	4.0	6.5(−10)	73, 81
$O^- + N_2 \rightarrow N_2O + e$	0.2	2.0(−19)	78
$O^- + O_3 \rightarrow O_3^- + O$		5.3(−10)	78, 81
$O^- + N_2O \rightarrow NO^- + NO$	>3.5	2.0(−10)	81
$\rightarrow N_2O^- + O$		2.0(−12)	81
$H^- + O_2 \rightarrow HO_2 + e$	0.3	1.5(−9)	81
$O_2^- + O_2 \rightarrow 2O_2 + e$		5.8(−19)	78
$O_2^- + O \rightarrow O_3 + e$	0.6	3.0(−10)	78, 81
$O_2^- + N_2 \rightarrow O_2 + N_2 + e$		1.1(−19)	78
$O_2^- + N \rightarrow NO_2 + e$	4.1	4.0(−10)	78, 81
$\rightarrow NO + O + e$	1.0	4.0(−10)	81
$O_2^- + O \rightarrow O_2 + O^-$		1.0(−11)	78
$O_2^- + O_3 \rightarrow O_3^- + O_2$		3.5(−10)	78, 81
$O_2^- + NO \rightarrow NO_2^- + O$		<1.0(−12)	81
$O_2^- + NO_3 \rightarrow O_2 + NO_3^-$		5.0(−10)	78
$O_2^- + N_2O \rightarrow O_3^- + N_2$		~0.1(−10)	81
$OH^- + O \rightarrow HO_2 + e$	0.9	2.0(−10)	81
$OH^- + NO_2 \rightarrow NO_2^- + OH$		1.0(−9)	81
$OH^- + H \rightarrow H_2O + e$	3.2	1.0(−9)	81
$OH^- + N \rightarrow HNO + e$	2.4	<1.0(−11)	81
$CN^- + H \rightarrow HCN + e$	1.6	~8.0(−10)	81
$NO^- + O_2 \rightarrow O_2^- + NO$		9.0(−10)	81
$NH_2^- + NO_2 \rightarrow NO_2^- + NH_3$		1.0(−9)	81
$O_3^- + NO_3 \rightarrow O_3 + NO_3^-$		5.0(−10)	78
$O_3^- + CO_2 \rightarrow CO_3^- + O_2$		4.0(−10)	78, 81
$O_3^- + NO \rightarrow NO_3^- + O$		1.0(−11)	78, 81
$\rightarrow NO_2^- + O_2$		≪1.0(−11)	81
$O_3^- + O \rightarrow O_2^- + O_2$		1.0(−10)	78
$CO_3^- + O \rightarrow O_2^- + CO_2$		0.8(−10)	78, 81
$CO_3^- + NO \rightarrow NO_2^- + CO_2$		9.0(−12)	78, 81
$NO_2^- + O_3 \rightarrow NO_3^- + O_2$		1.8(−11)	81
$NO_3^- + NO \rightarrow NO_2^- + NO_2$		<1.0(−12)	81
$NO_2^- + NO_2 \rightarrow NO_3^- + NO$		4.0(−12)	81
$NO_2^- + NO_3 \rightarrow NO_2 + NO_3^-$		5.0(−10)	78
$O_4^- + NO \rightarrow NO_3^- + O_2$		2.5(−10)	81
$O_4^- + CO_2 \rightarrow CO_4^- + O_2$		4.3(−10)	81

continued

Table 13. (continued)

Reaction	ΔE (eV)	k ($cm^3 sec^{-1}$)	Reference
$CO_4^- + NO \rightarrow NO_3^- + CO_2$		4.8(−11)	81
$O_4^- + O \rightarrow O_3^- + O_2$		4.0(−10)	81
$\rightarrow O^- + 2O_2$		<4.0(−10)	81
$CO_4^- + O \rightarrow CO_3^- + O_2$		1.5(−10)	81
$\rightarrow O_3^- + CO_2$		<1.5(−10)	81
$N_2O^- + O_2 \rightarrow O_3^- + N_2$		1.0(−9)	81
$O_2^- + O_2(^1\Delta) \rightarrow O_2 + e + O_2$		1.0(−9)	78
$O^- + O_2(^1\Delta) \rightarrow O_2^- + O$		1.0(−9)	78
$O^- + O_2(^1\Delta) \rightarrow O_3 + e$		1.0(−10)	78
$O^- + O_2(^1\Delta) \rightarrow O_2 + h\nu + O^-$		2.6(−4)	78
$CO_3^- + CO \rightarrow 2CO_2 + e$		5.0(−13)	73
$O_3^- + N_2 \rightarrow NO_2^- + NO$		<5.0(−14)	81
$O_3^- + NO_2 \rightarrow NO_2^- + O_3$		7.0(−11)	78, 81
$O_3^- + NO_2 \rightarrow NO_3^- + O_2$		2.0(−11)	78
$CO_3^- + NO_2 \rightarrow NO_3^- + CO_2$		1.0(−10)	73, 81

[a] Numbers in parentheses denote the power of 10.

Table 14. Binary Positive Ion–Molecule Reactions

Reaction	ΔE (eV)	k ($cm^3 sec^{-1}$)	Reference
$He^+ + N_2 \rightarrow N^+ + N + He$	0.3	1.7(−9)[a]	81
$\rightarrow N_2^+ + He$	9.0	1.5(−9)	81
$He^+ + CO \rightarrow C^+ + O + He$	2.2	1.7(−9)	81
$He^+ + NO \rightarrow N^+ + O + He$	3.6	1.8(−9)	81
$He^+ + O_2 \rightarrow O^+ + O + He$	5.9	2.1(−9)	81
$He^+ + CO_2 \rightarrow O^+ + CO + He$	4.5	1.2(−9)	81
$\rightarrow CO^+ + O + He$	4.1	1.2(−9)	81
$C^+ + O_2 \rightarrow CO^+ + O$	3.27	1.0(−9)	81
$C^+ + CO_2 \rightarrow CO^+ + CO$	2.96	1.9(−9)	81
$C^+ + N_2O \rightarrow NO^+ + CN$	5.54	9.1(−10)	81
$N^+ + H_2 \rightarrow NH^+ + H$		7.0(−10)	81
$N^+ + N \rightarrow N_2^+ + h\nu$		3.0(−17)	78
$N^+ + CO \rightarrow CO^+ + N$	0.53	5.0(−10)	81
$N^+ + NO \rightarrow NO^+ + N$	5.30	8.0(−10)	78, 81
$\rightarrow N_2^+ + O$		3.0(−12)	78
$\rightarrow O^+ + N_2$		1.0(−12)	78
$N^+ + O_2 \rightarrow O_2^+ + N$	2.47	4.5(−10)	81
$\rightarrow NO^+ + O$	6.70	5.0(−10)	78, 81
$\rightarrow O^+ + NO$		1.0(−12)	78
$N^+ + CO_2 \rightarrow CO_2^+ + N$	0.75	1.3(−9)	81
$\rightarrow NO^+ + CO$	6.40	1.8(−11)	81
$N^+ + O \rightarrow N + O^+$		1.0(−12)	78
$\rightarrow NO^+ + h\nu$		1.0(−17)	78
$N^+ + N_2O \rightarrow NO^+ + N_2$		5.5(−10)	81
$O^+ + H_2 \rightarrow OH^+ + H$	0.30	2.0(−9)	81

continued

Table 14. (continued)

Reaction	ΔE (eV)	k ($cm^3\ sec^{-1}$)	Reference
$O^+ + N_2 \rightarrow NO^+ + N$	1.10	2.4(−12)	81
$O^+ + NO \rightarrow NO^+ + O$	4.36	2.4(−11)	78
$O^+ + O_2 \rightarrow O_2^+ + O$	1.50	2.0(−11)	78, 81
$O^+ + CO_2 \rightarrow O_2^+ + O$	1.50	1.1(−9)	78, 81
$O^+ + N_2O \rightarrow N_2O^+ + O$	0.7	4.0(−10)	81
$\rightarrow O_2^+ + N_2$	5.0	2.0(−11)	81
$H_2^+ + H_2 \rightarrow H_3^+ + H$		2.1(−9)	81
$H_2^+ + O_2 \rightarrow HO_2^+ + H$		6.9(−9)	81
$He_2^+ + N_2 \rightarrow N_2^+ + 2He$	6.7	1.3(−9)	81
$N_2^+ + H_2 \rightarrow N_2H^+ + H$		1.9(−9)	81
$N_2^+ + N \rightarrow N^+ + N_2$	1.04	1.0(−12)	78
$N_2^+ + O \rightarrow NO^+ + N$	3.05	2.5(−10)	78, 81
$\rightarrow O^+ + N_2$	1.97	1.0(−12)	78
$N_2^+ + CO \rightarrow CO^+ + N_2$	1.6	7.0(−11)	81
$N_2^+ + O_2 \rightarrow O_2^+ + N_2$	3.5	1.1(−10)	81
$\rightarrow NO^+ + NO$	4.5	1.0(−17)	78
$N_2^+ + NO \rightarrow NO^+ + N_2$	6.3	4.9(−10)	81
$N_2^+ + CO_2 \rightarrow CO_2^+ + N_2$	1.75	9.0(−10)	81
$N_2^+ + N_2O \rightarrow NO^+ + N_2 + N$	1.8	4.0(−10)	81
$\rightarrow N_2O^+ + N_2$	2.6	5.0(−10)	81
$O_2^+ + N \rightarrow NO^+ + O$	4.0	1.8(−10)	78, 81
$O_2^+ + NO \rightarrow NO^+ + O_2$	2.8	8.0(−10)	78, 81
$O_2^+ + N_2 \rightarrow NO^+ + NO$		1.0(−16)	78
$CO^+ + H_2 \rightarrow COH^+ + H$		2.0(−9)	81
$CO^+ + O_2 \rightarrow O_2^+ + CO$	1.94	2.0(−10)	81
$CO^+ + CO_2 \rightarrow CO_2^+ + CO$	0.22	1.1(−9)	81
$H_3^+ + N_2 \rightarrow N_2H^+ + H_2$		1.0(−9)	81
$CO_2^+ + H_2 \rightarrow CO_2H^+ + H$		1.4(−9)	81
$CO_2^+ + O_2 \rightarrow O_2^+ + CO_2$	1.72	1.0(−10)	81
$CH_2^+ + H_2 \rightarrow CH_3^+ + H$		2.3(−10)	81
$N_2O^+ + H_2 \rightarrow N_2OH^+ + H$		4.0(−10)	81
$H_2O^+ + H_2O \rightarrow H_3O^+ + OH$		1.0(−9)	81
$O_2^+ \cdot N_2 + O_2 \rightarrow O_4^+ + N_2$		≥5(−11)	81
$O_4^+ + N_2O \rightarrow O_2^+ \cdot N_2O + O_2$		2.5(−10)	81
$O_2^+ \cdot N_2O + H_2O \rightarrow O_2^+ \cdot H_2O + N_2O$		≥1.0(−10)	81
$O_4^+ + H_2O \rightarrow O_2^+ \cdot H_2O + O_2$		1.2(−9)	81
$O_2^+ \cdot H_2O + H_2O \rightarrow H_3O^+ + OH + O_2$		3(−10)	81
$\rightarrow H_3O^+ \cdot OH + O_2$		1(−9)	81
$He_2^+ + NO \rightarrow NO^+ + 2He$		1.3(−9)	81
$He_2^+ + O_2 \rightarrow O_2^+ + 2He$		1.1(−9)	81
$He_2^+ + CO \rightarrow CO^+ + 2He$		1.4(−9)	81
$O^+ + O \rightarrow O_2^+ + h\nu$		1.0(−17)	78
$O^+ + NO \rightarrow O_2^+ + N$		3.0(−12)	78
$O^+ + N \rightarrow NO^+ + h\nu$		1.0(−18)	78
$O_2^+ + NO_2 \rightarrow NO^+ + O_3$		1.0(−11)	78

[a] Numbers in parentheses denote the power of 10.

Table 15a. Three-Molecule Reactions

Reaction	k ($cm^6\ sec^{-1}$)	Reference
$He^+ + 2He \rightarrow He_2^+ + He$	8.7(−32)[a]	81
$N_2^+ + 2N_2 \rightarrow N_4^+ + N_2$	7.2(−29)	81
$CO_2^+ + 2CO_2 \rightarrow C_2O_4^+ + CO_2$	3.0(−28)	81
$O^- + 2O_2 \rightarrow O_3^- + O_2$	8.6(−31)	81
$O^- + 2CO_2 \rightarrow CO^- + CO_2$	8.0(−29)	81
$O_2^- + 2CO_2 \rightarrow CO_4^- + CO_2$	9.0(−30)	81
$O_2^- + CO_2 + O_2 \rightarrow CO_4^- + O_2$	2.0(−29)	81
$H^+ + 2H_2 \rightarrow H_3^+ + H_2$	3.2(−29)	81
$N^+ + 2N_2 \rightarrow N_3^+ + N_2$	1.8(−29)	81
$O^- + CO_2 + He \rightarrow CO_3^- + He$	1.5(−28)	81
$N_2^+ + N_2 + He \rightarrow N_4^+ + He$	1.9(−29)	81
$N^+ + N_2 + He \rightarrow N_3^+ + He$	8.6(−30)	81
$O_2^+ + O_2 + He \rightarrow O_4^+ + He$	3.1(−29)	81
$O^+ + N_2 + He \rightarrow N_2O^+ + He$	5.4(−29)	81
$NO^+ + 2NO \rightarrow NO^+ \cdot NO + NO$	5.0(−30)	81
$O_2^+ + 2O_2 \rightarrow O_4^+ + O_2$	2.8(−30)	81
$O_2^- + 2O_2 \rightarrow O_4^- + O_2$	3.0(−31)	81
$O^- + O^+ + N \rightarrow O_2 + N$	2.0(−25)	78
$O^- + O^+ + N_2 \rightarrow O_2 + N_2$	2.0(−25)	78
$O^- + O^+ + O_2 \rightarrow 2O_2$	2.0(−25)	78
$O^- + O^+ + O \rightarrow O_2 + O$	2.0(−25)	78
$O_2^- + O^+ + M$[b] $\rightarrow O_3 + M$	2.0(−25)	78
$O^- + O_2^+ + M \rightarrow O_3 + M$	2.0(−25)	78
$O_2^- + O_2^+ + M \rightarrow 2O_2 + M$	2.0(−25)	78
$O^- + N_2^+ + M \rightarrow N_2O + M$	2.0(−25)	78
$O_2^- + N_2^+ + M \rightarrow O_2 + N_2 + M$	2.0(−25)	78
$O^- + NO^+ + M \rightarrow NO_2 + M$	2.0(−25)	78
$O_2^- + NO^+ + M \rightarrow O_2 + NO + M$	2.0(−25)	78
$O^+ + O + M \rightarrow O_2^+ + M$	1.0(−29)	78
$O^+ + N + M \rightarrow NO^+ + M$	1.0(−29)	78
$O^- + NO + M \rightarrow NO_2^- + M$	1.0(−29)	78
$O_2^- + N + M \rightarrow NO_2^- + M$	1.0(−29)	78
$O_2^- + O_2 + N_2 \rightarrow NO_2^- + NO_2$	5.8(−42)	78
$O + O + O_2 \rightarrow O_2 + O_2$	3.0(−33)	78
$O + O + O \rightarrow O_2 + O$	3.0(−33)	78
$O + O + N_2 \rightarrow O_2 + N_2$	3.0(−33)	78
$O + O_2 + O_2 \rightarrow O_3 + O_2$	5.5(−34)	78
$O + O_2 + N_2 \rightarrow O_3 + N_2$	5.5(−34)	78
$O + O_2 + O \rightarrow O_3 + O$	5.5(−34)	78
$N + O + M \rightarrow NO + M$	1.1(−32)	78
$O + N_2 + M \rightarrow N_2O + M$	1.4(−45)	78
$O + NO + O_2 \rightarrow NO_2 + O_2$	1.0(−31)	78
$O + NO + N_2 \rightarrow NO_2 + N_2$	1.0(−31)	78
$N + N + M \rightarrow N_2 + M$	5.0(−33)	78
$N + NO + M \rightarrow N_2O + M$	3.6(−36)	78
$NO + O_2 + NO \rightarrow NO_2 + NO_2$	1.6(−46)	78
$N^+ + O + M \rightarrow NO^+ + M$	1.0(−29)	78

continued

Table 15a. (continued)

Reaction	k ($cm^6\ sec^{-1}$)	Reference
$N^+ + N + M \rightarrow N_2^+ + M$	1.0(−29)	78
$NO^+ + H_2O + M \rightarrow NO^+ \cdot H_2O + M$	2.0(−28)	78
$NO^+ + NO + M \rightarrow NO^+ \cdot NO + M$	4.0(−30)	78
$NO_2^- + H_2O + M \rightarrow NO_2^- \cdot H_2O + M$	1.0(−28)	78
$O^- + CO_2 + O_2 \rightarrow CO_3^- + O_2$	3.1(−28)	73
$CO_3^- + H_2O + M \rightarrow CO_3^- \cdot H_2O + M$	1.0(−28)	73
$O_2^- + H_2O + M \rightarrow O_2^- \cdot H_2O + M$	2.2(−28)	73
$H + O_2 + M \rightarrow HO_2 + M$	3.7(−32)	77
$O + 2NO \rightarrow NO_2 + NO$	1.5(−31)	82
$O + NO + He \rightarrow NO_2 + He$	6.7(−32)	82
$O + O_2 + CO \rightarrow O_3 + CO$	6.7(−34)	82
$O + 2CO \rightarrow CO_2 + CO$	3.2(−32)	82
$O + CO + N_2 \rightarrow CO_2 + N_2$	2.2(−36)	82
$O + CO + CO_2 \rightarrow 2CO_2$	6.2(−36)	82
$H + H + M \rightarrow H_2 + M$	8.4(−33)	83

[a] Numbers in parentheses denote the power of 10.
[b] Symbol M represents spectator particle for energy/momentum conservation.

Table 15b. Rates with Temperature Dependence Known Explicitly

Reaction	n[a]	E/R[a] (°K)	A[a] ($cm^6\ mole^{-2}\ sec^{-1}$)	Temperature range (°K)	Reference
$N + N + N_2 \rightarrow N_2 + N_2$	0	−500	3.0(14)[b]	200–600	84
$O + N + N_2 \rightarrow NO + N_2$	−0.5	0	6.4(16)	200–400	84
$NO + NO + O_2 \rightarrow NO_2 + NO_2$	0	−530	1.2(9)	273–660	84
$NO_2 + NO + O_2 \rightarrow NO_2 + NO_3$	0	−400	2.9(7)	300–500	84
$O + NO + O_2 \rightarrow NO_2 + O_2$	0	−940	1.1(15)	200–500	84
$O_2 + H + He \rightarrow HO_2 + He$	0	−500	1.5(15)	300–2000	84
$O + CO + M^c \rightarrow CO_2 + M(CO)$	0	2184	2.4(15)	250–500	[d]
$O + O + M \rightarrow O_2 + M(Ar)$	0	900	1.9(13)	190–4000	[d]

[a] See Eq. (6.12).
[b] Numbers in parentheses denote the power of 10.
[c] Symbol M represents spectator particle for energy/momentum conservation.
[d] See reference 84, Volume 3 (to be published).

Table 16. Binary Positive-Ion–Negative-Ion Reactions

Reaction	k ($cm^3\ sec^{-1}$)	Reference
$O^- + O^+ \rightarrow O + O$	2.0(−7)[a]	78
$O_2^- + O^+ \rightarrow O_2 + O$	2.0(−7)	78
$NO_2^- + O^+ \rightarrow NO_2 + O$	2.0(−7)	78
$O_3^- + O^+ \rightarrow O_3 + O$	2.0(−7)	78
$O^- + O_2^+ \rightarrow O + O_2$	2.0(−7)	78
$O_2^- + O_2^+ \rightarrow O_2 + O_2$	2.0(−7)	78

continued

Table 16. (continued)

Reaction	k (cm^3 sec^{-1})	Reference
$NO_2^- + O_2^+ \rightarrow NO_2 + O_2$	2.0(−7)	78
$O_3^- + O_2^+ \rightarrow O_3 + O_2$	2.0(−7)	78
$O^- + N_2^+ \rightarrow O + N_2$	2.0(−7)	78
$O_2^- + N_2^+ \rightarrow O_2 + N_2$	2.0(−7)	78
$NO_2^- + N_2^+ \rightarrow NO_2 + N_2$	2.0(−7)	78
$O_3^- + NO_2^+ \rightarrow O_3 + NO_2$	2.0(−7)	78
$O_2^- + NO^+ \rightarrow O_2 + NO$	2.0(−7)	78
$O^- + NO^+ \rightarrow O + NO$	2.0(−7)	78
$NO_2^- + NO^+ \rightarrow NO_2 + NO$	2.0(−7)	78
$O_3^- + NO^+ \rightarrow O_3 + NO$	2.0(−7)	78
$NO_3^- + NO^+ \rightarrow NO_3 + N + O$	1.0(−7)	78
$O_2^- + NO^+ \cdot NO \rightarrow O_2 + NO + NO$	1.0(−7)	78
$O_3^- + NO^+ \cdot NO \rightarrow O_3 + NO + NO$	1.0(−7)	78
$NO_2^- + NO^+ \cdot NO \rightarrow NO_2 + NO + NO$	1.0(−7)	78
$NO_3^- + NO^+ \cdot NO \rightarrow NO_3 + NO + NO$	1.0(−7)	78
$O_2^- + NO^+ \cdot H_2O \rightarrow O_2 + NO + H_2O$	1.0(−7)	78
$O_3^- + NO^+ \cdot H_2O \rightarrow O_3 + NO + H_2O$	1.0(−7)	78
$NO_2^- + NO^+ \cdot H_2O \rightarrow NO_2 + NO + H_2O$	1.0(−7)	78
$O_2^- + NO^+ \rightarrow O_2 + N + O$	1.0(−7)	78
$NO_3^- + NO^+ \cdot H_2O \rightarrow NO_3 + NO + H_2O$	1.0(−7)	78
$O_2^- + O_2^+ \rightarrow O_2 + O + O$	1.0(−7)	78
$O_3^- + O_2^+ \rightarrow O_3 + O + O$	1.0(−7)	78
$NO_2^- + O_2^+ \rightarrow NO_2 + O + O$	1.0(−7)	78
$NO_3^- + O_2^+ \rightarrow NO_3 + O + O$	1.0(−7)	78
$NO_3^- + NO^+ \rightarrow NO_3 + NO$	2.0(−7)	78
$O_2^- + N_2^+ \rightarrow O_2 + N + N$	1.0(−7)	78
$NO_2^- + N_2^+ \rightarrow NO_2 + N + N$	1.0(−7)	78
$O_3^- + N_2^+ \rightarrow O_3 + N + N$	1.0(−7)	78
$NO_2^- + NO^+ \rightarrow NO_2 + N + O$	1.0(−7)	78
$O_3^- + NO^+ \rightarrow O_3 + N + O$	1.0(−7)	78
$N^+ + O^- \rightarrow N + O$	2.0(−7)	78
$O_2^- + O_2^+ \rightarrow O_2 + O_2(^1\Delta)$	1.0(−8)	78
$NO_2^- \cdot H_2O + NO^+ \rightarrow NO_2 + H_2O + NO$	1.0(−7)	78
$NO_2^- \cdot H_2O^+ O_2^+ \rightarrow NO_2 + H_2O + O_2$	1.0(−7)	78
$NO_2^- \cdot H_2O + NO^+ \cdot H_2O \rightarrow NO_2 + 2H_2O + NO$	1.0(−7)	78
$NO_2^- \cdot H_2O + NO^+ \cdot NO \rightarrow NO_2 + H_2O + 2NO$	1.0(−7)	78

[a] The numbers in parentheses denote the power of 10.

Table 17a. Binary Neutral Molecule Collisions

Reaction	k (cm^3 sec^{-1})	Reference
$N + O \rightarrow NO + h\nu$	2.0(−17)[a]	78
$O + NO \rightarrow NO_2 + h\nu$	6.2(−17)	78
$N + N \rightarrow N_2 + h\nu$	1.0(−17)	78
$O + N_2 \rightarrow NO + N$	3.2(−65)	78

continued

Table 17a. (*continued*)

Reaction	k (cm^3 sec^{-1})	Reference
$O+NO \rightarrow O_2+N$	6.7(−41)	78
$O+NO_2 \rightarrow NO+O_2$	4.5(−12)	78
$O+N_2O \rightarrow NO+NO$	8.1(−31)	78
$O+N_2O \rightarrow O_2+N_2$	2.7(−31)	78
$O+O_3 \rightarrow O_2+O_2$	9.4(−15)	78
$N+O_2 \rightarrow NO+O$	8.6(−17)	78
$N+NO \rightarrow N_2+O$	2.2(−11)	78
$N+NO_2 \rightarrow N_2+O_2$	1.5(−11)	78
$N+NO_2 \rightarrow NO+NO$	3.0(−12)	78
$N+NO_2 \rightarrow N_2O+O$	1.5(−13)	78
$NO+O_3 \rightarrow NO_2+O_2$	1.3(−14)	78
$NO_2+O_3 \rightarrow NO_3+O_2$	7.0(−17)	78
$O_3+O_2(^1\Delta) \rightarrow O+O_2+O_2$	1.9(−15)	78
$O_2(^1\Delta)+M^a \rightarrow O_2+M$	1.0(−19)	78
$O_2(^1\Delta)+N \rightarrow NO+O$	3.0(−12)	78
$H+O_2 \rightarrow OH+O$	3.0(−22)	77
$O+H_2 \rightarrow OH+H$	>9.0(−18)	77
$O+HO_2 \rightarrow OH+O_2$	9.0(−12)	77
$OH+CO \rightarrow CO_2+H$	1.8(−13)	77
$O+OH \rightarrow O_2+H$	3.6(−11)	77
$H_2+OH \rightarrow H_2O+H$	7.0(−15)	77
$OH+OH \rightarrow H_2O+O$	2.4(−13)	77
$H+OH \rightarrow H_2+O$	3.1(−17)	77

[a] Symbol M represents spectator particle for energy/momentum conservation.

Table 17b. Rates with Temperature Dependence Known Explicitly

Reaction	n[a]	E/R[a] (°K)	A[a] (cm^3 mole^{-1} s^{-1})	Temperature range (°K)	Reference
$O+OH \rightarrow O_2+H$	0	0	~2(13)[b]	300	85
$H_2+O \rightarrow H+OH$	1.0	4,480	1.8(10)	400–2000	84
$H+O_2 \rightarrow O+OH$	0.13	2,180	1.9(14)	300–2500	85
$H+OH \rightarrow H_2+O$	1.0	3,500	8.3(9)	400–2000	84
$H_2+OH \rightarrow H_2O+H$	0	2,590	2.2(13)	300–2500	84
$H_2O+H \rightarrow H_2+OH$	0	10,250	9.3(13)	300–2500	84
$H_2O+O \rightarrow OH+OH$	0	9,240	6.8(13)	300–2000	84
$OH+OH \rightarrow H_2O+O$	0	550	6.3(12)	300–2000	84
$H+HO_2 \rightarrow OH+OH$	0	950	2.5(14)	290–800	84
$OH+OH \rightarrow H+HO_2$	0	20,200	1.2(13)	290–800	84
$H+HO_2 \rightarrow H_2+O_2$	0	350	2.5(13)	290–800	84
$H_2+O_2 \rightarrow H+HO_2$	0	29,100	5.5(13)	290–800	84
$H_2O_2+OH \rightarrow H_2O+HO_2$	0	910	1.0(13)	300–800	84
$H_2O+HO_2 \rightarrow H_2O_2+OH$	0	16,500	2.8(13)	300–800	84
$H_2O_2+H \rightarrow H_2+HO_2$	0	1,900	1.7(12)	300–800	84
$H_2+HO_2 \rightarrow H_2O_2+H$	0	9,400	7.3(11)	300–800	84
$O_2+N \rightarrow NO+O$	1.0	3,150	6.4(9)	300–3000	84

continued

Table 17b. (continued)

Reaction	n^a	E/R^a (°K)	A^a (cm³ mole⁻¹ s⁻¹)	Temperature range (°K)	Reference
$NO_2+O \rightarrow NO+O_2$	0	300	1.0(13)	300–550	84
$NO+O_2 \rightarrow NO_2+O$	0	23,400	1.7(12)	300–550	84
$NO_2+NO_3 \rightarrow NO+NO_2+O_2$	0	1,600	1.4(11)	300–850	84
$NO+O_3 \rightarrow NO_2+O_2$	0	1,300	8.9(11)	200–350	84
$NO_2+H \rightarrow NO+OH$	0	740	3.5(14)	298–630	84
$N_2O+H \rightarrow N_2+OH$	0	7,600	7.6(13)	700–2500	84
$NO+N \rightarrow N_2+O$	0	0	1.6(13)	300–5000	84
$NO+O \rightarrow NO_2+h\nu$	−2.0	0	1.5(5)	230–3750	84
$O_3+NO_2 \rightarrow NO_3+O_2$	0	3,500	5.9(12)	285–300	84
$O_2+O_2 \rightarrow O+O_2+O_2$	−0.5	59,700	5.5(17)	300–7000	85
$O_2+O \rightarrow O+O+O$	−0.5	59,700	1.8(18)	300–7000	85
$O_2+M^c \rightarrow O+O+M$	−0.5	59,700	8.8(16)	300–7000	85
$M=(N_2, N, NO)$					
$HO_2+CO \rightarrow CO_2+OH$	0	11,900	1.5(14)	700–1000	84, Vol. 3
$O+CO \rightarrow CO_2+h\nu$	0	1,600	2.5(6)	200–2000	85, Vol. 3
$OH+CO \rightarrow CO_2+H$	$\log_{10} k = 10.83+3.94(-4)T$			250–2500	85, Vol. 3
$NO_2+NO_2 \rightarrow NO+NO$	0	13,500	4.0(12)	600–2000	83, Vol. 3

[a] See Eq. (6.12).
[b] Numbers in parentheses denote the power of 10.
[c] Symbol M represents spectator particle for energy/momentum conservation.

derived in detail in Chapter 8. For a spatially independent model, Eq. (3.75) of Haas may be written

$$\frac{d}{dt}\left(\frac{3}{2}n_e kT_e\right) = -en_e\bar{V}_e \cdot \mathbf{E} - n_e\nu_u(T_e)kT_e \tag{5.161a}$$

$\nu_u(T_e)$ is the electron energy-exchange collision frequency, defined by Eq. (5.162). $\bar{\mathbf{V}}_e$ is the electron drift velocity. A computer code (BOLTZ) (see Section 3.2) has been written which calculates $\nu_u(T_e)$ and $\bar{\mathbf{V}}_e$ via the definitions of Chapter 3. Notice that $-en_e\bar{\mathbf{V}}_e$ is the electron current density $\mathbf{j}_e$. Wiegand and Nighan[71] evaluate $\mathbf{j}_e$ using the appropriate circuit equation, assuming that the positive column is the load on a resistively ballasted DC power supply.

The system of coupled equations (6.10) and (5.161a) may be solved numerically by stepwise integration. Computational methods are discussed in Section 6.1.4. The number of differential equations may be reduced by one by consideration of charge balance. Since charge neutrality is preserved to a high degree within the plasma, we may calculate the number density of, e.g., the dominant negative ion, N^-, from the equation

$$\Delta n_{N^-} = \sum_{M^+} \Delta n_{M^+} - \sum_{M^- \neq N^-} \Delta n_{M^-} - \Delta n_e \tag{6.11}$$

where Δn_P is the change in the number density of species P during the time Δt.

6.1.2.2. *Notes on the Temperature-Dependent Rates of Tables 15b and 17b*

The temperature dependence of the rate coefficients is given in the general form

$$k = T^n A \exp(-E/RT) \tag{6.12}$$

For the reaction given in column 1 (Tables 15b and 17b), n is given in column 2, E/R (°K) in column 3, and A ($cm^3\, mole^{-1}\, sec^{-1}$ or $cm^6\, mole^{-2}\, sec^{-1}$) in column 4.

Formula (6.12) gives k in the same units as A. In order to arrive at the units employed for the time-independent rates, binary reactions should be divided by N_0 and ternary reactions by N_0^2, where

$$N_0 = 6.023 \times 10^{23} \text{ molecules/mole}$$

is Avogadro's number.

For reactions involving a structureless particle M, the actual species involved in the reaction measured has been included in parentheses after the reaction in column 1.

6.1.3. *Comparison with Other Models*

A plasma code has been developed by Niles,[78] who solved a set of 24 rate equations involving over 200 reactions of ionospheric importance. Most of the reactions he considers involve atoms, molecules, and ions of combinations of N and O. CO_2 and related ions and molecules are all treated as minority species. Wiegand and Nighan[71,72] extend the work of Niles by including the effect of CO_2 as a primary species. They solve rate equations for nearly 40 neutral and charged species involving approximately 300 reactions.

The calculations of Wiegand and Nighan[71] predict the temporal development of all species (primary and secondary) in electrically excited mixtures of CO_2, N_2, and He at pressures in the 10–100 Torr range. The temporal variations of the effective electron temperature, and the direct CO_2 ionization and CO_2 dissociative attachment coefficients, were also obtained.

Our code, developed mostly by S. A. Roberts, includes all the reactions considered by Wiegand and Nighan, and extends their results to include: (1) The addition of H_2 and CO in the initial gas mixture, (2) pressures up to atmospheric, and (3) the effect of varying gas-flow speeds.

The problems of dissociation of CO_2 and the buildup of negative ions are treated separately in the work of the Laser Group at St. Andrews University.[73,77] A plasma code is reported, but one which considers the production and loss of only the various negative ions. The assumption is made that the number densities of all neutral species (CO_2, O_2, CO, N_2O, NO_2, and NO) are constant during the discharge period. The densities of N_2O, NO_2, and NO are estimated from empirical knowledge, and the densities of CO_2, CO, and O_2 are obtained by a separate analysis that considers the dissociation and recombination of CO_2.[73,77] In the plasma code[73] no allowance is made for any neutral reactions of H_2, H_2O, or OH.

Our code is more general than this latter code in that it considers simultaneously all the reactions discussed in Reference 73 together with the production and loss of positive ions (including positive-ion–negative-ion reactions), and all reactions between neutral species.

Similar to the other codes mentioned, our code is based on a point model (spatially independent). The ambient temperature is assumed to remain constant during the discharge period. Unlike the plasma code of Niles,[78] the effects of diffusion are not included.

6.1.4. Computational Methodologies

The set of coupled differential equations (6.10) and (5.161a) exhibits the so-called stiffness property; that is, the various rate coefficients k have widely differing values, and special integration formulas are required for their efficient solution.

Techniques based on the Runge–Kutta scheme have been used by Niles[78] and by Nighan and Wiegand.[72] Niles based his computer code on that developed by Keneshea,[86] who employs the Kutta–Merson technique; this has an advantage in that it provides an error analysis at each integration step. Nighan and Wiegand based their original code on the Treanor method,[87] but more recently have used a modification of the work of Nosrati.[88]

A set of coupled first-order differential equations has been solved by Davies, Smith, and Thomson[19] using a modified Hamming predictor–corrector method, initial values at four time steps being computed using the Runge–Kutta method. This same method has been adapted to the present problem.

Since the plasma code solves over 30 equations involving over 300 reactions, it is desirable to avoid the need to write each equation in full within the code. The code has been written to generate the set of equations arising from the reactions contained in a data base of reaction rates.

6.1.5. PLASKEM—*A Program Used to Predict the Variation with Time of the Number Densities of Chemical Species within a Plasma*[89]

6.1.5.1. *Program Description*

The program consists of nine routines, MAIN, DEDWOD, FIND, FXY, DFEQS2, DO2AEF, PRNPL2, PLSCAL, PLSCL2, and a function subroutine, IICOMP. Figure 27 shows the program structure. The input data is read in by MAIN, and all variables are initialized. MAIN calls DEDWOD to discard any redundant reactions and/or redundant species, and then one of the routines DFEQS2 or DO2AEF to solve the set of differential equations. DFEQS2 or DO2AEF calls FXY to compute the right-hand side of the equations. PRNPL2 provides plots of number densities vs. time for six chemical species, two species on each of three graphs, on the off-line printer. A description of these routines follows.

MAIN

The input data is read in MAIN. The reactions, together with their rates, are read in. MAIN calls DEDWOD and then at each time step calls one of DFEQS2 or DO2AEF to solve the set of differential equations. MAIN prints out the solutions to the equations at every ten time steps and increases the time step throughout the run if required.

Subroutine DEDWOD (LAB1, LAB2, ND)

For a given set of reactions and given primary species, DEDWOD discards any reactions for which at least one of the initial particles is never present, and also discards any species that never acquire a nonzero number density. DEDWOD relabels the remaining reactions and species. The arguments of DEDWOD are listed and described in Table 18.

Subroutine FIND (N, LIST, NN1, NN2, ISW, INDEX)

Subroutine FIND effects a binary search on a list of ordered text variables and compares the elements in the list with the variable N, using

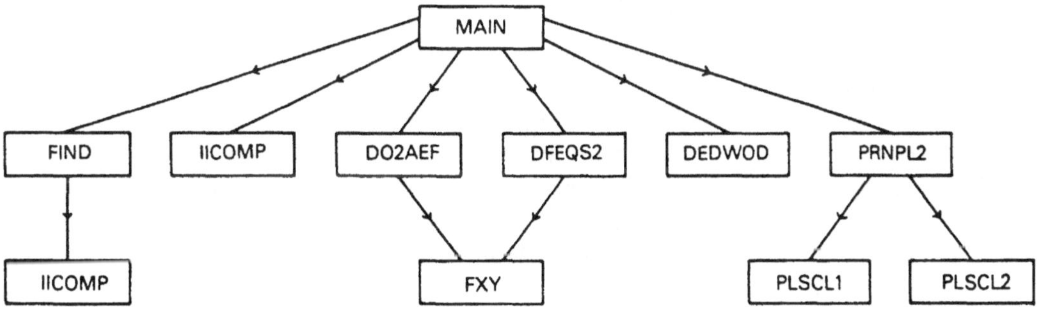

Fig. 27. Structure of PLASKEM.

Table 18. Arguments of DEDWOD

Argument	Description
LAB1	One-dimensional array, on exit LAB1(I) contains the old label of the species now labeled species I
LAB2	One-dimensional array, on exit LAB2(I) contains the old label of the reaction now labeled reaction *I*
ND	One-dimensional real array containing number densities of each species

the function IICOMP. The arguments of FIND are listed and described in Table 19.

Subroutine FXY (DE, E, T, N)

This subroutine computes the right-hand sides of the differential equations, FXY is equivalent to the routine EXTERN. The name of this routine is placed in an EXTERNAL statement. The arguments of FXY are listed and described in Table 20.

Subroutine DFEQS2 (K, NN, HH, XX, Y, BB, EXTERN)

This subroutine is described in Section 2.12.

Subroutine DO2AEF (X, Y, E, T, N, HO, H, EXTERN, A, AA, G, IP, B, F, K)

This subroutine is written by the Nottingham Algorithms Group and is described in their manual.[90]

Subroutine PRNPL2 (X, Y, Y2, XMAX, XINCR, YMAX, YINCR, ISX, ISY, NPTS)

Subroutine PLSCAL (V, VMAX, VINCR, NPTS, NDIVIS)
and
Subroutine PLSCL2 (V, V2, VMAX, VINCR, NPTS, NDIUIS)

These subroutines are each described in Section 3.2.1.

Table 19. Arguments of FIND

Argument	Description
N	Input text variable
LIST	One-dimensional ordered array of text variables
NN	Dimension of LIST
ISW	= 0 if N is not found in LIST = 1 if N is found in LIST
INDEX	Index of list such that LIST(INDEX) = N

Table 20. Arguments of FXY[a]

Argument	Description
DE	One-dimensional array of right-hand sides of differential equations
E	Array of dependent variables (number densities)
T	Independent variable (time)
N	Number of equations for solution

[a] FXY is the name of the routine that is equivalent to EXTERN.

Function subroutine IICOMP (I, A, J, B, K)

This subroutine compares two text variables A and B for equality. IICOMP is returned with value 0, −1, or 1 depending on whether A = B, A<B or A>B, respectively. The arguments of IICOMP are listed and described in Table 21.

6.1.5.2. Description of Input for PLASKEM

The input variables are listed and described in Table 22. The initial conditions in the plasma are read from the data file: these include gas temperature and pressure, primary species and their ratios, electron number density and the rates of in- and outflow of the gas.

The possible secondary species, and the reactions and their corresponding data base address, are all read from the direct access file store. If additional species and/or reactions are to be considered, these can be read from the data file. The debug switch IDEBUG is read and determines the amount of output from the program. The switch ODE determines the method to be used for solving the ODES according to: ODE = 1: solution by Runge–Kutta; ODE = 2: solution by Runge–Kutta and subsequently by the Hamming method; ODE = 3: solution by Gear's method.

Table 21. Arguments of IICOMP

Argument	Description
I	Number of characters to be compared
A	First text variable
J	Position in A where comparison is to start
B	Text variable to be compared to A
K	Position in B where comparison is to start

Table 22. Program Input for PLASKEM

Card	Variable	Description	Unit	Format
1	P	Gas pressure	Torr	E12.6
	TEMP	Gas ambient temperature	°K	E12.6
2	NSPEC	Number of primary species		I3
	IONS	= 0 if reactions involving electrons and ions are not included = 1 if negative ions and electrons are included = 2 if negative ions and positive ions are included		I3
3	XNAME(I), RP(I) I = 1, NSPEC	Symbol name and initial fraction of each primary species		4(A8, E10.4)
4	ENODEN	Electron number density	cm^{-3}	E12.6
	TLIM	Discharge time	nsec	E12.6
5	RATE2R } RATE3R }	= 0.0 if IONS = 1 or 3. Otherwise the generalized rate for $-ve$ ion, $+ve$ ion recombination (two and three-body)	$cm^3\ sec^{-1}$ $cm^6\ sec^{-1}$	E12.6 E12.6
6	S(I) I = 1, NSPECS	Rate of inflow of species *I*	$cm^{-3}\ sec^{-1}$	E12.6
	NU	Frequency of loss of all particles by processes other than reactions	sec^{-1}	E12.6
7	NEXTRA	Number of additional species to be included		I3
	MEXTRA	Number of additional reactions to be included		I3
8	SNAME(K)	Symbol name for extra species		10A8
9	ANAME(J), RATE(I) J = 1,6 I = 1, MEXTRA	Participant particles and rate of extra reactions	$cm^3\ sec^{-1}$ (or $cm^6\ sec^{-1}$)	6A8, E12.6
10	IDEBUG	= 0 Normal output = 1 Debug output		I3
11	ODE	= 1 (Runge–Kutta method of solution) = 2 (Hamming method of solution) = 3 (Gears method of solution)		I3

continued

Table 22. (continued)

Card	Variable	Description	Unit	Format
12	DISP(K)	Symbol names of species to be output on number density vs. time graphs		8A8
	Variables input from STORE1 (unformatted):			
	NSPECS	Number of allowed neutral molecules		
	SNAME(I)	Text symbol for molecule I		
	SNAME (I+NSPECS)	Text symbol for negative ion		
	SNAME (I+2*NSPECS)	Text symbol for positive ion		
	IP(J), FP(J), IADD J=1, 3	Initial and final particles of reaction, followed by data base address		

6.1.5.3. *Description of Output for* PLASKEM

The input variables are listed, followed by, depending on the value of IDEBUG:

Normal Output, IDEBUG = 0

The number density for each primary and secondary species is output each time step (every ten time steps for Runge–Kutta and Hamming methods of solution), together with the current–time value.

Debug Output, IDEBUG = 1

This is the same for normal output, plus the current TSTEP value and the gradients of the number densities after each call of FXY.

At the end of a run, log–log plots of number density vs. time are output for six chosen species, two species on each of three graphs.

6.1.6. *Results from* PLASKEM

Figure 28 shows some of the results (negative ions only) of a run of PLASKEM for a $CO_2:N_2:He = 1:1:8$ mixture at atmospheric pressure, with the discharge maintained by a source of electrons with density and duration.

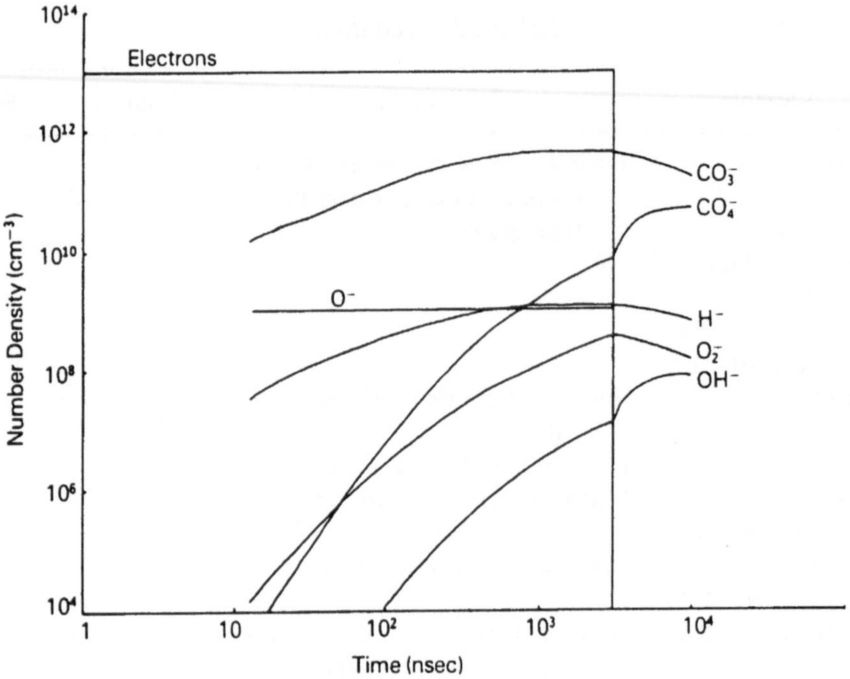

Fig. 28. Time history of negative-ion number densities in an atmospheric $CO_2:N_2:He = 1:1:8$ laser cavity.

6.2. *Energy Degradation of Primary Electrons*

Electron beams play a very important role in laser pumping. The objective of this section is to discuss a computer model to predict how many secondary electrons, ions, and excited states of the constituent atoms and molecules are produced when a high-energy electron beam passes through a gas mixture. A complete model of the electron-beam-controlled laser requires a knowledge of the plasma chemistry so that the time evolution of the concentrations of the products of the electron collisions could be followed as they react with each other. To determine how these products are distributed along the path of the primary electron it would be necessary to perform Monte Carlo calculations following individual electrons.

We do not intend to describe such an ambitious derivation of such a model here, although all the ingredients of the plasma chemistry model have been described in Section 6.1. In other words, we shall neglect the time and space dependence and restrict consideration to the energy degradation of a high-energy electron in a mixture of gases.

This section will be somewhat of a departure from the rest of the book in that the equations will not be derived from first principles but quoted from the original literature. For energies above 1 keV, we shall describe the continuous slowing-down approximations (CSDA) method in Section 6.2.1. For energies below 1 keV, we describe a discrete model.

6.2.1. CSDA Approximation

According to Peterson,[91] if we denote $N_{jt}^0(E_p)$ as the population of state j of species t, after degradation of the primary alone from energy E_p, then

$$N_{jt}^0(E_p)=\int_{W_{jt}}^{E_p}\frac{\sigma_{jt}(E)}{L(E)}\,dE \tag{6.13}$$

where W_{jt} is the threshold energy for exciting state j of species t, and $L(E)$ is the loss function:

$$L(E)=-\frac{1}{n}\frac{dE}{dx} \tag{6.14}$$

where n is the total number density of all particles:

$$n=\sum_t n_t \tag{6.15}$$

$\sigma_{jt}(E)$ is the cross section for the excitation of the j state of species t by an electron of energy E. If $\sigma_{it}(E, E_s)$ is the cross section for a primary electron of energy E to ionize a species t gas atom or molecule in state i and produce a secondary electron in the energy range (E_s, E_s+dE), then the secondary electron spectrum resulting from ionization of state i is given by

$$N_{it}(E_p, E_s)=\int_{2E_p+I_{it}}^{E_p}\frac{\sigma_{it}(E, E_t)}{L(E)}\,dE \tag{6.16}$$

where I_{it} is the ionization potential of species t.

According to Peterson, the total population of state j of species t, after secondaries are also degraded, is obtained from the integral equation

$$N_{jt}(E_p)=N_{jt}^0(E_p)+\sum_{r,i}\int_0^{(E_p-I_{ir})/2}N_{jt}(E_s)N_{ir}(E_p, E_s)\,dE_s \tag{6.17}$$

For an electron of energy E incident on a single species, Peterson and Green[92] give

$$\begin{aligned}L(E)=&\sum_i I_i\sigma_i(E)+\sum_d E_d\sigma_d(E)+\sum_{k,l} I_{k,l}\sigma_{k,l}(E)\\&+\sum_i\int_0^{(E-I_i)/2}E_s\sigma_i(E, E_s)\,dE_s+\sum_j W_j\sigma_j(E)+\cdots\end{aligned} \tag{6.18}$$

where the six terms represent: (i) the loss to primary ionization from which secondary electrons result; (ii) processes, like dissociative ionization, from which only low-energy secondaries emerge, unable to produce more ions;

(iii) double ionization; (iv) production of secondaries from primary ionizations; (v) discrete-state excitations; (vi) other loss mechanisms such as higher-order ionizations, recombinations, electron–electron, etc. The limits on the integral in Eq. (6.18) come from defining the higher-energy electron of the two secondaries as being the degraded primary.

For a mixture of gases, Eq. (6.18) must be generalized. If only processes (iv) and (v) are included, then

$$L(E)=\sum_t \frac{n_t}{n}\left[\sum_i \int_0^{(E-I_{it})/2} \sigma_{it}(E, E_s)E_s\, dE_s + \sum_j W_{jt}\sigma_{jt}(E)\right] \tag{6.19}$$

Schunk and Hays[93] have shown that the electron-energy loss rate to the ambient electrons, for $E \gg kT$ (where k is Boltzmann's constant and T is the electron temperature), can be described approximately by the formula

$$-\frac{dE}{dt}=\frac{w_p^2 e^2}{v}\begin{cases} \ln\left(\dfrac{mv^3}{\gamma e^2 w_p}\right) & \text{for } kT \ll E < \dfrac{me^4}{2\hbar^2} \\ \ln\left(\dfrac{mv^2}{\hbar w_p}\right) & \text{for } E > \dfrac{me^4}{2\hbar^2} \end{cases} \tag{6.20}$$

where v is the speed on an electron of energy E, the plasma frequency is

$$w_p = (4\pi n_e e^2/m)^{1/2} \tag{6.21}$$

$\ln\gamma$ is Euler's constant, m is the electron mass, and n_e is the electron number density. The contribution of this electron–electron loss term to Eq. (6.19) is

$$-\frac{1}{nv}\frac{dE}{dt} \tag{6.22}$$

Given $E_p \sim 1$ MeV, and the form of the cross sections, $L(E)$ can be calculated, which allows the calculation of $N_{jt}^0(E_p)$, $N_{it}(E_p, E_s)$, and then the solution of the integral equation for $N_{jt}(E_p)$.

6.2.2. Discrete Energy-Loss Method

Other than in electron–electron collisions, the primary electron will not lose energy continuously; instead it will lose energy in discrete quanta, depending on which state of the gas atom or molecule is excited or ionized and on the energy of the secondary electron. However, the discrete energy-loss method is so slow computationally that it is only used at energies below 1 keV, where most of the excitations occur in any case.

In carrying out the energy loss in discrete steps, Peterson[91] proposed dividing the energy scale 0 to E_p into a number of equally spaced points:

$$\Delta E < W$$

where W is the smallest inelastic threshold. These points define the centers of what are referred to as "bins." One electron is placed in the bin labeled p, and zero electrons in all other bins. As a result of the first inelastic collision, this one electron will redistribute itself fractionally in the lower bins, as controlled by the various cross sections. For ionization processes, the energy loss W equals $I_{it}+E_s$. Hence, the primary, which we defined earlier to be the electron with greater energy, falls into the bin corresponding to $E_p - W$, and the secondary is *created* in the bin corresponding to E_s. Now that the topmost bin is empty, the electrons, or fractional parts, in the next bin are redistributed and so the process is repeated until the primary is thermalized.

Dalgarno and Lejeune[94] give the collision frequency for a particular loss process with cross section σ, of an electron with velocity v and energy E, to be

$$\nu = nv\sigma(E) \qquad (\text{sec}^{-1}) \tag{6.23}$$

The collision frequency for inelastic process j for species t is then

$$\nu_{jt} = n_t v\sigma_{jt}(E) \tag{6.24}$$

and the ionization collision frequency will be

$$\nu_i(E) = \int_0^{(E-I_{it})/2} dE_s \nu_{it}(E, E_s) = n_t v \int_0^{(E-I_{it})/2} \sigma_{it}(E, E_s)\, dE_s \tag{6.25}$$

Dalgarno and Lejeune define a collision frequency for elastic energy loss in collision with thermal electrons by

$$\nu_e E_s = \frac{1}{\Delta E_s}\left(-\frac{dE}{dt}\right)_{E_s} \tag{6.26}$$

where ν_e^{-1} is the time taken for the electron to lose energy ΔE_s. The total collision frequency is given by

$$\nu(E) = \sum_t \left[\sum_j \nu_{jt}(E) + \sum_i \nu_{it}(E)\right] + \nu_e \tag{6.27}$$

The probability that an electron undergoes an inelastic collision with species t and loses energy W_{jt} is given by

$$P_{jt} \equiv \frac{\nu_{jt}}{\nu} \tag{6.28}$$

and the probability that an electron undergoes an ionizing collision losing energy $I_{it}+E_s$, creating an electron with energy E_s, will be

$$P_{it}(E, E_s) \equiv \frac{\nu_{it}(E, E_s)}{\nu} \tag{6.29}$$

From the preceding results it is possible to calculate:

(i) the total energy lost by an electron in bin k to electron–electron collisions;
(ii) the total number of all orders of secondary electrons produced;
(iii) the number of excitations of state n and ionizations to state i of species t.

The calculations are started at the lowest energy, $k=1$, and the discrete model is used for energies up to about 1 keV, and then the CSDA method is used for energies of 1 keV up to E_p.

7

Stability Analysis of Laser Plasmas

7.1. Introduction

In practice many physical systems are continually subject to small amplitude perturbations. In a plasma, for example, there will be local fluctuations of the plasma variables (species, number densities, etc.) as well as noise generated by flow turbulence and power-supply ripple.

We can envisage three possible responses of a physical system to these small perturbations:

(i) The perturbations are attenuated so as to decrease in amplitude, or are propagated through the system without attenuation. Under these conditions the system is stable.

(ii) The perturbations are amplified, but, at the same time, are "transported" out of the system (convective instability).

(iii) The perturbations are amplified and remain within the system (absolute, or nonconvective, instability).

If case (iii) applies, then the system is expected to be physically unstable and break down in some way, for example, the arcing of a plasma discharge.

Stability analysis sets out to predict the conditions for which systems are unstable. The basic ideas of the theory are introduced in the next section by performing a simplified and nonrigorous stability analysis on an equation of interest in electron-beam-sustained plasma discharges. In Section 7.3 a stability analysis is performed on a simple mechanical model that exhibits instability. The criteria that determine whether the instability is convective or absolute are determined for this case. In Section 7.4 and 7.5 examples are given of applications of stability analysis to laser plasma systems. In Section 7.6 a complete derivation of the theory of stability is given together with mathematical definitions of the concepts of convective and absolute instabilities.

In Fig. 29 we present a schematic diagram that summarizes the modes

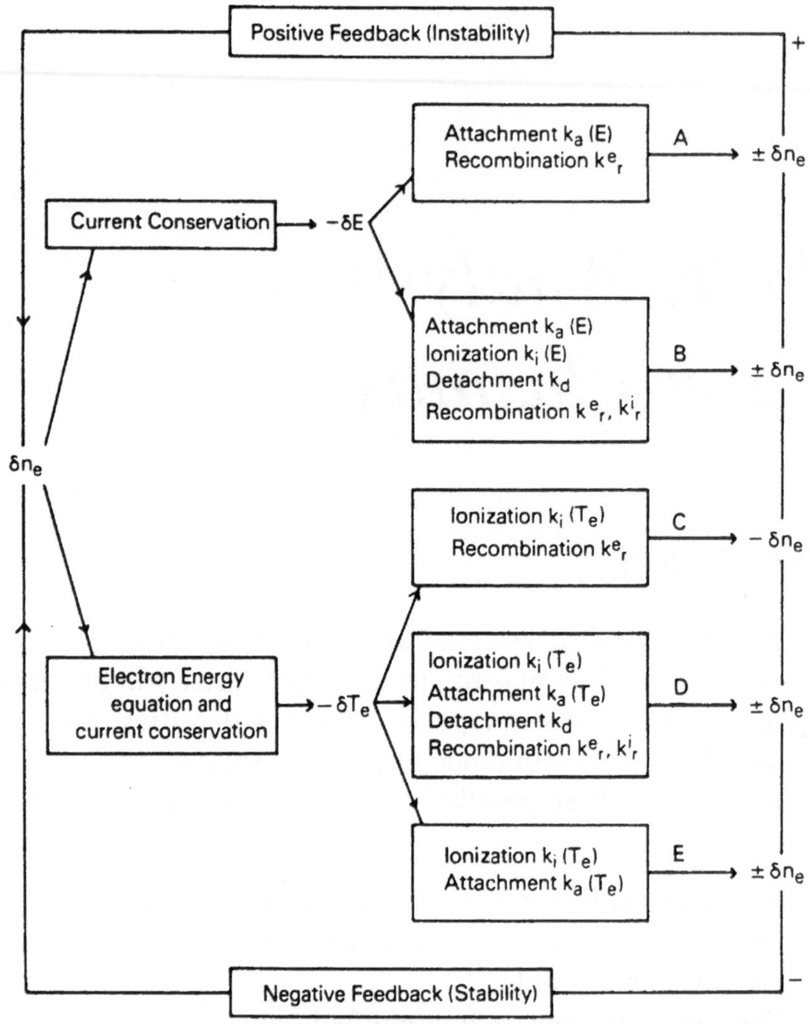

Fig. 29. Diagrammatic summary of the instabilities studied in Chapter 7. The initial disturbance in the electron number density n_e is shown at the left-hand side, followed by the resulting sequence of events for the various cases considered: (A) Douglas-Hamilton, Section 7.2; (B) Douglas-Hamilton, Section 7.2.1; (C) Nighan and Wiegand, Section 7.2.1; (D) Nighan and Wiegand, Section 7.5.3; and (E) Dyer and James, Section 7.4.

of instability which have been investigated in the literature. From this figure we see that a perturbation in the electron number density, δn_e, induces a change in either the local electric field, δE, or in the electron temperature, δT_e, which in turn changes the designated rate, k (E or T_e), hence δn_e completing the cycle, leading to stability or instability.

7.2. A Simplified Analysis of Attachment Instability (Douglas-Hamilton)

In this section we present a simplified derivation of the attachment instability of Douglas-Hamilton *et al.*[95,96] We consider the equation for the secondary electron number density, $n_e(t)$, in an electron-beam-controlled discharge [see also Eq. (2.89)]

$$\frac{dn_e}{dt} = S - \beta n_e - \alpha n_e^2 \tag{7.1}$$

which is derived by approximating Eq. (5.154), where S is the rate of production of secondary electrons by the primary electron beam, which is taken to be constant, β is the electron neutral attachment coefficient, and α the electron–ion recombination coefficient. The units of the left-hand side of Eq. (7.1) are $\text{cm}^{-3}\,\text{sec}^{-1}$. In Eq. (5.154) we neglect terms containing spatial derivatives and the loss frequency term, and the coefficients α and β are given in terms of rates by

$$\alpha = k_r^e \qquad (\text{cm}^3\,\text{sec}^{-1})$$

$$\beta = nk_a \qquad (\text{sec}^{-1})$$

where k_r^e is the electron positive-ion recombination rate, k_a the electron neutral attachment rate, and n the heavy-particle number density. An applied electric field $\mathbf{E}$ results in an electron current density $\mathbf{j}_e$ proportional to the product of n_e and $\mathbf{E}$:

$$\begin{aligned} \mathbf{j}_e &\approx e\mu_e n_e \mathbf{E} \\ &\approx e n_e v_d \qquad (\text{amp cm}^{-2}) \end{aligned} \tag{7.2}$$

where v_d is the drift velocity and μ_e the electron mobility. Equation (7.2) is an approximate form of the equation for the electron current density [see Eqs. (5.158) and (5.195)]. We see from these equations that if we neglect terms containing spatial derivatives, and if the mass-average velocity $\mathbf{v}_0$ is zero, assuming that there is no flow, then we obtain Eq. (7.2).

We suppose that n_e has reached its steady-state value, n_{e0}, given by Eq. (7.1) with $dn_{e0}/dt = 0$; hence

$$n_{e0} = \frac{-\beta \pm (\beta^2 + 4\alpha S)^{1/2}}{2\alpha} \tag{7.3}$$

where the plus sign is to be included since α, β, and S are positive and n_{e0} must be positive.

The attachment rate β is a function of the electric field $\mathbf{E}$ and we wish to know if this dependence of β on $\mathbf{E}$ can result in an instability. We suppose that there is a local fluctuation in the system that results in perturbations δn_e and $\delta \mathbf{E}$ in some region:

$$\left.\begin{aligned} n_e &\to n_{e0} + \delta n_e \\ \mathbf{E} &\to \mathbf{E} + \delta\mathbf{E} \\ \beta(\mathbf{E}) &\to \beta(\mathbf{E} + \delta\mathbf{E}) \approx \beta(\mathbf{E}) + \frac{\partial \beta}{\partial \mathbf{E}}\,\delta\mathbf{E} \end{aligned}\right\} \tag{7.4}$$

The perturbations (7.4) are substituted into Eq. (7.1) to obtain

$$\frac{d}{dt}(n_{e0}+\delta n_e)=S-\left(\beta+\frac{\partial\beta}{\partial \mathbf{E}}\,\delta\mathbf{E}\right)(n_{e0}+\delta n_e)-\alpha(n_{e0}+\delta n_e)^2 \qquad (7.5)$$

n_{e0} is an equilibrium value so that dn_{e0}/dt is zero and from Eq. (7.1) we have

$$S-\beta n_{e0}-\alpha n_{e0}^2=0 \qquad (7.6)$$

When second-order terms in perturbed quantities are neglected, Eq. (7.5) becomes

$$\frac{d}{dt}\,\delta n_e=-n_{e0}\frac{\partial\beta}{\partial\mathbf{E}}\,\delta\mathbf{E}-\beta\,\delta n_e-2\alpha n_{e0}\,\delta n_e \qquad (7.7)$$

Conservation of the electron current is implied by

$$\nabla\cdot\mathbf{j}_e+\frac{\partial\rho}{\partial t}=0 \qquad (7.8)$$

where

$$\rho=en_e$$

Hence, if the field **E** has only an x-component:

$$\frac{\partial}{\partial x}(e\mu_e n_e E)+e\frac{\partial n_e}{\partial t}=0$$

having used Eq. (7.2) for $\mathbf{j}_e$. At equilibrium, n_e is replaced by n_{e0} and $\partial n_{e0}/\partial t=0$; hence the first term only is nonzero, and we have

$$\mu_e\left(\frac{\partial n_{e0}}{\partial x}\right)E+\mu_e n_{e0}\frac{\partial E}{\partial x}=0$$

or

$$E\delta n_{e0}=-n_{e0}\delta E \qquad (7.9)$$

This result is substituted into the first term on the right-hand side of Eq. (7.7) to give

$$\frac{d}{dt}\,\delta n_e=E\frac{\partial\beta}{\partial E}\,\delta n_e-\beta\delta n_e-2\alpha n_{e0}\delta n_e \qquad (7.10)$$

We now *assume* that the time dependence of the perturbation $\delta n_e(t)$ can be characterized by a frequency ω:

$$\delta n_e(t)=\delta n_e(0)\,e^{-i\omega t} \qquad (7.11)$$

Substituting Eq. (7.11) into Eq. (7.10) and removing the common factor $\delta n_e(0)\, e^{-i\omega t}$ we obtain an equation for ω, namely,

$$-i\omega = E\frac{\partial \beta}{\partial E} - \beta - 2\alpha n_{e0} \tag{7.12}$$

the so called "dispersion relation." Now if Eq. (7.12) has a solution in which the imaginary part ω_I of $\omega = \omega_R + i\omega_I$ is positive, then, by Eq. (7.11), the perturbation will be amplified exponentially in time; that is,

$$\delta n_e(t) = \delta n_e(0)\, e^{-i\omega_R t}\, e^{\omega_I t} \tag{7.13}$$

Hence we take the criteria for instability to be

$$\omega_I \equiv \text{Im}\,\omega > 0 \tag{7.14}$$

Since all the quantities on the right-hand side of Eq. (7.12) are real, the criteria (7.14) applied to it gives

$$E\frac{\partial \beta}{\partial E} - \beta - 2\alpha n_{e0} > 0$$

or

$$E\frac{\partial \beta}{\partial E} - (\beta^2 + 4\alpha S)^{1/2} > 0 \tag{7.15}$$

where we have used Eq. (7.3) for n_{e0}. Equation (7.15) states that if β is a sufficiently strongly increasing function of E, then an instability will occur. Over a certain range of E the attachment rate β does increase, which gives rise to the "attachment instability" of Douglas-Hamilton *et al.*[95,96]

The physical reasoning behind the instability is that a local increase in the electric field results in a reduction in the electron density if β increases with E. To maintain a constant current density the electric field increases further and the attachment rate β is again enhanced, and so a "positive feedback" type of mechanism is established, which ends only when the system breaks down in some way.

7.2.1. *Stability Analysis of an Electron-Beam-Sustained Discharge*

In this section we present the full derivation of the conditions for instability in an electron-beam-sustained discharge, as given by Douglas-Hamilton *et al.*[96]

We consider rate equations for the number densities of electrons, n_e, one species of negative ion, n_-, and one species of positive ion, n_+:

$$\left.\begin{aligned}
\frac{\partial n_e}{\partial t}+\frac{\partial}{\partial x}(n_e\mu_e E) &= S-\beta n_e-\alpha_1 n_e n_+ +\gamma n_e \\
\frac{\partial n_-}{\partial t}+\frac{\partial}{\partial x}(n_-\mu_- E) &= \beta n_e-\alpha_2 n_- n_+ \\
\frac{\partial n_+}{\partial t}+\frac{\partial}{\partial x}(n_+\mu_+ E) &= S-(\alpha_1 n_e+\alpha_2 n_-)n_+ +\gamma n_e
\end{aligned}\right\} \tag{7.16}$$

The first of Eqs. (7.16) is an approximation to Eq. (5.154) for the electron number density. We are considering a stationary (nonflowing) plasma so that the mass-average velocity $\mathbf{v}_0$ in Eq. (5.154) is zero. The electron drift velocity $\bar{\mathbf{V}}_e$, on the left-hand side of Eq. (5.154), is expressed in terms of the electric field $\mathbf{E}$ and electron mobility μ_e, using Eq. (5.195) and neglecting diffusion terms since "the characteristic diffusion times at atmospheric pressures are several thousand times the characteristic discharge lifetime."[(96)] The second and third of Eqs. (7.16) are similar approximations to Eq. (5.35), which defines heavy-particle number densities. In Eqs. (7.16) the electric field is assumed to be in the x-direction.

The terms on the right-hand sides of Eqs. (7.16) represent the following processes: S, in $\text{cm}^{-3}\,\text{sec}^{-1}$, is the electron-source term describing the rate of production of secondary electrons due to neutral heavy-particle–primary-electron ionizing collisions. S is therefore also the positive-ion source term. β, in sec^{-1}, is the electron neutral-attachment coefficient. In terms of the electron neutral-attachment rate k_a ($\text{cm}^3\,\text{sec}^{-1}$), β is given by

$$\beta = nk_a$$

where n is the neutral-particle number density. α_1 and α_2, both in $\text{cm}^3\,\text{sec}^{-1}$, are the electron–ion and ion–ion recombination rates, k_r^e and k_r^i, respectively. γ, in sec^{-1}, is the electron-neutral-ionization coefficient, given in terms of the electron neutral-ionization rate k_i ($\text{cm}^3\,\text{sec}^{-1}$) by

$$\gamma = nk_i$$

From Eqs. (5.223) and (5.226) we can write down for the electric field E:

$$\frac{\partial E}{\partial x}=\frac{e}{\varepsilon_0}(n_+-n_--n_e) \tag{7.17}$$

In terms of the voltage V_0 applied across the discharge we deduce from the general relation

$$\mathbf{E} = -\nabla V$$

that

$$V_0 = -\int_0^L E\,dx \tag{7.18}$$

where L is the distance across the discharge.

The final equation required for the analysis is an expression for the total electric current density J. From Eqs. (5.158), (5.227), and (5.195) we have

$$J = eE(\mu_+ n_+ - \mu_- n_- - \mu_e n_e) \tag{7.19}$$

We shall perform the perturbation analysis around the equilibrium solutions of Eqs. (7.16). To find these equilibrium solutions, which we denote by n_{e0}, n_{-0}, and n_{+0}, we set the right-hand sides of Eqs. (7.16) equal to zero, namely,

$$\left.\begin{aligned} S - \beta n_{e0} - \alpha_1 n_{e0} n_{+0} + \gamma n_{e0} &= 0 \\ \beta n_{e0} - \alpha_2 n_{-0} n_{+0} &= 0 \\ S - (\alpha_1 n_{e0} + \alpha_2 n_{-0}) n_{+0} + \gamma n_{e0} &= 0 \end{aligned}\right\} \tag{7.20}$$

and conservation of charge gives

$$n_{+0} = n_{-0} + n_{e0} \tag{7.21}$$

Eliminating n_{+0} and n_{e0}, using Eq. (7.21) and the first two of Eqs. (7.20), straightforward algebra gives

$$\alpha_2[\alpha_2\gamma + (\alpha_1 - \alpha_2)\beta] n_{-0}^3 - \alpha_2[\beta(\gamma - \beta) + \alpha_2 S] n_{-0}^2 + 2\alpha_2\beta S n_{-0} - S\beta^2 = 0 \tag{7.22}$$

Now for low electric fields ($E < 10^3$ V/cm), γ is very much smaller than β, and if we also make the approximation

$$\alpha_1 \approx \alpha_2$$

then Eq. (7.22) becomes quadratic in n_{-0}:

$$\alpha_2[\beta(\gamma - \beta) + \alpha_2 S] n_{-0}^2 + 2\alpha_2\beta S n_{-0} - S\beta^2 = 0 \tag{7.23}$$

Equation (7.23) is readily solved for n_{-0} to obtain

$$n_{-0} = \frac{\beta(\alpha S)^{1/2}}{\alpha[\beta + (\alpha S)^{1/2}]} \tag{7.24}$$

the other root being negative and therefore unphysical. From Eqs. (7.20), (7.21), and (7.24) we can now obtain the equilibrium values n_{+0} and n_{e0}.

After a little algebra we obtain

$$n_{e0} = \frac{S}{\beta + (\alpha S)^{1/2}}$$
$$n_{+0} = (S/\alpha)^{1/2} \tag{7.25}$$

From Eq. (7.18) the equilibrium value, E_0, of the electric field is

$$E_0 = V_0/L \tag{7.26}$$

and from Eqs. (7.19), (7.21), (7.24) and (7.25) we can write the equilibrium current density, J_0, as

$$J_0 = -eE_0 \frac{S}{\beta + (\alpha S)^{1/2}} \left[\mu_e + \mu_- + \frac{2\beta}{(\alpha S)^{1/2}} \mu_- \right] \tag{7.27}$$

We now perturb the variables in Eqs. (7.16) and (7.17) about their equilibrium solutions:

$$n_e \rightarrow n_{e0} + \delta n_e$$
$$n_+ \rightarrow n_{+0} + \delta n_+$$
$$n_- \rightarrow n_{-0} + \delta n_- \tag{7.28}$$
$$E \rightarrow E_0 + \delta E$$

and allow for the dependence of the coefficients β and γ on E, so that

$$\beta(E) \rightarrow \beta(E_0 + \delta E) = \beta(E_0) + \frac{\partial \beta}{\partial E} \delta E$$
$$\gamma(E) \rightarrow \gamma(E_0 + \delta E) = \gamma(E_0) + \frac{\partial \gamma}{\partial E} \delta E \tag{7.29}$$

We put

$$\frac{\partial \beta}{\partial E} = \beta', \qquad \frac{\partial \gamma}{\partial E} = \gamma' \tag{7.30}$$

and henceforth we omit the subscripts 0. The perturbed forms of Eqs. (7.16) and (7.17) are, to first order in small quantities:

$$\frac{\partial \delta n_e}{\partial t} + \mu_e \frac{\partial}{\partial x}(n_e\, \delta E + E\, \delta n_e)$$
$$= -\beta' n_e\, \delta E - \beta\, \delta n_e - \alpha_1 n_e\, \delta n_+ - \alpha_1 n_+\, \delta n_e + \gamma\, \delta n_e + \gamma' n_e \delta E$$
$$\frac{\partial \delta n_-}{\partial t} + \mu_- \frac{\partial}{\partial x}(n_-\, \delta E + E\, \delta n_-) \tag{7.31a}$$
$$= \beta' n_e\, \delta E + \beta\, \delta n_e - \alpha_2 n_-\, \delta n_+ - \alpha_2 n_+\, \delta n_-$$

$$\frac{\partial \delta n_+}{\partial t} + \mu_+ \frac{\partial}{\partial x}(n_+\,\delta E + E\,\delta n_+)$$

$$= -(\alpha_1 n_e + \alpha_2 n_-)\,\delta n_+ - \alpha_1 n_+\,\delta n_e - \alpha_2 n_+\,\delta n_- + \gamma\,\delta n_e + \gamma' n_e\,\delta E \quad (7.31b)$$

$$\frac{\partial \delta E}{\partial x} = \frac{e}{\varepsilon_0}(\delta n_+ - \delta n_- - \delta n_e)$$

We now suppose that the perturbations δn_e, etc., have a space and time dependence given by

$$\delta n_e = \delta\hat{n}_e\, e^{i(kx-\omega t)}, \quad \text{etc.} \tag{7.32}$$

On substituting the form for the perturbations given by Eq. (7.32) into Eqs. (7.31) and noting that spatial derivatives of unperturbed (equilibrium) quantities are zero, we obtain

$$\begin{aligned}
&-i\omega\delta\hat{n}_e + ik\mu_e n_e\delta\hat{E} + ik\mu_e E\delta\hat{n}_e \\
&\qquad = -\beta' n_e\,\delta\hat{E} - \beta\,\delta\hat{n}_e - \alpha_1 n_e\,\delta\hat{n}_+ - \alpha_1 n_+\,\delta\hat{n}_e + \gamma\delta\hat{n}_e + \gamma' n_e\,\delta\hat{E} \\
&-i\omega\,\delta\hat{n}_- + ik\mu_- n_-\,\delta\hat{E} + ik\mu_-\,\delta\hat{n}_- E \\
&\qquad = \beta' n_e\,\delta\hat{E} + \beta\,\delta\hat{n}_e - \alpha_2 n_-\,\delta\hat{n}_+ - \alpha_2 n_+\,\delta\hat{n}_- \\
&-i\omega\delta\hat{n}_+ + \mu_+ n_+ ik\,\delta\hat{E} + \mu_+ ikE\,\delta\hat{n}_+ \\
&\qquad = -(\alpha_1 n_e + \alpha_2 n_-)\delta\hat{n}_+ - \alpha_1 n_+\delta\hat{n}_e - \alpha_2 n_+\delta\hat{n}_- + \gamma\,\delta\hat{n}_e + \gamma' n_e\delta\hat{E} \\
&ik\delta\hat{E} = \frac{e}{\varepsilon_0}(\delta\hat{n}_+ - \delta\hat{n}_- - \delta\hat{n}_e)
\end{aligned} \tag{7.33}$$

The last of Eqs. (7.33) may be used to eliminate $\delta\hat{E}$ from the first three of Eqs. (7.33) to give

$$\left(\omega - \frac{i\mu_e n_e e}{\varepsilon_0} - k\mu_e E - \frac{\beta' n_e e}{k\varepsilon_0} + i\beta + i\alpha_1 n_+ - i\gamma + \frac{\gamma' n_e e}{k\varepsilon_0}\right)\delta\hat{n}_e$$

$$+\left(-\frac{i\mu_e n_e e}{\varepsilon_0} - \frac{\beta' n_e e}{k\varepsilon_0} + \frac{\gamma' n_e e}{k\varepsilon_0}\right)\delta\hat{n}_-$$

$$+\left(\frac{i\mu_e n_e e}{\varepsilon_0} + \frac{\beta' n_e e}{k\varepsilon_0} + i\alpha_1 n_e - \frac{\gamma' n_e e}{k\varepsilon_0}\right)\delta\hat{n}_+ = 0$$

$$\left(\omega - \frac{i\mu_- n_- e}{\varepsilon_0} - k\mu_- E + i\alpha_2 n_+ + \frac{\beta' n_e e}{k\varepsilon_0}\right)\delta\hat{n}_-$$

$$+\left(-\frac{i\mu_- n_- e}{\varepsilon_0} - i\beta + \frac{\beta' n_e e}{k\varepsilon_0}\right)\delta\hat{n}_e$$

$$+\left(\frac{i\mu_- n_- e}{\varepsilon_0} + i\alpha_2 n_- - \frac{\beta' n_e e}{k\varepsilon_0}\right)\delta\hat{n}_+ = 0$$

$$\left[\omega+\frac{i\mu_+n_+e}{\varepsilon_0}-\mu_+kE+i(\alpha_1n_e+\alpha_2n_-)-\frac{\gamma'n_ee}{k\varepsilon_0}\right]\delta\hat{n}_+$$
$$+\left[-\frac{i\mu_+n_+e}{\varepsilon_0}+i\alpha_1n_+-i\gamma+\frac{\gamma'n_ee}{k\varepsilon_0}\right]\delta\hat{n}_e$$
$$+\left[-\frac{i\mu_+n_+e}{\varepsilon_0}+i\alpha_2n_++\frac{\gamma'n_ee}{k\varepsilon_0}\right]\delta\hat{n}_-=0 \tag{7.34}$$

We may express Eqs. (7.34) in matrix form

$$\left\{\begin{bmatrix}A_{11} & A_{12} & A_{13}\\ A_{21} & A_{22} & A_{33}\\ A_{31} & A_{32} & A_{33}\end{bmatrix}+\begin{bmatrix}\omega & 0 & 0\\ 0 & \omega & 0\\ 0 & 0 & \omega\end{bmatrix}\right\}\begin{bmatrix}\delta\hat{n}_e\\ \delta\hat{n}_-\\ \delta\hat{n}_+\end{bmatrix}=0 \tag{7.35}$$

We see from Eq. (7.35) that for nonzero $\delta\hat{n}_e$, $\delta\hat{n}_-$, and $\delta\hat{n}_+$ the determinant of the square matrix must be zero

$$\operatorname{Det}[\mathbf{A}+\omega\mathbf{1}]=0 \tag{7.36}$$

From Eqs. (7.34) and (7.35) the elements of the matrix A are given by

$$A_{11}=\left(-k\mu_eE-\frac{e}{\varepsilon_0k}\beta'n_e+\frac{e}{\varepsilon_0k}\gamma'n_e\right)+i\left(\beta-\gamma+\alpha_1n_+-\frac{e}{\varepsilon_0}n_e\mu_e\right)$$

$$A_{12}=\left(-\frac{e}{\varepsilon_0k}\beta'n_e+\frac{e}{\varepsilon_0k}\gamma'n_e\right)-i\left(\frac{e}{\varepsilon_0}n_e\mu_e\right)$$

$$A_{13}=\left(\frac{e}{\varepsilon_0k}\beta'n_e-\frac{e}{\varepsilon_0k}\gamma'n_e\right)+i\left(\alpha_1n_e+\frac{e}{\varepsilon_0}n_e\mu_e\right)$$

$$A_{21}=\left(\frac{e}{\varepsilon_0k}\beta'n_e\right)-i\left(\beta+\frac{e}{\varepsilon_0}\mu_-n_-\right)$$

$$A_{22}=\left(-k\mu_-E+\frac{e}{\varepsilon_0k}\beta'n_e\right)+i\left(\alpha_2n_+-\frac{e}{\varepsilon_0}\mu_-n_-\right) \tag{7.37}$$

$$A_{23}=\left(-\frac{e}{\varepsilon_0k}\beta'n_e\right)+i\left(\alpha_2n_-+\frac{e}{\varepsilon_0}\mu_-n_-\right)$$

$$A_{31}=\left(\frac{e}{\varepsilon_0k}\gamma'n_e\right)+i\left(\alpha_1n_+-\gamma-\frac{e}{\varepsilon_0}\mu_+n_+\right)$$

$$A_{32}=\left(\frac{e}{\varepsilon_0k}\gamma'n_e\right)+i\left(\alpha_2n_+-\frac{e}{\varepsilon_0}\mu_+n_+\right)$$

$$A_{33}=\left(-k\mu_+E-\frac{e}{\varepsilon_0k}\gamma'n_e\right)+i\left(\alpha_1n_e+\alpha_2n_-+\frac{e}{\varepsilon_0}\mu_+n_+\right)$$

From the perturbed form of Eq. (7.18) we have that, since V_0 is an externally applied voltage and does not suffer perturbations:

$$0 = \int_0^L \delta\hat{E}\, e^{i(kx-\omega t)}\, dx$$

$$= \frac{e^{-i\omega t}}{ik(e^{ikL}-1)}\delta\hat{E} \tag{7.38}$$

and therefore we have that

$$k = \frac{2\pi n}{L} \tag{7.39}$$

where n is some integer.

We can rewrite Eq. (7.36), the dispersion relation, in the form

$$\omega^3 + (p_1 + ip_2)\omega^2 + (p_3 + ip_4)\omega + (p_5 + ip_6) = 0 \tag{7.40}$$

where the p_i, $i = 1, 6$, are real and defined by

$$\begin{aligned}
p_1 + ip_2 &= (A_{11} + A_{22} + A_{33})\\
p_3 + ip_4 &= (A_{11}A_{22} + A_{11}A_{33} + A_{22}A_{33} - A_{32}A_{23}\\
&\quad - A_{21}A_{12} - A_{13}A_{31})\\
p_5 + ip_6 &= (A_{11}A_{22}A_{33} - A_{11}A_{32}A_{23} + A_{12}A_{23}A_{31} - A_{12}A_{21}A_{33}\\
&\quad + A_{13}A_{21}A_{32} - A_{13}A_{31}A_{22})
\end{aligned} \tag{7.41}$$

If we now write ω in terms of its real and imaginary parts:

$$\omega = \omega_R + i\omega_I$$

then Eq. (7.40) becomes

$$\begin{aligned}
&[\omega_R(\omega_R^2 - 3\omega_I^2) + p_1(\omega_R^2 - \omega_I^2) - 2p_2\omega_R\omega_I + p_3\omega_R - p_4\omega_I + p_5]\\
&\quad + i[\omega_I(3\omega_R^2 - \omega_I^2) + p_1 2\omega_R\omega_I + p_2(\omega_R^2 - \omega_I^2) + p_4\omega_R + p_3\omega_I + p_6] = 0
\end{aligned} \tag{7.42}$$

The real and imaginary parts of Eq. (7.42) must be separately zero. Now from Eq. (7.32) we have that, for instability

$$\text{Im}\,\omega > 0$$

Therefore, at the "instability boundary," defined by

$$\text{Im}\,\omega = 0$$

we have that $\omega_I = 0$. Setting ω_I at zero in Eq. (7.42) we obtain from the real and imaginary parts:

$$\begin{aligned}\omega_R^3+p_1\omega_R^2+p_3\omega_R+p_5&=0\\ p_2\omega_R^2+p_4\omega_R+p_6&=0\end{aligned}\tag{7.43}$$

From the second of Eqs. (7.43) we have

$$\omega_R=\frac{-p_4\pm(p_4^2-4p_2p_6)^{1/2}}{2p_2}\tag{7.44}$$

which is true at the instability boundary. Since ω_R is real we see from Eq. (7.44) that one of the conditions that must be satisfied for an instability to occur is

$$p_4^2\geqslant 4p_2p_6\tag{7.45}$$

To find the condition satisfied by the p's at the instability boundary as a result of the first of Eqs. (7.43), we write it in the form

$$\omega_R^2(\omega_R+p_1)+p_3\omega_R+p_5=0\tag{7.46}$$

Now from the second of Eqs. (7.43) we have

$$\omega_R^2=-\frac{p_4}{p_2}\omega_R-\frac{p_6}{p_2}\tag{7.47}$$

Eliminating ω_R^2 from the first of Eqs. (7.43) using Eq. (7.47) gives

$$\left(-\frac{p_4}{p_2}\omega_R+\frac{p_6}{p_2}\right)\left(\omega_R+p_1\right)+p_3\omega_R+p_5=0$$

or

$$\frac{p_4}{p_2}\left(\frac{p_4}{p_2}\omega_R+\frac{p_6}{p_2}\right)-\left(\frac{p_4p_1}{p_2}+\frac{p_6}{p_2}\right)\omega_R+\frac{p_6p_1}{p_2}+p_3\omega_R+p_5=0\tag{7.48}$$

where we have again used Eq. (7.47) to eliminate ω_R^2. We can rewrite Eq. (7.48) in the form

$$\omega_R=-\frac{p_4p_6+p_1p_2p_6+p_2^2p_5}{p_4^2-p_1p_2p_4-p_2p_6+p_2^2p_3}$$

and using the expression for ω_R given by Eq. (7.44) we obtain a relationship between the p's:

$$\frac{-p_4\pm(p_4^2-4p_2p_6)^{1/2}}{2p_2}=-\frac{p_4p_6+p_1p_2p_6+p_2^2p_5}{p_4^2-p_1p_2p_4-p_2p_6+p_2^2p_3}\tag{7.49}$$

If we rewrite Eq. (7.49) in the form

$$\pm(p_4^2-4p_2p_6)^{1/2}=p_4-2p_2\frac{p_4p_6+p_1p_2p_6+p_2^2p_5}{p_4^2-p_1p_2p_4-p_2p_6+p_2^2p_3}$$

and square both sides, then we obtain, after eliminating common factors on both sides:

$$-p_6 = -p_4\left(\frac{p_4p_6+p_1p_2p_6+p_2^2p_5}{p_4^2-p_1p_2p_4-p_2p_6+p_2^2p_3}\right)+\left(\frac{p_4p_6+p_1p_2p_6+p_2^2p_5}{p_4^2-p_1p_2p_4-p_2p_6+p_2^2p_3}\right)^2$$

which, after some straightforward algebra, can be written as

$$p_6 = -\frac{p_5(p_2^3p_5-p_4^3+p_2p_4p_6+p_1p_2p_4^2-p_2^2p_3p_4)}{(p_6-p_2p_3)^2-(p_1p_2-p_4)(p_3p_4-p_1p_6+2p_2p_5)} \tag{7.50}$$

as given by Douglas-Hamilton *et al.*[96] [their Eq. (28)].

Equations (7.45) and (7.50) are the conditions satisfied at the instability boundary

$$\text{Im}\ \omega = 0$$

Since the p's depend on the source term S and the applied field E from Eqs. (7.41), (7.37), (7.24), and (7.25), numerical values may be substituted into Eqs. (7.45) and (7.50) to obtain the attachment instability region in S–E space. The result of one such calculation by Douglas-Hamilton *et al.* is shown in Fig. 30.

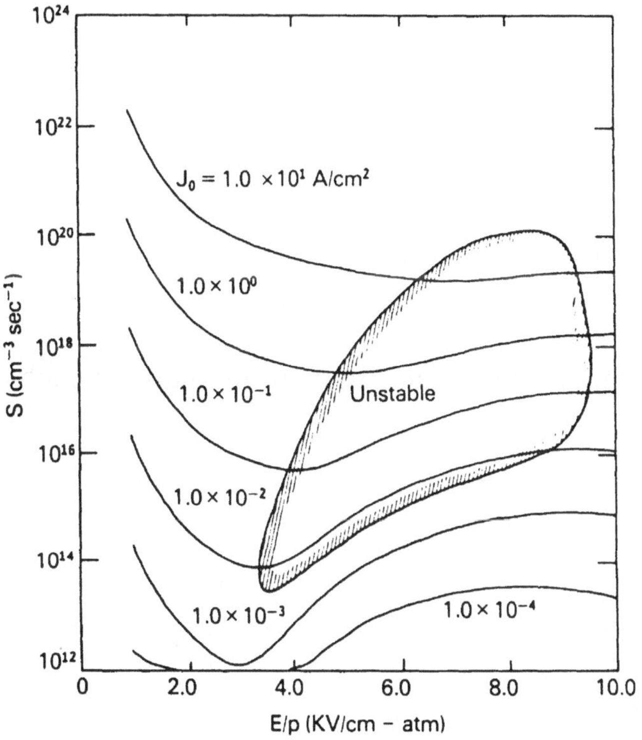

Fig. 30. Region of attachment instability in S–E/p space for $He:N_2:CO_2 = 3:2:1$ mixture at 760 Torr pressure, showing also the lines of constant equilibrium current density J_0 [see Eq. (7.27)]. (From Douglas-Hamilton *et al.*,[96] with permission.)

7.3. A Mechanical Model

Following Sturrock[97] we consider a linear array of oscillators, namely, a set of pendulums vibrating in planes transverse to the z-axis (Fig. 31). The equation of each decoupled oscillator is of the form

$$\frac{d^2\phi}{dt^2} = -\lambda\phi \tag{7.51}$$

where ϕ measures the angular displacement of a pendulum. Now suppose that each pendulum is connected to each of its neighbors by elastic strings. If the tension in each string is proportional to its length l', the proportionality constant being K, then the force F acting in the x-direction on the nth pendulum due to one of its neighbors is

$$F = Kl' \sin\theta$$

where θ is the angle to the z-axis subtended by the string. Figure 32, drawn in the x–z plane (assuming small displacements) illustrates the geometry. s is the length of the pendulum. Now

$$\sin\theta = \frac{s(\phi_{n+1} - \phi_n)}{l'}$$

and hence

$$\begin{aligned} F &= -Ks(\phi_{n+1} - \phi_n) \\ &= -\mu'(\phi_{n+1} - \phi_n) \end{aligned}$$

where we have put

$$\mu' = Ks$$

Therefore the total force on the nth pendulum due to each of its neighbors will be

$$F = \mu'(\phi_{n+1} - 2\phi_n + \phi_{n-1})$$

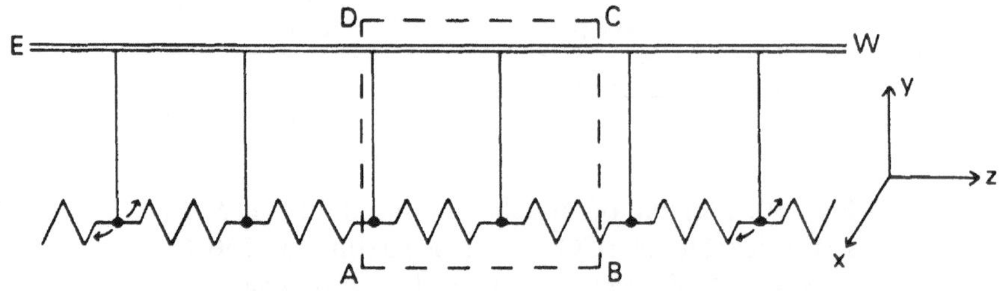

Fig. 31. The mechanical model, where $ABCD$ is the "restricted field of view."

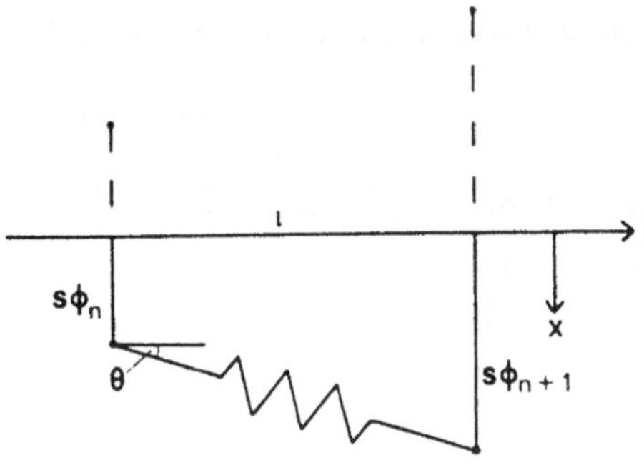

Fig. 32. The geometry in the x–z plane.

and the equation of motion (7.51) for the nth oscillator is now

$$\frac{d^2\phi_n}{dt^2} = -\lambda\phi_n - \mu'(\phi_{n+1} - 2\phi_n + \phi_{n-1}) \tag{7.52}$$

In the limiting case of infinitesimally small separation between oscillators, Eq. (7.52) becomes

$$\frac{\partial^2\phi(z,t)}{\partial t^2} = -\lambda\phi(z,t) + \mu\frac{\partial^2\phi(z,t)}{\partial z^2} \tag{7.53}$$

where $\mu \equiv -\mathrm{Lt}_{\Delta z\to 0}\,\Delta z^2\mu'$. If the whole assembly is translated in the z-direction with velocity v then we have, in terms of the moving coordinate system z', t' defined by

$$z' = z + vt$$

$$t' = t$$

that

$$\frac{\partial}{\partial z} = \frac{\partial z'}{\partial z}\frac{\partial}{\partial z'} + \frac{\partial t'}{\partial z}\frac{\partial}{\partial t'} = \frac{\partial}{\partial z'}$$

and

$$\frac{\partial}{\partial t} = \frac{\partial z'}{\partial t}\frac{\partial}{\partial z'} + \frac{\partial t'}{\partial t}\frac{\partial}{\partial t'}$$

$$= v\frac{\partial}{\partial z'} + \frac{\partial}{\partial t'}$$

Therefore, in terms of the coordinate system z', t', Eq. (7.53) is

$$\left(\frac{\partial}{\partial t} + v\frac{\partial}{\partial z}\right)^2 \phi = -\lambda\phi + \mu\frac{\partial^2\phi}{\partial z^2} \tag{7.54}$$

where we have now dropped the primes on t and z. We now suppose that the system is subject to fluctuations resulting in a perturbation $\delta\phi(z,t)$:

$$\phi(z,t) \to \phi(z,t) + \delta\phi(z,t) \tag{7.55}$$

where the z- and t-dependence of $\delta\phi(z,t)$ is assumed to be that of a traveling wave with wavenumber k and frequency ω:

$$\delta\phi(z,t)=\delta\phi_0\, e^{i(kz-\omega t)} \tag{7.56}$$

Replacing $\phi(z,t)$ in Eq. (7.54) by the perturbed form given by Eq. (7.55) and using Eq. (7.54) to eliminate terms depending on the unperturbed $\phi(z,t)$ we obtain

$$\left(\frac{\partial}{\partial t}+v\frac{\partial}{\partial z}\right)^2 \delta\phi = -\lambda\,\delta\phi+\mu\frac{\partial^2}{\partial z^2}\delta\phi \tag{7.57}$$

With the form for $\delta\phi(z,t)$ given by Eq. (7.56) we obtain, from Eq. (7.57), on removing the common factor $\delta\phi_0\, e^{i(kz-\omega t)}$ the dispersion relation

$$(\omega-vk)^2=\lambda+\mu k^2 \tag{7.58}$$

which has solutions

$$\omega=vk\pm(k^2\mu+\lambda)^{1/2} \tag{7.59}$$

If the pendulums are *hanging* in their stable positions and if the connecting links are under tension, then $\lambda>0$ and $\mu>0$. Therefore, from Eq. (7.59) we find that ω is real for all real values of k, so that the system is stable and no investigation of instability is required.

We now suppose that the pendulums are initially in their positions of unstable equilibrium. In this position all the pendulums have been rotated from their stable equilibrium positions shown in Fig. 31. The gravitational force now acts to increase any displacement of the pendulum from the unstable equilibrium position and so now we have $\lambda<0$ in Eq. (7.54). Therefore we write Eq. (7.54) as

$$\left(\frac{\partial}{\partial t}+v\frac{\partial}{\partial z}\right)^2 \phi = -b^2\phi+c^2\frac{\partial^2\phi}{\partial z^2} \tag{7.60}$$

We now suppose that the perturbation $\delta\phi(z,t)$ is not of a single wavelength, as in Eq. (7.56), but is a superposition of disturbances of slightly differing real wavelengths, k:

$$\delta\phi(z,\omega,t)=\int_{-\infty}^{\infty} dk\;\Phi(k)\, e^{i(kz-\omega t)} \tag{7.61}$$

where $\Phi(k)$ is sharply peaked around some value of k and is negligibly small elsewhere. The perturbed form of Eq. (7.60) is obtained in the same way as Eq. (7.57):

$$\left(\frac{\partial}{\partial t}+v\frac{\partial}{\partial z}\right)^2 \delta\phi = -b^2\,\delta\phi+c^2\frac{\partial^2\delta\phi}{\partial z^2} \tag{7.62}$$

Substituting the form for $\delta\phi(z, \omega, t)$, given by Eq. (7.61), into Eq. (7.62) we obtain

$$\int_{-\infty}^{+\infty} dk\, \Phi(k)[(\omega - vk)^2 + b^2 - c^2k^2]\, e^{i(kz-\omega t)} = 0 \tag{7.63}$$

Equation (7.63) must be true for any z, t, and hence we obtain the dispersion relation for the system

$$\omega^2 - 2vk\omega + [b^2 + k^2(v^2 - c^2)] = 0 \tag{7.64}$$

or

$$k^2(c^2 - v^2) + 2\omega vk - (b^2 + \omega^2) = 0 \tag{7.65}$$

When Eq. (7.64) is solved for ω we have

$$\omega = vk \pm (c^2k^2 - b^2)^{1/2} \tag{7.66}$$

a double-valued function of k, and so we deduce that an instability can occur for

$$k^2 < b^2/c^2 \tag{7.67}$$

since for values of k satisfying Eq. (7.67), ω is complex and the perturbation $\delta\phi(z, t)$, given by Eq. (7.61), then contains a real factor which increases exponentially with time.

The above treatment does not complete the stability analysis of the system. We recall that the (infinite) train of pendulums is moving in the pendulums' unstable equilibrium positions with a velocity v. Now imagine a stationary observer with a restricted field of view (*ABCD* in Fig. 31), who administers the initial perturbation by giving one of the pendulums a slight push as it moves past. This pendulum will begin to fall and will, because of the tensional coupling between the pendulums, pull the neighboring pendulums down after it. If the speed, measured along the z-axis, with which successive pendulums pull each other down is less than the velocity v of the system, then the observer with his restricted field of view will see the initial disturbance being amplified but carried out of his field of view, and after a finite time the system will appear to be in equilibrium again. On the other hand, if the pendulums pull each other down with a speed greater than the velocity of the system, then we expect the observer to see all the pendulums within his field of view fall down.

We wish to find the mathematical conditions that determine whether the instability will move out of the field of view (convective instability) or remain within it (absolute instability). If z_0 is a point within the field of view, then the instability will be convective if

$$\delta\phi(z_0, t) \to 0 \qquad \text{as } t \to \infty \tag{7.68}$$

Otherwise, the instability will be absolute, since if Eq. (7.68) is not satisfied we will have, because of the real exponential factor in Eq. (7.61) due to the complex ω:

$$\delta\phi(z_0, t) \to \infty \qquad \text{as } t \to \infty \tag{7.69}$$

Now substituting for ω, the solution of the dispersion relation given by Eq. (7.66), into Eq. (7.61), we have, for real k:

$$\delta\phi(z, t) = \int_{-\infty}^{\infty} dk\, \Phi(k) \exp(i\{kz - [vk \pm (c^2k^2 - b^2)^{1/2}]t\}) \tag{7.70}$$

If it is possible to transform Eq. (7.70) into an integral over all real values of ω, that is,

$$\delta\phi(z, t) = \int_{-\infty}^{\infty} d\omega\, X(\omega)\, e^{i[k(\omega)z - \omega t]} \tag{7.71}$$

then by Riemann's lemma* we would have

$$\delta\phi(z, t) \to 0 \qquad \text{as } t \to \pm\infty \tag{7.72}$$

Now if Eq. (7.72) is true, then we see that, for z_0, a particular value of z, the amplitude $\delta\phi$ of the disturbance goes to zero as t goes to infinity. Since the disturbance is being amplified (we know the system is unstable) it follows that the disturbance travels out of the system (considered as the region around z_0), and that therefore the instability is convective.

However, if Eq. (7.72) is not true, then $\delta\phi(z_0, t)$ will increase with time and the instability is absolute.

Therefore, if we can write Eq. (7.70) in the form given by Eq. (7.71), then the instability is convective; otherwise it is absolute.

The solution of the dispersion relation [Eq. (7.65)] for k in terms of real ω is

$$k(\omega) = \frac{v\omega}{v^2 - c^2} \pm \frac{[c^2\omega^2 - (v^2 - c^2)b^2]^{1/2}}{v^2 - c^2} \tag{7.73}$$

* Riemann's lemma[98,99] states that for a function $X(\omega)$ of ω, which is bounded in ω (nonnegligible only for some finite range of ω):

$$\int_{-\infty}^{+\infty} d\omega\, X(\omega) \sin \omega t \to 0 \qquad \text{as } t \to \infty$$

$$\int_{-\infty}^{+\infty} d\omega\, X(\omega) \cos \omega t \to 0 \qquad \text{as } t \to \infty$$

Therefore we obtain

$$\int d\omega\, X(\omega)\, e^{i\omega t} = \int d\omega\, X(\omega)(\cos \omega t + i \sin \omega t) \to 0 \qquad \text{as } t \to \infty$$

Using Eq. (7.73) and

$$dk\,\Phi(k) = d\omega \frac{dk(\omega)}{d\omega}\Phi[k(\omega)] \tag{7.74}$$

then Eq. (7.70) becomes

$$\delta\phi(z,t) = \int_{\substack{\mathrm{Lt}(v\pm c)k \\ k\to-\infty}}^{\substack{\mathrm{Lt}(v\pm c)k \\ k\to\infty}} d\omega \frac{dk(\omega)}{d\omega}\Phi[k(\omega)]$$

$$\times \exp\left[i\left(\left\{\frac{v\omega}{v^2-c^2} \pm \frac{[c^2\omega^2-(v^2-c^2)b^2]^{1/2}}{v^2-c^2}\right\}z - \omega t\right)\right] \tag{7.75}$$

Now from Eq. (7.67) we know that ω is complex on the real k-axis for

$$-b/c < k < b/c \tag{7.76}$$

In Figs. 33 and 34, which represent the complex k-plane, we have labeled these two points, $\pm b/c$, on the real k-axis. Now, only if the path of integration on the complex k-plane corresponding to real ω bridges this region of the real k-axis, will the integral over all values of real ω [Eq. (7.75)] correspond to the integral (7.70), over all values of real k. In other words, if there is a continuous path in the complex k-plane which goes from $k = +\infty$ to $k = -\infty$ along which ω is real as ω goes from $\omega = +\infty$ to $\omega = -\infty$, then the integral over all ω will correspond to integration over all k with the path of integration over k displaced into the complex plane.

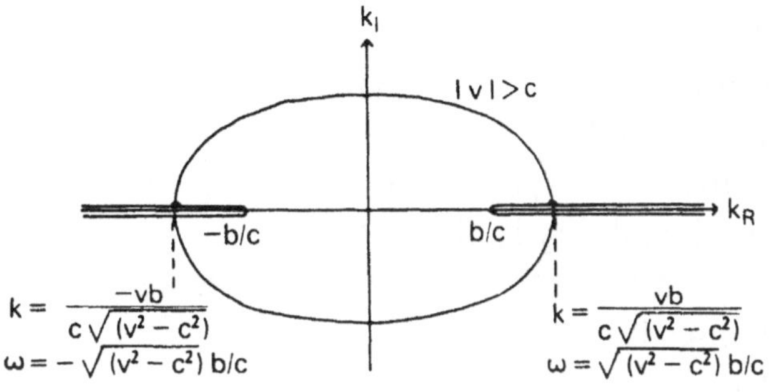

Fig. 33. Contour diagram of real ω for mechanical model corresponding to convective instability.

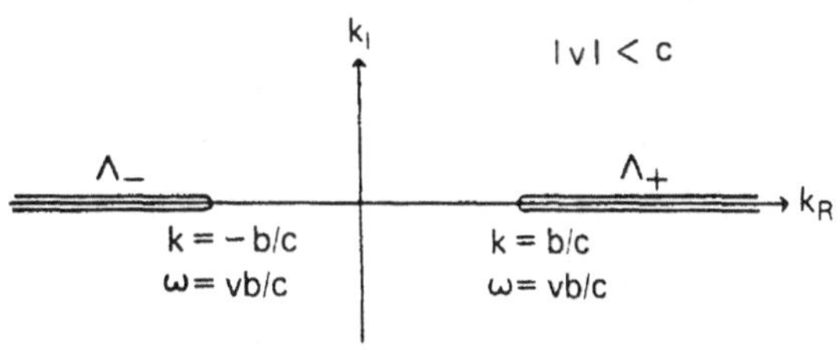

Fig. 34. Contour diagram of real ω for mechanical model corresponding to absolute instability.

This gives the same result[100] as integration over all real k provided that the integral in Eq. (7.75) is analytic in the region of the complex k-plane through which the path of integration is displaced.

From Eq. (7.73) the contours of real ω in the complex k-plane are determined by

$$\left.\begin{aligned} k_R &= \frac{v}{v^2-c^2}\omega \\ k_I &= \pm\frac{[(v^2-c^2)b^2-c^2\omega^2]^{1/2}}{v^2-c^2} \end{aligned}\right\} \quad \text{for } \omega^2<\left(\frac{v^2}{c^2}-1\right)b^2 \tag{7.77}$$

$$\left.\begin{aligned} k_R &= \frac{v\omega}{v^2-c^2}\pm\frac{[c^2\omega^2-(v^2-c^2)b^2]^{1/2}}{v^2-c^2} \\ k_I &= 0 \end{aligned}\right\} \quad \text{for } \omega^2>\left(\frac{v^2}{c^2}-1\right)b^2 \tag{7.78}$$

We now construct (Figs. 33 and 34) the curves on the complex k-plane parameterized by Eqs. (7.77) and (7.78). We know from Eq. (7.67) that ω is complex on the real k-axis within the range $-b/c<k<b/c$. We wish to know if the contours of real ω bridge this region of the real k-axis.

Two distinct cases emerge: If $|v|< c$, then the condition on ω for Eqs. (7.77), namely,

$$\omega^2<(v^2/c^2-1)b^2$$

cannot be satisfied since ω is taken to be real. Hence for $|v|<c$ the contours of k values related to real ω are given by Eqs. (7.78) only. These contours lie on the real k-axis and are shown in Fig. 34. The curves Λ_+ and Λ_- correspond to the $\pm$ sign in Eq. (7.78). We see that for $|v|<c$ the integral (7.75) over real ω does not correspond to integration over all real values of k. In fact it corresponds to the integral over k outside the region $-b/c<k<b/c$. Therefore Eq. (7.68) does not hold and we conclude that the instability is absolute.

In Fig. 35 we have plotted ω against k_R, where we have taken explicit values for v, b, and c, with $|v|<c$. The lines are the paths of real ω deduced from Eq. (7.78). Figure 35 may be regarded as three-dimensional, with the k_I-axis perpendicular to the k_R and ω axes. The paths of real ω are confined to the k_R–ω plane. We see that an integral of a function of ω over real ω from $\omega=+\infty$ to $\omega=-\infty$ does not correspond to an integral of the transformed integrand over all real values of k.

Alternatively if $|v|>c$, then both Eqs. (7.77) and (7.78) describe contours of real ω. The contours described by Eq. (7.78) will be of the same form as those shown in Fig. 35.

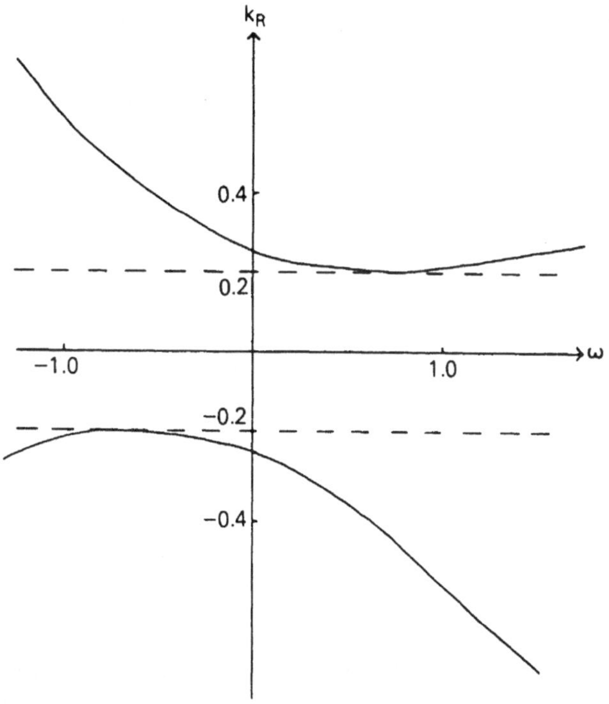

Fig. 35. Typical graph of real ω for absolute instability. The solid curves are the paths of real ω, and the k_I axis may be regarded as perpendicular to the figure. The paths of real ω are calculated from Eq. (7.78), where we have taken $v = 3$, $c = 5$, and $b = 1$. The condition for instability is given by Eq. (7.76): $-0.2 < k < 0.2$.

The contours described by Eq. (7.77) form an ellipse in the complex k-plane since the equation relating k_R and k_I, which is independent of ω, is

$$c^2k_R^2 + v^2k_I^2 = \frac{v^2b^2}{v^2 - c^2} \tag{7.79}$$

This ellipse meets the k_R-axis at the points given by $k_I = 0$ in Eq. (7.79), namely,

$$k_R = \pm\frac{vb}{c(v^2 - c^2)^{1/2}} \tag{7.80}$$

and, since for $|v| > c$ it follows that $vb/[c(v^2 - c^2)^{1/2}] > b/c$, the ellipse meets the contours that lie on the real k-axis. Hence the contour diagram is as shown in Fig. 36. The integral (7.75) over all values of real ω will correspond to the integral over all values of real k, and we conclude that the instability is convective.

In Fig. 36 we have plotted ω against k_R with explicit values for v, b, and c, and with $|v| > c$. The solid lines are the paths of real ω deduced from Eq. (7.78). The dashed line is the path of real ω deduced from Eq. (7.77); for these values of ω the values of k are complex and Fig. 37 shows how the paths of real ω are displaced from the real k_R axis. From these figures we see how a path over all real values of ω can correspond to a path over all real values of k.

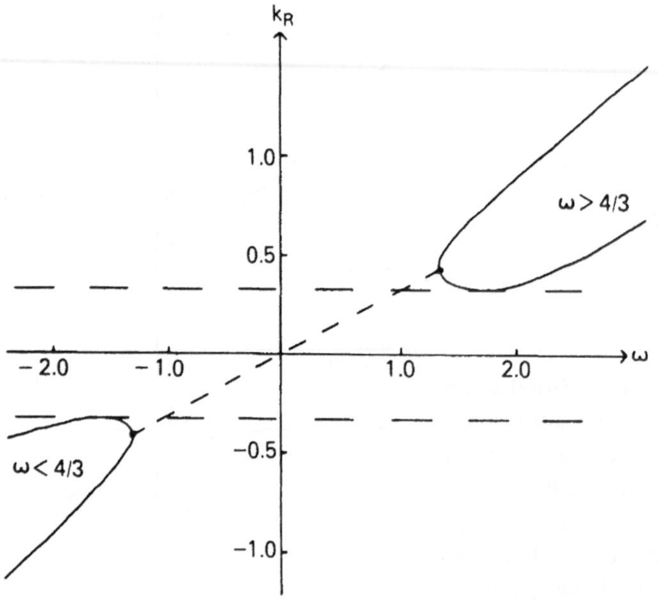

Fig. 36. Typical graph of real ω vs. k_R for convective instability. The solid curves lie in the k_R–ω plane and the dashed line which connects them represents those values of real ω for which k is complex. The solid curves are calculated from Eq. (7.78) and the dashed lines from Eq. (7.77), with $v=5$, $c=3$, and $b=1$. The condition for instability is given by Eq. (7.76): $-\frac{1}{3}<k<\frac{1}{3}$.

We have deduced that the system is unstable since the dispersion relation has solutions Im $\omega>0$. We have also deduced that the instability is absolute if the system of pendulums is moving with a velocity $|v|<c$, and it is convective if it is moving with a velocity greater than c.

The physical distinction between the absolute and convective instabilities in this case is that if the speed at which the pendulums "pull each other down" is less than the speed of the system then a stationary observer with a restricted field of view [see Fig. 31], who perturbs a pendulum from

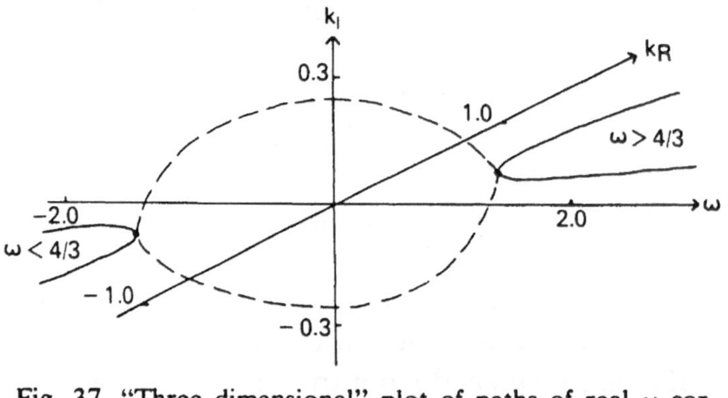

Fig. 37. "Three-dimensional" plot of paths of real ω corresponding to Fig. 36. The dashed curve is in a plane containing the k_I-axis, perpendicular to the k_R–ω plane.

its position of unstable equilibrium, will not see any pendulums "fall down." The instability is then convective. On the other hand, if the pendulums pull each other down faster than the system moves, then the same observer will see some pendulums fall down, even if the one that he pushed has moved out of his field of vision. The instability is absolute.

In this example it was obvious from the dynamics of the system that the system is unstable, but only with the above "kinetic analysis" is it possible to determine for what values of the parameters the instability is absolute or convective.

7.4. *Ionization and Attachment Instabilities in Self-Sustained Discharges (Dyer and James)*

7.4.1. *The Unperturbed Equations*

Dyer and James[(101)] consider equations for the number densities of electrons, n_e, and positive and negative ions, n_+, n_-, in a self-sustained discharge. From the heavy-particle number-density equation [Eq. (5.35)] applied to negative ions, and neglecting spatial derivatives, the source term (since the discharge is self-sustained), and the loss frequency term, we have

$$\frac{dn_-}{dt} = n_e n k_a - n_- n_+ k_r^i \tag{7.81a}$$

where we have assumed that the only processes affecting the negative-ion number density are electron neutral attachment, with rate k_a, and positive-ion–negative-ion recombination, with rate k_r^i. With number densities expressed in cm^{-3}, all rates are in $\text{cm}^3\,\text{sec}^{-1}$. With similar approximations applied to the electron number-density equation [Eq. (5.154)], we have

$$\frac{dn_e}{dt} = n_e n k_i - n_e n k_a - n_e n_+ k_r^e \tag{7.82a}$$

where k_i is the electron impact ionization rate and k_r^e is the electron–positive-ion recombination rate. We assume total charge conservation, so that the positive-ion number density is given by

$$n_+ = n_e + n_- \tag{7.83}$$

Following Dyer and James we rewrite Eqs. (7.81a) and (7.82a) with the attachment and ionization terms depending on the electron drift velocity

$\bar{V}_e$ [see Eqs. (5.152) and (5.195)]:

$$\frac{dn_-}{dt} = \bar{V}_e\bar{\beta}n_e - n_-n_+k_r^i \tag{7.81b}$$

$$\frac{dn_e}{dt} = \bar{V}_e\bar{\alpha}n_e - \bar{V}_e\bar{\beta}n_e - n_en_+k_r^e \tag{7.82b}$$

where

$$\bar{\alpha}(T_e) = \frac{nk_i(T_e)}{\bar{V}_e(T_e)}$$

$$\bar{\beta}(T_e) = \frac{nk_a(T_e)}{\bar{V}_e(T_e)}$$

are called the electron ionization and attachment coefficients, respectively, and are expressed in cm^{-1}. These coefficients are not the same as those used in Section 7.2, and so are distinguished by overbars. The electron drift velocity, $\bar{V}_e$, the coefficients $\bar{\alpha}$ and $\bar{\beta}$, and the rates k_r^i and k_r^e, are all functions of the electron temperature T_e. In order to perform a perturbation analysis on the above equations we first derive an expression relating fluctuations in the electron temperature, δT_e, to fluctuations in the electron number density, δn_e.

7.4.2. *Electron Number-Density Fluctuations in Terms of Electron Temperature Fluctuations*

As first pointed out by Haas,[70] and emphasized by Nighan and Wiegand,[72] it is possible, under certain simplifying assumptions, to relate directly the fluctuations in the electron number density, δn_e, to fluctuations of the electron temperature, δT_e.

From Eq. (5.161), the electron energy equation, we have, on neglecting terms containing spatial derivatives:

$$\frac{\partial}{\partial t}\left(\frac{3}{2}n_ekT_e\right) = \mathbf{j}_e \cdot \mathbf{E} - n_e\nu_u(T_e)kT_e \tag{7.84}$$

where we have used Eq. (5.162) to rewrite the final term in Eq. (5.161); $\mathbf{j}_e$ is the electron current density and is given in terms of the electron drift velocity $\bar{\mathbf{V}}_e$ by

$$\mathbf{j}_e = -en_e\bar{\mathbf{V}}_e \tag{5.158}$$

From Eq. (5.195) for the electron drift velocity we have, on neglecting diffusion terms (those containing spatial derivatives):

$$\bar{\mathbf{V}}_e = -\mu_e\mathbf{E} \tag{7.85}$$

where μ_e, the electron mobility, is given in terms of the electron momentum-transfer collision frequency $\nu_m(T_e)$ by

$$\mu_e = \frac{e}{m_e \nu_m(T_e)} \tag{5.193}$$

From Eqs. (5.158), (5.195), and (5.193) we have

$$\mathbf{j}_e = \frac{e^2}{m_e} \frac{n_e}{\nu_m} \mathbf{E} \tag{7.86}$$

The equation of conservation of the electron current is

$$\nabla \cdot \mathbf{j}_e + \frac{\partial (e n_e)}{\partial t} = 0 \tag{7.8}$$

We recall that, in Eq. (7.84), $\nu_u(T_e)$ is the electron energy-exchange collision frequency. Therefore ν_u^{-1} is a characteristic time for electron energy loss to the molecular gas. Now in mixtures of molecular gases of the type used in CO_2 gas lasers, ν_u^{-1} is very short compared to the time required for attachment, ionization, etc., owing to the large low-energy electron rates for vibrational excitation of the gas molecules. Therefore the time response of the electron energy density (and therefore electron temperature) is effectively instantaneous compared with the time required for changes in the charged-particle densities due to the processes (ionization, etc.) occurring in Eqs. (7.81b) and (7.82b).

The above considerations allow a quasi-steady relationship between the fluctuations δn_e, δT_e, $\delta \mathbf{E}$ to be derived from the steady-state forms of Eqs. (7.84) and (7.8), namely,

$$\mathbf{j}_e \cdot \mathbf{E} = n_e \nu_u k T_e \tag{7.87}$$

$$\nabla \cdot \mathbf{j}_e = 0 \tag{7.88}$$

Writing the perturbed form of Eq. (7.87) as

$$\delta \mathbf{j}_e \cdot \mathbf{E} + \mathbf{j}_e \cdot \delta \mathbf{E} = \delta(n_e \nu_u k T_e) \tag{7.89}$$

and of Eq. (7.86) as

$$\delta \mathbf{j}_e = \frac{e^2}{m_e} \delta\left(\frac{n_e}{\nu_m} \mathbf{E}\right) \tag{7.90}$$

then from Eqs. (7.86) and (7.90), Eq. (7.89) may be written

$$\frac{e^2}{m_e} \mathbf{E} \cdot \delta\left(\frac{n_e}{\nu_m} \mathbf{E}\right) + \frac{e^2}{m_e} \frac{n_e}{\nu_m} \mathbf{E} \cdot \delta \mathbf{E} = \delta(n_e \nu_u k T_e) \tag{7.91}$$

We now take the perturbations in $\delta\mathbf{E}$, $\delta\mathbf{j}_e$, δn_e, δT_e to have the periodic form

$$\delta\mathbf{E} = \delta\mathbf{E}_0\, e^{i(\mathbf{k}\cdot\mathbf{x}-\omega t)} \tag{7.92}$$

$$\delta\mathbf{j}_e = \delta\mathbf{j}_{e0}\, e^{i(\mathbf{k}\cdot\mathbf{x}-\omega t)}, \quad \text{etc.} \tag{7.93}$$

where $\mathbf{k}$, the wavenumber of the perturbation, is in the direction of $\delta\mathbf{E}$ and is at an angle ϕ to the direction of $\mathbf{E}$:

$$\mathbf{k}\cdot\mathbf{E} = kE\cos\phi \tag{7.94}$$

From Eq. (7.88) we have

$$\nabla\cdot\delta\mathbf{j}_e = 0 \tag{7.95}$$

and therefore, from Eqs. (7.90), (7.93), and (7.95):

$$0 = \mathbf{k}\cdot\delta\left(\frac{n_e}{\nu_m}\mathbf{E}\right)$$

or

$$\mathbf{k}\cdot\mathbf{E}\delta\left(\frac{n_e}{\nu_m}\right) + \frac{n_e}{\nu_m}\mathbf{k}\cdot\delta\mathbf{E} = 0$$

which becomes, using Eq. (7.94):

$$kE\cos\phi\cdot\delta\left(\frac{n_e}{\nu_m}\right) + \frac{n_e}{\nu_m}k\,\delta E = 0$$

or, on dividing through by k:

$$\frac{n_e}{\nu_m}\delta E = -E\cos\phi\,\delta\left(\frac{n_e}{\nu_m}\right)$$

and, multiplying through by $E\cos\phi$:

$$\frac{n_e}{\nu_m}E\,\delta E\cos\phi = -E^2\cos^2\phi\,\delta\left(\frac{n_e}{\nu_m}\right)$$

which may be written

$$\frac{n_e}{\nu_m}\mathbf{E}\cdot\delta\mathbf{E} = -E^2\cos^2\phi\,\delta\left(\frac{n_e}{\nu_m}\right) \tag{7.96}$$

Now expanding the first term in Eq. (7.91) and replacing $\mathbf{E}\cdot\delta\mathbf{E}$ by Eq. (7.96) we obtain

$$\frac{e^2}{m_e}E^2\delta\left(\frac{n_e}{\nu_m}\right) - 2\frac{e^2}{m_e}E^2\cos^2\phi\,\delta\left(\frac{n_e}{\nu_m}\right) = \delta(n_e\nu_u kT_e) \tag{7.97}$$

We can eliminate E^2 from the above equation by noting that substituting Eq. (7.86) for $\mathbf{j}_e$ into Eq. (7.87) gives

$$E^2 = \frac{m_e}{e^2} \nu_m \nu_u k T_e \tag{7.98}$$

From Eqs. (7.97) and (7.98) we have, on eliminating E^2:

$$\nu_m \nu_u k T_e \delta\left(\frac{n_e}{\nu_m}\right) - 2\nu_m \nu_u k T_e \delta\left(\frac{n_e}{\nu_m}\right) \cos^2 \phi = \delta(n_e \nu_u k T_e) \tag{7.99}$$

Now replacing $\delta\nu_m$ by $(\partial\nu_m/\partial T_e)\delta T_e$ and $\delta\nu_u$ by $(\partial\nu_u/\partial T_e)\delta T_e$, and defining

$$\hat{\nu}_m = \frac{T_e}{\nu_m}\frac{\partial \nu_m}{\partial T_e}, \qquad \hat{\nu}_u = \frac{T_e}{\nu_u}\frac{\partial \nu_u}{\partial T_e} \tag{7.100}$$

on expanding the terms in parentheses in Eq. (7.99), and using the definitions given by Eqs. (7.100), straightforward algebra gives

$$\frac{\delta T_e}{T_e} = \frac{-2\cos^2\phi}{1 + \hat{\nu}_u - \cos^2\phi\, \hat{\nu}_m} \frac{\delta n_e}{n_e} \tag{7.101}$$

which is the equation relating fluctuations δT_e and δn_e, as given by Nighan and Wiegand [Reference 72, Eq. (12)].

7.4.3. *Perturbation Analysis*

We now perform a perturbation analysis on Eqs. (7.81b) and (7.82b). Following Dyer and James we neglect the electron temperature dependence of k_r^e and k_r^i compared to that of $(\bar{V}_e\bar{\alpha})$ and $(\bar{V}_e\bar{\beta})$.

Eliminating n_+ in Eqs. (7.81b) and (7.82b) using Eq. (7.83) we obtain

$$\frac{dn_e}{dt} = (\bar{V}_e\bar{\alpha} - \bar{V}_e\bar{\beta})n_e - k_r^e n_e^2 - k_r^e n_e n_- \tag{7.102}$$

$$\frac{dn_-}{dt} = \bar{V}_e\bar{\beta} n_e - k_r^i n_e n_- - k_r^i n_-^2 \tag{7.103}$$

Subjecting Eqs. (7.102) and (7.103) to the perturbations

$$\begin{aligned} n_e &\to n_{e0} + \delta n_e \\ n_- &\to n_{-0} + \delta n_- \\ T_e &\to T_{e0} + \delta T_e \end{aligned} \tag{7.104}$$

and using Eq. (7.101) to eliminate δT_e, we obtain

$$\frac{d}{dt}\delta n_e = \left[(\bar{V}_e\bar{\alpha} - \bar{V}_e\bar{\beta}) + \frac{\partial(\bar{V}_e\bar{\alpha} - \bar{V}_e\bar{\beta})}{\partial T_e} F T_e - 2k_r^e n_{e0} - k_r^e n_{-0}\right]\delta n_e + (-k_r^e n_{e0})\,\delta n_- \tag{7.105}$$

$$\frac{d}{dt}\delta n_- = \left[-\bar{V}_e\bar{\beta} + \frac{\partial(\bar{V}_e\bar{\beta})}{\partial T_e} F T_e - k_r^i n_{-0}\right]\delta n_e + (-k_r^i n_{e0} - 2k_r^i n_{-0})\,\delta n_- \tag{7.106}$$

where we have defined

$$F \equiv \frac{2\cos^2\phi}{1 + \hat{\nu}_u - \cos 2\phi\,\hat{\nu}_m} \tag{7.107}$$

Rewriting Eqs. (7.105) and (7.106) as

$$\begin{aligned} \frac{d}{dt}\delta n_e &= a\,\delta n_e + b\,\delta n_- \\ \frac{d}{dt}\delta n_- &= c\,\delta n_e + d\,\delta n_- \end{aligned} \tag{7.108}$$

and assuming a periodic form for the perturbation,

$$\begin{aligned} \delta n_e &= \delta n_{e0}\, e^{-i\omega t} \\ \delta n_- &= \delta n_{-0}\, e^{-i\omega t} \end{aligned} \tag{7.109}$$

we obtain from Eqs. (7.108):

$$\begin{bmatrix} i\omega + a & b \\ c & i\omega + d \end{bmatrix} \begin{bmatrix} \delta n_{e0} \\ \delta n_{-0} \end{bmatrix} = 0 \tag{7.110}$$

For nonzero δn_{e0} and δn_{-0}, the determinant of the matrix in Eq. (7.110) must be zero, giving the dispersion relation as a quadratic equation in ω

$$\omega^2 - i\omega(a+d) - ad + bc = 0 \tag{7.111}$$

whose roots are

$$\begin{aligned} \omega &= \frac{1}{2}\{i(a+d) \pm [-(a+d)^2 - 4(bc - ad)]^{1/2}\} \\ &= \frac{i}{2}\{(a+d) \pm [(a-d)^2 + 4bc]^{1/2}\} \end{aligned} \tag{7.112}$$

From Eqs. (7.109) the condition for instability is

$$\operatorname{Im}\omega > 0 \tag{7.113}$$

and we see from Eq. (7.112) that a sufficient condition for the imaginary part of a root to be positive is

$$a+d>0 \tag{7.114}$$

irrespective of whether the square root in Eq. (7.112) is real or imaginary.

From Eqs. (7.105) amd (7.106), written in the form of (7.108), the instability condition given by Eq. (7.114) may be written as

$$\bar{V}_e(\bar{\alpha}-\bar{\beta})+\frac{\partial[\bar{V}_e(\bar{\alpha}-\bar{\beta})]}{\partial T_e}FT_e-[(2k_r^e+k_r^i)n_{e0}+(k_r^e+2k_r^i)n_{-0}]>0 \tag{7.115}$$

which is the condition derived by Dyer and James [Reference 101, Eq. (4)].

To evaluate Eq. (7.115), we need to know rates and other quantities as a function of gas mixture and E/N as well as the electron temperature dependence of $(\bar{\alpha}-\bar{\beta})\bar{V}_e$. If we did not use the equilibrium values for n_{e0} and n_{-0}, then we would have to evaluate Eq. (7.115) for each time t.

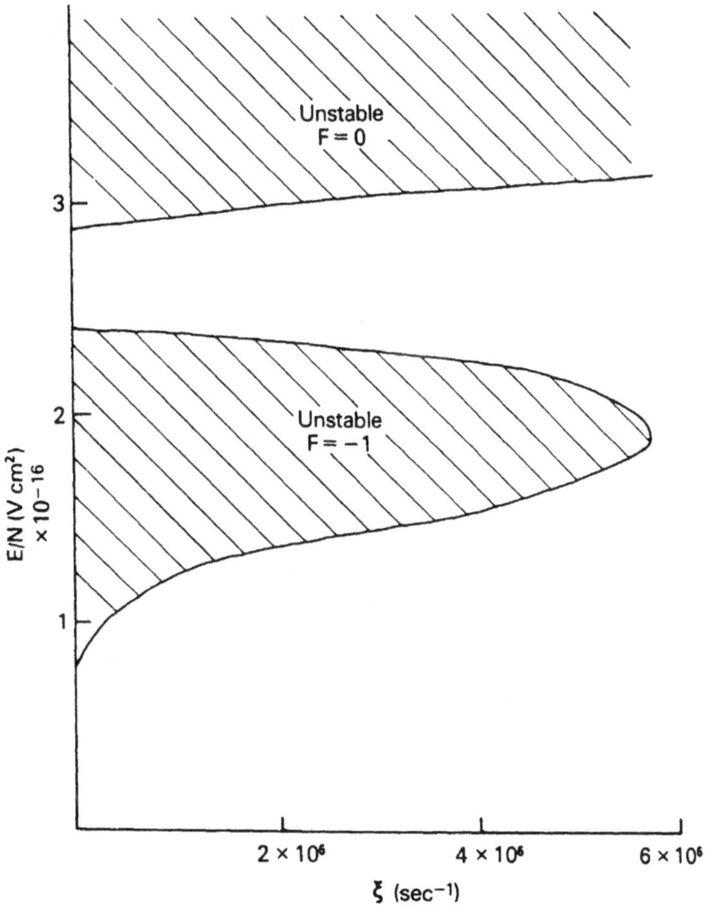

Fig. 38. Unstable regions in E/N-ξ space. (From Dyer and James,[101] with permission.)

Dyer and James consider two special cases of Eq. (7.115) corresponding to particular values of ϕ:

(i) For $\phi = \pi/2$ we have from Eq. (7.107) that $F = 0$ and hence Eq. (7.115) becomes

$$\bar{V}_e(\bar{\alpha} - \bar{\beta}) > [(2k_r^e + k_r^i)n_{e0} + (k_r^e + 2k_r^i)n_{-0}] \qquad (7.116)$$

(ii) For $\phi = 0$, $F \approx -1$. The unstable regions deduced from Eq. (7.115) by Dyer and James are shown in Fig. 38. In the figure, ξ is given by

$$\xi = [(2k_r^e + k_r^i)n_{e0} + (k_r^e + 2k_r^i)n_{-0}]$$

A more accurate plot of the regions of instability would be one in which all regions satisfying Eq. (7.115) were shown for all values of ϕ.

In Fig. 39 we have conjectured the domain of stability for all values of F by writing

$$\xi = A(E/N) + FB(E/N)$$

From Fig. 38 we see that at $F = 0$, $A(E/N)$ can be fit to a linear function of E/N, while for $F = 1$, we can fit $B(E/N)$ to a cubic function of E/N.

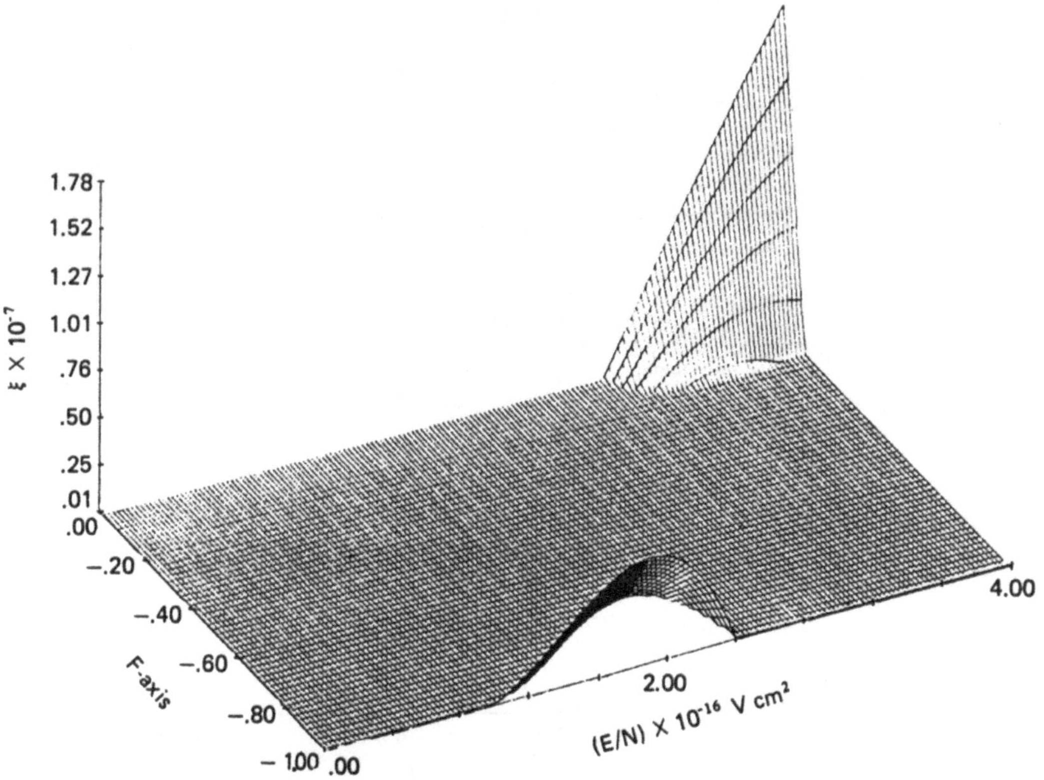

Fig. 39. Conjectured three-dimensional plot of regions of instability; operation inside either of the peaks is unstable.

7.5. Charged-Particle Modes of Instability (Nighan and Wiegand)

7.5.1. Rate Equations for Number Densities

In a study of the influence of negative-ion processes on steady-state properties in molecular gas discharges, Nighan and Wiegand[72] first consider the chemistry of the discharge (see Section 6.1.1). In order to carry out a stability analysis they then follow the approach of Haas,[70] and assume that there is only one species of negative ions and one of positive ions. Neglecting gradients and the loss frequency term, we may write Eq. (5.154) for the electron number density n_e in the form

$$\frac{\partial n_e}{\partial t} = n_e n k_i - n_e n_+ k_r^e - n_e n k_a + n_- n k_d + nS \tag{7.117}$$

using the Nighan and Wiegand definition of S, and the equation for the negative-ion number density, n_-, becomes

$$\frac{\partial n_-}{\partial t} = n_e n k_a - n_- n k_d - n_- n_+ k_r^i \tag{7.118}$$

where the k are the rates for the following processes: k_i is the electron impact ionization, $e + M \rightarrow M^+ + 2e$; k_r^e is the electron–positive-ion recombination, $e + M^+ \rightarrow M$; k_a is the electron neutral attachment, $e + M \rightarrow M^-$; k_d is the negative-ion detachment, $M^- + M \rightarrow 2M + e$; and k_r^i is the positive-ion–negative-ion recombination $M^- + M^+ \rightarrow 2M$. We recall that n is the total heavy-particle number density. We need not write down a separate equation for n_+, the number density of positive ions, if we assume that charge neutrality is preserved, since then

$$n_+ = n_e + n_- \tag{7.119}$$

Nighan and Wiegand point out that negative-ion clustering reactions of the type

$$O^- + CO_2 + M \rightarrow CO_3^- + M \tag{7.120}$$

are important for many circumstances of interest although they are not included in Eqs. (7.117)–(7.119). To include the effect of such reactions we would have to consider separate equations for different species of negative ions. Many important details of clustering processes are unknown at present, but nevertheless we shall see that (Section 7.5.4) the influence of the clustering process can be significant as regards plasma stability characteristics.

7.5.2. Charged-Particle Production and Loss Instability—No Negative Ions

In the absence of negative ions, and with no external source of ionization (i.e., self-sustained discharge), Eq. (7.117) becomes

$$\frac{\partial n_e}{\partial t} = n_e n k_i - n_e^2 k_r^e \tag{7.121}$$

where we have omitted the electron neutral-attachment term, since we are assuming that no negative ions are formed. Neglecting the electron temperature dependence of the recombination rate compared to that of the ionization rate, then the perturbed form of Eq. (7.121) is

$$\begin{aligned}\frac{\partial \delta n_e}{\partial t} &= \delta n_e n k_i + n_{e0} n \frac{\partial k_i}{\partial T_e} \delta T_e - 2n_{e0}\, \delta n_e k_r^e \\ &= \delta n_e n k_i + n_{e0} n k_i \hat{k}_i \frac{\delta T_e}{T_e} - 2n_{e0}\, \delta n_e k_r^e \end{aligned} \tag{7.122}$$

where

$$\hat{k}_i = \frac{T_e}{k_i}\frac{\partial k_i}{\partial T_e} = \frac{\partial \ln k_i}{\partial \ln T_e} \tag{7.123}$$

Equation (7.101) may be written

$$\frac{\delta T_e}{T_e} = \frac{-2\cos^2\phi}{\hat{\nu}_u'} \frac{\delta n_e}{n_e} \tag{7.124}$$

where

$$\hat{\nu}_u' = 1 + \hat{\nu}_u - \cos^2\phi \cdot \hat{\nu}_m \tag{7.125}$$

and using Eq. (7.124) to eliminate δT_e, Eq. (7.122) becomes

$$\frac{\partial}{\partial t}\delta n_e = \left(n k_i - n k_i \hat{k}_i \frac{2\cos^2\phi}{\hat{\nu}_u'} - 2n_{e0} k_r^e\right) \delta n_e \tag{7.126}$$

The electron number density n_e is given as the steady-state value of Eq. (7.121), namely,

$$n_{e0} n k_i - n_{e0}^2 k_r^e = 0$$

or

$$n_{e0} k_r^e = n k_i \tag{7.127}$$

Eliminating k_r^e from Eq. (7.126) we obtain

$$\frac{\partial}{\partial t}\delta n_e = n k_i \left(-\hat{k}_i \frac{2\cos^2\phi}{\hat{\nu}'} - 1\right) \delta n_e \tag{7.128}$$

and if we put

$$\delta n_e = \delta n_{e0}\, e^{-i\omega t}$$

then the instability criteria is Im $\omega > 0$, which, from Eq. (7.128) is

$$nk_i\left(-\hat{k}_i\frac{2\cos^2\phi}{\hat{\nu}'_u}-1\right)>0 \tag{7.129}$$

which is Eq. (14) of Nighan and Wiegand.[72]

Nighan and Wiegand have calculated $\hat{\nu}'_u$ for a wide variety of gas mixtures of present interest (CO_2 laser mixtures), and have found that it is positive except in very unusual cases. Since $\hat{k}_i$ is positive also, reflecting the strong increase in the electron production rate with increasing electron temperature, we see that the condition (7.129) is in general not satisfied. We deduce that the system is stable when ionization and electron–positive-ion recombination are the dominant processes. In the next section we investigate the stability of the discharge when negative-ion processes are important.

7.5.3. Charged-Particle Production and Loss Instability—Negative Ions Present

We now carry out a stability analysis on Eqs. (7.117) and (7.118), allowing for the electron temperature dependence of both the ionization and attachment rates, k_i and k_a.

The perturbed forms of Eqs. (7.117) and (7.118) are

$$\begin{aligned}\frac{\partial}{\partial t}\delta n_e &= \delta n_e(nk_i - n_+k_r^e - nk_a) + \delta n_-(nk_d)\\ &\quad + \delta n_+(-n_{e0}k_r^e) + \delta T_e\left(n_{e0}n\frac{\partial k_i}{\partial T_e} - n_{e0}n\frac{\partial k_a}{\partial T_e}\right)\\ \frac{\partial}{\partial t}\delta n_- &= \delta n_e(nk_a) + \delta n_-(-nk_d - n_{+0}k_r^i) + \delta n_+(-n_{-0}k_r^i)\\ &\quad + \delta T_e\left(n_{e0}n\frac{\partial k_a}{\partial T_e}\right)\end{aligned} \tag{7.130}$$

Using Eq. (7.119) to express δn_+ in terms of δn_e and δn_-, and Eq. (7.123) with a corresponding definition for k_a, Eqs. (7.130) become

$$\begin{aligned}\frac{\partial}{\partial t}\delta n_e &= \delta n_e(nk_i - n_{+0}k_r^e - nk_a - n_{e0}k_r^e)\\ &\quad + \delta n_-(nk_d - n_{e0}k_r^e) + \frac{\delta T_e}{T_e}(n_{e0}nk_i\hat{k}_i - n_{e0}nk_a\hat{k}_a)\end{aligned} \tag{7.130a}$$

$$\frac{\partial}{\partial t}\delta n_- = \delta n_e(nk_a - n_{-0}k_r^i) + \delta n_-(-nk_d - n_+k_r^i - n_{-0}k_r^i)$$
$$+\frac{\delta T_e}{T_e}(n_{e0}nk_a\hat{k}_a) \tag{7.130b}$$

Eliminating δT_e from Eqs. (7.130a and b) using Eq. (7.124) we obtain

$$\begin{aligned}\frac{\partial}{\partial t}\delta n_e &= \delta n_e\left[nk_i - n_{+0}k_r^e - nk_a - n_{e0}k_r^e - \frac{2\cos^2\phi}{\hat{\nu}'_u}(nk_i\hat{k}_i - nk_a\hat{k}_a)\right] \\ &\quad + \delta n_-(nk_d - n_{e0}k_r^e) \\ \frac{\partial}{\partial t}\delta n_- &= \delta n_e\left(nk_a - n_{-0}k_r^i - \frac{2\cos^2\phi}{\hat{\nu}'_u}nk_a\hat{k}_a\right) \\ &\quad + \delta n_-(-nk_d - n_{+0}k_r^i - n_{-0}k_r^i)\end{aligned} \tag{7.131}$$

Defining

$$\begin{aligned}A &= nk_i - n_{+0}k_r^e - nk_a - n_{e0}k_r^e - \frac{2\cos^2\phi}{\hat{\nu}'_u}(nk_1\hat{k}_i - nk_a\hat{k}_a) \\ B &= nk_d - n_{e0}k_r^e \\ C &= nk_a - n_{-0}k_r^i - \frac{2\cos^2\phi}{\hat{\nu}'_u}nk_a\hat{k}_a \\ D &= -nk_d - n_{+0}k_r^i - n_{-0}k_r^i\end{aligned} \tag{7.132}$$

and setting

$$\begin{aligned}\delta n_e &= \delta n_{e0}\, e^{+i\omega t} \\ \delta n_- &= \delta n_{-0}\, e^{+i\omega t}\end{aligned} \tag{7.133}$$

then Eqs. (7.131) may be written in the matrix form

$$\begin{bmatrix} -i\omega + A & B \\ C & -i\omega + D \end{bmatrix}\begin{bmatrix} \delta n_{e0} \\ \delta n_{-0} \end{bmatrix} = 0 \tag{7.134}$$

For nonzero δn_{e0} and δn_{-0} the determinant of the square matrix in Eq. (7.134) must be zero, and hence we have Eq. (7.111) again:

$$\omega^2 + i\omega(A+D) + (BC - AD) = 0$$

or

$$(i\omega)^2 - (i\omega)(A+D) + (AD - BC) = 0$$

which has solutions

$$\begin{aligned}i\omega &= \tfrac{1}{2}\{(A+D) \pm [(A+D)^2 - 4(AD-BC)]^{1/2}\} \\ &= \tfrac{1}{2}[-b \pm \tfrac{1}{2}(b^2 - 4c)^{1/2}]\end{aligned} \tag{7.135}$$

where we have defined real functions of the steady-state properties of the plasma by

$$\begin{aligned} -b &= A + D \\ c &= AD - BC \end{aligned} \tag{7.136}$$

The plus and minus signs in Eq. (7.135) do *not* distinguish between ionization and negative-ion (attachment) modes of instability. The relationship between the signs of b and c determines the mode of instability. From Eqs. (7.132) and (7.136) we have

$$\begin{aligned} -b &= -nk_i\hat{k}_i\left(\frac{2\cos^2\phi}{\hat{\nu}'_u}\right)\left(1 - \frac{k_a\hat{k}_a}{k_i\hat{k}_i}\right) \\ &\quad + nk_i - n_{+0}k_r^e - nk_a - n_{e0}k_r^e - nk_d - n_{+0}k_r^i - n_{-0}k_r^i \\ c &= -\left(nk_a - n_{-0}k_r^i - \frac{2\cos^2\phi}{\hat{\nu}'_u}nk_a\hat{k}_a\right)(nk_d - n_ek_r^e) \\ &\quad + \left[nk_i - n_{+0}k_r^e - nk_a - n_{e0}k_r^e - \frac{2\cos^2\phi}{\hat{\nu}'_u}(nk_i\hat{k}_i - nk_a\hat{k}_a)\right] \\ &\quad \times(-nk_d - n_{+0}k_r^i - n_{-0}k_r^i) \end{aligned} \tag{7.137}$$

To obtain the instability criteria given by Nighan and Wiegand we suppose that the number densities have their steady-state values, given by Eqs. (7.117) and (7.118), set equal to zero:

$$\begin{aligned} n_{e0}nk_i - n_{e0}n_{+0}k_r^e - n_{e0}nk_a + n_{-0}nk_d + nS &= 0 \\ n_{e0}nk_a - n_{-0}nk_d - n_{-0}n_{+0}k_r^i &= 0 \end{aligned} \tag{7.138}$$

Straightforward algebra using Eqs. (7.137) and (7.138) gives (dropping the subscript 0):

$$\begin{aligned} b &= -nk_i\hat{k}_i\left(\frac{2\cos^2\phi}{\hat{\nu}'_u}\right)\left(1 - \frac{k_a\hat{k}_a}{k_i\hat{k}_i}\right) \\ &\quad + \left(\frac{n_e}{n_+}n_+k_r^e + \frac{n_-}{n_+}n_+k_r^i + \frac{n_e}{n_-}nk_a + \frac{n_-}{n_e}nk_d + \frac{n}{n_e}S\right) \\ c &= nk_i\hat{k}_i\left(\frac{2\cos^2\phi}{\hat{\nu}'_u}\right)\left\{\left(\frac{n_-}{n_+}n_+k_r^i + \frac{n_e}{n}nk_a\right)\right. \\ &\quad \left. - \left[\left(1 + \frac{n_-}{n_+}\right)n_+k_r^i + \frac{n_e}{n_+}n_+k_r^e\right]\frac{k_a\hat{k}_a}{k_i\hat{k}_i}\right\} \\ &\quad + \left[\frac{n_e}{n_-}nk_an_+k_r^e + \frac{n_-}{n_e}k_r^enk_d + \frac{n_-}{n_+}n_+k_r^i + \frac{n_e}{n_-}nk_a\right)\frac{n}{n_e}S\right] \end{aligned} \tag{7.139}$$

Equations (7.135) and (7.139) are equivalent to those given by Nighan and Wiegand [Reference 72, Eqs. (15)–(17)] and Haas [Reference 70, Eqs. (52)–(54)].

Nighan and Wiegand and Haas have applied the stability conditions represented by Eqs. (7.135) and (7.139) to a wide range of gas mixtures in low-pressure self-sustained discharges and have concluded that:

(i) There are two modes of instability, one an ionization instability and the other a negative-ion (attachment) instability;

(ii) When b is negative the ionization mode is unstable, and when b and c are of opposite sign, the negative-ion mode is unstable;

(iii) For conditions of interest, c is always positive while b changes sign. Therefore, for $c>0$, the negative-ion mode will be unstable only for those conditions such that the ionization mode is also unstable, i.e., $b<0$;

(iv) Haas has shown that when the conditions for both types of instability are satisfied simultaneously the ionization instability has the fastest growth rate.

It should be emphasized here that both the considerations of Nighan and Wiegand and Haas are primarily directed toward the study of low-pressure glow discharges (20–100 Torr). At higher pressures some of the assumptions, such as charge neutrality and the neglect of gradients in plasma variables, may not be valid.

7.5.4. *Numerical Evaluation of Instability Regions*

Nighan and Wiegand evaluated the instability region given by $b>0$, with b given by Eq. (7.139). For a 20-Torr CO_2–N_2–He mixture they take the fractional concentration of the secondary species CO, regarded as an additive, as a variable parameter. The negative-ion species O^- is formed by dissociative attachment:

$$e+CO_2 \rightarrow CO+O^- \tag{7.140}$$

and the CO additive detaches electrons from O^-:

$$O^- + CO \rightarrow CO_2 + e \tag{7.141}$$

Given the initial CO_2–N_e–He–CO concentrations, the electron Boltzmann equation must be solved to obtain the electron temperature, upon which many of the rates depend [i.e., the rates in Eq. (7.139) and the rate for the reaction (7.140)]. The steady-state form of the rate equations (5.35) for the plasma chemistry can then be solved to determine the negative-ion (O^-) concentration, and the instability criteria can then be applied to determine the regions of instability. The result of such a calculation by Nighan and Wiegand is shown in Fig. 40.

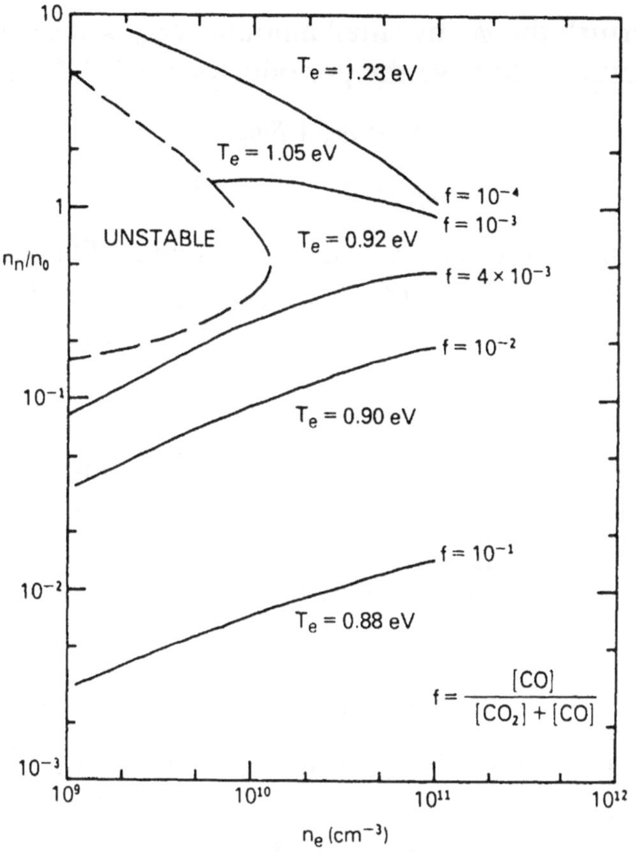

Fig. 40. Ionization instability boundary computed by Nighan and Wiegand[72] for a 20-Torr $CO_2:N_2:He = 1:7:12$ mixture. The electron density, n_e, and CO fraction, f, are the variable parameters from which the negative-ion concentration n_-/n, the electron temperature T_e, and the stability criteria are computed.

7.6. General Theory of Stability

7.6.1. Dispersion Relations for Perturbed Systems

Consider a physical system described by N variables ϕ_α, $\alpha = 1, \ldots, N$. We denote the equations for the system by

$$\begin{aligned} \hat{F}_1\phi_1 &= f_1(\phi_1, \phi_2, \ldots, \phi_N) \\ &\;\vdots \\ \hat{F}_N\phi_N &= f_N(\phi_1, \phi_2, \ldots, \phi_N) \end{aligned} \tag{7.142}$$

where the operator $\hat{F}_\alpha$ contains space and time derivatives of arbitrary orders with arbitrary coefficients, and the f_α are arbitrary functions of their arguments.

We now perturb the ϕ_α by small amounts $\delta\phi_\alpha$ whose space and time dependence is assumed to have the periodic form $e^{i(\mathbf{k}\cdot\mathbf{x}-\omega t)}$; hence

$$\begin{aligned}\phi_\alpha &\to \phi_\alpha + \delta\phi_\alpha \\ \delta\phi_\alpha &= \Phi_\alpha e^{i(\mathbf{k}\cdot\mathbf{x}-\omega t)}\end{aligned} \tag{7.143}$$

where the Φ_α are arbitrary but small constants. On substituting the perturbations of Eqs. (7.143) into Eqs. (7.142) we obtain a "dispersion relation," which is an algebraic relation in ω and $\mathbf{k}$, and is independent of the Φ_α:

$$D(\omega, \mathbf{k}) = 0 \tag{7.144}$$

The following example shows how the dispersion relation is obtained: We consider a hypothetical system described by the pair of equations

$$\begin{aligned}\frac{\partial\phi_1}{\partial t} + \mathbf{v}\cdot\nabla\phi_1 &= f_1(\phi_2)\phi_1 \\ \frac{\partial\phi_2}{\partial t} + \mathbf{v}\cdot\nabla\phi_2 &= f_2(\phi_1)\phi_2\end{aligned} \tag{7.145}$$

We make the perturbations (neglecting subscript 0 on equilibrium quantities)

$$\begin{aligned}\phi_1 &\to \phi_1 + \delta\phi_1 \\ \phi_2 &\to \phi_2 + \delta\phi_2\end{aligned}$$

Therefore

$$\begin{aligned}f_1(\phi_2) &\to f_1(\phi_2 + \delta\phi_2) \approx f_1(\phi_2) + \frac{\partial f_1}{\partial\phi_2}\delta\phi_2 \\ f_2(\phi_1) &\to f_2(\phi_1 + \delta\phi_1) \approx f_2(\phi_1) + \frac{\partial f_2}{\partial\phi_1}\delta\phi_1\end{aligned} \tag{7.146}$$

On substituting the perturbations (7.146) into Eqs. (7.145) neglecting terms of second order in perturbed quantities, and using Eqs. (7.145) to eliminate terms containing no perturbed quantities, we obtain

$$\begin{aligned}\left(\frac{\partial}{\partial t} + \mathbf{v}\cdot\nabla\right)\delta\phi_1 &= f_1\delta\phi_1 + \phi_1\frac{\partial f_1}{\partial\phi_2}\delta\phi_2 \\ \left(\frac{\partial}{\partial t} + \mathbf{v}\cdot\nabla\right)\delta\phi_2 &= f_2\delta\phi_2 + \phi_2\frac{\partial f_2}{\partial\phi_1}\delta\phi_1\end{aligned} \tag{7.147}$$

We now assume

$$\begin{aligned}\delta\phi_1 &= \Phi_1 e^{i(\mathbf{k}\cdot\mathbf{x}-\omega t)} \\ \delta\phi_2 &= \Phi_2 e^{i(\mathbf{k}\cdot\mathbf{x}-\omega t)}\end{aligned} \tag{7.148}$$

where Φ_1 and Φ_2 are arbitrary but small and do not depend on $\mathbf{x}$ or t. With the substitutions (7.148), Eqs. (7.147) become

$$\begin{aligned}(-i\omega+i\mathbf{v}\cdot\mathbf{k})\Phi_1&=f_1\Phi_1+\phi_1\frac{\partial f_1}{\partial\phi_2}\Phi_2\\(-i\omega+i\mathbf{v}\cdot\mathbf{k})\Phi_2&=f_2\Phi_2+\phi_2\frac{\partial f_2}{\partial\phi_1}\Phi_1\end{aligned}\tag{7.149}$$

or, in matrix notation

$$\begin{bmatrix}f_1+i\omega-i\mathbf{v}\cdot\mathbf{k} & \phi_1\dfrac{\partial f_1}{\partial\phi_2}\\ \phi_2\dfrac{\partial f_2}{\partial\phi_1} & f_2+i\omega-i\mathbf{v}\cdot\mathbf{k}\end{bmatrix}\begin{bmatrix}\Phi_1\\ \Phi_2\end{bmatrix}=0\tag{7.150}$$

Since Φ_1 and Φ_2 are arbitrary the determinant of the square matrix in Eq. (7.150) must be zero, and we have

$$(f_1+i\omega-i\mathbf{v}\cdot\mathbf{k})(f_2+i\omega-i\mathbf{v}\cdot\mathbf{k})-\phi_1\phi_2\frac{\partial f_1}{\partial\phi_2}\frac{\partial f_2}{\partial\phi_1}=0\tag{7.151}$$

which is the dispersion relation $D(\omega, \mathbf{k})=0$ corresponding to Eqs. (7.145).

We suppose that the general dispersion relation [Eq. (7.144)] for the system (7.142) has solutions

$$\omega=\Omega_j(\mathbf{k})\tag{7.152}$$

or, alternatively

$$\mathbf{k}=\mathbf{K}_j(\omega)\tag{7.153}$$

where j runs from unity to some integer N' and enumerates what we shall call the "modes" of the system. We can see from the above example that in general the number of modes will be equal to the order of the dispersion relation in ω, which in turn will equal the number of perturbed independent variables. If the dispersion relation (7.144) is solved for ω [Eq. (7.152)], and if the resulting functions Ω_j are real, then we see from Eq. (7.143) that the disturbance propagates through the system without attenuation. If, on the other hand, the solution gives $\operatorname{Im}\Omega_j>0$ for some j, then we see from Eq. (7.143) that the disturbance is amplified.

Here we come up against a difficulty. The fact that a disturbance is amplified is not a sufficient condition for instability to occur. The representation by Eq. (7.143) of a disturbance is in the form of a traveling wave and it is conceivable that a disturbance will move out of the system at the same time as it is being amplified. As a physical example we can consider flowing systems (e.g., gas dynamic lasers). Equations (7.145), for example, contain a "derivative following the motion" $\partial/\partial t+\mathbf{v}\cdot\boldsymbol{\nabla}$. In such systems we can easily envisage disturbances being carried along with the flow and out of the system. Alternatively, disturbances may remain stationary and amplify, resulting in instability. In the next section we establish criteria to distinguish between these cases.

7.6.2. Absolute and Convective Instabilities

In this section we derive the criteria for deciding whether perturbations that are amplified by the system remain within the system (absolute instability) or are transported out of the system (convective instability). We follow the approach of Sturrock.[97]

We first note that a general representation of a disturbance will be given by a superposition of disturbances of the form Eq. (7.143) with differing wavelengths:

$$\begin{aligned} \phi_\alpha &\to \phi_\alpha + \delta\phi_\alpha \\ \delta\phi_\alpha(\mathbf{x}, t) &= \int_{-\infty}^{\infty} d^3k\, \Phi_\alpha(\mathbf{k}) \exp\{i[\mathbf{k}\cdot\mathbf{x} - \Omega_j(\mathbf{k})t]\} \end{aligned} \tag{7.154}$$

where the integral is over all real values of $\mathbf{k}$; $\Omega_j(\mathbf{k})$ is the solution [Eq. (7.152)] of the dispersion relation (7.144) for the system (7.142). We now restrict our attention to one spatial dimension and we suppose that for some real value k_0 of k the solution $\Omega_j(k)$ of the dispersion relation is complex. We have, omitting for the purpose of clarity the subscripts α and j in Eq. (7.154):

$$\delta\phi = \int_{-\infty}^{\infty} dk\, \Phi(k)\, e^{i[kx - \Omega(k)t]} \tag{7.155}$$

We shall consider a quasi-monochromatic disturbance $\delta\phi$; that is, one whose wavelength is sharply peaked around k_0 and is negligibly small elsewhere. Explicitly, we could take

$$\Phi(k) = \exp\left[-\left(\frac{k-k_0}{\Delta k}\right)^{2n}\right] \tag{7.156}$$

where n is some positive integer and Δk is a measure of the width of the wavelength band of the disturbance, which may be made arbitrarily small.

We prove below that with a bounded $\Phi(k)$ [non-negligible only in some finite region of k-space, e.g., Eq. (7.156)]:

$$\delta\phi(x, t) \to 0 \qquad \text{as } x \to \pm\infty \tag{7.157}$$

This result is known as Riemann's lemma[98] (see also Section 7.3).

First we rewrite Eq. (7.155) in the form

$$\delta\phi(x) = \int_{-\infty}^{\infty} dk\, f(k)\, e^{ikx}$$
$$= \int_{-\infty}^{\infty} dk\, f(k) \cos kx + i \int_{-\infty}^{\infty} dk\, f(k) \sin kx \tag{7.158}$$

where $f(k) = \Phi(k)\, e^{-i\Omega(k)t}$, and we omit the t-dependence in Eq. (7.158). We consider the term containing the cosine on the right-hand side of Eq. (7.158). Now since $\Phi(k)$ is bounded, so is $f(k)$, and therefore for a given positive number ε we can choose a K so large that

$$\int_{K}^{\infty} dk|f(k)| < \varepsilon, \qquad \int_{-\infty}^{-K} dk|f(k)| < \varepsilon \tag{7.159}$$

and hence, for all x, we have

$$\int_{K}^{\infty} dk|f(k) \cos kx| < \varepsilon, \qquad \int_{-\infty}^{-K} dk|f(k) \cos kx| < \varepsilon \tag{7.160}$$

Now given $f(k)$, we can define a function $\chi(k)$, continuous in the interval $(-K, K)$, such that

$$\int_{-K}^{K} dk|f(k) - \chi(k)| < \varepsilon \tag{7.161}$$

and then, for all x,

$$\left| \int_{-K}^{K} dk[f(k) - \chi(k)] \cos kx \right| < \varepsilon \tag{7.162}$$

We now perform an integration by parts:

$$\int_{-K}^{K} dk\, \chi(k) \cos kx = \frac{\chi(K) \sin Kx}{x} + \frac{\chi(-K) \sin Kx}{x} - \frac{1}{x} \int_{-K}^{K} \sin kx \frac{d\chi(k)}{dk} dx \tag{7.163}$$

From Eq. (7.163) we see that for a fixed K we can choose an x_0 so large that the modulus of Eq. (7.163) is less than ε for $x > x_0$. Now, from Eqs. (7.160) and (7.162) we have

$$\begin{aligned} &\int_{-\infty}^{\infty} dk|f(k) \cos kx| < 3\varepsilon + \int_{-K}^{K} dk|x(k) \cos kx| \\ &\left| \int_{-\infty}^{\infty} dk\, f(k) \cos kx \right| < 4\varepsilon \qquad (x > x_0) \end{aligned} \tag{7.164}$$

and as $\varepsilon \to 0$, Eq. (7.164) becomes

$$\left| \int_{-\infty}^{\infty} dk\, f(k) \cos kx \right| \to 0 \qquad \text{as } x \to \infty$$

A similar proof applies to the sine term on the right-hand side of Eq. (7.158), and so Eq. (7.157) is proved.

Since by Eq. (7.157), $\delta\phi(x, t)$ is bounded in its spatial extent we shall, following Sturrock,[97] call $\delta\phi(x, t)$ a spacelike disturbance. Similarly we will call a disturbance timelike if it satisfies the condition

$$\delta\phi(x, t) \to 0 \qquad \text{as } t \to \pm\infty \tag{7.165}$$

Clearly a disturbance can be both spacelike and timelike. We can imagine typical "contour diagrams" for $\delta\phi$ for the various cases (Figs. 41–43). The qualitative features of the contours in Figs. 41–43 are readily deduced from the conditions [Eqs. (7.157) and (7.165)] that define spacelike and timelike disturbances.

We see from Fig. 43 that a disturbance that is both spacelike and timelike can be amplified but that the amplifying disturbance travels through the system. Only a disturbance that is spacelike but not timelike (Fig. 41) is amplified while remaining within the bounds of a finite physical

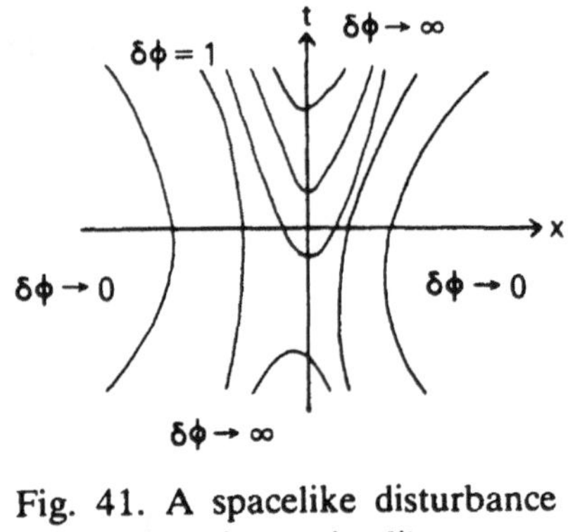

Fig. 41. A spacelike disturbance which is not timelike.

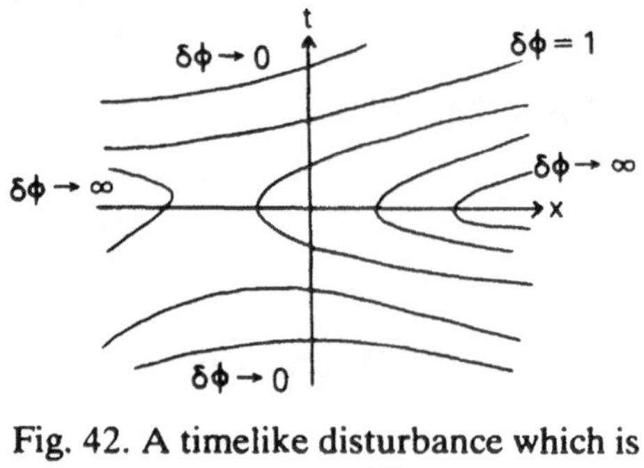

Fig. 42. A timelike disturbance which is not spacelike.

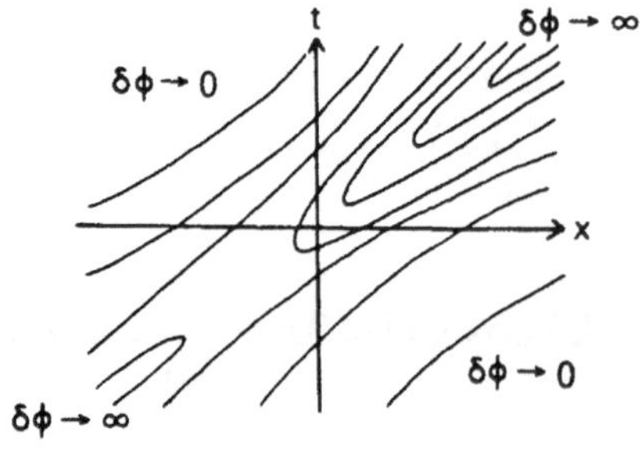

Fig. 43. A disturbance which is both spacelike and timelike.

system. We therefore arrive at the following definition:

> If a disturbance growing in time, which is composed of a narrow spectrum of wavenumbers and which therefore is a spacelike packet, is also a timelike packet, then the instability represented by this disturbance is convective; if, however, the spacelike disturbance is not a timelike disturbance then the instability is absolute.

This definition shows that a physical system that exhibits convective instability may exist in a stable state even in the presence of random disturbances, since these disturbances, although amplified, may be carried outside the physical bounds of the system. (They might be reflected by the boundaries of the system, however.) On the other hand, if a system exhibits absolute instability, then a random disturbance will grow in amplitude at the point where it originated, and we also expect that it will spread to extend over an arbitrarily large region of the system (Fig. 41). Such a physical system is said to be unstable or "self-oscillatory."

Having defined absolute and convective instabilities, it remains to be shown how solutions of the dispersion relations having $\operatorname{Im}\omega > 0$ can be tested to discover if the instability is absolute or convective. This shall be done in the next section.

7.6.3. Dispersion Relations: Identification of Instability Type

We recall that from a system of equations we obtain a dispersion relation

$$D(\omega, k) = 0 \tag{7.166}$$

which we can solve for ω or k, namely, either

$$\omega = \Omega(k) \tag{7.167}$$

or

$$k = K(\omega) \tag{7.168}$$

In the previous section we supposed that for some real value k_0 of k the solution [Eq. (7.167)] $\omega = \Omega(k_0)$ is complex. We now construct Figs. 44 and 45 in the k_R–k_I plane ($k = k_R + ik_I$). The curve Λ is the locus of the points given by Eq. (7.168), where ω takes on all real values. Since we are assuming that for $k = k_0$, ω is complex, the curve Λ does not pass through the point k_0. We suppose that $\omega = \Omega(k)$ is real for real k outside the range k_1 to k_2. The path of integration of the integral in Eq. (7.154), namely,

$$\delta\phi(x, t) = \int_{-\infty}^{\infty} dk\, \Phi(k)\, e^{i[kx - \Omega(k)t]} \tag{7.169}$$

therefore corresponds to integration over real values of ω, except for the contribution to the integral between the limits k_1 and k_2. Now by an

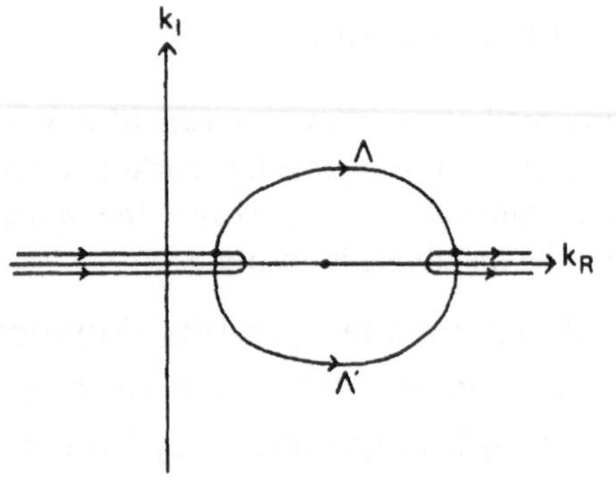

Fig. 44. Contour diagram of real ω for convective instability. A band of wavenumbers corresponding to instability is bridged by a locus of wavenumbers corresponding to real frequencies.

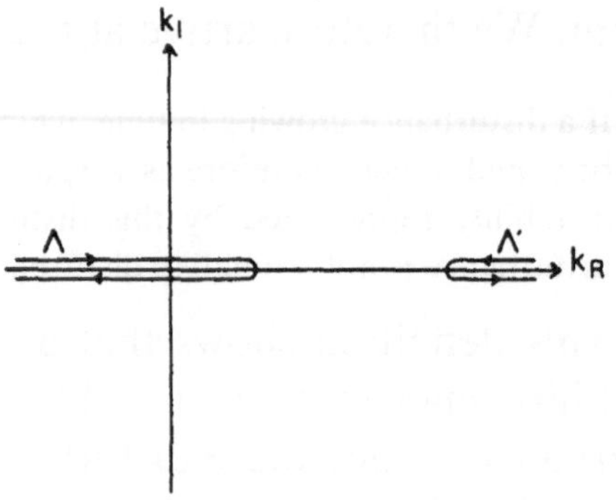

Fig. 45. Contour diagram of real ω for absolute instability. The band of wavenumbers corresponding to instability is not bridged by a locus of wavenumbers corresponding to real frequencies.

appropriate choice of $\Phi(k)$, for example, by making Δk in Eq. (7.156) arbitrarily small, we may make the contribution to the integral (7.169) negligibly small outside a small neighborhood of k_0. If the functions occurring in Eq. (7.169) are analytic we can distort the path of integration so that it runs from $k = -\infty$ to $k = \infty$ by some path other than the real axis. We ask if it is possible to displace the contour of integration in such a way that the integral (7.169) then corresponds to integration over all real values of ω. It will be possible to do this if and only if the curve "bridges" the gap between k_1 and k_2 (Fig. 44). Therefore, if Λ has one of the forms (Λ or Λ') shown in Fig. 44 we can use Eq. (7.168) to reexpress the integral (7.169) as

$$\delta\phi(x, t) = \int_{-\infty}^{+\infty} d\omega \frac{dK(\omega)}{d\omega} \Phi[K(\omega)]\, e^{i[K(\omega)x - \omega t]} \tag{7.170}$$

where ω runs over all real values.

By Riemann's lemma, if $\delta\phi$ can be expressed in the integral form of Eq. (7.170), then $\delta\phi$ is timelike. It is spacelike by Riemann's lemma and Eq. (7.157), and therefore the instability is convective. If, on the other hand the curves Λ and Λ' (representing a pair of modes) have the configuration shown in Fig. 45, then the integral (7.170) cannot be reexpressed as an integral for a timelike packet, and therefore the instability is absolute.

7.6.4. *Green's Function Formalism*

Haas[70] and, in greater detail, Briggs,[102] give equations for the response of a system to a localized source of disturbance. The equations are related to the Green's function formalism for a physical system, and in this section we review the connection between the Green's function formalism and the formalism of Sections 7.6.1–7.6.3.

We write the system of Eqs. (7.142) in the form

$$E_i[\phi_j(x, t)] = 0, \qquad i, j = 1, \ldots, N \tag{7.171}$$

and we suppose that with a small localized source $\bar{\delta J}_i(x, t)$ the system is described by the equations

$$E_i[\phi_j(x, t) + \delta\phi_j(x, t)] = \bar{\delta J}_i(x, t) \tag{7.172}$$

Comparing Eqs. (7.171) and (7.172) we see that we are regarding the $\bar{\delta J}_i$ as the "source" of the disturbances $\delta\phi_j$. Writing Eq. (7.172) in the form

$$E_i[\phi_j(x, t)] + \sum_{k=1}^{N} \frac{\delta E_i}{\delta \phi_k} \delta\phi_k(x, t) = \bar{\delta J}_i(x, t) \tag{7.173}$$

and using Eq. (7.171) we obtain

$$\sum_k \frac{\delta E_i}{\delta \phi_k} \delta\phi_k(x, t) = \bar{\delta J}_i(x, t) \tag{7.174}$$

In Eqs. (7.173) and (7.174), $\delta E_i/\delta\phi_k$ is the "functional derivative" of E_i with respect to ϕ_k defined by

$$\frac{\delta E_i}{\delta \phi_k} = \frac{\partial E_i}{\partial \phi_k} - \frac{\partial}{\partial x}\left[\frac{\partial E_i}{\partial(\partial \phi_k/\partial x)}\right] + \frac{\partial^2}{\partial x^2}\left[\frac{\partial E_i}{\partial(\partial^2 \phi_k/\partial x^2)}\right] - \cdots \tag{7.175}$$

From Eq. (7.174) we determine the responses $\delta\phi_k$ to the sources $\bar{\delta J}_i$:

$$\delta\phi_k(x, t) = \sum_i \left(\frac{\delta E_i}{\delta \phi_k}\right)^{-1} \bar{\delta J}_i(x, t) \tag{7.176}$$

where $(\delta E_i/\delta\phi_k)^{-1}$ is the inverse of $\delta E_i/\delta\phi_k$ regarded as a matrix. $(\delta E_i/\delta\phi_k)^{-1}$ is Green's function for the system.

If we now express $\delta\phi_k(x, t)$ as a Fourier–Laplace transform:

$$\delta\phi_j(x, t) = \int \frac{dk}{2\pi} \int \frac{d\omega}{2\pi} \Phi_j(k, \omega)\, e^{i(kx - \omega t)} \tag{7.177}$$

and substitute this expression for $\delta\phi_k$ into Eq. (7.176) we obtain

$$\sum_j \int \frac{dk}{2\pi} \int \frac{d\omega}{2\pi} \Phi_j(k, \omega) D_{ij}(k, \omega)\, e^{i(kx - \omega t)} = \int \frac{dk}{2\pi} \int \frac{d\omega}{2\pi} \bar{\delta J}_i(k, \omega t)\, e^{i(kx - \omega t)} \tag{7.178}$$

where $\bar{\delta J}_i(k, \omega)$ is the Fourier–Laplace transform of $\bar{\delta J}_i(x, t)$ and $D_{ij}(k, \omega)$ is defined by

$$\left(\frac{\delta E_i}{\delta \phi_j}\right) e^{i(kx - \omega t)} = D_{ij}(k, \omega)\, e^{i(kx - \omega t)} \tag{7.179}$$

where we recall that $\delta E_i/\delta\phi_j$ is a differential operator in x and t. Referring to Section 7.6.1 we see that

$$D_{ij}(k,\omega)=0 \tag{7.180}$$

is the dispersion relation for the system.

From Eq. (7.178) we obtain

$$\sum_j \Phi_j(k,\omega)D_{ij}(k,\omega)=\bar{\delta J}_i(k,\omega) \tag{7.181}$$

so that

$$\Phi_j(k,\omega)=D_{ij}^{-1}(k,\omega)\bar{\delta J}_i(k,\omega) \tag{7.182}$$

where D_{ij}^{-1} is the inverse of D_{ij}, regarded as a matrix. From Eqs. (7.177) and (7.182) we determine the responses $\delta\phi_j$ to the transforms of the sources of perturbation $\bar{\delta J}_i$:

$$\delta\phi_j(x,t)=\sum_i\int\frac{dk}{2\pi}\int\frac{d\omega}{2\pi}D_{ij}^{-1}(k,\omega)\bar{\delta J}_i(k,\omega)\,e^{i(kx-\omega t)} \tag{7.183}$$

as given by Haas [Reference 70, Eq. (36)]. We know that the physically allowed responses have ω and k related by the solutions of the dispersion relation (7.180). Since D_{ij}, regarded as a matrix, is singular, we see that the responses $\delta\phi_j$ "blow up" (absolute instability) unless there exists a path of integration in complex (k,ω) space resulting in a finite $\delta\phi_j(x,t)$ for any x as $t\to\infty$ (convective instability). This indicates the connection between the Green's function formalism and that of the previous sections.

8

The Haas Approximation to the General Model

In this chapter we shall derive Haas' plasma model from the general model of Chapter 5 and, following Haas,[70] obtain the perturbed equations required for a stability analysis.

A complete CO_2 laser plasma model would consist of differential equations for the following variables: (a) species number densities (59) [primary and neutral secondary species (18), positive ions (27), and negative ions (14) (see Table 11, Chapter 6)], (b) conservation of mass (1), (c) momentum (1), (d) vibrational energy (5), (e) translational energy (1), (f) electron number density (1), (g) electron energy (1), and (h) electric field (2), totaling 71 equations. A full stability analysis on this set of equations would therefore result in a polynomial in ω, the frequency of the perturbation, of order 71. Analytic solution of such a polynomial is intractable and Haas has suggested a simplified plasma model in which there are only two primary species, one atomic and one diatomic, whose number densities are assumed to remain constant (compared with the number densities of the secondary species). The model consists of equations for: (a) one species of positive ion (1), (b) one species of negative ion (1), (c) one electrically excited species (1), (d) conservation of mass (1), (e) momentum (1), (f) vibrational energy (1), (g) translational energy (1), (h) electron number density (1), (i) electron energy (1), and (j) electric field (2), making 11 equations in all. Under certain simplifying assumptions, Haas obtains instability criteria from this model.

The point plasma chemistry model of Chapter 6, together with the Haas model, provide a tractable, though approximate, means of determining the stability of CO_2 laser plasmas. The point plasma chemistry model will provide the species number densities, and vibrational and translational energy densities at any time during a discharge; the Haas stability criteria can be applied to these values to determine the operating regions of stability.

8.1. Equations of the Model

8.1.1. Dynamics of the Heavy Particles

We consider two primary species, one atomic and one diatomic, with number densities n_a and n_m, respectively. We assume:

(i) no chemical decomposition of the primary species, e.g., no $CO_2 \rightarrow CO$; i.e., appropriate k-rates vanish;

(ii) negligible *change* in n_a and n_m upon formation of secondary species (positive and negative ions and electrically excited species); these rates are zero;

(iii) negligible change in n_a and n_m due to diffusive processes (i.e., $\nu_j \approx 0$; and $\nabla \cdot (n_j \bar{\mathbf{V}}_j) = 0$.

With the above assumptions we can write, from Eq. (5.35):

$$Dn_a/Dt + n_a \nabla \cdot \mathbf{v}_0 = 0$$

$$Dn_m/Dt + n_m \nabla \cdot \mathbf{v}_0 = 0$$

which we can add to obtain the equation of continuity for the gas:

$$Dn/Dt + n \nabla \cdot \mathbf{v}_0 = 0 \tag{8.1}$$

where

$$n = n_a + n_m$$

We consider three secondary species: positive ions, negative ions, and an electronically excited species, with number densities n_+, n_-, and n_*, respectively. We again assume diffusion effects to be negligible ($\nu_j \approx 0$), and from Eq. (5.35) we write the conservation equations for the secondary species (where n_e is the secondary electron number density) as

$$\begin{aligned} &Dn_+/Dt + n_+ \nabla \cdot \mathbf{v}_0 + \nabla \cdot (n_+ \bar{\mathbf{V}}_+) \\ &\quad = S_+ + n_e n k_i(T_e) + n_e n_* k_i^*(T_e) - n_e n_+ k_r^e(T_e) + n_+ n_- k_r^i(T) \end{aligned} \tag{8.2}$$

where S_+ positive ions per second are created by the primary (high-energy) electron beam;

$$Dn_-/Dt + n_- \nabla \cdot \mathbf{v}_0 + \nabla \cdot (n_- \bar{\mathbf{V}}_-) = n_e n k_a(T_e) - n_+ n_- k_r^i(T) - n_- n k_d(T) \tag{8.3}$$

and

$$Dn_*/Dt + n_+ \nabla \cdot \mathbf{v}_0 = n_e n k_*(T_e) - n_e n_* k_i^*(T_e) - n_* n k_q(T) + S_* \tag{8.4}$$

Since the n_* are neutral, then $\bar{\mathbf{V}}_* = 0$. The rates occurring on the right-hand sides of Eqs. (8.2)–(8.4) describe processes of importance in low-pressure, weakly ionized discharges in molecular gases (Haas[70]). They are the rates for the following processes: $k_i(T_e)$, for electron impact ionization of neutral heavy particles; $k_i^*(T_e)$, for electron impact ionization of electronically excited species; $k_a(T_e)$, for electron heavy primary particle dissociative attachment; $k_*(T_e)$, for electron impact excitation of electronically excited species; $k_q(T)$, for deexcitation of electronically excited species by collisions with heavy particles; $k_r^e(T_e)$, for electron–positive-ion recombination; $k_r^i(T)$, for negative-ion–positive-ion recombination; $k_d(T)$, for detachment of a negative-ion electron by neutral impact. The effect of vibrational excitation on the processes in Eqs. (8.2)–(8.4) have been omitted but may be important. Since diffusive effects are neglected, only the contribution from the electric field is included in describing the average variance velocities $\bar{\mathbf{V}}_\pm$ of the ionic species. Therefore we have

$$\begin{aligned} \bar{\mathbf{V}}_+ &= \mu_+\mathbf{E} \\ \bar{\mathbf{V}}_- &= -\mu_-\mathbf{E} \end{aligned} \tag{8.5}$$

where μ_+ and μ_- are the positive- and negative-ion mobilities, which are related to the multicomponent diffusion coefficients D_{ij} from Eqs. (5.106) and (5.116). The mobilities may also be measured experimentally.

The momentum equation is given by

$$\boldsymbol{\nabla}\cdot\hat{P}-\sum_i n_i\mathbf{X}_i+\rho\frac{D\mathbf{v}_0}{Dt}=0 \tag{5.45}$$

From Eq. (5.7), neglecting the Vlasov field, we can write the second term as

$$\sum_i n_i\mathbf{X}_i=\sum_i n_i e_i\mathbf{E}=\rho_c\mathbf{E} \tag{8.6}$$

where

$$\rho_c(\mathbf{x}, t)=\sum_i n_i(\mathbf{x}, t)e_i \tag{8.7}$$

is the electric charge density. From Eqs. (5.129) and (5.131) we have the first term in Eq. (5.45) in the form

$$\nabla_a P_{ab}=\nabla_a\left[(p-\zeta)\,\delta_{ab}-\eta(\nabla_a v_{0b}+\nabla_b v_{0a})-\frac{2\eta}{3}\delta_{ab}\nabla_a v_{0a}-\eta_B\nabla\cdot\mathbf{v}_0\,\delta_{ab}\right]$$

and hence

$$\boldsymbol{\nabla}\cdot\hat{P}=\boldsymbol{\nabla}(p-\zeta)-\eta\nabla^2\mathbf{v}_0-\tfrac{2}{3}\eta\boldsymbol{\nabla}(\boldsymbol{\nabla}\cdot\mathbf{v}_0)-\eta_B\boldsymbol{\nabla}(\boldsymbol{\nabla}\cdot\mathbf{v}_0) \tag{8.8}$$

where η is the coefficient of shear viscosity, also called the kinematic viscosity coefficient, and η_B is the bulk viscosity coefficient which arises as a result of the rapid molecular rotational relaxation during dilatation of fluid elements. We shall follow Haas and neglect ζ; hence

$$\nabla \cdot \hat{P} = \nabla p - \eta \nabla^2 \mathbf{v}_0 - (\eta_B + \tfrac{2}{3}\eta)\nabla(\nabla \cdot \mathbf{v}_0) \tag{8.9}$$

Using Eqs. (8.6) and (8.9) we write the momentum equation, Eq. (5.45), in the form

$$\rho \frac{D\mathbf{v}_0}{Dt} = -\nabla p + \eta \nabla^2 \mathbf{v}_0 + (\eta_B + \frac{2}{3}\eta)\nabla(\nabla \cdot \mathbf{v}_0) + \rho_c \mathbf{E} \tag{8.10}$$

where

$$p = nkT \tag{5.98}$$

$$\rho = n_a m_a + n_m m_m \tag{8.11}$$

From Eq. (5.68) for the translational–rotational energy, and noting that for the mixture under consideration there is only one vibrational mode, we can write

$$\frac{DE^{(TR)}}{Dt} + E^{(TR)}\nabla \cdot \mathbf{v}_0 + \hat{P} : (\nabla \mathbf{v}_0)$$
$$= -\nabla \cdot \mathbf{q}^{(TR)} + \sum_{i=\pm} n_i e_i \mathbf{E} \cdot \bar{\mathbf{V}}_i + \frac{E^{(V)}(T^{(V)}) - E^{(V)}(T)}{\tau_{(VT)}(T)} + n_e \nu^{(TR)}(T_e)kT_e \tag{8.12}$$

From Eqs. (5.129) and (5.131) we have the third term in Eq. (8.12), when $\eta_B << \eta$:

$$\hat{P} : (\nabla \mathbf{v}_0) = p\nabla \cdot \mathbf{v}_0 - \eta \left[\sum_{ab} (\nabla_a v_{0b} + \nabla_b v_{0a})(\nabla_a v_{0b}) - \tfrac{2}{3}(\nabla \cdot \mathbf{v}_0)^2 \right]$$
$$= p\nabla \cdot \mathbf{v}_0 - \eta \Phi \tag{8.13}$$

defining Φ, the so-called viscous dissipation function. Neglecting the ion-Joule heating term (the term containing $\mathbf{E} \cdot \bar{\mathbf{V}}_i$) we can now write Eq. (8.12) as

$$\frac{DE^{(TR)}}{Dt} + (E^{(TR)} + p)\nabla \cdot \mathbf{v}_0$$
$$= -\nabla \mathbf{q}^{(TR)} + \eta \Phi + \frac{E^{(V)}(T^{(V)}) - E^{(V)}(T)}{\tau_{(VT)}(T)} + n_e \nu^{(TR)}(T_e)kT_e \tag{8.14}$$

In Eq. (8.14) we have from Eq. (5.58) that

$$E^{(TR)} = \tfrac{1}{2}(3n_a + 5n_m)kT \tag{8.15}$$

$\mathbf{q}^{(\mathrm{TR})}(\mathbf{x}, t)$ is the translation–rotation energy flux and is given by Eq. (5.136). Neglecting contributions from diffusion and vibrational temperature gradients, from Eqs. (5.136) we have

$$\mathbf{q}^{(\mathrm{TR})} = -\lambda'\nabla T \tag{8.16}$$

where λ' is the thermal conductivity of the mixture. $E^{(\mathrm{V})}(T^{(\mathrm{V})})$ is the vibrational energy density of the single vibrational mode:

$$E^{(\mathrm{V})}(T^{(\mathrm{V})}) = \frac{n_m\varepsilon}{\exp(\varepsilon/kT^{(\mathrm{V})}) - 1} \tag{1.69a}$$

where ε is the quantum of vibrational energy, $h\nu$. $\tau_{(\mathrm{VT})}(T)$ is the relaxation time for vibrational–translational energy transfer. We assume such energy transfer is due to predominantly to collisions between vibrationally excited molecules and the atomic diluent, and write [see also Eq. (2.6)]

$$\tau_{(\mathrm{VT})}^{-1} = n_a k_{10}(T)(1 - e^{-\varepsilon/kT}) \tag{8.17}$$

We recall that $\nu^{(\mathrm{TR})}$ is the translation–rotation contribution to the total electron energy-exchange collision frequency [see Eq. (5.65a)].

For the mode vibrational energy equation we have from Eq. (5.55):

$$\frac{DE^{(\mathrm{V})}(T^{(\mathrm{V})})}{Dt} + E^{(\mathrm{V})}(T^{(\mathrm{V})})\nabla\cdot\mathbf{v}_0 + \nabla\cdot\mathbf{q}^{(\mathrm{V})}$$

$$= \varepsilon n_e n_m X^{(\mathrm{V})} - \frac{E^{(\mathrm{V})}(T^{(\mathrm{V})}) - E^{(\mathrm{V})}(T)}{\tau_{(\mathrm{VT})}(T)} \tag{8.18}$$

where $X^{(\mathrm{V})}(T_e)$ is the effective vibrational electron excitation rate, which is related to $\nu^{(\mathrm{V})}(T_e)$, the vibrational contribution to the electron energy-exchange collision frequency, by

$$\varepsilon n_m X^{(\mathrm{V})} = \nu^{(\mathrm{V})} kT_e \tag{8.19}$$

$\mathbf{q}^{(\mathrm{V})}(\mathbf{x}, t)$ is the vibrational energy flux vector and is given by Eq. (5.142). Neglecting diffusion terms and the contribution to the vibrational energy flux due to the ambient temperature gradient, we have from Eq. (5.142) that

$$\mathbf{q}^{(\mathrm{V})} = -\lambda^{(\mathrm{V})}\nabla T^{(\mathrm{V})} \tag{8.20}$$

where $\lambda^{(\mathrm{V})}$ is the vibrational conductivity. We recall that there is only a single vibrational mode present in the Haas model.

8.1.2. Dynamics of the Electrons and Electromagnetic Field

Using Eq. (5.154) we can write down the electron number-density equation. On the right-hand side we allow for those processes already considered in Eqs. (8.2)–(8.4) in which electrons participate:

$$Dn_e/Dt + n_e\nabla\cdot\mathbf{v}_0 + \nabla\cdot(n_e\bar{\mathbf{V}}_e)$$
$$= n_e n k_i(T_e) + n_e n_* k_i^*(T_e) + n_- n k_d(T) - n_e n_+ k_r^e(T_e) - n_e n k_a(T_e) + S_e \tag{8.21}$$

$\bar{\mathbf{V}}_e(\mathbf{x}, t)$ is the electron drift velocity. From Eq. (5.195) we have

$$n_e\bar{\mathbf{V}}_e = -\nabla(n_e D_e) - n_e D_e \nabla[\ln(n)] - n_e\mu_e\mathbf{E} \tag{8.22}$$

where the electron diffusion coefficient D_e and mobility μ_e are given by

$$D_e = kT_e/m_e\nu_m \tag{5.192a}$$

$$\mu_e = e/m_e\nu_m \tag{5.193a}$$

where ν_m is the effective electron momentum-transfer collision frequency.

From Eqs. (5.161), (5.162), and (5.209) we can write down the electron energy equation in the form

$$\frac{D}{Dt}\left(\frac{3}{2}n_e kT_e\right) + \left(\frac{3}{2}n_e kT_e + p_e\right)\nabla\cdot\mathbf{v}_0 = \mathbf{j}_e\cdot\mathbf{E} - \nabla\cdot\mathbf{q}_e - n_e\nu_u(T_e)kT_e \tag{8.23}$$

where

$$p_e = n_e kT_e \tag{5.210a}$$

is the electron pressure and

$$\mathbf{j}_e = -en_e\bar{\mathbf{V}}_e \tag{5.158}$$

is the electron current density; ν_u is the electron energy-exchange collision frequency and $\mathbf{q}_e(\mathbf{x}, t)$ is the electron energy flux vector, given by

$$\mathbf{q}_e(\mathbf{x}, t) = -\lambda_e\nabla T_e - \frac{5}{2}\left(\frac{kT_e}{e}\right)\mathbf{j}_e \tag{5.221}$$

where

$$\lambda_e = \tfrac{5}{2}kn_e D_e \tag{5.222}$$

is the effective electron thermal conductivity.

We shall assume that the electric field $\mathbf{E}$ is constant in time (compared to the time scales in which other quantities change) so that, from Eqs. (5.223) and (5.224) we have

$$\nabla\cdot\mathbf{E} = \frac{\rho_c}{\varepsilon_0}$$
$$\nabla\times\mathbf{E} = 0 \tag{8.24}$$

8.2. The Steady-State Equations

Local steady-state relationships between the plasma variables may be obtained by setting time and space derivatives of plasma variables equal to zero. From the negative-ion number density [Eq. (8.3)] we obtain, on dividing through by n_-:

$$\frac{n_e n}{n_-} k_a - n_+ k_r^i - n k_d = 0 \tag{8.25}$$

We can define characteristic times τ (sec) for plasma reactions as follows:

$$\tau_a = (n k_a)^{-1} \tag{8.26}$$

$$\tau_r^i = (n_+ k_r^i)^{-1} \tag{8.27}$$

$$\tau_d = (n k_d)^{-1} \tag{8.28}$$

which are characteristic times for dissociative attachment, ion–ion recombination, and detachment, respectively. From Eqs. (8.25)–(8.28) we have

$$\frac{n_e}{n_-}\frac{1}{\tau_a} - \frac{1}{\tau_r^i} - \frac{1}{\tau_d} = 0 \tag{8.29}$$

From the electron number-density equation we have, on dividing through by n_e and using Eq. (8.25), that Eq. (8.21) becomes

$$n k_i + n_* k_i^* - \frac{n_+ n_-}{n_e} k_r^i - n_+ k_r^e + \frac{S_e}{n_e} = 0 \tag{8.30}$$

Defining the following characteristic times:

Direct ionization

$$\tau_i = (n k_i)^{-1} \tag{8.31}$$

Excited-state ionization

$$\tau_i^* = (n_* k_i^*)^{-1} \tag{8.32}$$

Electron–ion dissociative recombination

$$\tau_r^e = (n_+ k_r^e)^{-1} \tag{8.33}$$

we can then rewrite Eq. (8.30) in the form

$$\frac{1}{\tau_i} + \frac{1}{\tau_i^*} - \frac{1}{\tau_r^e} - \frac{n_-}{n_e}\frac{1}{\tau_r^i} + \frac{S_e}{n_e} = 0 \tag{8.34}$$

Eliminating τ_r^i between Eqs. (8.29) and (8.34) we obtain

$$\frac{1}{\tau_i}+\frac{1}{\tau_i^*}-\frac{1}{\tau_r^e}-\frac{1}{\tau_a}+\frac{n_-}{n_e}\frac{1}{\tau_d}+\frac{S_e}{n_e}=0 \tag{8.35}$$

From the steady-state form of Eq. (8.4), dividing by n_* we have

$$\frac{n_e n}{n_*}k_* = n_e k_i^* - nk_q + \frac{S_*}{n_*}=0 \tag{8.36}$$

Defining the following characteristic times:

Electron excitation

$$\tau_* = (nk^*)^{-1} \tag{8.37}$$

Electron quenching

$$\tau_q = (n_* k_q)^{-1} \tag{8.38}$$

we can then write Eq. (8.36) in the form

$$\frac{n_e}{n_*}\frac{1}{\tau_*}-\frac{n_e}{n_*}\frac{1}{\tau_i^*}-\frac{n}{n_*}\frac{1}{\tau_q}+\frac{S_*}{n_*}=0 \tag{8.39}$$

The characteristic times defined above are listed, together with typical values for low-pressure laser plasma mixtures, in Table 23.

From the vibrational energy equation (8.18) and Eq. (8.19), we obtain the steady-state relation

$$\frac{E^{(\mathrm{V})}(T^{(\mathrm{V})})-E^{(\mathrm{V})}(T)}{\tau_{(\mathrm{VT})}(T)} = n_e \nu^{(\mathrm{V})} kT_e \tag{8.40}$$

From the electron energy density equation (8.23), we obtain the steady-state relation

$$\mathbf{j}_e \cdot \mathbf{E} = n_e \nu_u kT_e \tag{8.41}$$

We obtain a steady-state expression for $\mathbf{j}_e$ by substituting Eq. (8.22) into

$$\mathbf{j}_e = -en_e\bar{\mathbf{V}}_e \tag{5.158}$$

$$= e\mu_e n_e \mathbf{E} \tag{8.42}$$

From Eqs. (8.41) and (8.42) we have

$$\frac{j_e^2}{\sigma_e} = n_e \nu_u kT_e \tag{8.43}$$

where we have set

$$\sigma_e = e\mu_e n_e \tag{8.44}$$

Table 23. Characteristic Times. Typical Discharge Conditions[a]

Process	Equation	Characteristic time	Range of values (sec)
I. Space-charge relaxation	(8.118b)	$\tau_C = \varepsilon_0/\sigma \sim \nu_m/\omega_{pe}^2$	10^{-10}–10^{-9}
II. Collisional energy transfer			
i. Electron heating	(8.135)	$\tau_e = \frac{3}{2}[n_e kT_e/(J_e^2/\sigma_e)] = \frac{3}{2}\nu_u^{-1}$	10^{-9}–10^{-8}
ii. Translation–rotation heating	(8.91)	$\tau_{(TR)} = nC_pT/(J_e^2/\sigma_e) \sim (n/n_e)(C_p/k)(T/T_e)\nu_u^{-1}$	10^{-3}–10^{-2}
iii. Vibration heating	(8.92)	$\tau_{(V)} = n_m C_v^{(V)}(T^{(V)})T^{(V)}/(J_e^2/\sigma_e) \sim (n_m/n_e)[C_v^{(V)}(T^{(V)})/k](T^{(V)}/T_e)\nu_u^{-1}$	10^{-3}–10^{-2}
iv. Vibrational relaxation	(8.17)	$\tau_{(V)} = [n_a k_{10}(1-e^{-\varepsilon/kT})]^{-1}$	10^{-4}–10^{-2}
III. Plasma kinetic processes			
i. Direct ionization	(8.31)	$\tau_i = [nk_i]^{-1}$	10^{-6}–10^{-5}
ii. Excited state ionization	(8.32)	$\tau_i^* = [n_* k_i^*]^{-1}$	10^{-6}–10^{-4}
iii. Attachment (dissociative)	(8.26)	$\tau_a = [nk_a]^{-1}$	10^{-6}–10^{-5}
iv. Detachment (direct neutral and associative)	(8.28)	$\tau_d = [nk_d]^{-1}$	10^{-6}–10^{-5}
v. Electronic excitation	(8.37)	$\tau_* = [nk_*]^{-1}$	10^{-6}–10^{-4}
vi. Electronic quenching	(8.38)	$\tau_q = [n_* k_q]^{-1}$	10^{-6}–10^{-4}
vii. Electron–ion dissociative recombination	(8.33)	$\tau_r^{(e)} = [n_p k_r^{(e)}]^{-1}$	10^{-6}–10^{-5}
viii. Ion–ion recombination	(8.27)	$\tau_r^{(i)} = [n_p k_r^{(i)}]^{-1}$	10^{-6}–10^{-5}
IV. Transport and other processes			
i. Sound propagation	(8.55)	$\tau_s = l/\alpha$	10^{-5}–10^{-4}
ii. Translation–rotation energy conduction	(8.91)	$\tau_{(TR)} = (nC_v l^2/\lambda^{(T)})$	10^{-3}–10^{-2}
iii. Viscous dissipation	(8.13)	$\tau_\mu = \rho l^2/\mu$	10^{-3}–10^{-2}
iv. Vibrational energy conduction	(8.92)	$\tau_{(V)} = n_m C_v^{(V)}(T^{(V)})l^2/\lambda^{(V)}$	10^{-2}–10^{-1}
v. Electron thermal conduction	(5.222)	$\tau_{T_e} = \frac{3}{2}(n_e kl^2/\lambda_e)$	10^{-7}–10^{-6}
vi. Ambipolar diffusion	(9.161)	$\tau_{amb} = l^2/[(1-\sigma_e/\sigma)D_e]$	10^{-5}–10^{-4}
vii. Negative-ion diffusion	(8.127)	$\tau_{neg} = \dfrac{l^2}{(\sigma_n/\sigma)(n_e/n_n)D_e}$	10^{-5}–10^{-4}

[a] Diatomic molecule–atom mixture, $p \sim 10$–100 Torr, $T \sim 300$–600°K, $T_e \sim 0.5$–2.0 eV, $T^{(V)} \sim 1000$–5000°K, $n_e/n \sim 10^{-8}$–10^{-6}, $l \sim 1$ cm; for a plane wave with wave-number k the characteristic length $l = k^{-1}$.

Equations (8.29), (8.35), (8.39), (8.40), and (8.43) are the steady-state equations given by Haas [Reference 70, Eqs. (15)–(19)], except Haas does not have the factor n/n_* in the $1/\tau_q$ term in Eq. (8.39).

8.3. The Perturbed Equations

8.3.1. The Primary Species Number-Density Equation

The primary species number-density equation is

$$Dn/Dt + n\nabla \cdot \mathbf{v}_0 = 0 \tag{8.1}$$

Following Haas we shall derive perturbed forms of the plasma equations in terms of a dimensionless spatial variable obtained by making the substitution

$$\mathbf{x} \rightarrow l\mathbf{x} \tag{8.45}$$

where l is expressed in centimeters and is to be identified with the characteristic dimension of a spontaneous disturbance within the plasma. Making the substitution of Eq. (8.45) into Eq. (8.1) we obtain

$$Dn/Dt + (n/l)\nabla \cdot \mathbf{v}_0 = 0 \tag{8.46}$$

We now subject the plasma variables in Eq. (8.46) to the perturbations

$$n \rightarrow n(1 + n') = n + nn' \tag{8.47}$$

$$\mathbf{v}_0 \rightarrow \mathbf{v}_0 + \alpha \mathbf{v}_0' \tag{8.48}$$

where n' and $\mathbf{v}_0'$ are dimensionless perturbations whose magnitudes are small compared to unity. α is the local frozen sound speed and is given by

$$\alpha = (\gamma kT/m)^{1/2} \tag{8.49}$$

where γ is the specific-heat ratio

$$\gamma = C_p/C_v \tag{8.50}$$

where C_p and C_v are, respectively, the specific heats at constant pressure and constant volume, and are given by Vincenti and Kruger (Reference 4, p. 11) to be

$$\begin{aligned} C_v &= \frac{(3n_a + 5n_m)k}{2n} \\ C_p &= C_v + k \end{aligned} \tag{8.51}$$

From Eqs. (8.47) and (8.48) the first-order perturbed form of Eq. (8.46) is

$$D(nn')/Dt+(1/l)[n\nabla\cdot(\alpha\mathbf{v}_0')+nn'\nabla\cdot\mathbf{v}_0]=0 \tag{8.52}$$

From Eqs. (8.46) and (8.52) we obtain

$$Dn'/Dt+(1/l)\nabla\cdot(\alpha\mathbf{v}_0')=0 \tag{8.53}$$

We now assume that the characteristic dimension l of a disturbance is small compared to the dimensions of the discharge and that the effects of steady-state gradients are small compared to the gradients of the perturbed variables. Then we can write Eq. (8.53) in the form

$$Dn'/Dt+(1/\tau_s)\nabla\cdot\mathbf{v}_0'=0 \tag{8.54}$$

where

$$\tau_s=l/\alpha \tag{8.55}$$

is a measure of the time for sound to cross the disturbance. Equation (8.54) is the perturbed form of the primary species number-density equation.

8.3.2. The Momentum Equation

In terms of the dimensionless spatial parameter the momentum equation (8.10) is given by

$$\rho\frac{D\mathbf{v}_0}{Dt}=-\frac{1}{l}\nabla p+\frac{\eta}{l^2}\nabla^2\mathbf{v}_0+\left(\eta_B+\frac{2}{3}\eta\right)\frac{1}{l^2}\nabla(\nabla\cdot\mathbf{v}_0)+\rho_c\mathbf{E} \tag{8.56}$$

Subjecting Eq. (8.56) to the perturbations given by Eqs. (8.47) and (8.48) together with

$$\rho\to\rho(1+\rho')=\rho+\rho\rho' \tag{8.57}$$

$$p=p(1+p')=p+pp' \tag{8.58}$$

$$\rho_c\to\rho_c+en_+\rho_c' \tag{8.59}$$

$$\mathbf{E}\to\mathbf{E}+E\mathbf{E}' \tag{8.60}$$

as defined by Haas, we obtain

$$\begin{aligned}(\rho+\rho\rho')\frac{D}{Dt}(\mathbf{v}_0+\alpha\mathbf{v}_0')&\\ =-\frac{1}{l}\nabla(p+pp')+\frac{\eta}{l^2}\nabla^2(\mathbf{v}_0+\alpha\mathbf{v}_0')&\\ +\left(\eta_B+\frac{2}{3}\eta\right)\frac{1}{l^2}\nabla[\nabla\cdot(\mathbf{v}_0+\alpha\mathbf{v}_0')]+(\rho_c+en_+\rho_c')(\mathbf{E}+E\mathbf{E}')&\end{aligned} \tag{8.61}$$

From Eqs. (8.56) and (8.61) we obtain the first-order perturbed equation

$$\alpha\rho\frac{D\mathbf{v}_0'}{Dt}=-\frac{p}{l}\nabla p'+\frac{\alpha\eta}{l^2}\nabla^2\mathbf{v}_0'+\frac{\alpha(\eta_B+\frac{2}{3}\eta)}{l^2}\nabla(\nabla\cdot\mathbf{v}_0')+\rho_c E\mathbf{E}'+en_+\mathbf{E}\rho_c' \quad (8.62)$$

where, as in the previous section, we have neglected gradients of steady-state plasma quantities.

Following Haas we shall replace Eq. (8.62) by equations for vorticity and pressure fluctuations. Defining the vorticity fluctuation $\boldsymbol{\Omega}'(\mathbf{x}, t)$ by

$$\boldsymbol{\Omega}'=\nabla\times\mathbf{v}_0' \quad (8.63)$$

we obtain, on dividing Eq. (8.62) by $\alpha\rho$ and taking the curl [Haas,[70] Eq. (21)]:

$$\frac{D\boldsymbol{\Omega}'}{Dt}-\frac{\eta}{(\rho l^2)}\nabla^2\boldsymbol{\Omega}'=\left(\frac{en_+}{\rho\alpha^2/l}\right)\frac{1}{\tau_s}(\nabla\rho_c')\mathbf{E} \quad (8.64)$$

where we have used Eqs. (8.55) and (8.148), the latter to be derived later.

To obtain the equation for pressure fluctuations we form the divergence of Eq. (8.62), and continue to neglect the gradient of steady-state plasma quantities; hence

$$\alpha\rho\frac{D}{Dt}\nabla\cdot\mathbf{v}_0'=-\frac{p}{l}\nabla^2 p'+\frac{\alpha}{l^2}\left(\eta_B+\frac{4}{3}\eta\right)\nabla^2\nabla\cdot\mathbf{v}_0'+\rho_c E\nabla\cdot\mathbf{E}'+en_+\mathbf{E}\cdot\nabla\rho_c' \quad (8.65)$$

Substituting for $\nabla\cdot\mathbf{v}_0'$ from Eq. (8.54), and using Eq. (8.148) for $\nabla\cdot\mathbf{E}'$ we obtain

$$-\alpha\rho\tau_s\frac{D^2n'}{Dt^2}=-\frac{p}{l}\nabla^2 p'-\frac{\alpha}{l^2}\tau_s\left(\eta_B+\frac{4}{3}\eta\right)\nabla^2\frac{Dn'}{Dt}+\frac{en_+l}{\varepsilon_0}\rho_c\rho_c'+en_+\nabla\cdot(\rho_c'\mathbf{E}) \quad (8.66)$$

The perturbed form of Eq. (5.98) is, to first order in small quantities using the definitions of Eqs. (8.77) (given below), (8.47), and (8.58):

$$p'=n'+T' \quad (8.67)$$

Using Eq. (8.67) to substitute for n' on the left-hand side of Eq. (8.66), and using Eq. (8.24) in terms of the dimensionless spatial parameter to combine the last two terms on the right-hand side of Eq. (8.66), we obtain, on dividing through by $\alpha\rho\tau_s$ and rearranging terms:

$$\frac{D^2p'}{Dt^2}-\frac{p}{\alpha\rho\tau_s l}\nabla^2 p'=\frac{1}{\rho l^2}\left(\eta_B+\frac{4}{3}\eta\right)\nabla^2\frac{Dn'}{Dt}+\frac{D^2T'}{Dt^2}-\frac{En_+l}{\alpha^2\rho\tau_s^2}\nabla\cdot\left(\rho_c'\frac{\mathbf{E}}{E}\right) \quad (8.68)$$

Equations (8.64) and (8.68) are equivalent to the perturbed form of Eq. (8.56), the momentum equation. They are in the form of wave equations with source terms for the vorticity and pressure fluctuations.

We shall now anticipate the result of the next section, which is an equation for DT'/Dt, the time derivative of the ambient temperature perturbation, in order to reexpress the right-hand side of Eq. (8.68). From Eq. (8.100) we obtain, on substituting for n' in the first term on the right-hand side, using Eq. (8.67) and operating on both sides by D/Dt:

$$\begin{aligned}\gamma\frac{D^2p'}{Dt^2} &= (\gamma-1)\frac{D^2p'}{Dt^2}+\left(\frac{\lambda'}{nC_v l^2}\right)\nabla^2\left(\frac{DT'}{Dt}\right)\\ &+\frac{\gamma}{\tau_{(T)}}\left[-P_{(V)}\hat{\tau}_{(VT)}-\frac{C_v^{(V)}(T)T}{C_v^{(V)}(T^{(V)})T^{(V)}}\left(\frac{\tau_v}{\tau_{(VT)}}\right)\right]\frac{DT'}{Dt}\\ &+\frac{1}{\tau_{(VT)}}\left(\frac{n_m C_v^{(V)}(T^{(V)})T^{(V)}}{nC_pT}\right)\frac{DT'^{(V)}}{Dt}+\frac{\gamma P_{(TR)}}{\tau_{(T)}}\frac{Dn_e'}{Dt}\\ &+\frac{\gamma P_{(TR)}}{\tau_{(T)}}(1+\hat{\nu}^{(TR)})\frac{DT_e'}{Dt}+\frac{\gamma}{\tau_{(T)}}(2P_{(V)}+P_{(TR)})\frac{Dn'}{Dt'}\end{aligned} \tag{8.69}$$

Using Eq. (8.69) to substitute for the D^2T'/Dt^2 term in Eq. (8.68) we obtain

$$\begin{aligned}\frac{D^2p'}{Dt^2}-\frac{p\gamma}{\alpha\rho\tau_s l}\nabla^2p' &= \frac{\gamma}{\rho l^2}\left(\eta_B+\frac{4}{3}\eta\right)\nabla^2\left(\frac{Dn'}{Dt}\right)+\left(\frac{\lambda'}{nC_v l^2}\right)\nabla^2\left(\frac{DT'}{Dt}\right)\\ &-\left(\frac{en_+El}{\rho\alpha^2}\right)\frac{1}{\tau_s^2}\nabla\cdot\left(\rho_c\frac{\mathbf{E}}{E}\right)+\frac{D\dot{Q}'_{(TR)}}{Dt}\end{aligned} \tag{8.70}$$

where

$$\begin{aligned}\frac{1}{\gamma}\frac{D\dot{Q}'_{(TR)}}{Dt} &= \frac{1}{\tau_{(T)}}\left[-P_{(V)}\hat{\tau}_{(VT)}-\frac{C_v^{(V)}(T)T}{C_v^{(V)}(T^{(V)})T^{(V)}}\left(\frac{\tau_v}{\tau_{(VT)}}\right)\right]\frac{DT'}{Dt}\\ &+\frac{1}{\tau_{(VT)}}\left(\frac{n_m C_v^{(V)}(T^{(V)})T^{(V)}}{nC_pT}\right)\frac{DT'^{(V)}}{Dt}\\ &+\frac{P_{(TR)}}{\tau_{(T)}}\frac{Dn_e'}{Dt}+\frac{P_{(TR)}}{\tau_{(T)}}(1+\hat{\nu}^{(TR)})\frac{DT_e'}{Dt}\\ &+\frac{1}{\tau_{(T)}}(2P_{(V)}+P_{(TR)})\frac{Dn'}{Dt}\end{aligned} \tag{8.71}$$

and $\dot{Q}'_{(TR)}$ is identified as the fluctuation in the volumetric heating of the translation–rotation degree of freedom of the neutral gas.

From Eqs. (8.49) and (8.55) we have, for the coefficient of the second term on the left-hand side of Eq. (8.70):

$$\frac{p\gamma}{\alpha\rho\tau_s l} = \frac{p\gamma}{\alpha^2\rho}\frac{1}{\tau_s^2}$$

$$= \frac{p\gamma m}{\gamma kT\rho}\frac{1}{\tau_s^2}$$

$$= \frac{nm}{\rho}\frac{1}{\tau_s^2}$$

$$= \frac{1}{\tau_s^2} \tag{8.72}$$

From Eqs. (8.70) and (8.72) we have

$$\frac{D^2p'}{Dt^2} - \frac{1}{\tau_s^2}\nabla^2 p' = \frac{\gamma}{\rho l^2}\left(\eta_B + \frac{4}{3}\eta\right)\nabla^2\left(\frac{Dn'}{Dt}\right) + \left(\frac{\lambda'}{nC_v l^2}\right)\nabla^2\left(\frac{DT'}{Dt}\right)$$
$$- \left(\frac{en_+El}{\rho\alpha^2}\right)\frac{1}{\tau_s^2}\nabla\cdot\left(\rho_c'\frac{\mathbf{E}}{E}\right) + \frac{DQ'_{(TR)}}{Dt} \tag{8.73}$$

which is an equivalent form of Eq. (8.68) for the pressure of fluctuation, p', and is the form given by Haas [Reference 70, Eq. (22)].

8.3.3. *The Translation–Rotation Energy Equation*

From Eqs. (8.15) and (8.51) the translation–rotation energy, $E^{(TR)}$, may be written in terms of the ambient temperature, T:

$$E^{(TR)} = nC_vT \tag{8.74}$$

In terms of the ambient temperature, T, and the dimensionless spatial variable [Eq. (8.45)], Eq. (8.14) for the translation–rotation energy becomes

$$\frac{D}{Dt}(nC_vT) + \frac{(nC_vT + p)}{l}\nabla\cdot\mathbf{v}_0$$
$$= \frac{1}{l^2}\nabla\cdot(\lambda'\nabla T) + \eta\Phi + \frac{n_m}{\tau_{(VT)}}[\varepsilon_{(V)}(T^{(V)}) - \varepsilon_{(V)}(T)] + n_e\nu^{(TR)}(T_e)kT_e \tag{8.75}$$

where, on the right-hand side of Eq. (8.75) we have used Eq. (8.16) to substitute for $q^{(TR)}$, and $\varepsilon_{(V)}$ is the vibrational energy per molecule, related to the vibrational energy per unit volume $E^{(V)}$ by

$$E^{(V)} = n_m\varepsilon_{(V)} \tag{8.76}$$

With the perturbations

$$T \to T(1+T') \tag{8.77}$$

$$T^{(V)} \to T^{(V)}(1+T'^{(V)}) \tag{8.78}$$

we have

$$\varepsilon_{(V)}(T) \to \varepsilon_{(V)}[T(1+T')] = \varepsilon_{(V)}(T) + \frac{\partial \varepsilon_{(V)}(T)}{\partial T} TT'$$

$$= \varepsilon_{(V)}(T) + C_v^{(V)}(T)TT' \tag{8.79}$$

defining the vibrational specific heat $C_v^{(V)}$. Similarly

$$\varepsilon_{(V)}(T^{(V)}) \to \varepsilon_{(V)}(T^{(V)}) + C_v^{(V)}(T^{(V)})T^{(V)}T'^{(V)}$$

where

$$C_v^{(V)}(T^{(V)}) = \frac{\partial \varepsilon_{(V)}(T^{(V)})}{\partial T^{(V)}} \tag{8.80}$$

We also have, for the temperature-dependent relaxation time $\tau_{(VT)}(T)$ in the third term on the right-hand side of Eq. (8.75):

$$\tau_{(VT)}(T) \to \tau_{(VT)}[T(1+T')]$$

and hence

$$\frac{1}{\tau_{(VT)}(T)} \to \frac{1}{\tau_{(VT)}(T)} + \left[\frac{\partial}{\partial T}\frac{1}{\tau_{(VT)}(T)}\right]TT'$$

$$= \frac{1}{\tau_{(VT)}(T)} - \frac{1}{\tau_{(VT)}^2(T)}\frac{\partial \tau_{(VT)}(T)}{\partial T}TT'$$

$$= \frac{1}{\tau_{(VT)}(T)} - \frac{\hat{\tau}_{(VT)}(T)}{\tau_{(VT)}(T)}T' \tag{8.81}$$

where

$$\hat{\tau}_{(VT)}(T) \equiv \frac{T}{\tau_{(VT)}(T)}\frac{\partial \tau_{(VT)}(T)}{\partial T} = \frac{\partial[\ln \tau_{(VT)}(T)]}{\partial[\ln(T)]} \tag{8.82}$$

Similarly we have for the final term on the right-hand side of Eq. (8.75):

$$\nu^{(TR)}(T_e) \to \nu^{(TR)}(T_e) + \frac{\partial \nu^{(TR)}}{\partial T_e}T_e T'_e = \nu^{(TR)}(T_e) + \nu^{(TR)}\hat{\nu}^{(TR)}T'_e \tag{8.83}$$

where we have defined

$$\hat{\nu}^{(T)} \equiv \frac{\partial[\ln \nu^{(TR)}(T_e)]}{\partial[\ln(T_e)]} \tag{8.84}$$

Noting that unprimed quantities are steady-state quantities, and that gradients only act on perturbed quantities, we can now write down the perturbed form of Eq. (8.75):

$$nC_vT\frac{DT'}{Dt}+nC_vT\frac{Dn'}{Dt}+\frac{\alpha(nC_vT+p)}{l}\nabla\cdot\mathbf{v}_0'$$
$$=\frac{\lambda'T}{l^2}\nabla^2T'-\hat{\tau}_{(\mathrm{VT})}(T)\frac{n_m}{\tau_{(\mathrm{VT})}(T)}[\varepsilon_{(\mathrm{V})}(T^{(\mathrm{V})})-\varepsilon_{(\mathrm{V})}(T)]T'$$
$$+\frac{n_m}{\tau_{(\mathrm{VT})}}[C_v^{(\mathrm{V})}(T^{(\mathrm{V})})T^{(\mathrm{V})}T'^{(\mathrm{V})}-C_v^{(\mathrm{V})}(T)TT']$$
$$+n_en_e'\nu^{(\mathrm{TR})}kT_e+n_e\nu^{(\mathrm{TR})}kT_eT_e'+\nu^{(\mathrm{TR})}\hat{\nu}^{(\mathrm{TR})}T_eT_e'kn_e \tag{8.85}$$

Dividing Eq. (8.85) through by nC_vT, and using Eq. (8.54) to eliminate $\nabla\cdot\mathbf{v}_0'$ we obtain

$$\frac{DT'}{Dt}-(\gamma-1)\frac{Dn'}{Dt}=\left(\frac{\lambda'}{nC_vl^2}\right)\nabla^2T'+\frac{1}{nC_vT}n_e\nu^{(\mathrm{TR})}kT_en_e'$$
$$+\left\{-\frac{\hat{\tau}_{(\mathrm{VT})}(T)}{nC_vT}\frac{n_m}{\tau_{(\mathrm{VT})}(T)}\right.$$
$$\left.\times[\varepsilon_{(\mathrm{V})}(T^{(\mathrm{V})})-\varepsilon_{(\mathrm{V})}(T)]\frac{n_m}{\tau_{(\mathrm{VT})}}\frac{C_v^{(\mathrm{V})}(T)}{nC_v}\right\}T'$$
$$+\frac{n_m}{\tau_{(\mathrm{VT})}}\frac{C_v^{(\mathrm{V})}(T^{(\mathrm{V})})T^{(\mathrm{V})}T'^{(\mathrm{V})}}{nC_vT}$$
$$+\frac{1}{nC_vT}n_e\nu^{(\mathrm{TR})}kT_e(1+\hat{\nu}^{(\mathrm{TR})})T_e' \tag{8.86}$$

where γ is the specific-heat ratio:

$$\gamma=\frac{C_p}{C_v} \tag{8.87}$$

Following Haas, we now define the quantities, $P_{(\mathrm{V})}$ and $P_{(\mathrm{TR})}$

$$P_{(\mathrm{V})}\equiv\frac{n_m\varepsilon_{(\mathrm{V})}(T^{(\mathrm{V})})-\varepsilon_{(\mathrm{V})}(T)}{\tau_{(\mathrm{VT})}j_e^2/\sigma_e} \tag{8.88}$$

$$P_{(\mathrm{TR})}\equiv\frac{n_en(\nu^{(\mathrm{TR})}/n)kT_e}{j_e^2/\sigma_e} \tag{8.89}$$

which are measures of the fractional power transfer from electrons into vibration and into translation–rotation excitation, respectively. The steady-state expression for j_e^2/σ_e is given by Eq. (8.43). From Eqs. (8.86), (8.88), and (8.89) we have

$$\frac{DT'}{Dt}-(\gamma-1)\frac{Dn'}{Dt}=\left(\frac{\lambda'}{nC_v l^2}\right)\nabla^2 T'$$
$$+\frac{\gamma}{\tau_{(\mathrm{T})}}\left[-P_{(\mathrm{V})}\hat{\tau}_{(\mathrm{VT})}-\frac{C_v^{(\mathrm{V})}(T)T}{C_v^{(\mathrm{V})}(T^{(\mathrm{V})})T^{(\mathrm{V})}}\left(\frac{\tau_{(\mathrm{V})}}{\tau_{(\mathrm{VT})}}\right)\right]T'$$
$$+\frac{1}{\tau_{(\mathrm{VT})}}\left[\frac{n_m C_v^{(\mathrm{V})}(T_{(\mathrm{V})})T_{(\mathrm{V})}}{nC_vT}\right]T'^{(\mathrm{V})}+\frac{\gamma P_{(\mathrm{TR})}}{\tau_{(\mathrm{T})}}n_e'$$
$$+\frac{\gamma P_{(\mathrm{TR})}}{\tau_{(\mathrm{T})}}(1+\hat{\nu}^{(\mathrm{TR})})T_e'+\frac{\gamma}{\tau_{(\mathrm{T})}}(2P_{(\mathrm{V})}+P_{(\mathrm{TR})})n' \tag{8.90}$$

where, following Haas, we have set

$$\tau_{(\mathrm{TR})}\equiv\frac{nC_pT}{j_e^2/\sigma_e} \tag{8.91}$$

$$\tau_{(\mathrm{V})}\equiv\frac{n_m C_v^{(\mathrm{V})}(T^{(\mathrm{V})})T^{(\mathrm{V})}}{j_e^2/\sigma_e} \tag{8.92}$$

$\tau_{(\mathrm{TR})}$ is identified with a characteristic time for translation–rotation heating, and $\tau_{(\mathrm{V})}$ with a characteristic time for vibration heating (Table 23). Equation (8.90) is Eq. (26) of Haas[70] with the exception of the n' term.

8.3.4. *The Vibrational Energy Equation*

Using Eqs. (8.19) and (8.20) we write the vibrational energy equation (8.18) in the form

$$\frac{DE^{(\mathrm{V})}(T^{(\mathrm{V})})}{Dt}+E^{(\mathrm{V})}(T^{(\mathrm{V})})\nabla\cdot\mathbf{v}_0-\nabla\cdot(\lambda^{(\mathrm{V})}\nabla T^{(\mathrm{V})})$$
$$=n_e\nu^{(\mathrm{V})}(T_e)kT_e-\frac{E^{(\mathrm{V})}(T^{(\mathrm{V})})-E^{(\mathrm{V})}(T)}{\tau_{(\mathrm{VT})}(T)} \tag{8.93}$$

Subjecting the first term of Eq. (8.93) to the perturbation

$$T^{(\mathrm{V})}\to T^{(\mathrm{V})}(1+T'^{(\mathrm{V})}) \tag{8.78}$$

we have

$$\frac{DE^{(\mathrm{V})}(T^{(\mathrm{V})})}{Dt}\to\frac{D}{Dt}[E^{(\mathrm{V})}(T^{(\mathrm{V})}+T^{(\mathrm{V})}T'^{(\mathrm{V})})]$$
$$=\frac{D}{Dt}\left[E^{(\mathrm{V})}(T^{(\mathrm{V})})+\frac{\partial E^{(\mathrm{V})}(T^{(\mathrm{V})})}{\partial T^{(\mathrm{V})}}T^{(\mathrm{V})}T'^{(\mathrm{V})}\right]$$
$$=\frac{DE^{(\mathrm{V})}(T^{(\mathrm{V})})}{Dt}+n_m C_v^{(\mathrm{V})}(T^{(\mathrm{V})})T^{(\mathrm{V})}\frac{DT'^{(\mathrm{V})}}{Dt} \tag{8.94}$$

To obtain the last term on the right-hand side of Eq. (8.94) we have used Eq. (8.76), and Eq. (8.80) for the vibrational specific heat, as well as making use of the fact that unprimed quantities are steady-state quantities.

Subjecting the second term of Eq. (8.93) to the perturbations given by Eqs. (8.78) and (8.48) we have, in terms of the dimensionless spatial variable defined by Eq. (8.45):

$$\frac{1}{l}E^{(V)}(T^{(V)})\nabla\cdot\mathbf{v}_0 \to \frac{1}{l}E^{(V)}(T^{(V)}+T^{(V)}T'^{(V)})\nabla\cdot(\mathbf{v}_0+\alpha\mathbf{v}_0')$$

$$=\frac{1}{l}\left[E^{(V)}(T^{(V)})+\frac{\partial E^{(V)}(T^{(V)})}{\partial T^{(V)}}T^{(V)}T'^{(V)}\right][\nabla\cdot\mathbf{v}_0+\nabla\cdot(\alpha\mathbf{v}_0')]$$

$$=\frac{1}{l}E^{(V)}(T^{(V)})\nabla\cdot\mathbf{v}_0+\frac{1}{\tau_s}E^{(V)}(T^{(V)})\nabla\cdot\mathbf{v}_0' \qquad (8.95)$$

where we have kept only first-order terms in perturbed quantities and have again used the fact that the unprimed variables are steady state to remove some terms. Using Eqs. (8.54) and (8.76) we can rewrite Eq. (8.95) in the form

$$\frac{1}{l}E^{(V)}(T^{(V)})\nabla\cdot\mathbf{v}_0 \to \frac{1}{l}E^{(V)}(T^{(V)})\nabla\cdot\mathbf{v}_0 - n_m\varepsilon_{(V)}(T^{(V)})\frac{Dn'}{DT} \qquad (8.96)$$

For the third term of Eq. (8.93) we have

$$\frac{1}{l^2}\nabla\cdot(\lambda^{(V)}\nabla T^{(V)}) \to \frac{1}{l^2}\nabla\cdot[\lambda^{(V)}\nabla(T^{(V)}+T^{(V)}T'^{(V)})$$

$$=\frac{1}{l^2}\nabla\cdot(\lambda^{(V)}\nabla T^{(V)})+\frac{\lambda^{(V)}}{l^2}T^{(V)}\nabla^2T'^{(V)} \qquad (8.97)$$

and for the first term on the left-hand side of Eq. (8.93) we have

$$n_e\nu^{(V)}(T_e)kT_e \to (n_e+n_en_e')\nu^{(V)}(T_e+T_eT_e')k(T_e+T_eT_e')$$

$$=n_e\nu^{(V)}(T_e)kT_e[1+n_e'+(1+\hat{\nu}^{(V)})T_e'] \qquad (8.98)$$

where we have kept only first-order terms in perturbed quantities, and where

$$\hat{\nu}^{(V)} \equiv \frac{T_e}{\nu^{(V)}(T_e)}\frac{\partial\nu^{(V)}(T_e)}{\partial T_e} \qquad (8.99)$$

Subjecting the last term of Eq. (8.93) to the perturbations given by Eqs. (8.77) and (8.78) we obtain

$$\begin{aligned}
\frac{E^{(V)}(T^{(V)})-E^{(V)}(T)}{\tau_{(VT)}(T)} &= \frac{n_m}{\tau_{(VT)}(T)}[\varepsilon_{(V)}(T^{(V)})-\varepsilon_{(V)}(T)] \\
&\to \frac{n_m}{\tau_{(VT)}(T+TT')} \\
&\quad \times [\varepsilon_{(V)}(T^{(V)}+T^{(V)}T'^{(V)})-\varepsilon_{(V)}(T+TT')] \\
&= \frac{n_m}{\tau_{(VT)}(T)}[1-\hat{\tau}_{(VT)}(T^{(V)})T'][\varepsilon_{(V)}(T^{(V)}) \\
&\quad +C_v^{(V)}(T^{(V)})T^{(V)}T'^{(V)}-\varepsilon_{(V)}(T)-C_v^{(V)}(T)TT'] \\
&= \frac{E^{(V)}(T^{(V)})-E^{(V)}(T)}{\tau_{(VT)}(T)} \\
&\quad -\hat{\tau}_{(VT)}(T^{(V)})\frac{n_m}{\tau_{(VT)}(T)}[\varepsilon_{(V)}(T^{(V)})-\varepsilon_{(V)}(T)]T' \\
&\quad +\frac{n_m}{\tau_{(VT)}(T)}[C_v^{(V)}(T^{(V)})T^{(V)}T'^{(V)}-C_v^{(V)}(T)TT')
\end{aligned} \tag{8.100}$$

where we have used Eqs. (8.76), (8.79), (8.80), and (8.82) and have kept only first-order terms in perturbed quantities.

From Eqs. (8.94), (8.96), (8.97), (8.98), and (8.100) and using Eq. (8.93) to eliminate terms not containing perturbed variables, we obtain the perturbed form of Eq. (8.93) as

$$\begin{aligned}
n_m C_v^{(V)}(T^{(V)})T^{(V)}\frac{DT'^{(V)}}{Dt} - n_m\varepsilon_{(V)}(T^{(V)})\frac{D_{n'}}{Dt}-\frac{\lambda^{(V)}}{l^2}T^{(V)}\nabla^2 T'^{(V)} \\
= n_e\nu^{(V)}(T_e)kT_e n_e' + n_e\nu^{(V)}(T_e)kT_e(1+\hat{\nu}^{(V)})T_e' \\
+\hat{\tau}_{(VT)}(T^{(V)})\frac{n_m}{\tau_{(VT)}(T)}[\varepsilon_{(V)}(T^{(V)})-\varepsilon_{(V)}(T)]T' \\
-\frac{n_m}{\tau_{(VT)}(T)}[C_v^{(V)}(T^{(V)})T^{(V)}T'^{(V)}-C_v^{(V)}(T)TT']
\end{aligned} \tag{8.101}$$

On dividing Eq. (8.101) through by $n_m C_v^{(V)}(T^{(V)})T^{(V)}$ and rearranging terms we obtain

$$\frac{DT'^{(\mathrm{V})}}{Dt}=\frac{\varepsilon_{(\mathrm{V})}(T^{(\mathrm{V})})}{C_v^{(\mathrm{V})}(T^{(\mathrm{V})})T^{(\mathrm{V})}}\frac{Dn'}{Dt}+\left[\frac{\lambda^{(\mathrm{V})}}{n_m C_v^{(\mathrm{V})}(T^{(\mathrm{V})})l^2}\right]\nabla^2 T'^{(\mathrm{V})}$$

$$+\frac{n_e\nu^{(\mathrm{V})}(T_e)kT_e}{n_m C_v^{(\mathrm{V})}(T^{(\mathrm{V})})T^{(\mathrm{V})}}n_e'+\frac{n_e\nu^{(\mathrm{V})}(T_e)kT_e}{n_m C_v^{(\mathrm{V})}(T^{(\mathrm{V})})T^{(\mathrm{V})}}(1+\hat{\nu}^{(\mathrm{V})})T_e'-\frac{T'^{(\mathrm{V})}}{\tau_{(\mathrm{VT})}(T)}$$

$$+\frac{1}{T^{(\mathrm{V})}C_v^{(\mathrm{V})}(T^{(\mathrm{V})})}\left\{\frac{\hat{\tau}_{(\mathrm{VT})}(T^{(\mathrm{V})})}{\tau_{(\mathrm{VT})}(T)}[\varepsilon_{(\mathrm{V})}(T^{(\mathrm{V})})-\varepsilon_{(\mathrm{V})}(T)]\right.$$

$$\left.+\frac{1}{\tau_{(\mathrm{VT})}(T)}C_v^{(\mathrm{V})}(T)T\right\}T' \tag{8.102}$$

Using Eqs. (8.40), (8.76), (8.88), and (8.92) we obtain, for the third term on the right-hand side of Eq. (8.102):

$$\frac{n_e\nu^{(\mathrm{V})}(T_e)kT_e}{n_m C_v^{(\mathrm{V})}(T^{(\mathrm{V})})T^{(\mathrm{V})}}n_e'=\frac{\varepsilon_{(\mathrm{V})}(T^{(\mathrm{V})})-\varepsilon_{(\mathrm{V})}(T)}{\tau_{(\mathrm{VT})}C_v^{(\mathrm{V})}(T^{(\mathrm{V})})T^{(\mathrm{V})}}n_e'$$

$$=\frac{C_v^{(\mathrm{V})}(T^{(\mathrm{V})})T^{(\mathrm{V})}P_{(\mathrm{V})}}{n_m}\frac{j_e^2}{\sigma_e}n_e'$$

$$=\frac{P_{(\mathrm{V})}}{\tau_{(\mathrm{V})}}n_e' \tag{8.103}$$

Similarly, for the fourth term on the right-hand side of Eq. (8.102) we have

$$\frac{n_e\nu^{(\mathrm{V})}(T_e)kT_e}{n_m C_v^{(\mathrm{V})}(T^{(\mathrm{V})})T^{(\mathrm{V})}}(1+\hat{\nu}^{(\mathrm{V})})T_e'=\frac{P_{(\mathrm{V})}}{\tau_{(\mathrm{V})}}(1+\hat{\nu}^{(\mathrm{V})})T_e' \tag{8.104}$$

For the last term of Eq. (8.102) we have, from Eqs. (8.88) and (8.92):

$$\frac{1}{T^{(\mathrm{V})}C_v^{(\mathrm{V})}(T_{(\mathrm{V})})}\left\{\frac{\hat{\tau}_{(\mathrm{VT})}(T^{(\mathrm{V})})}{\tau_{(\mathrm{VT})}}[\varepsilon_{(\mathrm{V})}(T^{(\mathrm{V})})-\varepsilon_{(\mathrm{V})}(T)]+\frac{1}{\tau_{(\mathrm{VT})}(T)}C_v^{(\mathrm{V})}(T)T\right\}T'$$

$$=\left[\frac{P_{(\mathrm{V})}j_e^2/\sigma_e}{n_m T^{(\mathrm{V})}C_v^{(\mathrm{V})}(T^{(\mathrm{V})})}\hat{\tau}_{(\mathrm{VT})}(T^{(\mathrm{V})})+\frac{1}{\tau_{(\mathrm{VT})}(T)}\frac{C_v^{(\mathrm{V})}(T)T}{C_v^{(\mathrm{V})}(T^{(\mathrm{V})})T^{(\mathrm{V})}}\right]T'$$

$$=\frac{1}{\tau_{(\mathrm{V})}}\left[P_{(\mathrm{V})}\hat{\tau}_{(\mathrm{VT})}(T^{(\mathrm{V})})+\frac{\tau_{(\mathrm{V})}}{\tau_{(\mathrm{VT})}(T)}\frac{C_v^{(\mathrm{V})}(T)}{C_v^{(\mathrm{V})}(T^{(\mathrm{V})})}\frac{T}{T^{(\mathrm{V})}}\right]T' \tag{8.105}$$

From Eqs. (8.103) and (8.105) we can rewrite Eq. (8.102) as

$$\frac{DT'^{(V)}}{Dt}=\frac{\varepsilon_{(V)}(T^{(V)})}{C_v^{(V)}(T^{(V)})T^{(V)}}\frac{Dn'}{Dt}+\left[\frac{\lambda^{(V)}}{n_m C_v^{(V)}(T^{(V)})l^2}\right]\nabla^2 T'^{(V)}+\frac{P_{(V)}}{\tau_{(V)}}n_e'$$
$$+\frac{P_{(V)}}{\tau_{(V)}}(1+\hat{\nu}^{(V)})T_e'$$
$$-\frac{T'^{(V)}}{\tau_{(VT)}(T)}+\frac{1}{\tau_{(V)}}\left[P_{(V)}\hat{\tau}_{(VT)}(T^{(V)})+\frac{\tau_{(V)}}{\tau_{(VT)}(T)}\frac{C_v^{(V)}(T)T}{C_v^{(V)}(T^{(V)})T^{(V)}}\right]T' \tag{8.106}$$

which is the form of the equation for the vibrational temperature fluctuations given by Haas except that for the coefficient of n' he has

$$\frac{\gamma}{\tau_{(T)}}(2P_{(V)}+P_{(TR)})\quad\text{instead of}\quad\frac{\varepsilon_{(V)}(T^{(V)})}{C_v^{(V)}(T^{(V)})T^{(V)}}\frac{D}{Dt}$$

8.3.5. *The Charge Conservation Equation*

From Eq. (8.7) we have for the charge density $\rho_c(\mathbf{x}, t)$:

$$\rho_c=e(n_+-n_--n_e) \tag{8.107}$$

Following Haas we shall use Eq. (8.107) to write down equations for the perturbed quantities ρ_c', n_-', and n_e'. From Eq. (8.107) we have

$$\frac{1}{e}\frac{D\rho_c}{Dt}=\frac{Dn_+}{Dt}-\frac{Dn_-}{Dt}-\frac{Dn_e}{Dt} \tag{8.108}$$

Substituting for the quantities on the right-hand side of Eq. (8.108) using Eqs. (8.2), (8.3), and (8.21), we obtain after rearrangement of terms:

$$\frac{D\rho_c}{Dt}+\rho_c\nabla\cdot\mathbf{v}_0+e\nabla\cdot(n_+\bar{\mathbf{V}}_+-n_-\bar{\mathbf{V}}_--n_e\bar{\mathbf{V}}_e)-eS_++eS_e=0 \tag{8.109}$$

Equations (8.5) and (8.22) for the charged-particle drift velocities allow Eq. (8.109) to be written in the form

$$\frac{D\rho_c}{Dt}+\rho_c\nabla\cdot\mathbf{v}_0+e\nabla\cdot(n_+\mu_+\mathbf{E}+n_-\mu_-\mathbf{E}+\nabla(n_eD_e)+n_eD_e\nabla[\ln(n)]+n_e\mu_e\mathbf{E})$$
$$-eS_++eS_e=0 \tag{8.110}$$

Subjecting the plasma variables to the perturbations of Eqs. (8.47), (8.48), and (8.57)–(8.60) and replacing the spatial variable by the dimensionless variable, we write the perturbed form of Eq. (8.110) as

$$en_+\frac{D\rho_c'}{Dt}+\frac{\rho_c\alpha}{l}\nabla\cdot\mathbf{v}_0'+\frac{1}{l}\nabla\cdot\Big[\sigma_+\mathbf{E}n_+'+\sigma_+E\mathbf{E}'+\sigma_-\mathbf{E}n_-'+\sigma_-E\mathbf{E}'$$
$$+\hat{\sigma}_+\sigma_+\mathbf{E}T'+\hat{\sigma}_-\sigma_-\mathbf{E}T'+\frac{e}{l}\nabla(n_eD_eT_e'-n_eD_e\hat{\nu}_mT_e')+\frac{en_eD_e}{l}\nabla n'$$
$$+\sigma_e\mathbf{E}n_e'+\sigma_eE\mathbf{E}'-\sigma_e\mathbf{E}\hat{\nu}_mT_e'\Big]=0 \tag{8.111}$$

where we have defined electrical conductivities by

$$\begin{aligned}\sigma_+&=en_+\mu_+\\ \sigma_-&=en_-\mu_-\\ \sigma_e&=en_e\mu_e\end{aligned} \tag{8.112}$$

and put

$$\hat{\sigma}_+=\frac{T}{\sigma_+}\frac{\partial\sigma_+}{\partial T},\qquad \hat{\sigma}_-=\frac{T}{\sigma_-}\frac{\partial\sigma_-}{\partial T}$$

and we have used Eq. (5.192) for the electron diffusion coefficient D_e to obtain

$$(n_eD_e)'=n_eD_eT_e'-n_eD_e\hat{\nu}_mT_e \tag{8.113}$$

where

$$\hat{\nu}_m=\frac{T_e}{\nu_m}\frac{\partial\nu_m}{\partial T_e} \tag{8.114}$$

Defining

$$\begin{aligned}\sigma_i&=\sigma_++\sigma_-\\ \sigma&=\sigma_i+\sigma_e\end{aligned}$$

and dividing Eq. (8.111) by en_+, we obtain

$$\frac{D\rho_c'}{Dt}-\left(\frac{\rho_c}{en_+}\right)\frac{Dn'}{Dt}+\frac{\sigma}{en_+l}E\nabla\cdot\mathbf{E}'$$
$$+\frac{\mathbf{E}\cdot\nabla}{en_+l}(\sigma_+n_+'+\sigma_-n_-'+\sigma_en_e'-\sigma_e\hat{\nu}_mT_e')+\frac{n_eD_e}{n_+l^2}\nabla^2(T_e'-\hat{\nu}_mT_e+n')=0 \tag{8.115}$$

where we have used Eq. (8.54) as well as the fact that gradients of unperturbed quantities are zero. The perturbed form of Eq. (8.107) is

$$en_+\rho_c'=e(n_+n_+'-n_-n_-'-n_en_e') \tag{8.116a}$$

Therefore

$$n'_+ = \rho'_c + \frac{n_-}{n_+} n'_- + \frac{n_e}{n_+} n'_e \tag{8.116b}$$

Using Eq. (8.116b) to eliminate n'_+ from Eq. (8.115) and using Eq. (8.148) for $\mathbf{\nabla} \cdot \mathbf{E}'$, we obtain the final form of the charge conservation equation:

$$\frac{D\rho'_c}{Dt} - \left(\frac{\rho_c}{en_+}\right)\frac{D_{n'}}{Dt} + \frac{1}{\tau_c}\rho'_c + \frac{1}{\tau_c}\left(\frac{\varepsilon_0 \mathbf{E}}{en_+ l}\right) \cdot \mathbf{\nabla}\left[\frac{\sigma_+}{\sigma}\rho'_c + \left(\frac{\sigma_e}{\sigma} + \frac{\sigma_+}{\sigma}\frac{n_e}{n_+}\right)n'_e \right.$$
$$\left. + \frac{\sigma_i}{\sigma}\hat{\sigma}_i T' + \left(\frac{\sigma_-}{\sigma} + \frac{\sigma_+}{\sigma}\frac{n_-}{n_+}\right)n'_- - \frac{\sigma_e}{\sigma}\hat{\nu}_m T'_e\right] + \frac{n_e}{n_+}\frac{D_e}{l^2}\nabla^2 \eta'_e = 0 \tag{8.117}$$

where we have defined

$$\eta'_e \equiv n' + (1 - \hat{\nu}_m)T'_e \tag{8.118a}$$

and space–charge relaxation

$$\tau_c^{-1} \equiv \varepsilon_0/\sigma \tag{8.118b}$$

8.3.6. *The Electron, Negative-Ion, and Electronically Excited Species Number-Density Equations*

From Eqs. (8.21) and (8.22), the electron number-density equation is

$$\frac{Dn_e}{Dt} + n_e \mathbf{\nabla} \cdot \mathbf{v}_0 - \mathbf{\nabla} \cdot \{\mathbf{\nabla}(n_e D_e) + n_e D_e \mathbf{\nabla}[\ln(n)] + n_e \mu_e \mathbf{E}\}$$
$$= n_e n k_i(T_e) + n_e n_* k_i^*(T_e) + n_- n k_d(T) - n_e n_+ k_r^e(T_e) - n_e n k_a(T_e) + S_e \tag{8.119}$$

In the same way that Eq. (8.117) was derived from Eq. (8.110) we obtain the perturbed form of Eq. (8.119):

$$\frac{Dn'_e}{Dt} + \frac{1}{\tau_s}\mathbf{\nabla} \cdot \mathbf{v}'_0 - \frac{D_e}{l^2}\nabla^2 \eta'_e - \frac{\sigma_e}{n_e}\frac{n_+}{\varepsilon_0}\rho'_c - \frac{\sigma_e}{\varepsilon_0}\left(\frac{\varepsilon_0 \mathbf{E}}{en_e l}\right) \cdot \mathbf{\nabla}(n'_e - \hat{\nu}_m T'_e)$$
$$= (nk_i + n_* k_i^* - n_+ k_r^e - nk_a)n'_e + \left(nk_i + \frac{n_- n}{n_e}k_d - nk_a\right)n'$$
$$+ n_* k_i^* n'_* + \frac{n_- n}{n_e} k_d n'_- - n_+ k_r^e n'_+$$
$$+ (nk_i \hat{k}_i + n_* k_i^* \hat{k}_i^* - n_+ k_r^e \hat{k}_r^e - nk_a \hat{k}_a)T'_e + \frac{n_- n}{n_e} k_d \hat{k}_d T' \tag{8.120}$$

where

$$\hat{k}_j(T_e) \equiv \frac{T_e}{k_j(T_e)}\frac{\partial k_j(T_e)}{\partial T_e} \tag{8.121}$$

for all the electron temperature-dependent rates and

$$\hat{k}_d(T) = \frac{T}{k_d(T)} \frac{\partial k_d(T)}{\partial T} \tag{8.122}$$

To obtain the electron density perturbation equation in the form given by Haas, we use Eq. (8.117) for the charge perturbation to eliminate the fourth term on the left-hand side of Eq. (8.120), and Eq. (8.116b) to substitute for n'_+ on the right-hand side of Eq. (8.120):

$$\begin{aligned}
&\frac{Dn'_e}{Dt} + \frac{1}{\tau_s}\nabla \cdot \mathbf{v}'_0 - \left(1 - \frac{\sigma_e}{\sigma}\right)\frac{D_e}{l^2}\nabla^2 \eta'_e \\
&\quad - \frac{\sigma_e}{\varepsilon_0}\left(\frac{\varepsilon_0 \mathbf{E}}{e n_e l}\right) \cdot \nabla\left[-\frac{\sigma_+}{\sigma}\rho'_c + \left(1 - \frac{\sigma_e}{\sigma} - \frac{n_e}{n_+}\frac{\sigma_+}{\sigma}\right)n'_e\right. \\
&\quad \left. - \left(\frac{\sigma_-}{\sigma} + \frac{\sigma_+}{\sigma}\frac{n_-}{n_+}\right)n'_- - \left(1 - \frac{\sigma_e}{\sigma}\right)\hat{\nu}_m T'_e - \frac{\sigma_i}{\sigma}\hat{\sigma}_i T'\right] \\
&= -\frac{\sigma_e}{\sigma}\frac{n_+}{n_e}\frac{D\rho'_c}{Dt} - n_+ k^e_r \rho'_c + (nk_i + n_* k^*_i - (n_+ + n_e)k^e_r - nk_a)n'_e \\
&\quad + \left(nk_i - \frac{n_- n}{n_e}k_d - nk_a\right)n' + n_* k^*_i n'_* + \left(\frac{n_- n}{n_e}k_d - n_- k^e_r\right)n'_- \\
&\quad + (nk_i\hat{k}_i + n_* k^*_i \hat{k}^*_i - n_+ k^e_r \hat{k}^e_r - nk_a\hat{k}_a)T'_e + \frac{n_- n}{n_e}k_d\hat{k}_d T'
\end{aligned} \tag{8.123a}$$

Finally we rewrite the right-hand side of Eq. (8.123a) using the characteristic times given in Table 23 and the steady-state relationship, Eq. (8.34), to obtain [using $\tau_r^{(e)} \equiv (n + k_r^e)^{-1}$]

$$\begin{aligned}
&\frac{Dn'_e}{Dt} + \frac{1}{\tau_s}\nabla \cdot \mathbf{v}'_0 - \left(1 - \frac{\sigma_e}{\sigma}\right)\frac{D_e}{l^2}\nabla^2 \eta'_e \\
&\quad - \frac{\sigma_e}{\varepsilon_0}\left(\frac{\varepsilon_0 \mathbf{E}}{e n_e l}\right) \cdot \nabla\left[-\frac{\sigma_+}{\sigma}\rho'_c + \left(1 - \frac{\sigma_e}{\sigma} - \frac{n_e}{n_+}\frac{\sigma_+}{\sigma}\right)n'_e\right. \\
&\quad \left. - \left(\frac{\sigma_-}{\sigma} + \frac{\sigma_+}{\sigma}\frac{n_-}{n_+}\right)n'_- - \left(1 - \frac{\sigma_e}{\sigma}\right)\hat{\nu}_m T'_e - \frac{\sigma_i}{\sigma}\hat{\sigma}_i T'\right] \\
&= -\frac{\sigma_e}{\sigma}\frac{n_+}{n_e}\frac{D\rho'_c}{Dt} - \frac{1}{\tau_r^{(e)}}\rho'_c - \left(\frac{n_e}{n_+}\frac{1}{\tau_r^{(e)}} + \frac{n_-}{n_e}\frac{1}{\tau_d} + \frac{S_e}{n_e}\right)n'_e
\end{aligned}$$

$$+\left(\frac{n_-}{n_e}\frac{1}{\tau_d}-\frac{n_-}{n_+}\frac{1}{\tau_r^{(e)}}\right)n'_-+\frac{1}{\tau_i^*}n'_*+\left(\frac{1}{\tau_r^{(e)}}-\frac{1}{\tau_i^*}\right)n'$$

$$+\frac{1}{\tau_i}\left(\hat{k}_i+\frac{\tau_i}{\tau_i^*}\hat{k}_i^*-\frac{\tau_i}{\tau_r^{(e)}}\hat{k}_r^{(e)}-\frac{\tau_i}{\tau_a}\hat{k}_a\right)T'_e+\frac{n_-}{n_e}\frac{1}{\tau_d}\hat{k}_d T \tag{8.123b}$$

which is the equation given by Haas [Reference 70, Eq. (29)].

The equation for the negative-ion density perturbation, n'_-, is obtained in the same way. From Eqs. (8.3) and (8.5) we have

$$\frac{Dn_-}{Dt}+n_-\nabla\cdot\mathbf{v}_0-\nabla\cdot(n_-\mu_-\mathbf{E})=n_e n k_a(T_e)-n_+n_-k_r^i(T)-n_-nk_d(T) \tag{8.124}$$

The perturbed form of Eq. (5.124) is

$$\frac{Dn'_-}{Dt}+\frac{1}{\tau_s}\nabla\cdot\mathbf{v}'_0-\frac{\sigma_-}{n_-}\frac{n_+}{\varepsilon_0}\rho'_c\frac{\sigma_-}{\varepsilon_0}\left(\frac{\varepsilon_0\mathbf{E}}{en_-l}\right)\cdot\nabla(n'_-+\sigma_-T')$$

$$=\frac{n_en}{n_-}k_an'_e+\left(\frac{n_en}{n_-}k_a-nk_d\right)n'-n_+k_r^in'_+$$

$$-(n_+k_r^i+nk_d)n'_--\frac{n_en}{n_-}k_a\hat{k}_aT'_e-(n_+k_r^i\hat{k}_r^i+nk_d\hat{k}_d)T' \tag{8.125}$$

Now using Eq. (8.117) for the charge perturbation to eliminate the third term on the left-hand side of Eq. (8.125), and Eq. (8.116b) to substitute for n'_+ on the right-hand side, we obtain

$$\frac{Dn'_-}{Dt}+\frac{1}{\tau_s}\nabla\cdot\mathbf{v}'_0+\frac{\sigma_-}{\sigma}\frac{n_e}{n_-}\frac{D_e}{l^2}\nabla^2\eta'_e$$

$$-\frac{\sigma_-}{\varepsilon_0}\left(\frac{\varepsilon_0\mathbf{E}}{en_-l}\right)\cdot\nabla\left[-\frac{\sigma_+}{\sigma}\rho'_c-\left(\frac{\sigma_e}{\sigma}+\frac{\sigma_+}{\sigma}\frac{n_e}{n}\right)n'_e\right.$$

$$\left.+\left(1-\frac{\sigma_-}{\sigma}-\frac{\sigma_+}{\sigma}\frac{n_-}{n_+}\right)n'_-+\frac{\sigma_e}{\sigma}\hat{\nu}_mT'_e+\left(\sigma_--\frac{\sigma_i}{\sigma}\hat{\sigma}_i\right)T'\right]$$

$$=-\frac{\sigma_i}{\sigma}\frac{n_+}{n_-}\frac{D\rho'_c}{Dt}-n_+k_r^i\rho'_c+\left(\frac{n_en}{n_-}k_a-n_ek_r^i\right)n'_e+\left(\frac{n_en}{n_-}k_a-nk_d\right)n'$$

$$-(n_+k_r^i+nk_d+n_-k_r^i)n'_-+\frac{n_en}{n_-}k_a\hat{k}_aT'_e-(n_+k_r^i\hat{k}_r^i+nk_d\hat{k}_d)T' \tag{8.126}$$

The final form of the equation for n'_- is obtained by rewriting the right-hand side of Eq. (8.126), using the characteristic times of Table 23 and the steady-state relationship (8.29) with $\tau_{\text{neg}} \equiv l^2/[(\sigma_-/\sigma)(n_e/n_-)D_e]$

$$
\begin{aligned}
&\frac{Dn'_-}{Dt} + \frac{1}{\tau_s}\nabla \cdot \mathbf{v}'_0 + \frac{1}{\tau_{\text{neg}}}\nabla^2 \eta'_e \\
&\quad - \frac{\sigma_-}{\varepsilon_0}\left(\frac{\varepsilon_0 \mathbf{E}}{en_- l}\right) \cdot \nabla\left[-\frac{\sigma_+}{\sigma}\rho'_c - \left(\frac{\sigma_e}{\sigma} + \frac{\sigma_+}{\sigma}\frac{n_e}{n_+}\right)n'_e\right. \\
&\quad \left. + \left(1 - \frac{\sigma_-}{\sigma} - \frac{\sigma_+}{\sigma}\frac{n_-}{n_+}\right)n'_- + \frac{\sigma_e}{\sigma}\hat{\nu}_m T'_e + \left(\hat{\sigma}_- - \frac{\hat{\sigma}_i}{\sigma}\sigma_i\right)T'\right] \\
&= -\frac{\sigma_i}{\sigma}\frac{n_+}{n_-}\frac{D\rho'_c}{Dt} - \frac{1}{\tau_r^{(i)}}\rho'_c + \left(\frac{n_e}{n_-}\frac{1}{\tau_a} - \frac{n_e}{n_+}\frac{1}{\tau_r^{(i)}}\right)n'_e + \frac{1}{\tau_r^{(i)}}n' \cdot \\
&\quad + \frac{n_e}{n_-}\frac{1}{\tau_a}\hat{k}_a T'_e - \left(\frac{1}{\tau_r^{(i)}}\hat{k}^i_r + \frac{1}{\tau_d}\hat{k}_d\right)T'
\end{aligned}
\tag{8.127}
$$

which is the equation by Haas [Reference 70, Eq. (30)].

The equation for the electronically excited species number-density perturbation, n'_*, is obtained from Eq. (8.4):

$$
\frac{Dn_*}{Dt} + n_*\nabla \cdot \mathbf{v}_0 = n_e n k_*(T_e) - n_e n_* k_i^*(T_e) - n_* n k_q(T) + S_* \tag{8.4}
$$

Perturbing the variables in Eq. (8.4) and dividing through by n_* we have

$$
\begin{aligned}
\frac{Dn'_*}{Dt} + \frac{1}{\tau_s}\nabla \cdot \mathbf{v}'_0 &= \left(\frac{n_e n}{n_*}k_* - n_e k_i^*\right)n'_e + \left(\frac{n_e n}{n_*}k_* - nk_q\right)n' \\
&\quad - (n_e k_i^* + nk_q)n'_* + \left(\frac{n_e n}{n_*}k_*\hat{k}_* - n_e k_i^* \hat{k}_i^*\right)T'_e - nk_q\hat{k}_q T'
\end{aligned}
\tag{8.128}
$$

In terms of the characteristic times of Table 23 and using the steady-state relationship (8.39), Eq. (8.128) becomes

$$
\begin{aligned}
\frac{Dn'_*}{Dt} + \frac{1}{\tau_s}\nabla \cdot \mathbf{v}'_0 &= \left(\frac{n}{n_*}\frac{1}{\tau_q} - \frac{S_*}{n_*}\right)n'_e + \frac{n_e}{n_*}\left(\frac{1}{\tau_*} - \frac{1}{\tau_i^*}\right)n' \\
&\quad - \left(\frac{n_e}{n_*}\frac{1}{\tau_i^*} + \frac{n}{n_*}\frac{1}{\tau_q}\right)n'_* - \frac{n}{n_*}\frac{1}{\tau_q}\hat{k}_q T' \\
&\quad + \frac{n_e}{n_*}\frac{1}{\tau_*}\left(\hat{k}_* - \frac{\tau_*}{\tau_i^*}\hat{k}_i^*\right)T'_e
\end{aligned}
\tag{8.129}
$$

which is the equation given by Haas [Reference 70, Eq. (31)] except that some terms on the right-hand side are not in agreement.

8.3.7. The Electron Energy Equation

From Eqs. (8.23) and (8.23b) the unperturbed form of the electron energy equation is

$$\frac{D}{Dt}(\tfrac{3}{2}n_e kT_e)+\tfrac{5}{2}n_e kT_e \boldsymbol{\nabla}\cdot\mathbf{v}_0'=\mathbf{j}_e\cdot\mathbf{E}-\boldsymbol{\nabla}\cdot\mathbf{q}_e-n_e\nu_u kT_e \qquad (8.130)$$

Perturbing the plasma variables in Eq. (8.130) and dividing the resulting equation by $\frac{3}{2}n_e kT_e$, we obtain

$$\frac{DT_e'}{Dt}+\frac{Dn_e'}{Dt}+\frac{5}{3}\frac{1}{\tau_s}\boldsymbol{\nabla}\cdot\mathbf{v}_0'$$

$$=\frac{\mathbf{j}_e'\cdot\mathbf{E}}{\frac{3}{2}n_e kT_e}+\frac{\mathbf{j}_e\cdot\mathbf{E}'E}{\frac{3}{2}n_e kT_e}-\frac{(1/l)\boldsymbol{\nabla}\cdot\mathbf{q}_e'}{\frac{3}{2}n_e kT_e}-\frac{2}{3}\nu_u(n_e'+T_e'+\hat{\nu}_u T_e') \qquad (8.131)$$

where $\mathbf{j}_e'$ and $\mathbf{q}_e'$ are the perturbed forms of $\mathbf{j}_e$ and $\mathbf{q}_e$ and are to be evaluated in terms of the perturbed plasma variables. From Eq. (5.221) we have

$$\boldsymbol{\nabla}\cdot\mathbf{q}_e'=-\frac{\lambda_e T_e}{l}\nabla^2 T_e'-\frac{5}{2}\frac{k}{e}T_e\mathbf{j}_e\cdot\boldsymbol{\nabla}T_e'-\frac{5}{2}\frac{k}{e}T_e\boldsymbol{\nabla}\cdot\mathbf{j}_e' \qquad (8.132)$$

From Eqs. (8.42)–(8.44) we have the steady-state relationships

$$\mathbf{j}_e=\sigma_e\mathbf{E} \qquad (8.133)$$

$$j_e^2/\sigma_e=n_e\nu_u kT_e \qquad (8.134)$$

Following Haas we define an "electron heating" relaxation time τ_e by

$$\tau_e=\tfrac{3}{2}\nu_u^{-1} \qquad (8.135)$$

Then, from Eqs. (8.133)–(8.135) we have

$$\frac{3}{2}n_e kT_e=\frac{\tau_e j_e^2}{\sigma_e}=\tau_e\sigma_e E^2 \qquad (8.136)$$

Using Eqs. (8.132) and (8.136) we write Eq. (8.131) in the form

$$\frac{DT_e'}{Dt}+\frac{Dn_e'}{Dt}+\frac{5}{3}\frac{1}{\tau_s}\boldsymbol{\nabla}\cdot\mathbf{v}_0'$$

$$=\frac{\mathbf{j}_e'\cdot\mathbf{E}}{\frac{3}{2}n_e kT_e}+\frac{1}{\tau_e}\left(\frac{\mathbf{E}}{E}\right)\cdot\mathbf{E}'-\frac{1}{\tau_e}n_e'-\frac{1}{\tau_e}(1+\hat{\nu}_u)T_e'+\frac{2e\lambda_e}{3n_e kl^2}\nabla^2 T_e'$$

$$+\frac{5}{3}\frac{1}{ln_e e}\boldsymbol{\nabla}\cdot\mathbf{j}_e'+\frac{5}{3}\frac{\sigma_e}{\varepsilon_0}\left(\frac{\varepsilon_0\mathbf{E}}{en_e l}\right)\cdot\boldsymbol{\nabla}T_e' \qquad (8.137)$$

It remains to evaluate $\mathbf{j}_e'$. From Eqs. (5.158) and (5.195) we have

$$\mathbf{j}_e = e\nabla(n_eD_e) + en_eD_e\nabla\ln(n) + en_e\mu_e\mathbf{E} \tag{8.138}$$

Using Eq. (8.113), and continuing to neglect the spatial derivatives of steady-state quantities, we obtain from Eq. (8.138):

$$\begin{aligned}\mathbf{j}_e' &= \frac{e}{l}\nabla(n_eD_eT_e - n_eD_e\hat{\nu}_mT_e') + \frac{en_eD_en'}{l}\\ &\quad +\sigma_eE\mathbf{E}' - \sigma_e\hat{\nu}_m\mathbf{E}T_e' + \sigma_e\mathbf{E}n_e'\\ &= \frac{en_eD_e}{l}\nabla\eta_e' + \sigma_eE\mathbf{E}' - \sigma_e\hat{\nu}_m\mathbf{E}T_e' + \sigma_e\mathbf{E}n_e'\end{aligned} \tag{8.139}$$

where we have used Eq. (8.118a) for η_e'. From Eqs. (8.139) and (8.136) we have, for the first term on the right-hand side of Eq. (8.137):

$$\frac{\mathbf{j}_e'\cdot\mathbf{E}}{\frac{3}{2}n_ekT_e} = \frac{en_eD_e\mathbf{E}}{l\frac{3}{2}n_ekT_e}\nabla\eta_e' + \frac{1}{\tau_e}\left(\frac{\mathbf{E}}{E}\right)\cdot\mathbf{E}' - \frac{1}{\tau_e}\hat{\nu}_mT_e' + \frac{1}{\tau_e}n_e' \tag{8.140}$$

From Eqs. (5.192) and (5.193) we have

$$eD_e = \mu_ekT_e \tag{8.141}$$

which allows Eq. (8.140) to be rewritten in the form

$$\frac{\mathbf{j}_e'\cdot\mathbf{E}}{\frac{3}{2}n_ekT_e} = \frac{2}{3}\frac{\sigma_e}{\varepsilon_0}\left(\frac{\varepsilon_0\mathbf{E}}{ln_ee}\right)\cdot\nabla\eta_e' + \frac{1}{\tau_e}\left(\frac{\mathbf{E}}{E}\right)\cdot\mathbf{E}' - \frac{1}{\tau_e}\hat{\nu}_mT_e' + \frac{1}{\tau_e}n_e' \tag{8.142}$$

Similarly, we have for the penultimate term on the right-hand side of Eq. (8.137) using Eq. (8.139):

$$\begin{aligned}\frac{5}{3}\frac{1}{ln_ee}\nabla\cdot\mathbf{j}_e' &= \frac{5}{3}\frac{1}{ln_ee}\left(\frac{en_eD_e}{l}\nabla^2\eta_e' + \sigma_eE\nabla\cdot\mathbf{E}' - \sigma_e\hat{\nu}_m\mathbf{E}\cdot\nabla T_e' + \sigma_e\mathbf{E}\cdot\nabla n_e'\right)\\ &= \frac{5}{3}\frac{D_e}{l^2}\nabla^2\eta_e' + \frac{5}{3}\frac{1}{ln_ee}\sigma_eE\rho_c'\frac{en_+l}{\varepsilon_0E} - \frac{5}{3}\frac{1}{ln_ee}\sigma_e\hat{\nu}_m\mathbf{E}\cdot\nabla\cdot T_e'\\ &\quad + \frac{5}{3}\frac{1}{ln_ee}\sigma_e\mathbf{E}\cdot\nabla n_e'\\ &= \frac{5}{3}\frac{D_e}{l^2}\nabla^2\eta_e' + \frac{5}{3}\frac{n_+}{n_e}\frac{\sigma_e}{\varepsilon_0}\rho_c' - \frac{5}{3}\frac{\sigma_e}{\varepsilon_0}\left(\frac{\varepsilon_0\mathbf{E}}{en_el}\right)\cdot\nabla(\hat{\nu}_mT_e' - n_e')\end{aligned} \tag{8.143}$$

where we have used Eq. (8.116) for n_e'.

Using Eqs. (8.142) and (8.143), Eq. (8.137) becomes

$$\frac{DT'_e}{Dt}+\frac{Dn'_e}{Dt}+\frac{5}{3}\frac{1}{\tau_s}\nabla\cdot\mathbf{v}'_0$$
$$=\frac{2}{3}\frac{\sigma_e}{\varepsilon_0}\left(\frac{\varepsilon_0\mathbf{E}}{ln_e}\right)\cdot\nabla\eta'_e+\frac{2}{\tau_e}\left(\frac{\mathbf{E}}{E}\right)\cdot\mathbf{E}'-\frac{1}{\tau_e}\hat{\nu}_m T'_e+\frac{1}{\tau_e}n'_e-\frac{1}{\tau_e}n'_e$$
$$-\frac{1}{\tau_e}(1+\hat{\nu}_u)T'_e+\frac{2e\lambda_e}{3n_e kl^2}\nabla^2 T'_e+\frac{5}{3}\frac{\sigma_e}{\varepsilon_0}\left(\frac{\varepsilon_0\mathbf{E}}{en_e l}\right)\cdot\nabla T'_e$$
$$+\frac{5}{3}\frac{D_e}{l^2}\nabla^2\eta'_e+\frac{5}{3}\frac{n_+}{n_e}\frac{\sigma_e}{\varepsilon_0}\rho'_c-\frac{5}{3}\frac{\sigma_e}{\varepsilon_0}\left(\frac{\varepsilon_0\mathbf{E}}{en_e l}\right)\cdot\nabla(\hat{\nu}_m T'_e-n'_e) \tag{8.144}$$

Following Haas we now define Ω'_e by

$$\Omega'_e\equiv 2\left(\frac{\mathbf{E}}{E}\right)\cdot\mathbf{E}'+n'_e-n'-\hat{\nu}_m T'_e+\frac{2}{3}\tau_e\left(\frac{\sigma_e}{\varepsilon_0}\right)\frac{\varepsilon_0\mathbf{E}}{en_e l}\nabla\eta'_e \tag{8.145}$$

so that Eq. (8.144) becomes

$$\frac{DT'_e}{Dt}+\frac{Dn'_e}{Dt}+\frac{5}{3}\frac{1}{\tau_s}\nabla\cdot\mathbf{v}'_0$$
$$=\frac{1}{\tau_e}\Omega'_e-\frac{1}{\tau_e}n'_e+\frac{1}{\tau_e}n'-\frac{1}{\tau_e}(1+\nu_u)T'_e$$
$$+\frac{2}{3}\frac{e\lambda_e}{n_e kl^2}\nabla^2 T'_e+\frac{5}{3}\frac{\sigma_e}{\varepsilon_0}\left(\frac{\varepsilon_0\mathbf{E}}{en_e l}\right)\cdot\nabla(T'_e-\hat{\nu}_m T'_e+n'_e)$$
$$+\frac{5}{3}\frac{D_e}{l^2}\nabla^2\eta'_e+\frac{5}{3}\frac{n_+}{n_e}\frac{\sigma_e}{\varepsilon_0}\rho'_c \tag{8.146}$$

Finally, we use Eq. (8.117) for the charge fluctuation to eliminate ρ'_c in the last term in Eq. (8.146):

$$\frac{DT'_e}{Dt}+\frac{Dn'_e}{Dt}+\frac{5}{3}\frac{1}{\tau_s}\nabla\cdot\mathbf{v}_0$$
$$=\frac{1}{\tau_e}\Omega'_e+\frac{2}{3}\frac{e\lambda_e}{n_e kl^2}\nabla^2 T'_e+\frac{5}{3}\left(1-\frac{\sigma_e}{\sigma}\right)\frac{D_e}{l^2}\nabla^2\eta'_e$$
$$-\frac{5}{3}\frac{n_+}{n_e}\frac{\sigma_e}{\sigma}\frac{D\rho'_c}{Dt}-\frac{1}{\tau_e}n'_e+\frac{1}{\tau_e}n'-\frac{1}{\tau_e}(1+\hat{\nu}_u)T'_e$$

$$+\frac{5}{3}\frac{\sigma_e}{\sigma}\left(\frac{\varepsilon_0\mathbf{E}}{en_e l}\right)\cdot\boldsymbol{\nabla}\left\{-\frac{\sigma_+}{\sigma}\rho_c'+\left(1-\frac{\sigma_e}{\sigma}-\frac{n_e}{n_+}\frac{\sigma_+}{\sigma}\right)n_e'\right.$$

$$\left.-\left(\frac{\sigma_-}{\sigma}+\frac{n_-}{n_+}\frac{\sigma_+}{\sigma}\right)n_-'+\left[1-\left(1-\frac{\sigma_e}{\sigma}\right)\hat{\nu}_m\right]T_e'-\frac{\sigma_i}{\sigma}\hat{\sigma}_i T'\right\} \tag{8.147}$$

which is the equation given by Haas [his Eq. (32)].

8.3.8. The Electromagnetic Field Equations

Subjecting the equations (8.24) for the electromagnetic field to the perturbations given by Eqs. (8.59) and (8.60), and neglecting the spatial derivatives of steady-state variables, we have, in terms of the dimensionless spatial parameter defined by Eq. (8.45):

$$\frac{\varepsilon_0 E}{en_+ l}\boldsymbol{\nabla}\cdot\mathbf{E}'=\rho_e' \tag{8.148}$$

$$\boldsymbol{\nabla}\times\mathbf{E}'=0 \tag{8.149}$$

8.4. Stability Analysis

Having obtained the equations for the perturbed variables, a stability analysis may be performed as described in Chapter 7. Haas notes that the characteristic times (Table 23) are spread over a large range, and consequently the general dispersion relation can be partitioned into a set of three relatively independent dispersion relations of order three or less. These can be solved to obtain stability criteria. We shall not present the detailed analysis required to obtain the stability criteria; the results of the analysis are presented and discussed by Haas. We remark that a computer solution of the full dispersion relation would be possible. We also remark here that a possible alternative approach to the stability problem, especially for realistic gases, would be to solve the equations of the general plasma model numerically and then introduce numerical perturbations to determine stability stemming from whether or not the perturbations are amplified during the continued solution of the equations.

9

Devices

9.1. Unstable Optical Resonators

9.1.1. Nomenclature*

Consider an optical wave propagating along the $+z$-direction, so that it may be written as

$$u(x, y, z, t) = \mathrm{Re}[\tilde{u}(x, y)\, e^{i(\omega t - \beta z)}] \tag{9.1}$$

The basic problem of optical-beam propagation is: Given the complex wave amplitude $\tilde{u}_0(x_0, y_0)$ across an input plane z_0, find the complex amplitude and phase $\tilde{u}(x, y)$ of the wave across any later output plane, z. The most common wave used in analysis is one having a Gaussian variation in amplitude across the wavefront:

$$\begin{aligned} \tilde{u}(x, y) &= \left(\frac{2}{\pi}\right)^{1/2} \frac{1}{\omega} \exp\left(-\frac{i\pi}{\lambda} \frac{x^2 + y^2}{\tilde{q}}\right) \\ \frac{1}{\tilde{q}} &= \frac{1}{R} - i\frac{\lambda}{\pi\omega^2} \end{aligned} \tag{9.2}$$

where R is the radius of the spherical wave and ω is the spot size.

Consider a Gaussian light beam with spot size ω_0 at $z = 0$, spot sizes ω_1 and ω_2, and radii of curvature R_1 and R_2 at two other arbitrary planes $z = z_1$ and $z = z_2$ (see Fig. 46). If two curved mirrors are placed at the planes z_1 and z_2, with radii of curvature exactly matching the beam wavefronts, but with diameters considerably larger than the beam spot sizes, these mirrors will reflect the Gaussian beam directly back on itself at each end. In this way, the beam will be forced to bounce back and forth between the two mirrors forming a standing wave. That is to say, this trapped Gaussian beam will form a standing wave *resonant mode* in the optical cavity formed by the two mirrors. Equation (9.2) is the simplest

* See A. E. Siegman,[16] Chapter 8.

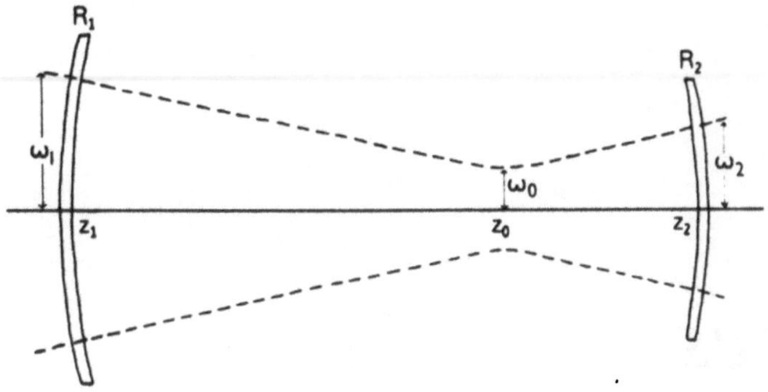

Fig. 46. A stable optical resonator, with mirror separation L.

waveform calculated by Huygens' principle; in general there is a doubly infinite set of waves, the higher-order Hermite–Gaussian waves, which will give rise to higher-order transverse modes in optical resonators.[16] We note that the wavefront curvature and spot-size parameters of Eq. (9.2) are independent of the mode indices m and n. Most laser devices tend to oscillate not only in higher-order modes, but in many such modes at once. Single-mode oscillation, which is limited to the usually desired lowest-order mode, can be obtained either by making the gain strongest on the axis, or by increasing the losses for the higher-order modes. If geometrical ray optics predicts that a ray, not parallel to the z-axis, is more greatly displaced from the z-axis every round-trip (so that the ray eventually leaves the sides of the resonator system), then the resonator is said to be unstable.

The mode in an unstable resonator consists, in general, of two opposite diverging spherical waves, as shown in Fig. 47. The resonator loses energy in propagating out past the outer edges of one or both mirrors.

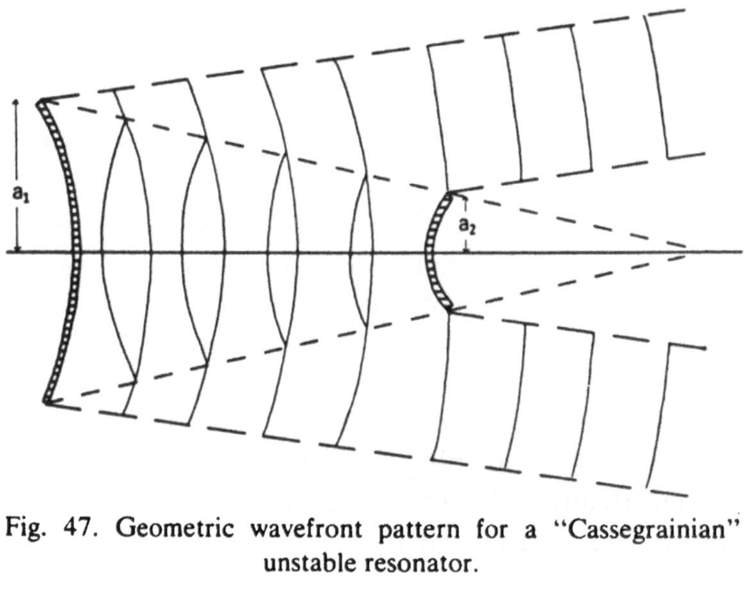

Fig. 47. Geometric wavefront pattern for a "Cassegrainian" unstable resonator.

This loss mechanism can be regarded as the useful laser output beam; it is called "diffraction coupling." Although the output mode is an annular ring in the *near field*, this transforms in the *far field*[103] into a single main beam with no hole on the axis. With proper design, the losses for the higher-order modes in unstable optical resonators can be made larger than the losses for the lowest-order mode. Then a laser with an unstable resonator can have a good transverse mode selection even with a large Fresnel number, something difficult to obtain with the usual stable Gaussian modes.

For a resonator to be unstable, its g-parameters,

$$g_i \equiv 1 - L/R_i, \qquad i = 1, 2 \tag{9.3}$$

must lie in the regions[16]

$$g_1 g_2 > 1 \qquad \text{(positive branch)} \tag{9.4}$$

$$g_1 g_2 < 0 \qquad \text{(negative branch)} \tag{9.5}$$

In Fig. 47 we present the geometric wavefront pattern for the type I ($g_1 g_2 > 1$, $g_1 > 0$), positive-branch unstable resonator. In a confocal resonator, the focal points of the mirrors coincide (recall $R = 2f$). The Fresnel number, N_F, of a mirror of radius "a," which is reflecting radiation of wavelength λ, is

$$N_F = a_1^2 / L\lambda \tag{9.6}$$

9.1.2. *Wave Propagation in Free Space Using Finite-Difference Methods*[18]

Under stationary conditions the electromagnetic-field distribution between the resonator mirrors is time independent. Consequently, subject to boundary conditions, the field distribution can be represented by the equation

$$(\nabla^2 + k^2) v(x, y, z) = 0 \tag{9.7}$$

where z is normal to the resonator mirrors, $k = 2\pi/\lambda$, where λ is the wavelength, and v is the complex amplitude. In order to reduce Eq. (9.7) to a form suitable for a finite-difference scheme, a constant phase, which depends only on the value of z, is removed from the distribution:

$$v(x, y, z) = u(x, y, z)\, e^{ikz} \tag{9.8}$$

Both the amplitude distribution and the phase variation in the x–y plane remain unchanged. The phasor u is again a complex quantity satisfying

$$\frac{\partial^2 u}{\partial z^2} e^{ikz} + 2ik\, e^{ikz} \frac{\partial u}{\partial z} + e^{ikz} \left(\frac{\partial^2}{\partial x^2} + \frac{\partial^2}{\partial y^2} \right) u = 0 \tag{9.9}$$

It is found that u is very weakly dependent on z, and so we can neglect the first term; hence Eq. (9.9) becomes

$$2ik\frac{\partial u}{\partial z}+\left(\frac{\partial^2}{\partial x^2}+\frac{\partial^2}{\partial y^2}\right)u=0 \tag{9.10}$$

which is the form of the diffusion equation. We note that neglecting u_{zz} has changed the wave equation from elliptic to parabolic. The stability condition of the usual explicit method[104]

$$0<\Delta t/[k(\Delta x)^2]<\tfrac{1}{2}$$

imposes a burdensome limitation on the numerical treatment of parabolic differential equations, since a moderately fine spatial division may require an immoderately fine "temporal" division.

Du Fort and Frankel[105] have introduced an alternative difference equation, making use of a diagonal lattice (see Fig. 48) in the z–x and z–y planes, by omitting those mesh points for which $m+j$ and $m+l$, say, are odd, where

$$z=m\,\Delta z, \qquad x=j\,\Delta x, \qquad \text{and} \quad y=l\,\Delta u \tag{9.11}$$

The term u_z is defined by

$$u_z\equiv\frac{1}{2\,\Delta z}(u_{j,l}^{m+1}-u_{j,l}^{m-1})+T_z \tag{9.12}$$

where T_z is the truncation error of the equation. In the representation of u_{xx} by a second difference with respect to j, the end terms $u_{j-1,l}^{m}$ and $u_{j+1,l}^{m}$ may be used as in the usual explicit finite-difference formula, but the midpoint term, $-2u_{j,l}^{m}$, is now omitted from the lattice and is replaced by

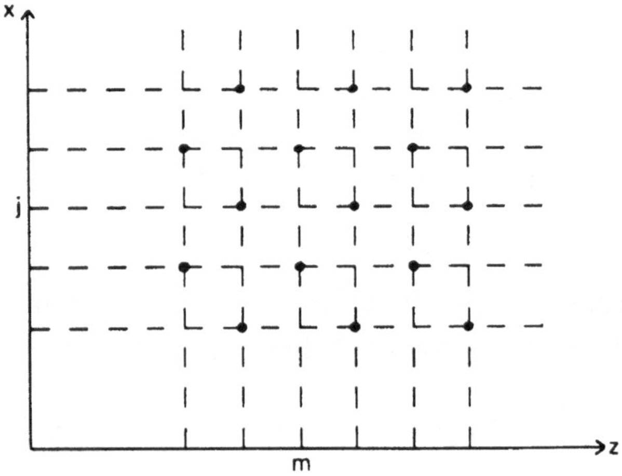

Fig. 48. Du Fort and Frankel diagonal lattice.

the mean of the two terms of neighboring j values, that is,

$$-2u_{j,l}^{m} \approx -u_{j,l}^{m+1} - u_{j,l}^{m-1} \tag{9.13}$$

Hence

$$u_{xx} = \frac{1}{(\Delta x)^2}(u_{j+1,l}^{m} - u_{j,l}^{m+1} - u_{j,l}^{m-1} + u_{j-1,l}^{m}) + T_x \tag{9.14}$$

and similarly,

$$u_{yy} = \frac{1}{(\Delta y)^2}(u_{j,l+1}^{m} - u_{j,l}^{m+1} - u_{j,l}^{m-1} + u_{j,l-1}^{m}) + T_y \tag{9.15}$$

Equations (9.12), (9.14), and (9.15) are substituted into Eq. (9.10) to give the difference equation

$$u_{j,l}^{m+1} - u_{j,l}^{m-1} + \frac{2\,\Delta z}{2ik(\Delta x)^2}(u_{j+1,l}^{m} - u_{j,l}^{m+1} - u_{j,l}^{m-1} + u_{j-1,l}^{m})$$
$$+ \frac{2\,\Delta z}{2ik(\Delta x)^2}(u_{j,l+1}^{m} - u_{j,l}^{m+1} - u_{j,l}^{m-1} + u_{j,l-1}^{m}) = 0$$

If we define

$$\beta \equiv \frac{\Delta z}{k(\Delta x)^2} \quad \text{and} \quad \gamma \equiv \frac{\Delta z}{k(\Delta y)^2} \tag{9.16}$$

then the difference equation becomes

$$u_{j,l}^{m+1}(1 + i\beta + i\gamma) = u_{j,l}^{m-1}(1 - i\beta - i\gamma) + i\beta(u_{j+1,l}^{m} + u_{j-1,l}^{m})$$
$$+ i\gamma(u_{j,l}^{m} + u_{j,l-1}^{m})$$

which, upon rearrangement, becomes

$$\begin{aligned} &u_{j,l}^{m+1}(1 + \beta^2 + \gamma^2 + 2\beta\gamma) \\ &\quad = u_{j,l}^{m-1} + i\beta(u_{j+1,l}^{m} - 2u_{j,l}^{m-1} + u_{j-1,l}^{m}) + \beta^2(u_{j+1,l}^{m} - u_{j,l}^{m-1} + u_{j-1,l}^{m}) \\ &\quad + \beta\gamma(u_{j+1,l}^{m} - u_{j,l}^{m-1} + u_{j-1,l}^{m}) + \beta\gamma(u_{j,l+1}^{m} - u_{j,l}^{m-1} + u_{j,l-1}^{m}) \\ &\quad + i\gamma(u_{j,l+1}^{m} - 2u_{j,l}^{m-1} + u_{j,l-1}^{m}) + \gamma^2(u_{j,l+1}^{m} - u_{j,l}^{m-1} + u_{j,l-1}^{m}) \\ &\quad + \text{truncation error terms} \end{aligned} \tag{9.17}$$

Equation (9.17) allows the calculation of the field at $z + \Delta z$ from known field values at z and $z - \Delta z$.

In each cycle of the calculation, $u_{j,l}^m$ may be evaluated for all j and l values associated with one m value, even j or l occurring in one cycle, odd j or l in the next—the "leap-frog" method. In the "pyramid" method, the $u_{j,l}^m$ may be evaluated in each cycle for all j along a diagonal line $m+j=$ const, or $m-j=$ const.

Problem 8. Show that

$$T_z + T_x + T_y = \frac{i}{k} u_{zz}\left[\left(\frac{\Delta z}{\Delta x}\right)^2 + \left(\frac{\Delta z}{\Delta y}\right)^2\right] + \frac{1}{3} u_{zzz}(\Delta z)^2 - \frac{i}{12k}\left[\frac{\partial^4 u}{\partial x^4}(\Delta x)^2 + \frac{\partial^4 u}{\partial y^4}(\Delta y)^2\right] \tag{9.18}$$

The errors due to the partial derivatives in Eq. (9.18) can be made arbitrarily small by decreasing the values of Δx, Δy, and Δz as required. For the second and third terms on the right-hand side, the three increments can be reduced independently of each other, but in the first term Δz must decrease faster than Δx and Δy. In other words,

$$\Delta z(\Delta x, \Delta y)$$

If, on the other hand, $\Delta z/\Delta x$ and $\Delta z/\Delta y$ are kept constant, say equal to ε, then Eq. (9.17) is consistent, not with the diffusion equation (9.10) but with the hyperbolic equation

$$2ik\frac{\partial u}{\partial z} + \left(\frac{\partial^2}{\partial x^2} + \frac{\partial^2}{\partial y^2}\right)u + \frac{2i\varepsilon^2}{k}\frac{\partial^2 u}{\partial z^2} = 0 \tag{9.19}$$

We can ensure that the first term will not contribute significantly if we impose

$$\Delta z < k(\Delta x)^2 \tag{9.20}$$

One drawback of this type of finite-difference technique is its inability to treat discontinuous distributions. Such a distribution is obtained in the reflectivity of a mirror possessing sharp edges. What the finite difference scheme sees at the mirror edge is a ramp of width Δx, say, where the reflection coefficient has dropped from a value of about unity to zero across one segment. As Δx is decreased, then at a particular value instabilities will be introduced into the solution. In order to overcome this difficulty Rensch has tapered the mirror edges with a Gaussian profile, where the intensity reflection coefficient varies as

$$R = R_0 \exp\{-[(x-x_0)^2 + (y-y_0)^2]/\tau^2\} \tag{9.21a}$$

for

$$|x| \geqslant x_0, \quad -a \leqslant y \leqslant a; \qquad |y| \geqslant y_0, \quad -a \leqslant x \leqslant a \tag{9.21b}$$

where R_0 is the uniform reflectivity of the mirror, x_0 and y_0 are the mirror radii measured from the mirror center for the portion of the mirror with uniform reflectivity R_0, and τ is the truncation distance. The position of x_0 is determined by

$$a = x_0 + (0.7)^{1/2}\tau \tag{9.22}$$

where $2a$ is the actual mirror diameter and $\tau(0.7)^{1/2}$ is added so that the effective mirror diameter is measured between the half-intensity points of the Gaussian edges. For

$$\tau/\Delta x = \tau/\Delta y \geqslant 2 \tag{9.23}$$

Rensch asserts that the error terms can be made arbitrarily small in the region of the mirror edge.

From Eq. (9.17), in order to calculate the value of u at $z+\Delta z$, values are needed at z and $z-\Delta z$. Thus a starting procedure has to be used to deduce conditions at $z=\Delta z$ from those at $z=0$. We can replace Eq. (9.12) by the forward-difference formula

$$u_z = \frac{1}{\Delta z}(u_{j,l}^{1} - u_{j,l}^{0}) \tag{9.24}$$

and replace Eqs. (9.14) and (9.15) by the standard explicit formula

$$u_{xx} = \frac{1}{(\Delta x)^2}(u_{j+1,l}^{0} - 2u_{j,l}^{0} + u_{j-1,l}^{0}) \tag{9.25}$$

so that for the first step the finite-difference representation of the diffusion equation will be

$$u_{j,l}^{1} = u_{j,l}^{0} + \frac{1}{2}[\beta(u_{j+1,l}^{0} - 2u_{j,l}^{0} + u_{j-1,l}^{0}) + \gamma(u_{j,l+1}^{0} - 2u_{j,l}^{0} + u_{j,l-1}^{0})] \tag{9.26}$$

9.1.3. Coordinate Systems

In Fig. 49 we present a schematic of an unstable confocal resonator. The dotted curves represent spherical wavefronts appearing to emanate from the common focus F, which are reflected at the back mirror, radius a_1, into plane waves represented by dot–dash lines. The field width PQ is subdivided into mesh points Δx apart; typical dimensions are $PQ \sim 15$ cm, $L \sim 166$ cm, and $R_1 \sim 630$ cm.

If the electric field on PQ has constant amplitude and phase,

$$u = A\,e^{i\theta}$$

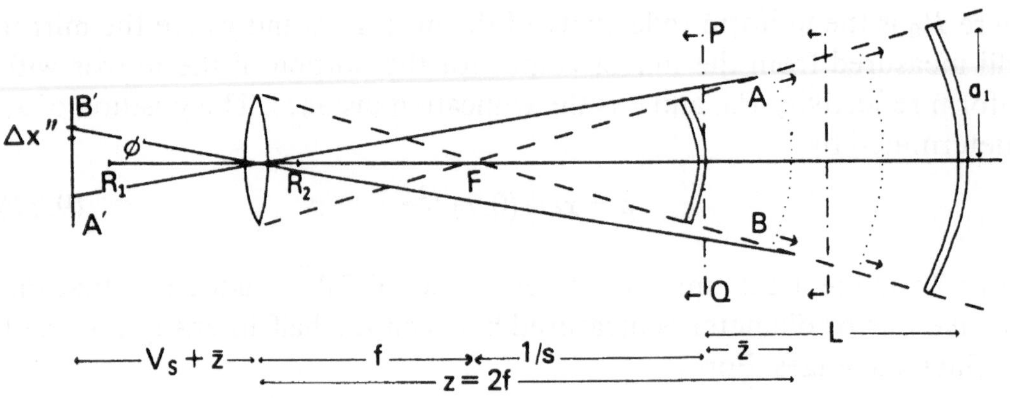

Fig. 49. Wavefronts in an unstable confocal resonator.

then upon reflection the phase is changed by varying amounts depending on the distance from the axis (see Fig. 50):

$$\theta_i = (2z_i \cdot 2\pi)/\lambda \tag{9.27}$$

Consequently, there will be increasing variation in u from one mesh step to the next as we move away from the axis. In practice it is found that for low Fresnel numbers, $N_F \sim 10$, the number of mesh points across the field width should be $N_x \sim 100$.

As N_F increases [see Eq. (9.6)], which is equivalent to increasing a_1, the pathlengths z_i are increased, and so Δx must be decreased to maintain the same order of accuracy in u across the field width as for $N_F \sim 10$. It is found[106] that

$$\text{for} \quad N_F \sim 40, \quad N_x \sim 200+$$

Alternatively, increasing N_F is equivalent to decreasing L, which implies decreasing R_1 since

$$R_1 = 2mL/(m-1) \tag{9.28}$$

where m is the magnification imposing a decrease on R_2, leading to an increase in the z_i of Fig. 50 and requiring a smaller Δx, or equivalently

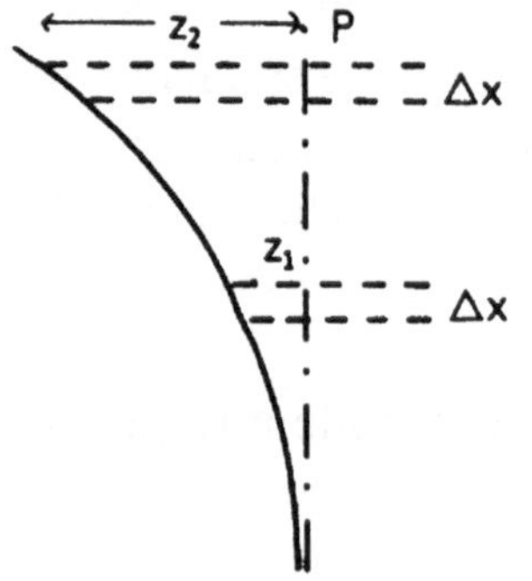

Fig. 50. Pathlengths at the output mirror.

higher N_x. Since lasers with $N_F \sim 360$ have been constructed, it is evident that three-dimensional computer modeling using a Cartesian mesh in the physical frame of reference would require huge N_x and N_y.

Rensch[18] has proposed a variable coordinate system, S', to remove the dependence of the number of mesh points on the mirror curvature. The objective of Rensch's method is to transform the propagation of the spherical wave at each point from the output mirror to the back mirror into the propagation of a plane wave in the variable coordinate frame. To derive this transformation, we imagine a lens placed to the left of F in Fig. 49, with F as its focus. Let z be the distance of an "object" in the physical space between the mirrors from the lens, and let z' be the distance of the image from the lens. In the paraxial approximation the spherical wave AB will be transformed into the plane wave $A'B'$.

From the thin-lens formula,

$$1/z + 1/z' = 1/f$$

we have

$$z' = zf/(z-f) \tag{9.29}$$

Hence

$$x/z = x'/z'$$

When Eq. (9.29) is differentiated,

$$dz'/dz = f^2/(z-f)^2$$

which leads to

$$\frac{c}{dz/dt} = \frac{f^2}{(z-f)^2}\frac{c}{dz'/dt}$$

where c is the velocity of light *in vacuo*. If n' denotes the refractive index in frame S', then

$$n' \equiv \frac{c}{dz'/dt} = \frac{(z-f)^2}{f^2}n = \frac{f^2}{(z'-f)^2}n \tag{9.30a}$$

which shows that there exists a nonisotropic medium in the S' frame, since we have $n'(z')$.

A new coordinate frame S'' is defined in which the medium is isotropic:

$$n'' = \frac{(z'-f)^2}{f^2}n' = n \tag{9.30b}$$

which implies

$$\frac{c}{dz''/dt} = \frac{z'-f}{f} \frac{c}{dz'/dt}$$

or

$$dz'' = \left(\frac{f}{z'-f}\right)^2 dz'$$

Hence

$$z'' = -\frac{fz'}{z'-f} = -z$$

$$x'' = x' \quad \text{and} \quad y'' = y' \tag{9.31}$$

Rensch postulates that the electric field in the S'' frame, namely v'', is given as the solution of

$$\left(\frac{\partial^2}{\partial x''^2} + \frac{\partial^2}{\partial y''^2} + \frac{\partial^2}{\partial z''^2} + n''k^2\right) v'' = 0 \tag{9.32}$$

The transformation law for the electric-field amplitude is found from the conservation of beam power:

$$\iint |u''|^2 \, dx'' \, dy'' = \iint |u'|^2 \, dx' \, dy' = \iint |u|^2 \, dx \, dy \tag{9.33}$$

From Eq. (9.29) we have

$$dx' = f/(z-f)\, dx \quad \text{and} \quad dy' = f/(z-f)\, dy \tag{9.34}$$

Hence

$$\iint |u'|^2 \, dx' \, dy' = \iint |u'|^2 \left(\frac{f}{z-f}\right)^2 dx \, dy$$

which upon comparing with Eq. (9.33) gives

$$u = u' \frac{f}{z-f} = u'' \frac{f}{z-f} \tag{9.35}$$

having used Eq. (9.33). Alternatively, we can write Eq. (9.35) as

$$u'' = \frac{z-f}{f} u \tag{9.36}$$

The procedure for using S'' is to extract the phase $e^{ikz''}$ from the solution to Eq. (9.32), thus leading to the analog of the parabolic equation (9.10) with the double-primed variables whose finite-difference representation will be Eq. (9.17) with u'' instead of u. We remove the double prime from the quantities in the analog to Eq. (9.17) by noting that at z and $z \pm \Delta z$, where $z = 2f$:

$$(u'')^m = \frac{2f-f}{f} u^m = u^m$$
$$(u'')^{m\pm 1} = \frac{2f \pm \Delta z - f}{f} u^{m\pm 1} \tag{9.37}$$
$$= \frac{f \pm \Delta z}{z} u^{m\pm 1}$$

If we define

$$A \equiv \frac{f-\Delta z}{f} \tag{9.38}$$

then

$$(u'')^{m-1} = A u^{m-1} \tag{9.39}$$

and

$$(u'')^{m+1} = \left(1 + \frac{\Delta z}{f}\right) u^{m+1} \tag{9.40}$$

When Eqs. (9.37), (9.39), and (9.40) are substituted into the double-primed analog of Eq. (9.17), we have

$$\begin{aligned}\left(1+\frac{\Delta z}{f}\right) & u_{j,l}^{m+1}(1+\beta^2+\gamma^2+2\beta\gamma) \\ &= Au_{j,l}^{m-1} + i\beta(u_{j+1,l}^m - 2Au_{j,l}^{m-1} + u_{j-1,l}^m) + \beta^2(u_{j+1,l}^m - Au_{j,l}^{m-1} + u_{j-1,l}^m) \\ &\quad + \beta\gamma(u_{j+1,l}^m - Au_{j,l}^{m-1} + u_{j-1,l}^m) + \beta\gamma(u_{j,l+1}^m - Au_{j,l}^{m-1} + u_{j,l-1}^m) \\ &\quad + i\gamma(u_{j,l+1}^m - 2Au_{j,l}^{m-1} + u_{j,l-1}^m) + \gamma^2(u_{j,l+1}^m - Au_{j,l}^{m-1} + u_{j,l-1}^m)\end{aligned} \tag{9.41}$$

When the factor multiplying $u_{j,l}^{m+1}$ is taken to the right-hand side of the equation, and we use

$$\left(1+\frac{\Delta z}{f}\right)^{-1} \approx 1 - \frac{\Delta z}{f} = A \tag{9.42}$$

we have precisely Rensch's form [Reference 18, Eq. (26)] of the difference equation for the step-by-step integration from the output mirror to the back mirror.

In the variable coordinate system, f, $\Delta x''$, and $\Delta y''$ are functions of z. Let V_s be the distance measured from the output mirror to the virtual source (see Fig. 49); then

$$f = V_s + \bar{z} \tag{9.43}$$

From Fig. 49 we see that

$$\Delta x'' \approx (V_s + \bar{z})\,\Delta\phi = (V_s + \bar{z})\frac{\overline{\Delta x}}{V_s}$$

$$\approx \left(1 + \frac{\bar{z}}{V_s}\right)\overline{\Delta x} \tag{9.44}$$

where $\overline{\Delta x}$ is the mesh size at the output mirror when $\bar{z} = 0$. Similarly we have

$$\Delta y'' \approx \left(1 + \frac{\bar{z}}{V_s}\right)\overline{\Delta y} \tag{9.45}$$

After reflection has occurred at the back mirror the return propagation is performed using Eq. (9.17), with the final mesh sizes $m\,\Delta x$ and $m\,\Delta y$, where m is the geometric magnification of the resonator. At the end of the return trip at the output mirror in the S frame, the N_xN_y field values that define the beam at each cell of width $m\,\Delta x$ and $m\,\Delta y$ must be replaced with those representing the beam at each mesh of widths $\overline{\Delta x}$ and $\overline{\Delta y}$. This is done by interpolation before Eq. (9.41) is used to propagate from the output mirror to the back mirror in the frame S''.

9.1.4. Gas Dynamic Lasers

To take into account the interaction of the propagating electric field with a laser medium, the finite-difference representation of the preceding sections must be corrected for the local gain and index of refraction at each step Δz. Consider an electric field propagating through a medium with complex refractive index η; then Eq. (9.7) is replaced by (in one dimension)

$$\left(\frac{\partial}{\partial z^2} + \eta^2 k^2\right) v(z) = 0 \tag{9.46}$$

where we can demonstrate the real and imaginary (absorption and emission) parts of the refractive-index explicitly:

$$\eta k = nk + ig/2 \tag{9.47}$$

The solution of Eq. (9.46) is

$$v(z) = v_0\, e^{-i(nk + ig/2)z}$$

where v_0 is the initial value of the electric field at $z = 0$. After one mesh step,

$$v(z + \Delta z) = v(z)\, e^{-ink\Delta z + g\Delta z/2} \tag{9.48}$$

From Eq. (9.8), in one dimension, we have

$$v_f(z+\Delta z) = v_f(z)\, e^{-ik\Delta z}$$

where the subscript f denotes the free-space solution. When we take the ratio of the two preceding equations and

$$v(z)/v_f(z) = 1$$

then

$$v(z+\Delta z) = v_f(z+\Delta z)\, e^{-i(n-1)k\Delta z + g\Delta z/2}$$

which in three dimensions is written as

$$v(x, y, z+\Delta z) = v_f(x, y, z+\Delta z)\, e^{-i[n(x,y)-1]k\Delta z + g(x,y)\Delta z/2} \tag{9.49}$$

To perform the calculation one begins with an arbitrary initial amplitude. Since the process of integrating back and forth between the mirrors is an iterative procedure, the closer one can guess the initial amplitude to the final result the faster will be the convergence. In Fig. 51 we present a schematic of the axial segments along the direction of propagation of the light, but perpendicular to the direction of gas flow. The initial complex wave $v(x, y, 0)$ located between segment 1 and the output mirror is reflected from the output mirror and then propagated across segment 1 using Eq. (9.41), the free-space propagation algorithm, to give $v_f(x, y, \Delta z)$. The gain $g(x, y)$ and the refractive index $n(x, y)$ are calculated in segment 1 and the medium-corrected complex field $v(x, y, \Delta z)$ is calculated using Eq. (9.49). The intensity of the corrected complex field is stored for use in calculating the gain in segment 1 on the return pass. In other words in the calculation of $g(x, y)$ one uses the sum of the right and left traveling field

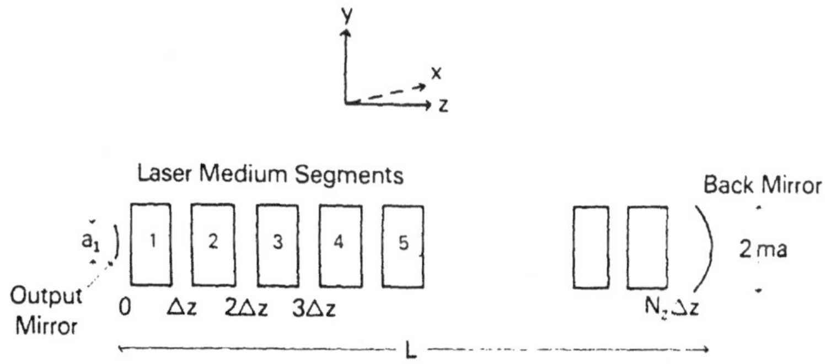

Fig. 51. Schematic of the N_z segments across each of which the gain and refractive index are assumed uniform.

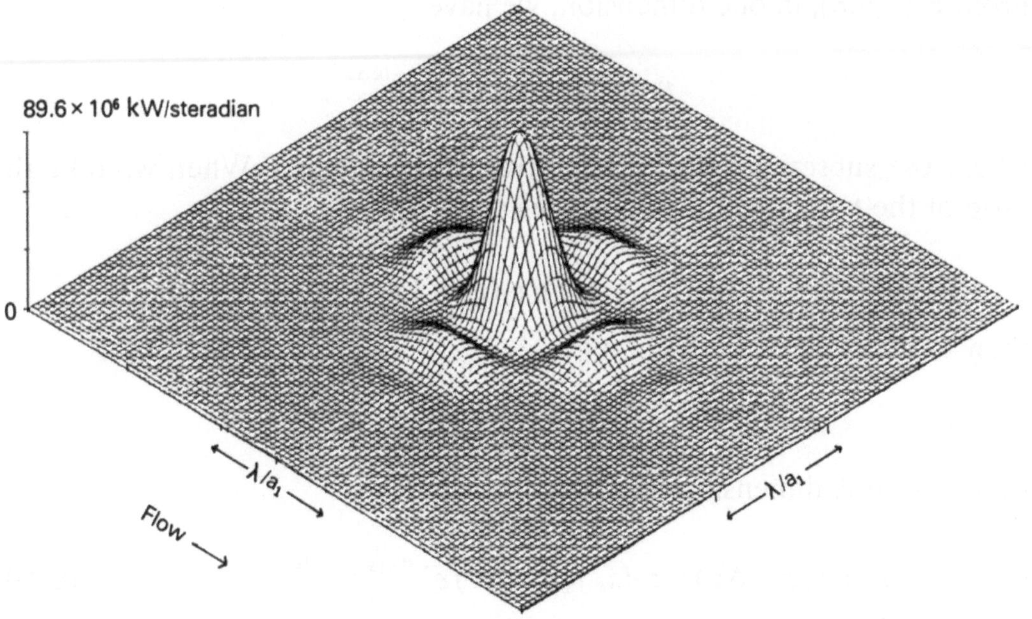

Fig. 52. Far-field intensity for the following parameters: $N_f = 60$, $M = 2.5$, $N_x = 61$, $N_y = 31$, $N_z = 88$, $\Delta x/\tau = 32$, $\beta = 0.1003$, $a_1 = 2.5219$ cm, $L = 100.0$ cm, $\lambda = 0.00106$ cm, for square mirrors.

intensities. The procedure for determining the complex field at the end of segment 2 is to use the medium corrected complex field in the free-field propagation finite-difference equation (9.41).

To calculate the index of refraction for each segment one needs a model; neither Rensch[18] nor Jones[106] describe their model. In gas dynamic lasers, where one can expect substantial density changes in the medium, then one must ensure that the refractive index accurately represents such changes (see Born and Wolf[107]).

In Section 2.7 we described the kinetic model used by Rensch. For example, Eq. (2.70b) can be written as

$$\frac{dE_1}{dt} = \frac{dx}{dt}\frac{dE_1}{dx} = V\frac{dE_1}{dx} = \nu_1 \,\Delta N\, WI_{\nu_0}(x) + \frac{\nu_1}{\nu_3}\frac{E_3 - E_3(T, T_1, T_2)}{\tau_3(T, T_1, T_2)} - \frac{E_1 - E_1(T)}{\tau_{10}(T)} - \frac{E_1 - E_1(T_2)}{\tau_{12}(T_2)} \qquad (9.50)$$

which is precisely the form used by Jones.[108] We have neglected the electron-pumping term of Eq. (2.70b) as the population inversion is brought about by supersonic expansion through a nozzle. The local gain of the medium is given by Eq. (2.75), where $\Delta N(x)$ is the population inversion.

In Fig. 52 we present Jones' results[108] for the far-field intensity, calculated as the two-dimensional Fourier transform of the near field.[103]

9.2. *Waveguide Lasers*[109]

9.2.1. *Introduction*

The waveguide laser (Fig. 53) is constructed from a capillary tube made, for example,[110] from Pyrex and containing a $CO_2:N_2:He$ mixture. An applied electric field **E**, in the direction of the tube axis, maintains a self-sustained discharge. The free electrons formed in this discharge excite the CO_2 and N_2 vibrational modes, resulting in population inversion between the CO_2 laser levels.

The energy that the electrons lose while vibrationally exciting the CO_2 and N_2 molecules is eventually transferred (through a series of V–V and V–T collisions) to the translational energy of the gas (apart from any energy lost to lasing). Gain decreases with increasing gas temperature, and therefore the capillary tube is immersed in a coolant of flowing liquid which removes heat from the tube. This heat-transfer process results in a temperature distribution across the tube.

The ionization rate at which electrons are created in the self-sustained discharge is a function of E/N and gas temperature. Electrons are lost to the walls of the tube by diffusion. The resulting radial density distribution of electrons affects both the radial temperature distribution and the radial distribution of molecular excited states.

The objective of the model developed in this section is to determine gain as a function of tube radius by solving the kinetic equations. The problem is to determine the gas temperature and electron density profiles. A comprehensive description of the use of these lasers for near-earth space applications has been given by McElroy *et al.*[111]

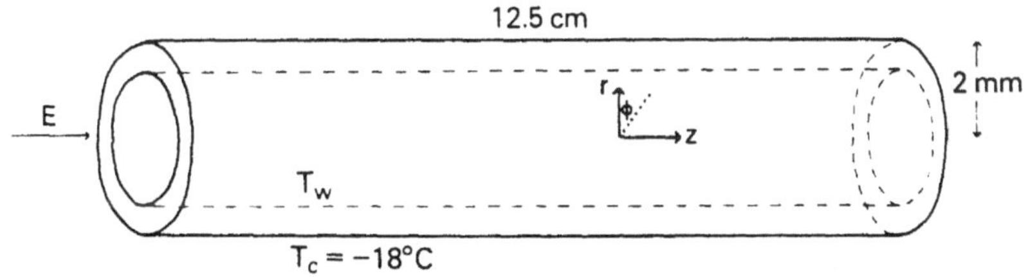

Fig. 53. Schematic of the waveguide laser with representative quantities. T_w, temperature of the inner wall; T_c, temperature of the coolant.

9.2.2. Translational–Rotational and Vibrational Energy Equations

9.2.2.1. Steady-State Form of the Translational–Rotational and Vibrational Energy Equations

The general equation for the translational–rotational energy, $E^{(TR)}(T)$, of a $CO_2:N_2:He$ gas mixture is given by Eq. (5.68) to be

$$\frac{DE^{(TR)}}{Dt}+E^{(TR)}\nabla\cdot\mathbf{v}_0+\nabla\cdot\mathbf{q}^{(TR)}+\hat{P}:(\nabla\mathbf{v}_0)-\sum_i\mathbf{X}_i\cdot\bar{\mathbf{V}}_i$$
$$=n_e\nu^{(TR)}kT_e+\frac{E_1-E_1(T)}{\tau_{10}(T)}+\frac{E_2-E_2(T)}{\tau_{20}(T)}+\left(1-\frac{h\nu_2}{h\nu_3}-\frac{h\nu_1}{h\nu_3}\right)\frac{E_3-E_3(T,T_1^{(V)},T_2^{(V)})}{\tau_3(T,T_1^{(V)},T_2^{(V)})} \tag{9.51}$$

where, on the right-hand side, we have used the explicit form for the rate of change of translational energy in a $CO_2:N_2:He$ mixture due to V–V and V–T collisions, as given by Eq. (2.123). Equation (9.51) contains six unknown dependent variables $E^{(TR)}$, $\mathbf{q}^{(TR)}$, and E_i; alternatively we can consider T, $\mathbf{q}^{(TR)}$, and $T_i^{(V)}$ as the six dependent variables.

The general equations for the energy of the CO_2 and N_2 vibrational modes are given by Eq. (5.55) and the vibrational kinetic terms are given by the right-hand sides of Eqs. (2.70)–(2.73); hence

$$\frac{DE_1}{Dt}+E_1\nabla\cdot\mathbf{v}_0+\nabla\cdot\mathbf{q}_1^{(V)}$$
$$=h\nu_1n_en_{CO_2}X_1-\frac{E_1-E_1^e(T)}{\tau_{10}(T)}-\frac{E_1-E_1^e(T_2^{(V)})}{\tau_{12}(T_2^{(V)})}+\frac{h\nu_1}{h\nu_3}\frac{E_3-E_3^e(T,T_1^{(V)},T_2^{(V)})}{\tau_3(T,T_1^{(V)},T_2^{(V)})}+h\nu_1\,\Delta N\,WI_\nu \tag{9.52}$$

$$\frac{DE_2}{Dt}+E_2\nabla\cdot\mathbf{v}_0+\nabla\cdot\mathbf{q}_2^{(V)}$$
$$=h\nu_2n_en_{CO_2}X_2-\frac{E_2-E_2^e(T)}{\tau_{20}(T)}+\frac{E_1-E_1^e(T_2^{(V)})}{\tau_{12}(T_2^{(V)})}+\frac{h\nu_2}{h\nu_3}\frac{E_3-E_3^e(T,T_1^{(V)},T_2^{(V)})}{\tau_3(T,T_1^{(V)},T_2^{(V)})} \tag{9.53}$$

$$\frac{DE_3}{Dt}+E_3\nabla\cdot\mathbf{v}_0+\nabla\cdot\mathbf{q}_3^{(V)}$$
$$=h\nu_3n_en_{CO_2}X_3-\frac{E_3-E_3^e(T,T_1^{(V)},T_2^{(V)})}{\tau_3(T,T_1^{(V)},T_2^{(V)})}+\frac{E_4-E_4^e(T_3^{(V)})}{\tau_{43}(T)}-h\nu_3\,\Delta N\,WI_\nu \tag{9.54}$$

$$\frac{DE_4}{Dt}+E_4\nabla\cdot\mathbf{v}_0+\nabla\cdot\mathbf{q}_4^{(V)}=h\nu_4 n_e n_{N_2}X_4-\frac{E_4-E_4^e(T_3)}{\tau_{43}(T)} \tag{9.55}$$

The four equations (9.52)–(9.55) provide relationships between $T_i^{(V)}$, $\mathbf{q}_i^{(V)}$, and T. Consequently we now have ten dependent variables T, $T_i^{(V)}$, $\mathbf{q}_i^{(V)}$, and $\mathbf{q}^{(TR)}$ but, as yet, only five equations. The five remaining equations for $\mathbf{q}^{(TR)}$ and $\mathbf{q}_i^{(V)}$ are given by Eqs. (5.136) and (5.142). For the moment we are assuming n_e and I_ν are known.

In Eqs. (9.51)–(9.55), $\mathbf{v}_0$ is the mass-average velocity of a flowing gas and D/Dt is the "time derivative following the motion," given by Eq. (5.14a). We are considering a nonflowing gas in a steady-state condition; therefore we remove time derivatives and set $\mathbf{v}_0$ equal to zero in Eq. (9.51) to obtain

$$\nabla\cdot\mathbf{q}^{(TR)}=\frac{E_1-E_1(T)}{\tau_{10}(T)}+\frac{E_2-E_2(T)}{\tau_{20}(T)}+\left(1-\frac{h\nu_2}{h\nu_3}-\frac{h\nu_1}{h\nu_3}\right)\frac{E_3-E_3^e(T,T_1^{(V)},T_2^{(V)})}{\tau_3(T,T_1^{(V)},T_2^{(V)})} \tag{9.56}$$

We have neglected the last term on the left-hand side of Eq. (9.51), which represents the change in translational energy due to the acceleration of positive and negative ions. We have also neglected the first term on the right-hand side of Eq. (9.51), which represents the gain in the gas translational energy due to elastic electron–molecule collisions. We are justified in making the latter approximation since the large mass difference between electrons and molecules results in a negligible energy transfer in elastic collisions, compared to the energy transfer in inelastic electron–molecule collisions and the energy transfer in molecular V–V and V–T collisions. In numerical calculations using the Boltzmann code (see Section 3.2), the energy transfer due to elastic collisions is found to be typically less than 1% of the energy transfer due to elastic collisions.

With the assumption of a steady state and no flow, the left-hand sides of Eqs. (9.52)–(9.55) become $\nabla\cdot\mathbf{q}_i^{(V)}$, $i=1,2,3,4$. If we now add these equations, we obtain

$$\begin{aligned}\nabla\cdot(\mathbf{q}_1^{(V)}+\mathbf{q}_2^{(V)}+\mathbf{q}_3^{(V)}+\mathbf{q}_4^{(V)})=&\ n_e h(\nu_1 n_{CO_2}X_1+\nu_2 n_{CO_2}X_2+\nu_3 n_{CO_2}X_3\\&+\nu_4 n_{N_2}X_4)\\&-\frac{E_1-E_1^e(T)}{\tau_{10}(T)}-\frac{E_2-E_2^e(T)}{\tau_{20}(T)}-h\nu_L\,\Delta N\,WI_\nu\\&-\left(1-\frac{h\nu_1}{h\nu_3}-\frac{h\nu_2}{h\nu_3}\right)\frac{E_3-E_3^e(T,T_1^{(V)},T_2^{(V)})}{\tau_3(T,T_1^{(V)},T_2^{(V)})}\end{aligned} \tag{9.57}$$

where ν_L is the laser frequency, given by

$$\nu_L = \nu_3 - \nu_1 \tag{9.58}$$

The V–V and V–T terms in Eq. (9.57) and the translational energy (9.56) are eliminated by adding these two equations together to give

$$\nabla \cdot (\mathbf{q}^{(TR)} + \mathbf{q}_1^{(V)} + \mathbf{q}_2^{(V)} + \mathbf{q}_3^{(V)} + \mathbf{q}_4^{(V)})$$
$$= n_e h(\nu_1 n_{CO_2} X_1 + \nu_2 n_{CO_2} X_2 + \nu_3 n_{CO_2} X_3 + \nu_4 n_{N_2} X_4) - h\nu_L \,\Delta N\, W I_\nu \tag{9.59}$$

The sum of the $\mathbf{q}$ on the left-hand side is the total translational–rotational and vibrational energy flux. The first term on the right-hand side of Eq. (9.59) is the rate of energy gain by all four vibrational modes due to electron excitation of vibrational states. The second term on the right-hand side is the rate of loss of energy by the vibrational modes due to lasing. Equation (9.59) therefore states that at any point $\mathbf{x}$, the divergence of the total flux of vibrational and translational–rotational energy equals the rate of gain of vibrational energy by electron excitation minus the rate of loss of vibrational energy due to lasing.

Equations (9.52)–(9.55), with only $\nabla \cdot \mathbf{q}_i^{(V)}$ on their left-hand sides, and Eq. (9.59), which replaces Eq. (9.51), are a set of five equations for the spatial distributions of vibrational and translational energy in terms of divergence of the translational and rotational energy fluxes. The model of the waveguide is completed by providing equations for: (i) the translational and vibrational energy fluxes (Section 9.2.2.2); (ii) the electron number density $n_e(\mathbf{x})$ (Section 9.2.2.3); and (iii) in the case of a laser oscillator, for the cavity-field intensity, $I_\nu(\mathbf{x})$.

9.2.2.2. The Translational–Rotational and Vibrational Energy Fluxes, $\mathbf{q}^{(TR)}(x)$, $\mathbf{q}_i^{(V)}(x)$

9.2.2.2.1. General Considerations Concerning the Transport of Translational and Vibrational Energy by Conduction and Diffusion. We shall see in Section 9.2.2.3 that an approximation to the steady-state radial electron distribution, $n_e(r)$, is a zero-order Bessel function. At the center of the tube the electron number density is at a maximum and therefore it is here that the rate of electron excitation of the vibrational modes will be greatest. There will result a radial distribution of effective vibrational temperatures $T_i^{(V)}(r)$ as shown in Fig. 54. The higher the vibrational temperature $T_i^{(V)}$, the greater the rates of V–T energy transfer from the vibrational modes to the translational energy of the gas. There will result a radial ambient temperature distribution as shown in Fig. 54. Steady-state radial temperature distributions will be attained when, at any point r, the

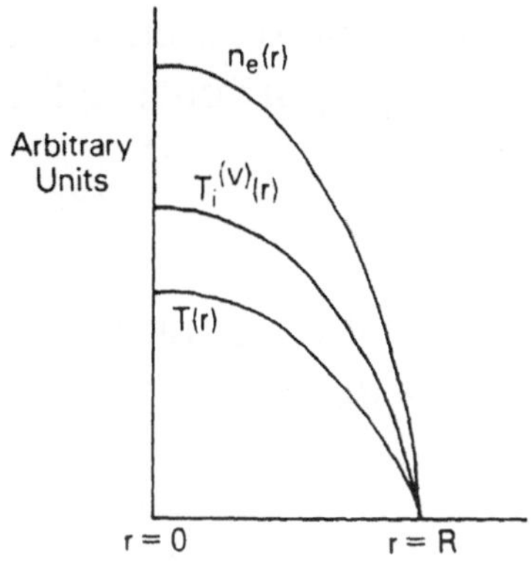

Fig. 54. Radial distributions of n_e, $T_i^{(V)}$, and T.

rate of transport of vibrational and translational energy radially outwards equals the rate at which electrons are transferring vibrational energy to the gas.

We shall now discuss the various mechanisms by which translational and vibrational energy are transported through a gas:

i. Conduction. An ambient temperature gradient in any material (solid as well as gas) results in a flow of heat through the material. The rate of heat flow is proportional to the temperature gradient, and in a gas it is a flux of molecular translational energy due to conduction. Similarly a vibrational temperature gradient in any material (solid or gas) means that in some region there are more vibrationally excited molecules than in a neighboring region. There results a flow of vibrational energy through the material in a direction, which tends to equalize the vibrational temperatures of the two regions. The rate of flow of vibrational energy is the vibrational energy flux due to conduction.

Two such neighboring regions are shown schematically in Fig. 55 with vibrational and ambient temperatures as indicated. The translational–rotational energy flux, $q^{(TR)}$, due to the ambient temperature gradient, is proportional to the ambient temperature gradient:

$$q^{(TR)} = -\lambda\left(\frac{T^u - T^l}{\Delta x}\right) \tag{9.60}$$

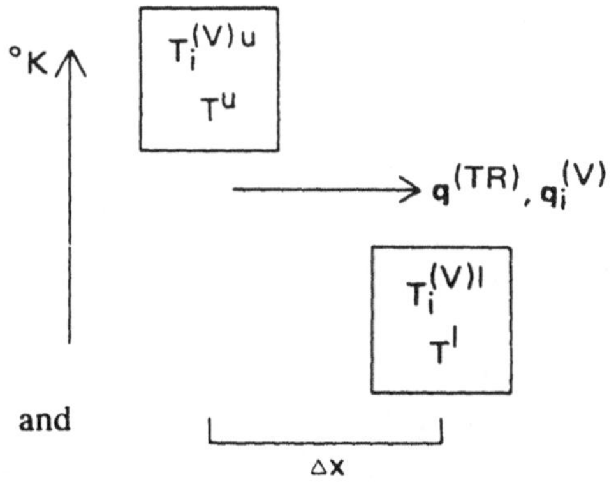

Fig. 55. Schematic of translational–rotational and vibrational fluxes.

while the vibrational energy flux, $q_{iV}^{(V)}$ due to the vibrational temperature gradient is proportional to the vibrational temperature gradient itself

$$q_{iV}^{(V)} = -\lambda_i^{(V)}\left(\frac{T_i^{(V)u} - T_i^{(V)l}}{\Delta x}\right) \tag{9.61}$$

A vibrational temperature gradient also causes a translational–rotational energy flux $q_V^{(TR)}$:

$$q_V^{(TR)} = -\sum_i \lambda_i^{(VT)}\left(\frac{T_i^{(V)u} - T_i^{(V)l}}{\Delta x}\right) \tag{9.62}$$

while, conversely, an ambient temperature gradient causes a vibrational energy flux $q_i^{(V)}(T)$:

$$q_{iT}^{(V)} = -\lambda_i^{(TV)}\left(\frac{T^u - T^l}{\Delta x}\right) \tag{9.63}$$

For example, if the ambient temperature gradient is zero and the vibrational temperature gradient is nonzero, there will still be a net translational energy flux due to the transport of vibrational energy which is then changed into translational energy by V–T processes.

ii. Diffusion. The four conduction processes described above can take place in any material. In a gas, the molecules are free to move and carry translational and vibrational energy with them to other regions of the gas. These are diffusion processes. In the absence of pressure and concentration gradients the rate of diffusion depends on temperature gradients, and four diffusion coefficients for the diffusion of vibrational and translational energy are required in the same way as four conduction coefficients are required to describe the conduction processes.

Having described the physical processes contributing to the transport of vibrational and translational–rotational energy we now derive expressions for the flux vectors $\mathbf{q}_i^{(V)}$ and $\mathbf{q}^{(TR)}$.

9.2.2.2.2. The Translational–Rotational Energy Flux $\mathbf{q}^{(TR)}(\mathbf{x})$. The translational energy flux $\mathbf{q}^{(TR)}(\mathbf{x})$ is given by Eq. (5.136). We now make the following assumptions that simplify the expression for the translational energy flux:

(i) We continue to neglect the presence of ions so that the last two terms on the right-hand side of Eq. (5.106a) are zero;

(ii) we assume that there no pressure gradients, i.e., that density variations are manifest solely as variations in gas temperature. With this assumption the second term on the right-hand side of Eq. (5.106a) will be zero;

(iii) we assume that the fractional composition of the various species of the gas mixture is constant throughout the cavity so that the first term on the right-hand side of Eq. (5.106a) is zero.

With the above assumptions $\mathbf{d}_i$ is zero, and from Eqs. (5.136) and (5.116) we have

$$\mathbf{q}^{(TR)} = -\frac{5}{2}kT\sum_i \frac{1}{m_i}D_i^{(T)}\nabla \ln T - \lambda'\nabla T - \sum_i \lambda_i^{(VT)}\nabla T_i^{(V)}$$
$$-\frac{5}{2}kT\sum_i \frac{1}{m_i}D_i^{(V)}\nabla \ln T_i^{(V)}$$

or

$$\mathbf{q}^{(TR)} = -\left(\lambda' + \frac{5}{2}k\sum_i \frac{D_i^{(T)}}{m_i}\right)\nabla T - \sum_i \left[\left(\lambda_i^{(VT)} + \frac{5}{2}k\frac{T}{T_i^{(V)}}\frac{D_i^{(V)}}{m_i}\right)\nabla T_i^{(V)}\right] \tag{9.64}$$

When Eq. (9.64) is substituted into Eq. (9.59) we have

$$\nabla\cdot\left[\left(\lambda' + \frac{5}{2}k\sum_i \frac{D_i^{(T)}}{m_i}\right)\nabla T\right] = -\nabla\cdot\left[\sum_i\left(\lambda_i^{(VT)} + \frac{5}{2}k\frac{TD_i^{(V)}}{T_i^{(V)}m_i}\right)\nabla T_i^{(V)}\right]$$
$$+\sum_i \nabla\cdot\mathbf{q}_i^{(V)} - n_e h\sum_i \nu_i n_i X_i + h\nu_L\,\Delta N\,WI_\nu \tag{9.65}$$

If we now make the assumption of axial symmetry so that for any function $f(x)$ we have

$$f(r,\phi,z) = f(r) \qquad \text{for all } \phi, z$$

then Eq. (9.65) becomes

$$\frac{1}{r}\frac{d}{dr}\left[r\left(\lambda' + \frac{5}{2}k\sum_i \frac{D_i^{(T)}}{m_i}\right)\frac{d}{dr}T\right] = -\frac{1}{r}\frac{d}{dr}\left[r\sum_i\left(\lambda_i^{(VT)} + \frac{5}{2}k\frac{T}{T_i^{(V)}}\frac{D_i^{(V)}}{m_i}\right)\frac{d}{dr}T_i^{(V)}\right]$$
$$+h\nu_L\Delta N\,WI_\nu + \sum_i \frac{dq_i^{(V)}}{dr} - n_e h\sum_i \nu_i n_i X_i \tag{9.66}$$

which is our form of the equation for the radial ambient temperature profile $T(r)$.

9.2.2.2.3. The Vibrational Energy Fluxes $\mathbf{q}_i^{(V)}(x)$. The vibrational energy flux $\mathbf{q}_j^{(V)}$ is given by Eq. (5.142). As a result of assumptions (i)–(iii) in the previous section, $\mathbf{d}_i$ is zero and we have Eq. (5.116) in the form

$$\bar{\mathbf{V}}_i = -\frac{1}{m_i n_i}D_i^{(T)}\nabla \ln T - \frac{1}{m_i n_i}D_i^{(V)}\nabla \ln T_i^{(V)} \tag{9.67}$$

and therefore Eq. (5.142) becomes under these assumptions

$$\mathbf{q}_j^{(V)} = -\frac{\bar{\varepsilon}_j}{m_j}D_j^{(T)}\nabla \ln T - \lambda_j^{(VT)}\nabla T - \frac{\bar{\varepsilon}_j}{m_j}D_j^{(V)}\nabla \ln T_j^{(V)} - \lambda_j^{(V)}\nabla T_j^{(V)} \tag{9.68}$$

which can be written in the equivalent form

$$\mathbf{q}_i^{(\mathrm{V})} = -\left\{\lambda_i^{(\mathrm{VT})} + \left[\frac{1}{2}h\nu_i + \frac{E_i(T_i^{(\mathrm{V})})}{N_i m_i}\right]\frac{D_i^{(\mathrm{T})}}{T}\right\}\nabla T$$

$$-\left\{\lambda_i^{(\mathrm{V})} + \left[\frac{1}{2}h\nu_i + \frac{E_i(T_i^{(\mathrm{V})})}{N_i m_i}\right]\frac{D_i^{(\mathrm{V})}}{T_i^{(\mathrm{V})}}\right\}\nabla T_i^{(\mathrm{V})} \tag{9.69}$$

where $D_i^{(\mathrm{T})}$ is the multicomponent thermal diffusion coefficient [Eq. (5.117a)], $\lambda_i^{(\mathrm{V})}$ is the vibrational conductivity [Eq. (5.143a)], and we have used Eq. (5.96b) for $\bar{\varepsilon}_j$.

9.2.2.3. Boundary Conditions

The axially symmetric form (r-dependence only) of Eqs. (9.52)–(9.55), with $\nabla \cdot \mathbf{q}_i^{(\mathrm{V})}$ only on their left-hand sides, and Eqs. (9.64), (9.66), and (9.69), are a set of ten coupled first-order ordinary differential equations for T, $T_i^{(\mathrm{V})}$ $(i = 1, 4)$, $q^{(\mathrm{TR})}$, and $q_i^{(\mathrm{V})}$ $(i = 1, 4)$. We require ten boundary conditions for the solution of these equations. Cylindrical symmetry imposes five boundary conditions since the gradient of each temperature distribution is zero at the center of the cylinder:

$$\left.\frac{dT}{dr}\right|_{r=0} = 0, \qquad \left.\frac{dT_i^{(\mathrm{V})}}{dr}\right|_{r=0} = 0 \tag{9.70}$$

For the remaining five boundary conditions we shall suppose that the ambient temperature at the inner wall of the cylinder is known. (See Section 9.2.6 for a derivation of the inner wall temperature in terms of the outer wall temperature and the heat flux through the wall.) So we have

$$T(R) = T_{\mathrm{wall}} \tag{9.71}$$

and all the vibrational temperatures equal the ambient temperature at the inner wall:

$$T_i^{(\mathrm{V})}(R) = T(R), \qquad i = 1, 4 \tag{9.72}$$

9.2.3. Electron Density Profile

9.2.3.1. Electron Number-Density Equation

The general equation for the electron number density, $n_e(\mathbf{x}, t)$, is given by Eq. (5.154).

On the right-hand side of Eq. (5.154), S_e is a source term denoting the rate of production of electrons by external means (i.e., a primary electron beam), which is not usually present in the waveguide laser; therefore $S = 0$.

ν_e is the loss frequency, which we assume to be negligibly small, of electrons by means other than reactions, and the quantities $k_{ab;cd}$ are rates for reactions in which free electrons are produced or lost, the indices referring to particular atomic and molecular species. Examples of such reactions are:

Ionization

$$e + CO_2 \to CO_2^+ + 2e$$

Dissociative Attachment

$$e + CO_2 \to CO + O^-$$

Dissociative Recombination

$$e + CO_2^+ \to CO + O$$

Associative Detachment

$$O^- + CO \to CO_2 + e$$

For the CO_2 waveguide laser, we make the following assumptions to simplify Eqs. (5.154) and (5.195) for the electron drift velocity:

(i) As in Section 9.2.2.1 we are considering the steady-state distribution and we set the time derivative in Eq. (5.154) equal to zero;

(ii) As in Section 9.2.2.1 we are considering a nonflowing gas and so we set the mass-average velocity $\mathbf{v}_0$ in Eqs. (5.154) and (5.195) equal to zero;

(iii) In Eq. (5.195) we assume that gradients of the heavy-particle number density are negligible compared to gradients of the electron number density. We therefore neglect the second term on the right-hand side of Eq. (5.195).

With the above assumptions we can write Eq. (5.154) for the electron number density in the form

$$\nabla \cdot [n_e(\mathbf{x})\bar{\mathbf{V}}_e(E/N, \mathbf{x})] = Z(E/N, \mathbf{x})n_e(\mathbf{x}) \tag{9.73}$$

where we have set

$$n_e Z = \sum_{ab} n_e n_b k_{ab;e\gamma} - \sum_{c} n_e n_c k_{ec;\gamma} + \sum_{abc} n_a n_b n_c k_{abc;e\gamma} - \sum_{ed} n_e n_c n_d k_{ecd;\gamma} \tag{9.74}$$

where Z is the net rate of electron production. Equation (5.195), with assumptions (i)–(iii) above, may be written in the form

$$n_e(\mathbf{x})\bar{\mathbf{V}}_e(E/N, \mathbf{x}) = -\nabla[n_e(\mathbf{x})D_e(E/N, \mathbf{x})] - n_e(\mathbf{x})\mu_e(E/N, \mathbf{x})\mathbf{E}(\mathbf{x}) \tag{9.75}$$

Equations (9.73) and (9.75) are combined to obtain the steady-state electron number-density equation appropriate to waveguide lasers:

$$\nabla^2(n_e D_e) + \nabla \cdot (n_e \mu_e \mathbf{E}) + Z n_e = 0 \tag{9.76}$$

The third term in this equation is the net rate of production of electrons by reactions at some point **x**. This rate is balanced by a loss of electrons due to diffusion (first term) and a loss of electrons due to drift under the influence of the electric field (second term).

If we now make the assumption of radial symmetry then Eq. (9.76) becomes

$$\frac{1}{r}\frac{d}{dr}\left[r\frac{d}{dr}(n_e D_e)\right] + \frac{d}{dr}(n_e \mu_e E_r) + Z n_e = 0 \tag{9.77}$$

where E_r is the radial component of the electric field. Radial symmetry implies that the axial component E_z of the electric field is a constant and that the angular component E_ϕ is zero. The electron diffusion coefficient $D_e(r)$ is given by Eq. (5.187).

9.2.3.2. Boundary Conditions

Equation (9.77) for the electron number-density radial distribution is a second-order ordinary differential equation and requires two boundary conditions. In the same way as for the temperature equations, radial symmetry imposes the boundary condition

$$\left.\frac{dn_e(r)}{dr}\right|_{r=0} = 0 \tag{9.78}$$

For the second boundary condition we assume the electron number density to be zero at the wall:

$$n_e(R) = 0 \tag{9.79}$$

9.2.4. Equation for the Electric Field

The steady-state form of Maxwell's equation for the electric field **E** is given by Eq. (5.223). As in previous sections we neglect the effect of positive and negative ions and obtain from Eqs. (5.223) and (5.226):

$$\nabla \cdot \mathbf{E} = -\frac{e}{\varepsilon} n_e(\mathbf{x}) \tag{9.80}$$

If we now assume a radially symmetric electron distribution, then the axial component E_z of the electric field is constant and given by the applied field, while the angular component E_ϕ is zero; then we obtain an equation for the radial component E_r of the electric field:

$$\frac{dE_r}{dr} = -\frac{e}{\varepsilon} n_e \tag{9.81}$$

The boundary condition for this equation is imposed by radial symmetry and is

$$E_r(r = 0) = 0 \tag{9.82}$$

The radial component E_r of the electric field is due entirely to the presence of electrons. The direction of the applied field is along the axis of the cylinder and therefore it has no radial component.

In quantities which depend on the ratio E/N, i.e., $D_e(E/N)$, E is the total electric-field strength; therefore

$$E(r) = [E_r(r)^2 + E_z^2]^{1/2} \tag{9.83}$$

9.2.5. *Small-Signal Gain, $\alpha(\mathbf{x})$*

In the point model of the laser, the small-signal gain was given by Eq. (2.75). For a spatial variation of population inversion we have

$$\alpha(\mathbf{x}) = c\nu_L \, \Delta N(\mathbf{x}) W(\mathbf{x}) \tag{9.84}$$

where $\Delta N(\mathbf{x})$ is the population inversion and is given by the spatial form of Eq. (2.42e), namely,

$$\Delta N(\mathbf{x}) = N_{001}(\mathbf{x}) P(J, \mathbf{x}) - (\theta_J/\theta_{J+1}) N_{100}(\mathbf{x}) P(J+1, \mathbf{x}) \tag{9.85}$$

where the excited-state populations $N_{001}(\mathbf{x})$ and $N_{100}(\mathbf{x})$ are given by the spatial forms of Eq. (2.42b), that is,

$$N_{001}(\mathbf{x}) = n_{CO_2} \exp[-h\nu_3/kT_3(\mathbf{x})] Z(\mathbf{x}) \tag{9.86}$$

$$N_{100}(\mathbf{x}) = n_{CO_2} \exp[-h\nu_1/kT_1(\mathbf{x})] Z(\mathbf{x}) \tag{9.87}$$

where

$$Z(\mathbf{x}) = \{1 - \exp[-h\nu_1/kT_1(\mathbf{x})]\} \times \{1 - \exp[-h\nu_2/kT_2(\mathbf{x})]\}^2 \{1 - \exp[-h\nu_3/kT_3(\mathbf{x})]\} \tag{9.88}$$

In Eq. (9.85), θ_J is given by Eq. (2.42c), and from Eq. (2.42a) we have

$$P(J, \mathbf{x}) = \left[\frac{2hcB}{kT(\mathbf{x})}\right] \theta_J \exp\left[\frac{-hcBJ(J+1)}{kT(\mathbf{x})}\right] \tag{9.89}$$

In Eq. (9.84), $W(\mathbf{x})$ is the stimulated-emission rate on line center and is given by the spatial form of Eq. (2.58) as

$$W(\mathbf{x}) = F\lambda^2 [4\pi^2 \nu_L \tau_{sp} \, \Delta\nu_L(\mathbf{x})]^{-1} \tag{9.90}$$

where we have included a filling factor F. τ_{sp} is the lifetime for spontaneous emission and $\Delta\nu_L(\mathbf{x})$ is the laser-transition homogeneous line width given as the spatial generalization of Eq. (2.86):

$$\Delta\nu_L(\mathbf{x}) = \sum_j \frac{n_j Q_j}{\pi}\left[\frac{8kT(\mathbf{x})}{\pi\mu_j}\right]^{1/2} \tag{9.91}$$

9.2.6. *Inside Wall Temperature, T_{in}, in Terms of Outside Wall Temperature, T_{out}*

In Section 9.2.2.3 we specified boundary conditions for the energy-flux equations in terms of the temperature T_{in} at the inside wall of the cylinder. Since it is the temperature T_{out} at the outside wall of the cylinder that is the experimentally measurable quantity we shall now relate T_{in} to T_{out}.[112] The heat flux $\mathbf{q}(\mathbf{r})$ in the wall of the capillary, between R and $\bar{R}$ in Fig. 56, is given by Eq. (9.64); ignoring the diffusion terms and the vibrational temperature term since the wall is a solid:

$$\mathbf{q}(\mathbf{r}) = -\lambda_c(T)\nabla T \tag{9.92}$$

where $\lambda_c(T)$ is the thermal conductivity of the material. With the assumption of cylindrical symmetry this equation becomes

$$q(r) = -\lambda_c(T)\, dT/dr \tag{9.93}$$

The heat flux $q(r)$ is the rate of flow of heat per unit area. Therefore the total heat flux Q, through a cylindrical shell of radius r and length L, is

$$Q = -2\pi r L \lambda_c(T)\, dT/dr \tag{9.94}$$

and hence

$$dT = -\frac{Q}{2\pi L\lambda_c(T)}\frac{dr}{r} \tag{9.95}$$

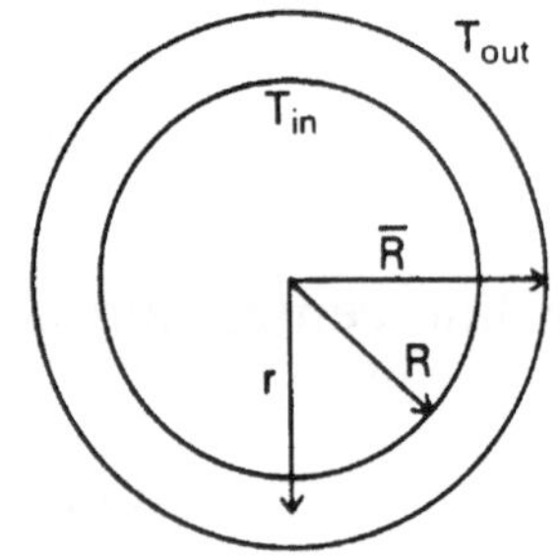

Fig. 56. Cross-sectional view of the waveguide capillary and its wall.

If we now assume that the thermal conductivity λ_c is temperature independent, then we can integrate Eq. (9.95) to obtain

$$T_{\text{out}} - T_{\text{in}} = -\frac{Q}{2\pi L\lambda_c}(\ln \bar{R} - \ln R)$$
$$= -\frac{Q}{2\pi L\lambda_c}\ln\frac{\bar{R}}{R} \tag{9.96}$$

In the steady state the total heat flux Q must equal the total radial flux of translational and vibrational energy in the gas given by

$$Q = 2\pi rL[q^{(\text{TR})}(r) + \sum_i q_i^{(\text{V})}(r)] \tag{9.97}$$

and the right-hand side of this equation should be constant for all r in Fig. 56 given by $0 < r < R$. (This will be an energy-balance check on the numerical solution of the equation). From Eqs. (9.96) and (9.94) we have the equation relating the inner and outer wall temperatures, namely,

$$T_{\text{in}} = T_{\text{out}} + \frac{r}{\lambda_c}\left[q^{(\text{TR})}(r) + \sum_i q_i^{(\text{V})}(r)\right]\ln\frac{\bar{R}}{R} \tag{9.98}$$

9.2.7. *Comparison with Other Formulations*

9.2.7.1. *Equations of D. J. Day*[113]

9.2.7.1.1. The Vibrational Energy Equations. When the cavity-field intensity terms and the first two terms are neglected in Eqs. (9.52)–(9.55), we have, on dividing each equation by $h\nu_i$, the following alternative forms:

$$\nabla\cdot\mathbf{I}_1 = \frac{p_{\text{CO}_2}}{kT}n_eX_1 - \frac{N_1 - N_1^e(T)}{\tau_{10}(T)} - \frac{N_1 - N_1^e(T_2^{(\text{V})})}{\tau_{12}(T_2^{(\text{V})})} + \frac{N_3 - N_3^e(T, T_1^{(\text{V})}, T_2^{(\text{V})})}{\tau_3(T, T_1^{(\text{V})}, T_2^{(\text{V})})} \tag{9.99}$$

$$\nabla\cdot\mathbf{I}_2 = \frac{p_{\text{CO}_2}}{kT}n_eX_2 - \frac{N_2 - N_2^e(T)}{\tau_{20}(T)} + \frac{N_1 - N_1^e(T_2^{(\text{V})})}{\tau_{12}(T_2^{(\text{V})})} + \frac{N_3 - N_3^e(T, T_1^{(\text{V})}, T_2^{(\text{V})})}{\tau_3(T, T_1^{(\text{V})}, T_2^{(\text{V})})} \tag{9.100}$$

$$\nabla\cdot\mathbf{I}_3 = \frac{p_{\text{CO}_2}}{kT}n_eX_3 - \frac{N_3 - N_3^e(T, T_1^{(\text{V})}, T_2^{(\text{V})})}{\tau_3(T, T_1^{(\text{V})}, T_2^{(\text{V})})} + \frac{N_4 - N_4^e(T_3^{(\text{V})})}{\tau_{43}(T)} \tag{9.101}$$

$$\nabla\cdot\mathbf{I}_4 = \frac{p_{\text{N}_2}}{kT}n_eX_4 - \frac{N_4 - N_4^e(T_3^{(\text{V})})}{\tau_{43}(T)} \tag{9.102}$$

where we have defined

$$\mathbf{I}_i = \frac{\mathbf{q}_i^{(V)}}{h\nu_i}, \qquad N_i = \frac{E_i}{h\nu_i}, \qquad N_i^e = \frac{E_i^e}{h\nu_i} \tag{9.103}$$

and where we have used

$$n_j = \frac{p_j}{kT} \tag{9.104}$$

where p_j is the partial pressure of species j. Equations (9.99)–(9.102) are the same as those given by Day.

If in Eq. (9.69) it is assumed that the contribution to the vibrational energy fluxes $\mathbf{q}_i^{(V)}$ due to the ambient temperature gradient ∇T is negligible compared to the contribution from the vibrational temperature gradient, then we have

$$\mathbf{q}_j^{(V)} = -\left\{\lambda_j^{(V)} + \left[\frac{1}{2}h\nu_j + \frac{E_j(T_j^{(V)})}{n_j m_j}\right]\frac{D_j^{(V)}}{T_j^{(V)}}\right\}\nabla T_j^{(V)} \tag{9.105}$$

If we now assume that the vibrational conduction term in Eq. (9.105) is negligible compared with the vibrational diffusion term, then we have (also neglecting $h\nu_j/2$)

$$\mathbf{q}_j^{(V)} = -\frac{E_j(T_j^{(V)})D_j^{(V)}}{n_j m_j T_j^{(V)}}\nabla T_j^{(V)} \tag{9.106}$$

Dividing this equation by $h\nu_j$ and using the definitions for $\mathbf{I}_j$ and N_j given by Eqs. (9.103), we obtain

$$\mathbf{I}_j = -\frac{N_j(T_j^{(V)})D_j^{(V)}}{n_j m_j T_j^{(V)}}\nabla T_j^{(V)} \tag{9.107}$$

We now use the relations

$$\nabla\left[\frac{N_j(T_j^{(V)})}{n_j}\right] = \left[\frac{\partial(N_j(T_j^{(V)})/n_j)}{\partial T_j^{(V)}}\right]\nabla T_j^{(V)} \tag{9.108}$$

and, from Eq. (1.69),

$$\begin{aligned}\frac{\partial}{\partial T_j^{(V)}}\left[\frac{N_j(T_j^{(V)})}{n_j}\right] &= \frac{\partial}{\partial T_j^{(V)}}\left[\frac{1}{\exp(h\nu_j/kT_j^{(V)})-1}\right]\\ &= \frac{h\nu_j}{kT_j^2[\exp(h\nu_j/kT_j^{(V)})-1]^2}\\ &= \frac{h\nu_j}{kT_j^{(V)2}}\left[\frac{N_j(T_j^{(V)})}{n_j}\right]^2\end{aligned} \tag{9.109}$$

we write Eq. (9.107) in the form

$$\mathbf{I}_j = -\frac{kT_j^{(V)2}}{h\nu_j} \frac{n_j^2}{N_j(T_j^{(V)})^2} \frac{N_j(T_j^{(V)})D_j^{(V)}}{n_j m_j T_j^{(V)}} \boldsymbol{\nabla}\left[\frac{N_j(T_j^{(V)})}{n_j}\right]$$

$$= -\frac{kT_j^{(V)}D_j^{(V)}}{h\nu_j N_j(T_j^{(V)})m_j} \boldsymbol{\nabla} N_j(T_j^{(V)}) \tag{9.110}$$

which is to be compared with the equation given by Day:

$$\mathbf{I}_i = D_{CO_2}\boldsymbol{\nabla} N_i, \qquad i = 1, 2, 3 \tag{9.111}$$

$$\mathbf{I}_4 = D_{N_2}\boldsymbol{\nabla} N_4 \tag{9.112}$$

9.2.7.1.2.   Translational Energy Equation. If in Eq. (9.65) we neglect translational diffusion, vibrational diffusion, and vibrational conduction, then we have (omitting the cavity-field intensity term)

$$\boldsymbol{\nabla} \cdot (\lambda'\boldsymbol{\nabla} T) = \sum_i \boldsymbol{\nabla} \cdot \mathbf{q}_i^{(V)} - n_e h \sum_i \nu_i n_i X_i$$

$$= \sum_i h\nu_i \boldsymbol{\nabla} \cdot \mathbf{I}_i - n_e h \sum_i \nu_i n_i X_i \tag{9.113}$$

which is the same as the equation given by Day except that for the second term on the right-hand side he has

$$-v_z E_z n_e$$

The derivation of this term shall be given in Section 9.2.7.2.

9.2.7.1.3.   The Electron Number-Density Equation. If we make the assumption that the net electron production rate Z in Eq. (9.76) is zero, then we obtain

$$\nabla^2(n_e D_e) + \boldsymbol{\nabla} \cdot (n_e \mu_e \mathbf{E}) = 0 \tag{9.114}$$

or, on integration:

$$\boldsymbol{\nabla}(n_e D_e) + n_e \mu_e \mathbf{E} = 0 \tag{9.115}$$

If we now make the assumption that the electron diffusion coefficient D_e is constant, then we obtain

$$\boldsymbol{\nabla} n_e = -\frac{n_e \mu_e \mathbf{E}}{D_e} \tag{9.116}$$

If we now use Einstein's relation (see Hochstim,[50] page 153; this relation only holds if the electrons have a near Maxwell velocity distribution):

$$\frac{\mu_e}{D_e} = \frac{e}{kT_e} \tag{9.117}$$

then Eq. (9.116) becomes

$$\nabla n_e = -\frac{n_e e \mathbf{E}}{kT_e} \tag{9.118}$$

which is the same as the equation given by Day except that he has a factor e^2 instead of our factor e.

9.2.7.1.4. The Electric Field Equation. Equation (9.80) for the electric field is the same as the equation given by Day, except that he has a factor of $\varepsilon\varepsilon_0$ instead of our factor ε.

9.2.7.2. Model of Laderman and Byron

9.2.7.2.1. Translational Energy Equation. Laderman and Byron[114] have presented equations for the ambient temperature radial profile and electron number-density radial profile in CO_2 lasers.

Laderman and Byron do not consider the effect of CO_2 vibrational modes and the corresponding vibrational fluxes. If we ignore all terms containing effective vibrational temperatures, then Eq. (9.66) becomes

$$\frac{1}{r}\frac{d}{dr}\left[r\left(\lambda' + \frac{5}{2}k\sum_i \frac{D_i^{(T)}}{m_i}\right)\frac{dT}{dr}\right] + n_e h \sum_i \nu_i n_i X_i = 0 \tag{9.119}$$

If we now assume that the radial transport of translational energy is by conduction, rather than diffusion, then we can write this equation in the form

$$\frac{1}{r}\frac{d}{dr}\left(r\lambda'\frac{dT}{dr}\right) + G(r) = 0 \tag{9.120}$$

where we have defined

$$G(r) \equiv n_e h \sum_i \nu_i n_i X_i \tag{9.121}$$

In Eq. (9.120), $G(r)$ is the rate at which heat is generated at position r due to electron–molecule collisions, and the first term is the rate at which heat is conducted away from position r. Equation (9.120) is the same as the equation given by Laderman and Byron except that their formula for G is

$$G_L = en_e v_D E \tag{9.122}$$

where v_D is the electron drift velocity. We shall now derive this expression.

If a current I (amps) is flowing along a conductor across which there is a potential V (volts), then energy is lost to the conductor and appears as heat (ohmic heating). The rate of heat generation P is

$$P = IV \qquad \text{(watts)} \tag{9.123}$$

If the conductor has length L (cm) and cross-sectional area A (cm^2), then the current density j is

$$j = I/A \qquad (\text{amp cm}^{-2}) \tag{9.124}$$

and the electric field E is given by

$$E = V/L \qquad (\text{volt cm}^{-1}) \tag{9.125}$$

From Eqs. (9.123)–(9.125) the rate of heat generation per unit volume is given by

$$G_L = P/LA = jE \tag{9.126}$$

The current density j is given by Eq. (2.152b), which when substituted into Eq. (9.126), gives us Eq. (9.122). It may be that Eq. (9.122) is a more accurate expression for the rate of heat generation than Eq. (9.121). Equation (9.121) calculates the rate of heat generation due to electrons losing energy to the CO_2 and N_2 vibrational modes during inelastic collisions. It does not include the rate of heat generation due to electrons losing energy to other excited states of CO_2, N_2, and He. The calculation of the drift velocity performed by BOLTZ (see Section 3.2) includes the effect of all inelastic electron–molecule collisions. Following Laderman and Byron we now define dimensionless parameters

$$\rho = r/R \tag{9.127}$$

$$\hat{T}(\rho) = T(r)/T(R) \tag{9.128}$$

$$\Lambda(\rho) = \lambda'(r)/\lambda'(R) \tag{9.129}$$

where we recall that R is the inside radius of the cylinder, and the coefficient of thermal conductivity $\lambda[T(r)]$ is spatially dependent through its temperature dependence. We now introduce the variable θ by the transformation

$$\theta = \int_0^T \Lambda \, d\hat{T} \tag{9.130}$$

Dividing Eq. (9.120) by $\lambda(R)$ and $T(R)$ and multiplying through by R^2 we obtain, on using Eqs. (9.127)–(9.129):

$$\frac{1}{\rho}\frac{d}{d\rho}\left(\rho\Lambda\frac{d\hat{T}}{d\rho}\right) + \frac{R^2}{\lambda(R)T(R)}G = 0 \tag{9.131}$$

Then, using Eq. (9.130) we obtain

$$\frac{1}{\rho}\frac{d}{d\rho}\left(\rho\frac{d\theta}{d\rho}\right) + \frac{R^2}{\lambda(R)T(R)}G = 0 \tag{9.132}$$

or

$$\frac{d^2\theta}{d\rho^2}+\frac{1}{\rho}\frac{d\theta}{d\rho}+\frac{R^2}{\lambda(R)T(R)}G=0 \tag{9.133}$$

which is the same as the equation given by Laderman and Byron [Reference 114, Eq. (2)] and is the Poisson equation in cylindrical coordinates.

9.2.7.2.2. Electron Number-Density Equation. Our form of the electron number-density equation is given in Eq. (9.77). The third term in Eq. (9.77) is the net rate of production of electrons by reactions. This rate is balanced by a loss of electrons due to diffusion (first term) and drift due to the electric field (second term). If the assumption is made that the electron loss by drift is negligible compared to the loss by diffusion, then Eq. (9.77) becomes

$$\frac{1}{r}\frac{d}{dr}\left[r\frac{d}{dr}(n_eD_e)\right]+Zn_e=0 \tag{9.134}$$

We recall that in Section 9.2.7.1.3 the assumption was made that the third term in Eq. (9.77) is negligible compared to the other two terms. Following Laderman and Byron we now define a dimensionless electron number density $\tilde{N}$ by

$$n_e=n_{e0}\tilde{N} \tag{9.135}$$

where n_{e0} is the electron number density on the axis of the cylinder. Using Eqs. (9.127) and (9.135) we have, for Eq. (9.134):

$$\frac{1}{R^2}\frac{d}{d\rho}\left[\rho\frac{d}{d\rho}(n_{e0}\tilde{N}D_e)\right]+Zn_{e0}\tilde{N}\rho=0$$

which becomes

$$\frac{d^2\tilde{N}}{d\rho^2}+\frac{1}{\rho}\frac{d\tilde{N}}{d\rho}\left(\frac{2\rho}{D_e}\frac{dD_e}{d\rho}+1\right)+\frac{\tilde{N}}{D_e}\left(\frac{d^2D_e}{d\rho^2}+\frac{1}{\rho}\frac{dD_e}{d\rho}\right)+\frac{Z\tilde{N}R^2}{D_e}=0 \tag{9.136}$$

Equation (9.136) is the same as the equation given by Laderman and Byron [Reference 114, Eq. (6)] except they do not have the third term or the factor 2 in the second term. Therefore if we make the assumption that the variations $dD_e/d\rho$ and $d^2D_e/d\rho^2$ are negligible compared to the differentials of $\tilde{N}$, then we obtain from Eq. (9.136) the equation given by Laderman and Byron (except for the factor of 2 in the second term):

$$\frac{d^2\tilde{N}}{d\rho^2}+\frac{1}{\rho}\frac{d\tilde{N}}{d\rho}\left(\frac{2\rho}{D_e}\frac{dD_e}{d\rho}+1\right)+\frac{ZR^2}{D_e}\hat{N}=0 \tag{9.137}$$

We note that for constant Z and D_e, this equation is Bessel's equation.

9.2.7.2.3. Solution of the Equations. The model of Laderman and Byron consists of an equation for the normalized temperature $\theta(\rho)$ [Eq. (9.133)], and an equation for the normalized electron number density $\tilde{N}(\rho)$ [Eq. (9.137)].

The electron production rate Z and the electron diffusion coefficient D_e are both functions of E/N. They are therefore both functions of the normalized temperature θ because of the relation

$$p = NkT \tag{9.138}$$

(we recall we are assuming constant pressures). Therefore Eqs. (9.133) and (9.137) are a pair of coupled ordinary differential equations. Laderman and Byron solved these equations numerically with the boundary conditions

$$\left.\frac{d\hat{T}}{d\rho}\right|_{\rho=0} = 0 \tag{9.139a}$$

$$\left.\frac{d\tilde{N}}{d\rho}\right|_{\rho=0} = 0 \tag{9.139b}$$

$$\hat{T}(\rho = 1) = 1 \tag{9.139c}$$

$$\tilde{N}(\rho = 1) = 0 \tag{9.139d}$$

Since the boundary conditions are given at two points, an iterative procedure ("shooting method") was used in which the value of $\hat{T}$ $(\rho = 0)$ was adjusted until the boundary conditions given by Eqs. (9.139) were satisfied.

Laderman and Byron also solved the equations of their model assuming constant Z and D_e. Equations (9.133) and (9.137) are then uncoupled and the solution of Eq. (9.137) is the Bessel function distribution; Equation (9.133) may be solved numerically using this solution for the electron distribution. The results of Laderman and Byron's calculations are shown in Fig. 57, together with the results obtained if a constant electron density across the tube is assumed.

9.2.7.3. AuYeung and Haus

If in Eq. (9.66) we neglect the translational diffusion term, vibrational temperature gradients, vibrational energy fluxes, and the cavity-field intensity term, then we obtain

$$\frac{1}{r}\frac{d}{dr}\left(r\lambda'\frac{dT}{dr}\right) + Q = 0 \tag{9.140}$$

where we have defined

$$Q \equiv n_e h \sum_i \nu_i n_i X_i \tag{9.141}$$

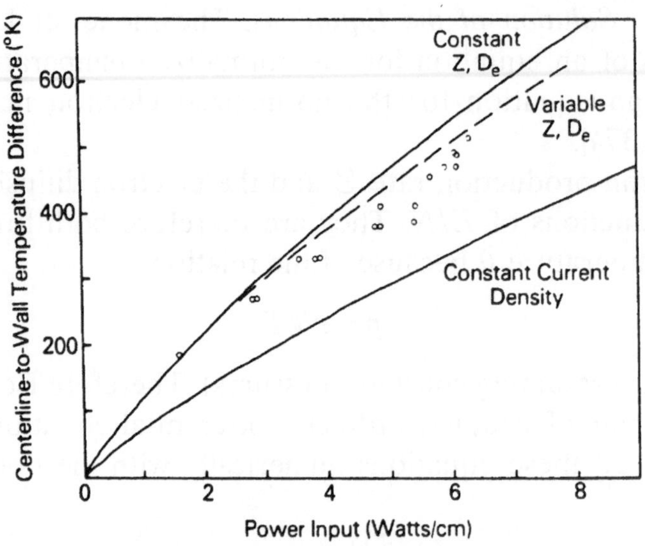

Fig. 57. Centerline temperature as a function of power input for a $CO_2:N_2:He = 1:3:16$ mixture, from Laderman and Byron,[114] showing also experimental results for 13–31 Torr devices with flow rates in the range 0.02–0.08 gm/sec.

Equation (9.140) is the equation solved by AuYeung and Haus [Reference 115, Eq. (35)] to determine the radial ambient temperature profile. Their expression for Q is

$$Q = en_e(\nu_a V_a + \nu_b V_b) \tag{9.142}$$

and the relationships between the quantities in Eqs. (9.141) and (9.142) are

$$\begin{aligned} eV_a &= h\nu_3 \approx h\nu_4 \\ eV_b &= h\nu_1 \approx 2h\nu_2 \\ \nu_a &= X_3 + X_4 \\ \nu_b &= X_1 + \tfrac{1}{2}X_2 \end{aligned} \tag{9.143}$$

AuYeung and Haus make the approximations $h\nu_3 = h\nu_4$ and $h\nu_1 = 2h\nu_2$.

9.2.7.4. *Cohen's Treatment of Waveguide Laser Kinetics*

We derive Cohen's[116] equations for waveguide CO_2 laser kinetics and show how, under Cohen's approximations, steady-state solutions to these equations are obtained.

In the preceding sections we have derived the waveguide equations of other authors from the plasma model equations developed in Chapter 5. Cohen's equations are rate equations for the energies of the first excited state of vibrational modes (higher excited states of the modes are neglected) rather than mode vibrational energy equations.

9.2.7.4.1. Kinetic Equations. Cohen's kinetic model consists of a set of three rate equations for:

(a) the number density, N_2^*, of nitrogen molecules in the first ($v = 1$) excited vibrational state;

(b) the number density n_2 of CO_2 molecules in the (0, 0, 1) state (i.e., the upper laser level);

(c) the number density n_1 of CO_2 molecules in the (1, 0, 0) and (0, 1, 0) excited states of CO_2. The strong Fermi resonance between the (1, 0, 0) and (0, 2, 0) states allows the approximation that these two states are in equilibrium with each other (i.e., they each have the same effective vibrational temperature).

We denote the number density of molecules in the ground [N_2 ($v = 0$)] state by N_2^0 and the number density of CO_2 molecules in the ground (0, 0, 0) state by n_0. We define f_1 to be the fraction of the CO_2 molecules in the combined (1, 0, 0) and (0, 1, 0) state that are in the lower laser level [the (1, 0, 0) state].

Cohen considers the following processes in forming his kinetic equations, shown schematically in Fig. 58:

(i) Electron excitation of the CO_2 (000) state to the CO_2 (001) state. The time rate of change of n_2 due to this process is given by

$$n_e(t)X_{001}n_0 \tag{9.144}$$

where $n_e(t)$ is the electron number density and X_{001} is the electron vibrational excitation rate for the first excited state of the asymmetric stretch mode [see Eq. (3.101)]. Following Cohen we define

$$\alpha \equiv n_e X_{001} \tag{9.145}$$

(ii) Electron deexcitation of the CO_2 (001) state to the CO_2 (000) state. The time rate of change of n_2 due to this process is given by

$$n_e(t)\bar{X}_{001}n_2 \tag{9.146}$$

where $\bar{X}_{001}$ is the electron vibrational deexcitation rate for the CO_2 (001) state. We define

$$\rho \equiv n_e \bar{X}_{001} \tag{9.147}$$

From Eq. (3.75) we see that the deexcitation rate of state i to the

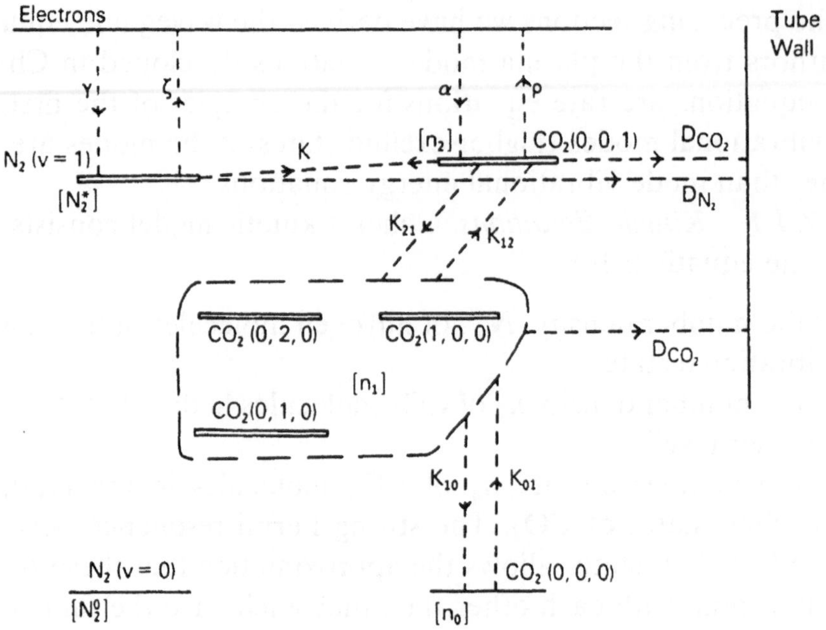

Fig. 58. Schematic energy-level diagram for Cohen's rate equations, where $[i]$ denotes the number densities of the various states.

ground state of species s will be

$$\bar{X}_s^i = (2e/m)^{1/2} \int_0^{\infty} Q_{-i}^s(u) f(u, E/N, T) u \, du$$

This integral is evaluated using the reciprocity relation, Eq. (3.72).

(iii) Electron excitation and deexcitation of the N_2 $(v = 1)$ state. The time rate of change of N_2^* due to these processes is given by

$$n_e(t) X_{N_2} N_2^0 - n_e(t) \bar{X}_{N_2} N_2^* \equiv \gamma N_2^0 - \zeta N_2^* \tag{9.148}$$

where we have defined

$$\gamma \equiv n_e X_{N_2} \tag{9.149}$$

$$\zeta \equiv n_e \bar{X}_{N_2} \tag{9.150}$$

Cohen does not consider electron excitation and deexcitation of the CO_2 (010) and (100) states.

(iv) V–V energy exchange between the N_2 $(v = 1)$ and CO_2 (001) states (see Section 2.3). The time rate of change of n_2 due to this process is given by

$$n_0 N_2^* k_{000,1;001,0}^{CO_2\text{-}N_2} - n_2 N_2^0 k_{001,0;000,1}^{CO_2\text{-}N_2} \tag{9.151}$$

With the approximation that this is a resonant process the two rates are equal, and we define

$$K = k_{000,1;001,0}^{CO_2\text{-}N_2} = k_{001,0;000,1}^{CO_2\text{-}N_2} \tag{9.152}$$

The time rate of change of N_2^* due to this process is then given by

$$n_2 N_2^0 K - n_0 N_2^* K \tag{9.153}$$

(v) V–V energy exchange between the CO_2 (001) and CO_2 (100) plus (010) states (see Section 2.4). The time rate of change of n_2 due to this process is given by [see Eq. (2.22)]

$$\sum_t n_1 n^t k^t_{110,001} - \sum_t n_2 n^t k^t_{001,110} \tag{9.154}$$

where n^t/n are the fractional particle concentrations [see Eq. (1.38)]. We define

$$K_{12} = \sum_t n^t k^t_{110,001} \tag{9.155}$$

$$K_{21} = \sum_t n^t k^t_{001,110} \tag{9.156}$$

The time rate of change of n_1 due to this process will be given by

$$wK_{21}n_2 - wK_{12}n_1 \tag{9.157}$$

where the factor w takes into account the fact that V–V energy exchange between the upper and lower laser levels results in excitation of more than one quanta of the lower laser level (see Section 2.4).

(vi) V–T energy exchange between the CO_2 (0, 1, 0) state and the translational energy of the gas (see Section 2.2). The time rate of change of n_1 due to this process is

$$\tfrac{1}{2} n_0 \sum_t n^t k^{t\text{-}CO_2}_{000,010} - \tfrac{1}{2} n_1 \sum_t n^t k^{t\text{-}CO_2}_{010,000} \tag{9.158}$$

where the factor $\frac{1}{2}$ is introduced because the excitation of one quanta of the (0, 1, 0) state is equivalent to the excitation of half a quanta of the (1, 0, 0) state [the energy of the (0, 1, 0) state is half the energy of the (1, 0, 0) state]. In Eq. (9.158) we define

$$\sum_t n^t k^{t\text{-}CO_2}_{000,010} = K_{01} \tag{9.159}$$

$$\sum_t n^t k^{t\text{-}CO_2}_{010,000} = K_{10} \tag{9.160}$$

(vii) Finally we consider diffusive loss to the walls: vibrationally excited molecules colliding with the wall and suffering deexcitation. The contributions to the rate equations from this process are obtained from the diffusion equation (see Bond *et al.*,[36] p. 255)

$$\frac{\partial n_i}{\partial t} = D_i \nabla^2 n_i \tag{9.161}$$

by putting $\nabla^2 \simeq 1/\Lambda^2$, where Λ is a dimension over which diffusion occurs. For a capillary tube of radius r_0, Λ is given by [102,103]

$$\Lambda = r_0/2.405 \tag{9.162}$$

The D_i, expressed in $\text{cm}^2\,\text{sec}^{-1}$, are ambipolar diffusion coefficients.[79] The time rate of change of each excited-state population due to this diffusion loss is given by

$$\frac{-\beta_i D_i}{\Lambda^2} n_i \tag{9.163}$$

where the β_i coefficients are numerical coefficients and estimates of the average number of quanta lost per wall deexcitation of the ith mode. For example, Gordietz *et al.*[117] have estimated β_2 to be 1.5.

With the above considerations we may now write down the rate equation for each excited state:

$$\frac{dn_2}{dt} = \alpha n_0 - \rho n_2 + n_0 N_2^* K - n_2 N_2^0 K + n_1 K_{12} - n_2 K_{21} - \frac{\beta_2 D_{CO_2}}{\Lambda^2} n_2 \tag{9.164}$$

$$\frac{dn_1}{dt} = w n_2 K_{21} - w n_1 K_{12} + \frac{1}{2} n_0 K_{01} - \frac{1}{2} n_1 K_{10} - \frac{\beta_1 D_{CO_2}}{\Lambda^2} n_1 \tag{9.165}$$

$$\frac{dN_2^*}{dt} = \gamma N_e^0 - \zeta N_2^* - n_0 N_2^* K + n_2 N_2^0 K - \frac{\beta_{N_2} D_{N_2}}{\Lambda^2} N_2^* \tag{9.166}$$

9.2.7.4.2. Steady-State Solution of the Kinetic Equations. Equations (9.164)–(9.166) are kinetic equations for a CO_2 laser amplifier considered as a three-level system. All variables are a function of position and time. We now assume cylindrical symmetry, so that the variables have r-dependence only, and consider the steady-state condition, so that the time derivatives in these equations are zero. Furthermore, Cohen assumes that the electron number density $n_e(r)$ and the excited-state number densities $n_2(r)$, $n_1(r)$, and $N_2^*(r)$ have zero-order Bessel function radial distributions:

$$\begin{aligned} n_e(r) &= n_e J_0(r/\Lambda) \\ n_2(r) &= n_2 J_0(r/\Lambda) \\ n_1(r) &= n_1 J_0(r/\Lambda) \\ N_2^*(r) &= N_2^* J_0(r/\Lambda) \end{aligned} \tag{9.167}$$

where, on the right-hand sides, n_e, n_2, n_1, and N_2^* now refer to the number densities at the tube center ($r=0$), and $J_0(r/\Lambda)$ is the zero-order Bessel function and

$$J_0(r_0/\Lambda)=J_0(2.405)=0 \tag{9.168}$$

where r_0 is the inner radius of the capillary tube (R in Fig. 56). We now substitute Eqs. (9.167) into Eqs. (9.164)–(9.166), set the time derivatives equal to zero, multiply the resulting equation by r and integrate over r between $r=0$ and $r=r_0$, and over ϕ between 0 and 2π. We obtain

$$\alpha n_0 a_1 - \rho n_2 a_2 + K n_0 N_2^* a_1 - K n_2 N_2^0 a_1 + K_{12} n_1 a_1 - K_{21} n_2 a_1 - \beta_2 \frac{D_{CO_2}}{\Lambda^2} n_2 a_1 = 0 \tag{9.169}$$

$$w K_{21} n_2 a_1 - w K_{12} n_1 a_1 + \frac{1}{2} K_{01} n_0 - \frac{1}{2} K_{10} n_1 a_1 - \beta_1 \frac{D_{CO_2}}{\Lambda^2} n_1 a_1 = 0 \tag{9.170}$$

$$\gamma N_2^0 a_1 - \zeta N_2^* a_2 - K n_0 N_2^* a_1 + K n_2 N_2^0 a_1 - \beta_{N_2} \frac{D_{N_2}}{\Lambda^2} N_2^* a_1 = 0 \tag{9.171}$$

where

$$a_1 = \frac{1}{\pi r_0^2} \int_0^{2\pi} d\phi \int_0^{r_0} r J_0\left(\frac{r}{\Lambda}\right) dr \approx 0.434 \tag{9.172}$$

$$a_2 = \frac{1}{\pi r_0^2} \int_0^{\pi} d\phi \int_0^{r_0} r J_0^2\left(\frac{r}{\Lambda}\right) dr \approx 0.270 \tag{9.173}$$

We now divide Eqs. (9.169)–(9.171) through by a_1 and define

$$\begin{aligned} \rho' &\equiv \rho a_2/a_1 \\ K_{01}'' &\equiv K_{01}/a_1 \\ \zeta' &\equiv \zeta a_2/a_1 \end{aligned} \tag{9.174}$$

to obtain

$$\alpha n_0 - \rho' n_2 + K n_0 N_2^* - K n_2 N_2^0 + K_{12} n_1 - K_{21} n_2 - \beta_2 \frac{D_{CO_2}}{\Lambda^2} n_2 = 0 \tag{9.175}$$

$$w K_{21} n_2 - w K_{12} n_1 + \frac{1}{2} K_{01}'' n_0 - \frac{1}{2} K_{10} n_1 - \beta_1 \frac{D_{CO_2}}{\Lambda^2} n_1 = 0 \tag{9.176}$$

$$\gamma N_2^0 - \zeta' N_2^* - K n_0 N_2^* + K n_2 N_2^0 - \beta_{N_2} \frac{D_{N_2}}{\Lambda^2} N_2^* = 0 \tag{9.177}$$

The separate conservation of the total number of N_2 and CO_2 molecules implies that

$$n(r) = n_0(r) + n_1(r) + n_2(r) \tag{9.178}$$

$$N_2(r) = N_2^0(r) + N_2^*(r) \tag{9.179}$$

or

$$n(r) = n_0 + n_1 J_0(r/\Lambda) + n_2 J_0(r/\Lambda) \tag{9.180}$$

$$N_2(r) = N_2^0 + N_2^* J_0(r/\Lambda) \tag{9.181}$$

where we have used Eqs. (9.167). Integrating Eqs. (9.180) and (9.181) over ϕ and r, we obtain

$$n = n_0 + n_1 a_1 + n_2 a_1 \tag{9.182}$$

$$N_2 = N_2^0 + N_2^* a_1 \tag{9.183}$$

where n is the total number of CO_2 molecules and N_2 denotes the number of nitrogen molecules per unit length of the tube.

We now eliminate n_0 and N_2^0 from Eqs. (9.175)–(9.177) using Eqs. (9.182) and (9.183) to obtain

$$\alpha(n - n_1 a_1 - n_2 a_1) - \rho' n_2 + K(n - n_1 a_1) N_2^* - K n_2 N_2 + K_{12} n_1 - K_{21} n_2 - \beta_2 \frac{D_{CO_2}}{\Lambda^2} n_2 = 0 \tag{9.184}$$

$$w K_{21} n_2 - w K_{12} n_1 + \frac{1}{2} K_{01}''(n - n_1 a_1 - n_2 a_1) - \frac{1}{2} K_{10} n_1 - \beta_1 \frac{D_{CO_2}}{\Lambda^2} n_1 = 0 \tag{9.185}$$

$$\left[-\gamma a_1 - \zeta' - K(n - n_1 a_1) - \beta_{N_2} \frac{D_{N_2}}{\Lambda^2}\right] N_2^* + \gamma N_2 + K n_2 N_2 = 0 \tag{9.186}$$

where, in Eqs. (9.184) and (9.186) we have collected together terms in N_2^*. Eliminating N_2^* from these two equations we obtain

$$N_2^* = \frac{-\alpha(n - n_1 a_1 - n_2 a_1) + \rho' n_2 + K n_2 N_2 + K_{21} n_2 + \beta_2 (D_{CO_2}/\Lambda^2) n_2}{K(n - n_1 a_1)}$$

$$= \frac{-\gamma N_2 - K n_2 N_2}{-\gamma a_1 - \zeta' - K(n - n_1 a_1) - \beta N_2 (D_{N_2}/\Lambda^2)}$$

or

$$\left[\gamma a_1 + \zeta' + K(n - n_1 a_1) + \beta_{N_2} \frac{D_{N_2}}{\Lambda^2}\right]\left[-\alpha(n - n_1 a_1 - n_2 a_1) + \rho' n_2 + K n_2 N_2 + K_{21} n_2 + \beta_2 \frac{D_{CO_2}}{\Lambda^2} n_2\right] - K(n - n_1 a_1)(\gamma N_2 + K n_2 N_2) = 0 \tag{9.187}$$

where we are now neglecting the excitation of the ν_3 mode from the ν_{12} mode (the K_{12} terms).

We now use Eq. (9.185) to express n_2 in terms of n_1:

$$n_2 = \frac{-\frac{1}{2}K''_{01}(n - n_1 a_1) + \frac{1}{2}K_{10}n_1 + \beta_1(D_{CO_2}/\Lambda^2)n_1}{wK_{21} - \frac{1}{2}K''_{01}a_1}$$

$$= \frac{[\frac{1}{2}K_{01} + \frac{1}{2}K_{10} + \beta_1(D_{CO_2}/\Lambda^2)]n_1}{wK_{21} - \frac{1}{2}K_{01}} + \frac{-\frac{1}{2}K''_{01}n}{wK_{21} - \frac{1}{2}K_{01}} \tag{9.188}$$

where we have used the second of Eqs. (9.174). We now define

$$\bar{s} \equiv \frac{\frac{1}{2}K_{01} + \frac{1}{2}K_{10} + \beta_1(D_{CO_2}/\Lambda^2)}{wK_{21} - \frac{1}{2}K_{01}} \tag{9.189}$$

$$\delta \equiv \frac{-\frac{1}{2}K''_{01}n}{wK_{21} - \frac{1}{2}K_{01}} \tag{9.190}$$

and Eq. (9.188) may then be written

$$n_2 = \bar{s}n_1 + \delta \tag{9.191}$$

Substituting Eq. (9.191) for n_2 into Eq. (9.187) we obtain

$$-\left[\gamma a_1 + \zeta + K(n - n_1 a_1) + \beta_{N_2}\frac{D_{N_2}}{\Lambda^2}\right]$$

$$\times\left[-\alpha n + \alpha n_1 a_1 + (\bar{s}n_1 + \delta)\left(\alpha a_1 + \rho' + KN_2 + K_{21} + \beta_2\frac{D_{CO_2}}{\Lambda^2}\right)\right]$$

$$+K(n - n_1 a_1)[\gamma N_2 + KN_2(\bar{s}n_1 + \delta)] = 0 \tag{9.192}$$

Collecting together the coefficients of the powers of n_1 we have

$$n_1^2 K''(\alpha'' + \bar{s}\Omega) - n_1[K''(\gamma N_2 + \alpha n - \delta\Omega) + K\bar{s}\theta N_2 + (\alpha'' + \bar{s}\Omega)(\theta + Kn)]$$

$$+(\theta + Kn)(\alpha n - \delta\Omega + \gamma N_2) - \theta N_2(\gamma + \delta K) = 0 \tag{9.193}$$

where

$$K'' = Ka_1, \qquad \alpha'' = \alpha a_1, \qquad \gamma'' = \gamma a_1 \tag{9.194}$$

$$\Omega = \alpha'' + \rho' + K_{21} + \beta_2\frac{D_{CO_2}}{\Lambda^2}$$

$$\theta = \gamma'' + \zeta' + \beta_{N_2}\frac{D_{N_2}}{\Lambda^2} \tag{9.195}$$

Equations (9.189), (9.190), and (9.193)–(9.195) agree with Eqs. (8)–(12) of Cohen[116] *except for the fact that* Cohen's sign for the fourth term of Eq. (9.193) is negative, and instead of Eqs. (9.194) he has

$$K'' = K/a_1, \qquad \alpha'' = \alpha/a_1, \qquad \gamma'' = \gamma/a_1$$

9.3. Far-Infrared (FIR) Gas Lasers

9.3.1. Transitions in Symmetric-Top Molecules

The development of CO_2 laser-pumped FIR lasers by Chang and Bridges[118] has permitted the construction of CW laser sources throughout the FIR spectral region.[119] These sources are capable of very stable milliwatt level CW outputs at wavelengths from 40 μm to 1.8 mm. However, most experimental studies have used pulsed-infrared pump sources which are capable of generating FIR pulses of several kilowatts.[120]

The principal objective in developing a theoretical model is to examine the dependence of laser performance on gas pressure, pump intensity, molecular relaxation rates, and buffer gases. Most theoretical attention has been given to symmetric-top molecules such as methyl fluoride, CH_3F. The lowest frequency vibrational mode of the CH_3F molecule is the ν_3 mode (C–F bond stretching) at 1048.6 cm^{-1}. If all other vibrational modes of the molecule are ignored, then the energy levels have been given by Townes and Schawlow[121] to be

$$E/h = \nu_3(v+\tfrac{1}{2}) + B_vJ(J+1) + (C_v - B_v)K^2 - D_{Jv}J^2(J+1)^2$$
$$- D_{JKv}J(J+1)K^2 - D_{Kv}K^4 \qquad (9.196)$$

where v is the vibrational quantum number, and J and K are the rotational quantum numbers related to the total angular momentum and the axial component of the angular momentum, respectively, with $K = 0, \pm 1, \pm 2, \ldots, \pm J$. The coefficients $C_v > B_v \gg D_{Jv}$, D_{JKv}, and D_{Kv} are rotational constants. Vibrational–rotational transitions among these levels follow the selection rules $\Delta v = 1$, $\Delta J = 0$, or ± 1 and $\Delta K = 0$, and give rise to an absorption band near 9.5 μm. Since the molecule has a permanent electric dipole moment (1.849742 D), pure rotational transitions are also allowed with selection rules $\Delta v = 0$, $\Delta J = \pm 1$, and $\Delta K = 0$. These latter transitions are the ones which occur in the millimeter to submillimeter region.

A partial energy-level diagram is presented in Fig. 59 showing the resonant absorption of the $P(20)$ line, at 9.55 μm, of the CO_2 laser by the

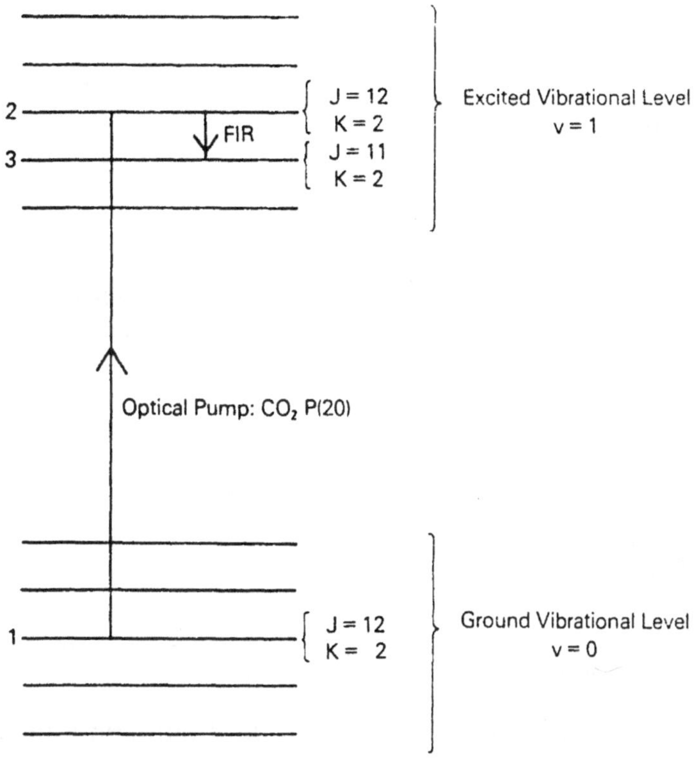

Fig. 59. Energy-level diagram of CH_3F showing various collision processes and level notation.

$Q(12, 2)$ ($v=0\to 1$, $J=12\to 12$, $K=2\to 2$) vibrational–rotational transition of CH_3F, thereby causing population inversion between the $J=12$ and 11 levels in the $v=1$ vibrational state. Actually the 9.55 μm line falls between the two sub-branches of the $Q(12, 1)$ and $Q(12, 2)$ which are separated from each other by $\simeq$80 MHz, but since the strength of these lines is proportional to K^2, the $Q(12, 2)$ transition is expected to be the most strongly excited.

Since the CO_2 $P(20)$ line (frequency ν_p) is offset by a little from the absorption line center (frequency ν_0), then for the pump radiation traveling in one direction, $+z$ say, *resonance* absorption will only take place with those molecules with velocity v such that

$$\nu_p = \nu_0(1+v/c)$$

while for pump radiation traveling in the opposite direction, $-z$, resonance absorption will only take place with those molecules with velocity $-v$. Consequently, the absorbed pump radiation produces two narrow spikes of population in the upper FIR level as shown in Fig. 60.

Let $I_\nu^\pm$ (cm^{-2} sec^{-1}) be the forward and backward traveling wave-pump intensities and $\sigma^\pm$ (cm^2) be the forward and backward stimulated-absorption cross sections; then the induced transition rate for optical

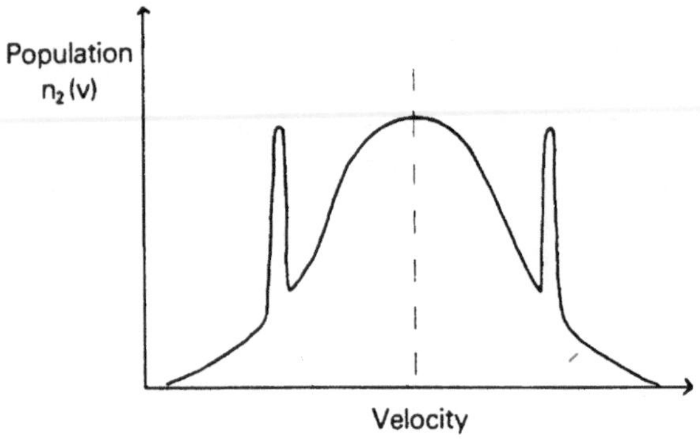

Fig. 60. Maxwellian profile and velocity-selected excitation of the upper level.

pumping will be

$$W(\nu, v)=[\sigma^{+}(\nu, v)I_{\nu}^{+}+\sigma^{-}(\nu, v)I_{\nu}^{-}] \qquad (\mathrm{sec}^{-1}) \tag{9.197}$$

where, from Eqs. (2.64d) and (2.58),

$$\sigma^{\pm}(\nu, v)=\frac{\lambda^{2}}{8\pi\tau_{\mathrm{sp}}}\cdot\frac{1}{2\pi}\frac{\Delta\nu}{[\nu-\nu^{0}(1\pm v/c)]^{2}+[(\Delta\nu)/2]^{2}} \qquad (\mathrm{cm}^{2}) \tag{9.198}$$

for Doppler broadened lines, where $\Delta\nu$ is the homogeneous line width.

To calculate the FIR output power we adopt the arguments of Section 2.6. Let ρ (cm^{-3}) be the number of photons per unit volume; then the energy density will be $h\nu\rho$ ($\mathrm{erg}\cdot\mathrm{cm}^{-3}$), and the energy flux will be $ch\nu\rho$ ($\mathrm{erg}\cdot\mathrm{cm}^{-2}\,\mathrm{sec}^{-1}$). Consequently, the one-way energy flux will be $\frac{1}{2}ch\nu\rho$. For an output mirror of area A, where t is the fractional coupling loss per double pass (round-trip), then

$$P_{\mathrm{out}}=\tfrac{1}{2}Atch\nu\rho \tag{9.199a}$$

If α_0 (cm^{-1}) is the small-signal gain, and the length of a round-trip is $2L$, then the total gain is $2L\alpha_0$. If a is the loss per double pass other than coupling losses, then the total loss is $(t+a)$ and

$$\text{net gain}=2L\alpha_0-(t+a)$$

The saturated photon density, ρ_s, is defined by

$$\rho\equiv\rho_s\frac{\text{net gain}}{\text{total loss}} \qquad (\mathrm{cm}^{-3}) \tag{9.200}$$

which leads to, introducing the saturated power P_s:

$$P_{\mathrm{out}}=\frac{1}{2}Atch\nu\rho_s\left(\frac{2L\alpha_0}{t+a}-1\right)=\frac{t}{2}P_s\left(\frac{2L\alpha_0}{t+a}-1\right) \tag{9.199b}$$

This formula is used by Henningsen and Jensen[122] and Hodges *et al.*[123] The problem now is to develop an explicit model for P_s and α_0.

9.3.2. Tucker's Theory[124]

9.3.2.1. Rate Equations

The CO_2 pumping photon is absorbed by the rotational sublevel n_1 of the ground vibrational level thereby promoting the molecule to n_2 of a low-lying excited vibrational state. The level n_2 is the upper laser level which is stimulated to give up an FIR photon and drop to the neighboring n_3 level. Within each rotational sublevel the pumping is only effective for molecules inside a narrow velocity range, and collisions tend to restore the velocity distribution toward Maxwellian, $f(v)$. The objective of this section is to derive rate equations for $n_i(v, t)$, whose degeneracies are denoted by

$$g_i = 2J_i + 1, \qquad i = 1, 2, 3 \tag{9.201}$$

There is a large number of rotational sublevels within k_BT of the active levels in both upper and lower vibrational manifolds, and those *not* explicitly labled are considered together with population M and N, respectively, where

$$M \gg n_2(v) \quad \text{or} \quad n_3(v) \qquad \text{and} \qquad N \gg n_1(v) \tag{9.202}$$

Consequently, the equilibrium populations of the three levels will be

$$n_1^e(v) = f(v)f_1N, \qquad n_2^e(v) = f(v)f_2M, \qquad n_3^e(v) = f(v)f_3M \tag{9.203}$$

where f_i, the Boltzmann occupation factors, denote the fraction of vibrational manifold population contained in each of these rotational sublevels in thermal equilibrium [see Eq. (4.19) and its approximation, Eq. (4.23)].

Let τ_c be the relaxation time for establishing equilibrium among the rotational levels of a vibrational level; then the rate equations will include the usual (see Chapter 1) cross-relaxation terms

$$\text{(A)} \qquad (n_i - n_i^e)/\tau_c, \ldots, \quad \text{etc.}$$

The relaxation time, τ_c, is the reciprocal of the collision frequency

$$W_c = \tau_c^{-1} \qquad (\text{sec}^{-1}) \tag{9.204}$$

From Eqs. (2.43) and (2.44) we showed that the net stimulated increase in radiation energy density per unit time is [see also Eq. (9.197)]

$$\Delta N \lambda^2 I_\nu g(\nu, v)/8\pi\tau_{sp} = \Delta N \sigma(\nu, v) I_\nu \tag{9.205}$$

where we recall from Eq. (2.49) that $g(\nu, v)$ is a line-width function:

$$g(\nu, v) = \frac{1}{2\pi} \frac{\Delta\nu}{[\nu - \nu^0(1 + v/c)]^2 + [(\Delta\nu)/2]^2} \tag{9.206}$$

We note that

$$I_\nu^{\pm} = I_\nu/2h\nu \tag{9.207}$$

and so the net stimulated increase in photon density will be Eq. (9.205) divided by $h\nu$, which we will write in the form

(B) $$\Delta n\, W(\nu, v)$$

where $W_p(\nu_p, v)$ and $W_f(\nu_f, v)$ are the induced transition rates for optical pumping and stimulated FIR emission, respectively, given by Eq. (9.197). We note that Tucker uses I_ν and takes $+v$ only in Eq. (9.197). For the pump transition we have

$$\Delta n_{21} \equiv n_2 - \frac{g_2 n_1}{g_1} \tag{9.208}$$

and similarly for Δn_{23}. The number of molecules in level i, whatever their velocity, is given by integrating over all velocities u

$$\int n_i(u)\, du$$

and the equilibrium fraction of them with velocity v will be

$$f(v) \int n_i(u)\, du \equiv \bar{n}_i^e(v) \tag{9.209}$$

Within each rotational sublevel, collisions will take place, at frequency W_K, which will tend to restore the velocity distribution toward its equilibrium Maxwellian profile; consequently, the rate equations will include velocity cross-relaxation terms of the form

(C) $$W_K[n_i(v) - \bar{n}_i^e], \ldots, \quad \text{etc.}$$

The rate equations for the levels' occupation numbers per unit volume are constructed from (A), (B), and (C) above:

$$\dot{n}_1(v) = W_p(\nu_p, v)\left[n_2(v) - \frac{g_2 n_1(v)}{g_1}\right] - W_c[n_1(v) - n_1^e(v)] - W_K[n_1(v) - \bar{n}_1^e] \tag{9.210}$$

$$\dot{n}_2(v) = -W_p(\nu_p, v)\left[n_2(v) - \frac{g_2 n_1(v)}{g_1}\right] - W_f(\nu_f, v)\left[n_2(v) - \frac{g_2 n_3(v)}{g_3}\right] - W_c[n_2(v) - n_2^e(v)] - W_K[n_2(v) - \bar{n}_2^e] \tag{9.211}$$

$$\dot{n}_3(v) = W_f(\nu_f, v)\left[n_2(v) - \frac{g_2 n_3(v)}{g_3}\right] - W_c[n_3(v) - n_3^e(v)] - W_K[n_3(v) - \bar{n}_3^e(v)] \tag{9.212}$$

If N_0 represents the total density of active molecules, then

$$N+M \approx N_0 \tag{9.213}$$

If Γ denotes the rate at which molecules in the upper vibrational manifold relax to the ground-state manifold, then the number of such transitions is

(D) $$\Gamma M$$

The number of molecules which are taken out of the "bath" N into level "1," independent of velocity, is given by

$$W_c f_1 N$$

Hence we have the additional pair of rate equations

$$\dot{N} = W_c\left[\int n_1(u)\,du - f_1 N\right] + \Gamma M \tag{9.214}$$

$$\dot{M} = W_c\left\{\int [n_2(u)+n_3(u)]\,du - (f_2+f_3)M\right\} - \Gamma M \tag{9.215}$$

where the left-hand terms of these equations allow for the designated levels populating the baths N and M.

9.3.2.2. *CW Solution of the Rate Equations*

The saturated gain at the FIR frequency, ν_f, is given by Eq. (2.64b) to be in the (+) direction of propagation:

$$\alpha(\nu_f, v) = \sigma^+(\nu_f, v)\,\Delta n_{23}(v) \qquad (\text{cm}^{-1}) \tag{9.216}$$

where the effective radiative cross section is given by Eq. (9.198). With the superscripts $\pm$ being dropped at line center we have

$$\sigma(\nu_f) = [\lambda_f^2 g(\nu_f)/8\pi\tau_{\text{sp}}^f] \qquad (\text{cm}^2)$$

where g is given in Eq. (9.206) and the rate equations are solved for Δn_{23} with $\dot{n}_i = 0$, but $I_f \neq 0$ for saturated gain.

It is assumed that λ_f is so long that Doppler broadening is unimportant at all reasonable pressures; therefore the induced transition rate for stimulated FIR emission is velocity-independent:

$$W_f(\nu_f, v) \to W_f(0) \tag{9.217}$$

We now define the deviations of the populations from their thermalized values,

$$\begin{aligned} \Delta n_1(v) &\equiv n_1(v) - n_1^e(v) \\ \Delta n_2(v) &\equiv n_2(v) - n_2^e(v) \\ \Delta n_3(v) &\equiv n_3(v) - n_3^e(v) \end{aligned} \tag{9.218}$$

and for steady-state operation the first of the rate equations becomes

$$0 = W_p(\nu_p, v)\Delta n_2(v) + W_p(\nu_p, v)n_2^e(v) - \left[\frac{g_2 W_p(\nu_p, v)}{g_1} + W_c\right]\Delta n_1(v)$$
$$+\frac{g_2 W_p(\nu_p, v)n_1^e}{g_1} - W_K \Delta n_1(v) - W_K n_1^e(v) + W_K \bar{n}_1^e$$

which can be rewritten as, dropping ν_p in $W_p(\nu_p, v)$ henceforth,

$$0 = -\left[W_K + W_c + \frac{g_2}{g_1} W_p(v)\right] \Delta n_1(v) + W_p(v)\Delta n_2(v)$$
$$-W_p(v)f(v)\left[\frac{g_2}{g_1} f_1 N - f_2 M\right] + W_K\left[f(v)\int n_1(u)\,du - f(v)f_1 N\right] \qquad (9.219)$$

where

$$f_1 N \equiv \int n_1^e(u)\,du \qquad (9.220)$$

can be substituted for the last term in Eq. (9.219).

Equation (9.214) can be written in the steady state as

$$0 = W_c\left\{\int [n_1(u) - n_1^e(u)]\,du\right\} + \Gamma M$$

or

$$M = -\frac{W_c}{\Gamma}\int \Delta n_1(u)\,du \qquad (9.221)$$

Equation (9.211) becomes

$$0 = -W_p(v)\Delta n_2(v) - W_p(v)n_2^e(v) + W_p(v)\frac{g_2}{g_1}\Delta n_1(v) + W_p(v)\frac{g_2}{g_1}n_1^e(v)$$
$$-W_f(0)\Delta n_2(v) - W_f(0)n_2^e(v) + W_f(0)\frac{g_2}{g_1}\Delta n_3(v) + W_f(0)\frac{g_2}{g_3}n_3^e(v)$$
$$-W_c\Delta n_2(v) - W_K\Delta n_2(v) - W_K n_2^e(v) + W_K \bar{n}_2^e$$

which can be rearranged into the form

$$0 = -[W_K + W_c + W_p(v) + W_f(0)]\,\Delta n_2(v) + \frac{g_2}{g_1}W_p(v)\Delta n_1(v)$$
$$+ W_f(0)\frac{g_2}{g_1}\Delta n_3(v) + W_p(v)f(v)\left(-f_2 M + \frac{g_2}{g_1}f_1 N\right)$$
$$+ W_f(0)f(v)\left(-f_2 + \frac{g_2}{g_3}f_3\right)M + W_K f(v)\int \Delta n_2(u)\,du \qquad (9.222)$$

Under steady-state conditions Eq. (9.212) becomes

$$0 = W_f(0)\,\Delta n_2(v) + W_f(0) n_2^e(v) - W_f(0)\frac{g_2}{g_3}\,\Delta n_3(v) - W_f(0)\frac{g_2}{g_3} n_3^e(v)$$

$$-W_c\,\Delta n_3(v) - W_K\,\Delta n_3(v) - W_K n_3^e(v) + W_K \bar{n}_3^e$$

which can be rearranged into the form

$$0 = -\left[W_K + W_c + \frac{g_2}{g_3} W_f(0)\right]\Delta n_3(v) + W_f(0)\,\Delta n_2(v)$$

$$-W_f(0) f(v)\left[-f_2 + \frac{g_2}{g_3} f_3\right] M + W_K f(v)\int \Delta n_3(u)\,du \qquad (9.223)$$

Equation (9.215) becomes

$$0 = W_c\left\{\int [n_2(u) + n_3(u) - n_2^e(u) - n_3^e(u)]\,du\right\} - \Gamma M$$

or

$$M = \frac{W_c}{\Gamma}\int [\Delta n_2(u) - \Delta n_3(u)]\,du \qquad (9.224)$$

which together with Eq. (9.213) completes the time-independent system of equations.

To simplify the system of equations we add Eqs. (9.222) and (9.223) to obtain

$$0 = -(W_K + W_c)[\Delta n_2(v) + \Delta n_3(v)] + \frac{g_2}{g_1} W_p(v)\,\Delta n_1(v) - W_p(v)\,\Delta n_2(v)$$

$$+ W_p(v) f(v)\left(\frac{g_2}{g_1} f_1 N - f_2 M\right) + W_K f(v)\int [\Delta n_2(u) + \Delta n_3(u)]\,du$$

When this result is compared with Eq. (9.219) we are able to make the identification

$$\Delta n_2(v) + \Delta n_3(v) = -\Delta n_1(v) \qquad (9.225)$$

Equation (9.225) is used to eliminate Δn_2 from the steady-state system of equations to obtain Eq. (9.219) as

$$0 = -\left[W_K + W_c + \left(1 + \frac{g_2}{g_1}\right) W_p(v)\right]\Delta n_1(v) - W_p(v)\,\Delta n_3(v)$$

$$- W_p(v) f(v)\left(\frac{g_2}{g_1} f_1 N - f_2 M\right) - W_K f(v)\frac{\Gamma M}{W_c} \qquad (9.226)$$

having used Eq. (9.221). Equation (9.223) becomes

$$0=-\left[W_K+W_c+\left(1+\frac{g_2}{g_3}\right)W_f(0)\right]\Delta n_3(v)-W_f(0)\,\Delta n_1(v)$$

$$-W_f(0)f(v)\left(\frac{g_2f_3}{g_3}-f_2\right)M+W_Kf(v)\int\Delta n_3(u)\,du \qquad (9.227)$$

Equations (9.226) and (9.227) together with (9.221), (9.225), and (9.213) provide a system of five equations for the five unknowns Δn_1, Δn_2, Δn_3, N, and M. We begin by integrating Eq. (9.227) over velocity, using Eq. (9.221), to obtain

$$0=-\left[W_K+W_c+\left(1+\frac{g_2}{g_3}\right)W_f(0)\right]\int\Delta n_3(u)\,du+W_f(0)\frac{\Gamma M}{W_c}$$

$$-W_f(0)\left[\frac{g_2}{g_3}f_3-f_2\right]M+W_K\int\Delta n_3(u)\,du$$

from which we obtain

$$\int\Delta n_3(u)\,du=\frac{\Gamma MW_f(0)}{W_c[W_c+(1+g_2/g_3)W_f(0)]}\left[1-\frac{W_c}{\Gamma}\left(\frac{g_2f_3}{g_3}-f_2\right)\right] \qquad (9.228)$$

This result is substituted back into Eq. (9.227) to give

$$\Delta n_3(v)=\frac{-W_f(0)\,\Delta n_1(v)}{[W_K+W_c+(1+g_2/g_3)W_f(0)]}+\frac{1}{[W_K+W_c+(1+g_2/g_3)W_f(0)]}$$

$$\times\left\{-W_f(0)f(v)M\left[\frac{g_2f_3}{g_3}-f_2\right]\right.$$

$$\left.+\frac{W_Kf(v)\Gamma MW_f(0)}{W_c[W_c+(1+g_2/g_3)W_f(0)]}\left[1-\frac{W_c}{\Gamma}\left(\frac{g_2f_3}{g_3}-f_2\right)\right]\right\}$$

$$=\frac{-W_f(0)\,\Delta n_1(v)}{[W_K+W_c+(1+g_2/g_3)W_f(0)]}+\frac{1}{[W_K+W_c+(1+g_2/g_3)W_f(0)]}$$

$$\times\left(\frac{W_Kf(v)\Gamma MW_f(0)}{W_c[W_c+(1+g_2/g_3)W_f(0)]}-W_f(0)f(v)M\left(\frac{g_2f_3}{g_3}-f_2\right)\right.$$

$$\left.\times\left\{1+\frac{W_K}{[W_c+(1+g_2/g_3)W_f(0)]}\right\}\right)$$

$$=\frac{-W_f(0)\,\Delta n_1(v)}{[W_K+W_c+(1+g_2/g_3)W_f(0)]}+\frac{[f(v)\Gamma MW_f(0)]/W_c}{W_c+(1+g_2/g_3)W_f(0)}$$

$$\times\left[\frac{W_K}{W_K+W_c+(1+g_2/g_3)W_f(0)}-\left(\frac{g_2f_3}{g_3}-f_2\right)\frac{W_c}{\Gamma}\right] \qquad (9.229)$$

To simplify the algebra, it is convenient to introduce the numbers

$$C_4 \equiv W_f(0)\left[W_c + \left(1+\frac{g_2}{g_1}\right)W_f(0)\right]^{-1} \tag{9.230}$$

and C_1 defined by

$$(1-C_1)\left(1+\frac{g_2}{g_1}\right) \equiv W_f(0)\left[W_K + W_c + \left(1+\frac{g_2}{g_3}\right)W_f(0)\right]^{-1} \tag{9.231}$$

With these definitions, Δn_3 can be written as

$$\Delta n_3(v) = -(1-C_1)\left(1+\frac{g_2}{g_1}\right)\Delta n_1(v) + \frac{f(v)\Gamma M C_4}{W_c}\left[\frac{W_K}{W_K + W_c + (1+g_2/g_3)W_f(0)} - \left(\frac{g_2 f_3}{g_3} - f_2\right)\frac{W_c}{\Gamma}\right] \tag{9.232}$$

which gives $\Delta n_3(v)$ in terms of $\Delta n_1(v)$ and M. To derive an expression for $\Delta n_1(v)$ in terms of the M and N_0, we substitute Eq. (9.232) into Eq. (9.226) to give, writing W_p for $W_p(\nu_p, v)$,

$$0 = -\left[W_K + W_c + C_1\left(1+\frac{g_2}{g_1}\right)W_p\right]\Delta n_1 - \frac{W_p f(v)\Gamma M C_4}{W_c}\left[\frac{W_K}{W_K + W_c + (1+g_2/g_3)W_f(0)} - \left(\frac{g_2 f_3}{g_3} - f_2\right)\frac{W_c}{\Gamma}\right] - W_p f(v)\frac{g_2 f_1}{g_1}(N_0 - M) + W_p f(v) f_2 M - W_K f(v)\frac{\Gamma M}{W_c}$$

which can be rearranged into the form

$$\Delta n_1\left[W_K + W_c + C_1\left(1+\frac{g_2}{g_1}\right)W_p\right] = -W_K\frac{\Gamma M}{W_c}f(v) - W_p f(v)\left\{\frac{g_2 f_1 N_0}{g_1} + \frac{\Gamma M}{W_c}W_K C_4\left[W_K + W_c + \left(1+\frac{g_2}{g_3}\right)W_f\right]^{-1} - M\left[\left(\frac{g_2 f_1}{g_1} + f_2\right) + \left(\frac{g_2 f_3}{g_3} - f_2\right)C_4\right]\right\}$$

Hence

$$\Delta n_1(v) \equiv -\left[W_K\frac{\Gamma M}{W_c}f(v) + C_0 W_p(v) f(v)\right]\left[W_K + W_c + C_1\left(1+\frac{g_2}{g_1}\right)W_p(v)\right] \tag{9.233}$$

which defines the number C_0. The velocity dependence of the various factors in Eq. (9.233) is shown explicitly.

We now define the number

$$C_p(\nu_p) \equiv \int dv f(v) \frac{C_1(1+g_2/g_1)W_p(\nu_p, v)}{W_K + W_c + C_1(1+g_2/g_1)W_p(\nu_p, v)} \tag{9.234}$$

and note the result

$$\begin{aligned} 1 - C_p &= \int dv\, f(v) - C_p \\ &= \int dv f(v) \frac{W_K + W_c}{W_K + W_c + C_1(1+g_2/g_1)W_p(v)} \end{aligned} \tag{9.235}$$

Equation (9.233) is integrated with respect to the velocity to obtain, using Eq. (9.221):

$$-\frac{\Gamma M}{W_c} = -\frac{W_K \Gamma M(1-C_p)}{W_c(W_K + W_c)} - \frac{C_0 C_p}{C_1(1+g_2/g_1)} \tag{9.236}$$

Since C_0 depends on M [see Eq. (9.233)], its explicit form must be substituted into Eq. (9.236) to solve for M. We begin by naming the last factor in C_0 as

$$C_3\left(1+\frac{g_2}{g_1}\right) \equiv \left(1+\frac{g_2 f_1}{g_1 f_2}\right) + \left(\frac{g_2 f_3}{g_3 f_2} - 1\right) C_4 \tag{9.237}$$

and so

$$C_0 = \frac{g_2 f_1 N_0}{g_1} + \frac{\Gamma M W_K}{W_c} C_4 \left[W_K + W_c + \left(1+\frac{g_2}{g_3}\right) W_f \right]^{-1} - M f_2 C_3 \left(1+\frac{g_2}{g_1}\right) \tag{9.238}$$

This result is substituted into Eq. (9.236) to give

$$\begin{aligned} M\Bigg[1 + C_p\frac{W_K}{W_c} - \left(\frac{W_K + W_c}{W_c}\right)\left(\frac{C_p}{C_1}\right)\frac{W_K C_4}{(1+g_2/g_1)[W_K + W_c + (1+g_2/g_3)W_f(0)]} \\ + \left(\frac{W_K + W_c}{W_c}\right)\left(\frac{C_p}{C_1}\right) C_3 f_2 \frac{W_c}{\Gamma}\Bigg] = \left(\frac{W_K + W_c}{W_c}\right)\left(\frac{C_p}{C_1}\right)\frac{g_2}{g_1 + g_2} N_0 f_1 \frac{W_c}{\Gamma} \end{aligned} \tag{9.239}$$

We now define the number C_2 by

$$\frac{C_4}{1+g_2/g_1} \equiv 1 - C_2 \tag{9.240}$$

and use the defining equations for C_1 and C_4 to eliminate

$$W_c + \left(1+\frac{g_2}{g_3}\right) W_f(0) \equiv \beta$$

In other words, taking the ratio of Eqs. (9.226) and (9.227) we obtain

$$\frac{C_4}{(1-C_1)(1+g_2/g_1)}=\frac{W_K+\beta}{\beta}$$

from which we obtain

$$\frac{W_K}{\beta}=\frac{C_1-C_2}{1-C_1}$$

which is substituted into Eq. (9.239) to obtain

$$M\left\{1+C_p\frac{W_K}{W_c}-\left(1+\frac{W_K}{W_c}\right)\left(\frac{C_p}{C_1}\right)(1-C_2)\left[1+\frac{1-C_1}{C_1-C_2}\right]^{-1}\right.$$
$$\left.+\left(\frac{W_K+W_c}{W_c}\right)\left(\frac{C_p}{C_1}\right)C_3f_2\frac{W_c}{\Gamma}\right\}=\frac{g_2}{g_1+g_2}N_0f_1\frac{W_c}{\Gamma}\left(\frac{W_K+W_c}{W_c}\right)\left(\frac{C_p}{C_1}\right)$$

Finally, simplifying the terms in the curly brackets, we obtain

$$M\left[1-C_p+\left(\frac{W_K+W_c}{W_c}\right)\left(\frac{C_p}{C_1}\right)(C_2+C_3f_2\frac{W_c}{\Gamma}\right]$$
$$=\frac{g_2}{g_1+g_2}N_0f_1\frac{W_c}{\Gamma}\left(\frac{W_K+W_c}{W_c}\right)\left(\frac{C_p}{C_1}\right) \tag{9.241}$$

In this result, we see that C_1 and C_4 are defined in terms of the degeneracies, g_i, and the various rates, W [see Eqs. (9.231) and (9.230) respectively], which in turn define C_p, C_3, and C_2. Hence we can calculate M knowing N_0 and f_1. Now that we know M we calculate C_0 from Eq. (9.236):

$$\frac{\Gamma M}{W_c}\left[1-\frac{W_K}{W_K+W_c}(1-C_p)\right]\left(\frac{C_1}{C_p}\right)\left(1+\frac{g_2}{g_1}\right)=C_0$$

which is substituted into Eq. (9.233) to give

$$-\Delta n_1(v)=\left[W_K+W_c+C_1\left(1+\frac{g_2}{g_1}\right)W_p(\nu)\right]^{-1}\left(\frac{\Gamma Mf(v)}{W_c}\right)$$
$$\times\left[W_K+W_p(v)\left(\frac{W_c+W_KC_p}{W_K+W_c}\right)\left(\frac{C_1}{C_p}\right)\left(1+\frac{g_2}{g_1}\right)\right]$$
$$=\frac{\Gamma Mf(v)}{W_c}\left\{\frac{W_K}{W_K+W_c}+\frac{W_c}{W_K+W_c}\left(\frac{C_1}{C_p}\right)\left(1+\frac{g_2}{g_1}\right)\right.$$
$$\left.\times\left[W_K+W_c+C_1\left(1+\frac{g_2}{g_1}\right)W_p(v)\right]^{-1}W_p(v)\right\} \tag{9.242}$$

which expresses $\Delta n_1(v)$ in terms of known quantities, and so the steady-state result has been obtained since Δn_3 is given by Eq. (9.232) and Δn_2 by Eq. (9.225). The population inversion for the FIR transition is defined by

$$\begin{aligned}
\Delta n_{23}(v) &\equiv n_2(v)-\frac{g_2}{g_3}n_3(v) \\
&= \Delta n_2(v)+f_2 f(v)M-\frac{g_2}{g_3}[\Delta n_3(v)+f_3 f(v)M] \\
&= \Delta n_2-\frac{g_2}{g_3}\Delta n_3-\left(\frac{g_2 f_3}{g_3}-f_2\right)f(v)M \\
&= \frac{\Gamma M}{W_c}f(v)\left\{\left[1-\left(\frac{g_2 f_3}{g_3}-f_2\right)\frac{W_c}{\Gamma}\right]\frac{W_c}{W_c+(1+g_2/g_3)W_f(0)}\right. \\
&\quad+\frac{W_c}{W_K+W_c+(1+g_2/g_3)W_f(0)} \\
&\quad\left.\times\left[\left(\frac{C_1}{C_p}\right)\frac{(1+g_2/g_1)W_p(v)}{W_K+W_c+C_1(1+g_2/g_1)W_p(v)}-1\right]\right\}
\end{aligned} \tag{9.243}$$

When this result is substituted into Eq. (9.216) for the gain, *and the velocity average is taken* for a homogeneously broadened FIR transition for which $v=0$ in Eq. (9.206), we have the saturated gain in an oscillator given by

$$\alpha(\nu_f)=\frac{\Gamma}{W_c}M\left[1-\left(\frac{g_2 f_3}{g_3}-f_2\right)\frac{W_c}{\Gamma}\right]\frac{\lambda_f^2}{8\pi\tau_{sp}^f}g(\nu_f)\frac{W_c}{W_c+(1+g_2/g_3)W_f(0)} \tag{9.244}$$

since the velocity average of the second term in Eq. (9.243) vanishes when we use the defining equation for C_p. When Eq. (9.241) is substituted for M in Eq. (9.244), we define the function

$$F(I_p)\equiv\frac{(1+W_K/W_c)(C_p/C_1)}{(1-C_p)+(1+W_K/W_c)(C_p/C_1)[C_2+C_3 f_2(W_c/\Gamma)]} \tag{9.245}$$

where we have shown the explicit dependence on the pump intensity I_p, since C_p depends on W_p, which in turn [see Eq. (9.197)] depends on I_p. Consequently, the saturated gain is given by

$$\begin{aligned}
\alpha(\nu_f) &= \frac{g_2}{g_1+g_2}f_1 N_0 \\
&\quad\times\left[1-\left(\frac{g_2 f_3}{g_3}-f_2\right)\frac{W_c}{\Gamma}\right]F\frac{\lambda_f^2}{8\pi\tau_{sp}^f}g(\nu_f)\frac{1}{1+(1+g_2/g_3)(W_f(0)/W_c)}
\end{aligned} \tag{9.246}$$

From Eqs. (9.197) and (9.207) we have the induced transition rate for the FIR transition, from which we have the FIR intensity given by

$$I_f = \frac{8\pi h\nu_f \tau_{sp}^f}{\lambda_f^2} \frac{W_f(0)}{g(\nu_f)} \tag{9.247}$$

This expression is substituted for $W_f(0)$ in the final factor of Eq. (9.246) to give

$$\alpha(\nu_f) = \frac{g_2}{g_1+g_2} f_1 N_0 \times \left[1 - \left(\frac{g_2 f_3}{g_3} - f_2\right)\frac{W_c}{\Gamma}\right] F \frac{\lambda_f^2}{8\pi\tau_{sp}^f} g(\nu_f) \frac{1}{1+[I_f/I_s(\nu_f)]} \quad (\text{cm}^{-1}) \tag{9.248}$$

where we have defined the saturation parameter for the FIR intensity by

$$I_s(\nu_f) \equiv \frac{8\pi h\nu_f \tau_{sp}^f}{\lambda_f^2} \frac{W_c}{g(\nu_f)} \left(1 + \frac{g_2}{g_3}\right)^{-1} \times 10^{-7} (\text{W cm}^{-2}) \tag{9.249}$$

9.3.2.3. *FIR Gain*

The interpretation of Eq. (9.246) can be understood as follows. The density of molecules in level n_1 in thermal equilibrium is $f_1 N_0$. A sufficiently strong pump field will result in a fraction $g_2/(g_1+g_2)$ being promoted to the upper FIR level. The saturation of the inversion with pump power is contained in $F(I_p)$ which includes the ratio W_c/Γ. This ratio is also found in the factor

$$1 - \frac{W_c}{\Gamma}\left(\frac{g_2 f_3}{g_3} - f_2\right) > 0 \qquad \text{for positive gain} \tag{9.250}$$

If the rotational thermalization rate W_c is too large compared to the vibrational relaxation rate Γ, then a molecule initially pumped into n_2 will undergo so many rotational collisions before decaying back to the ground state that population of the rotational sublevels in the upper manifold approaches a thermal distribution, and there will be no inversion for the FIR transition. However, if Γ is sufficiently large, population of the lower FIR level n_3 by collision is avoided and a net inversion is achieved.

The function $F(I_p)$ contains the dependence of FIR gain on pump intensity.

We now collect together the various other quantities required in the calculation of $\alpha(\nu_f)$. First we require $F(I_p)$, which is given by Eq. (9.245) and which requires C_1, C_2, C_3, and C_p. From Eqs. (9.231), (9.247), and (9.249) we have

$$C_1 = 1 - \frac{g_1}{g_1+g_2}\frac{W_f(0)/W_c}{(1+W_K/W_c)+(W_f(0)/W_c)(1+g_2/g_3)}$$

$$= 1 - \frac{g_1}{g_1+g_2}\frac{g_3}{g_2+g_3}\frac{I_f/I_s(\nu_f)}{(1+W_K/W_c)+[I_f/I_s(\nu_f)]} \tag{9.251}$$

All the quantities required to calculate I_f and $I_s(\nu_f)$ [see Eqs. (9.247) and (9.249), respectively] are given in Table 24 in Section 9.3.4. From Eqs. (9.240), (9.247), (9.249), and (9.230) we have

$$C_2 = 1 - \frac{g_1}{g_1+g_2}\frac{g_3}{g_2+g_3}\frac{I_f/I_s(\nu_f)}{1+[I_f/I_s(\nu_f)]} \tag{9.252}$$

From Eq. (9.237) we have

$$C_3 = \frac{g_1}{g_1+g_2}\left\{\left(1+\frac{g_2 f_1}{g_1 f_2}\right)+\left(\frac{g_2 f_3}{g_3 f_2}-1\right)\frac{I_f/I_s(\nu_f)}{1+[I_f/I_s(\nu_f)]}\right\}$$

$$= 1 + \frac{g_2}{g_1+g_2}\left(\frac{f_1}{f_2}-1\right) + \frac{g_1}{g_1+g_2}\frac{g_3}{g_2+g_3}\left(\frac{g_2 f_3}{g_3 f_2}-1\right)\frac{I_f/I_s(\nu_f)}{1+[I_f/I_s(\nu_f)]} \tag{9.253}$$

Finally, C_p is defined by Eq. (9.234), with $W_p(v)$ given by Eq. (9.197) and

$$f(v) = (\pi^{1/2}\,\Delta v)^{-1}\exp(-v^2/\Delta v^2), \qquad \Delta v = (2k_B T/M)^{1/2}$$

Hence, we have

$$C_p = \pi^{-1/2}\int_{-\infty}^{\infty}\frac{dv}{\Delta v}\exp(-v^2/\Delta v^2)\left[1+\frac{W_K+W_c}{C_1(1+g_2/g_1)W_p(v)}\right]^{-1}$$

The factor in the denominator becomes, using Eqs. (9.197), (9.198), and (9.207),

$$\left(1+\frac{g_2}{g_1}\right)\frac{W_p(v)}{W_K+W_c} \to \frac{(1+g_2/g_1)\lambda_p^2}{8\pi h\nu_p\tau_{\text{sp}}^p}\frac{2}{(W_K+W_c)\pi\Delta\nu_h}\frac{I_p}{2}$$

$$\times\left\{\frac{(\Delta\nu_h/2)^2}{[\nu_p-\nu_p^0(1+v/c)]^2+(\Delta\nu_h/2)^2}+\frac{(\Delta\nu_h/2)^2}{[\nu_p-\nu_p^0(1-v/c)]^2+(\Delta\nu_h/2)^2}\right\}$$

$$\equiv I_0^{-1}\left(\left\{1+\left[\frac{2\nu_p-2\nu_p^0(1+v/c)}{\Delta\nu_h}\right]^2\right\}^{-1}+\left\{1+\left[\frac{2\nu_p-2\nu_p^0(1-v/c)}{\Delta\nu_h}\right]^2\right\}^{-1}\right)\frac{I_p}{2} \tag{9.254}$$

which defines the saturation parameter for the pump intensity, I_0. Hence, C_p is given by

$$C_p(\nu_p, I_p) = \pi^{-1/2} \int_{-\infty}^{\infty} \frac{dv}{\Delta v} \exp(-v^2/\Delta v^2)$$

$$\times\left[1 + \frac{2I_0}{C_1 I_p}\left(\sum_{\&=\pm}\left\{1 + \left[\frac{2\nu_p - 2\nu_p^0(1 \,\&\, v/c)}{\Delta\nu_h}\right]^2\right\}^{-1}\right)^{-1}\right]^{-1} \quad (9.255)$$

where the & means the sum of the two terms, one with a + and one with a −. To simplify the notation we define

$$\alpha \equiv (2\nu_p - 2\nu_p^0)/\Delta\nu_h, \qquad \beta \equiv 2\nu_p^0/c\,\Delta\nu_h, \qquad \gamma \equiv 2I_0/C_1 I_p \quad (9.256)$$

and readily show the denominator to be

$$1 + \gamma\left[\frac{1 + 2(\alpha^2 + v^2\beta^2) + (\alpha^2 - v^2\beta^2)^2}{2(1 + \alpha^2 + v^2\beta^2)}\right]$$

When this factor is regrouped according to v^2 we have

$$C_p = \pi^{-1/2} \int_{-\infty}^{\infty} \frac{dv}{\Delta v} \exp(-v^2/\Delta v^2) \frac{d + ev^2}{av^4 + bv^2 + c} \quad (9.257)$$

where

$$a \equiv \beta^4\gamma, \qquad b \equiv 2\beta^2(1 + \gamma - \gamma\alpha^2), \qquad c \equiv 2(1 + \alpha^2) + \gamma + 2\gamma\alpha^2 + \gamma\alpha^4$$
$$d \equiv 2(1 + \alpha^2), \qquad e \equiv 2\beta^2 \quad (9.258)$$

In Tucker's theory, only the first term in included in the sum & in Eq. (9.245) and he introduces the parameter

$$\varepsilon_p \equiv (\beta\,\Delta v)^{-1} = \frac{\Delta\nu_h}{[2\nu_p^0(2k_BT/Mc^2)^{1/2}]}$$
$$= (\ln 2)^{1/2} \frac{\Delta\nu_h}{\Delta\nu_D^p} \quad (9.259)$$

where we have introduced the Doppler width at the pump frequency, $\Delta\nu_D^p$. Consequently, ε_p is proportional to the ratio of the two widths, and we can rewrite C_p in the form

$$C_p(\nu_p, I_p) = \pi^{-1/2} \int_{-\infty}^{\infty} dx \exp(-x^2) \frac{d + 2(x/\varepsilon_p)^2}{\gamma(x/\varepsilon_p)^4 + 2(1 + \gamma - \gamma\alpha^2)(x/\varepsilon_p)^2 + c}$$

where $x \equiv v/\Delta v$. This integral can be evaluated by the quadrature formula given in Krylov.[125]

To calculate the small-signal gain $\alpha_0(\nu_p)$ we also use Eq. (9.248), but $I_f = 0$ in that formula as well as in C_1, C_2, and C_3. In other words we solve the steady-state rate equations with I_f set to zero in those equations.

9.3.3. Absorption at the Pump Frequency

The absorption coefficient γ (cm^{-1}) at the pump frequency is defined by

$$\gamma(\nu_p) \equiv \int dv[n_2(v) - g_2 n_1(v)/g_1]\left\{\frac{1}{2}[\sigma^+(\nu_p, v) + \sigma^-(\nu_p, v)]\right\} \tag{9.260}$$

There are two cases of interest: first, where there is no FIR present, and second, in the presence of FIR. In Tucker's theory, he obtained formulas for the absorption coefficient in the absence of FIR, taking only σ^+ into account.

In the first case, no FIR, we have

$$W_f(0) = 0 \qquad \text{and} \qquad \Delta n_3 = 0$$

and from Eqs. (9.225) and (9.242)

$$\Delta n_2 = -\Delta n_1$$

$$= \frac{\Gamma M f(v)}{W_c}\left[\frac{W_K}{W_K + W_c} + \frac{W_c}{W_K + W_c}\frac{1}{C_p^0}\frac{(1 + g_2/g_1)W_p(v)}{W_K + W_c + (1 + g_2/g_1)W_p(v)}\right] \tag{9.261}$$

since from Eq. (9.231) we have $C_1^0 = 1$. The superscript 0 denotes no FIR. From Eq. (9.241) we have

$$M = \frac{g_2}{g_1 + g_2} N_0 f_1 \frac{W_c}{\Gamma} F^0(I_p) \tag{9.262}$$

where F^0 is given by Eq. (9.245) with, from Eq. (9.237):

$$C_3^0 = 1 + \frac{g_2}{g_1 + g_2}\left(\frac{f_1}{f_2} - 1\right) \tag{9.263}$$

since $C_4^0 = 0$ from Eq. (9.230).

The inversion on the pump transition is

$$[n_2(v) - g_2 n_1(v)/g_1] = \Delta n_2(v) + f(v) f_2 M - \frac{g_2}{g_1}\Delta n_1(v) - \frac{g_2}{g_1} f(v) f_1 N$$

$$= \left(1 + \frac{g_2}{g_1}\right)[-\Delta n_1(v)] - \frac{g_2}{g_1} f_1 N_0 f(v) + \left(f_2 + \frac{g_2}{g_1} f_1\right) M f(v) \tag{9.264}$$

When Eqs. (9.261) and (9.262), using the explicit form of $F^0(I_p)$, are substituted into this result we obtain

$$\left[n_2(v)-\frac{g_2}{g_1}n_1(v)\right]=-\frac{g_2}{g_1}f_1N_0f(v)\left[(1-C_p^0)+\frac{W_K+W_c}{W_c}C_p^0\left(1+C_3^0f_2\frac{W_c}{\Gamma}\right)\right]^{-1}$$
$$\times(W_K+W_c)\left[W_K+W_c+\left(1+\frac{g_2}{g_1}\right)W(\nu_p,v)\right]^{-1} \qquad (9.265)$$

When this result is substituted into Eq. (9.260) and we use Eqs. (9.197) and (9.207) we obtain

$$\gamma(\nu_p)=-\frac{g_2}{g_1}f_1N_0\left[(1-C_p^0)+\frac{W_K+W_c}{W_c}C_p^0\left(1+C_3^0f_2\frac{W_c}{\Gamma}\right)\right]^{-1}$$
$$\times\int dv\frac{1}{2}\frac{W(\nu_p,v)}{I_p/2h\nu_p}\frac{W_K+W_c}{W_K+W_c+(1+g_2/g_1)W(\nu_p,v)}$$
$$=-\frac{g_2f_1N_0h\nu_p(W_K+W_c)}{g_1I_p(1+g_2/g_1)}C_p^0(\nu_p)\left[(1-C_p^0)\right.$$
$$\left.+\frac{W_K+W_c}{W_c}C_p^0\left(1+C_3^0f_2\frac{W_c}{\Gamma}\right)\right]^{-1}=-\frac{g_2}{g_1}f_1N_0\cdot\frac{\lambda_p^2}{8\pi\tau_{sp}^p}\frac{2}{\pi\,\Delta\nu_h}\frac{I_0}{I_p}$$
$$\times\frac{C_p^0(\nu_p)}{(1-C_p^0)+[(1+W_K)/W_c]C_p^0[1+C_3^0f_2(W_c/\Gamma)]} \qquad (9.266)$$

where we have used the definition of I_0 [see Eq. (9.254)].

In the presence of FIR radiation, the inversion on the pump transition is also given by Eq. (9.264), but $\Delta n_2(v)$ must include the effect of $\Delta n_3(v)$ [see Eq. (9.232)] leading to a considerably more complicated expression for $\gamma(\nu_p)$.

9.3.4. Input Data for a CW–FIR Code

In Table 24 we present a list of the physical constants required to calculate the small-signal gain and the absorption coefficient. We have included $\bar{N}_0$ at $T=300°K$ and 1 Torr in this table. To calculate the total number density at other temperatures and pressures we use

$$N_0=\bar{N}_0(T_{300}/T)p \quad \text{cm}^{-3} \qquad (9.267)$$

where p is given in Torr. According to Hodges *et al.*,[123]

$$W_c=\pi\,\Delta\nu_h, \qquad \Delta\nu_h\approx 40\text{ MHz Torr}^{-1}$$
$$=125.663706p\text{ MHz}=1.256637\times10^8p \quad \text{sec}^{-1} \qquad (9.268)$$

Table 24. Physical Constants for CW–FIR Calculation

Constant	Value	Units	Reference
g_i	$(2J_i+1)$		
h	6.6252×10^{-27}	erg sec	
c	2.998×10^{10}	cm sec^{-1}	
$\bar{N}_0$	3.223684×10^{16}	cm^{-3} Torr^{-1}	
B_0	25536.1466	MHz	Freund *et al.*[127]
B_1	25197.57	MHz	Freund *et al.*[127]
C_1-C_0	−294.09	MHz	Freund *et al.*[127]
C_0	152328.38	MHz	Andersen *et al.*[128]
λ_f	496	μm	
$(\tau_{sp}^f)^{-1}$	4.2×10^{-3}	sec^{-1}	Chang and Bridges[118]
k_B	1.38044×10^{-16}	erg $(°\text{K})^{-1}$	
ν_p^0	31383945.8	MHz	Chang and Bridges[118]
τ_{sp}^p	4.2	sec	Smith[126]
$\Delta\nu_D^p$	79.4	MHz	Eq. (9.249)

For low pressures and small diameter (d) beams, Hodges *et al.* give the vibrational relaxation to be diffusion-dominated:

$$\Gamma\approx1.8\times10^3/[p\ (\text{Torr})d^2\ (\text{cm})]\quad \text{sec}^{-1} \tag{9.269}$$

The FIR line shape will be given by

$$g(\nu_f)\approx(2/\pi)\,\Delta\nu_h\approx1.591549\times10^{-8}/p\quad \text{sec} \tag{9.270}$$

Hodges *et al.* set I_f to zero in their evaluation of this formula; an estimate can be obtained from

$$P_{\text{out}}\sim\frac{t}{2}I_fA,\qquad I_f\equiv\rho ch\nu \tag{9.199a}$$

where $t\sim0.2$. When the values of the constants given in Table 24 and the value of W_c given above are substituted into Eq. (9.249) we obtain

$$I_s(\nu_f)=3.685p^2\ (\text{Torr})\quad \text{W cm}^{-2} \tag{9.271}$$

The calculation of $F(I_p,\nu_p)$, using Eq. (9.245), requires a value for W_K, which is neglected in Hodges *et al.* [compare their Eq. (2) with the definition of I_0 given in Eq. (9.255)]. We have been unable to find a value in the literature. Smith[126] carried out calculations for both $W_K=0$ and $W_K=W_C$, thereby hoping to bracket the effect of velocity relaxation. We are now in a position to calculate C_1, C_2, and C_3 using Eqs. (9.251)–(9.253).

The fraction of molecules in a particular state, f, is the product of the fraction f_v in the vibrational state of interest (we have set $f_0 = N$ and $f_1 = M$) and the fraction of these, f_{JK}, in a particular rotational state. If statistical weight due to nuclear spin is neglected, and Bh and Ch are small compared with kT, then Townes and Schawlow (Reference 121, p. 75) give

$$f_{JK} = S(I, K)\frac{(2J+1)}{4I^2+4I+1}\exp\{[-BJ(J+1) \\ -(C-B)K^2]h/kT\}[B^2Ch^3/\pi(k_BT)^3]^{1/2} \quad (9.272)$$

where B and C are in cycles per second, I is the nuclear spin of hydrogen, and for CH_3F:

$$S(I, K) = 2(4I^2+4I)$$

The values of the rotational constants were taken from Freund *et al.*[127] and Andersen *et al.*[128]

9.3.5. SUBMMW: A Code Used to Predict CW Submillimeter Wave Laser Performance

The objectives of the code SUBMMW (see Smith[126]) are to calculate the saturated and small-signal gains for two-way and one-way pump propagation, the output powers, and absorption coefficients. Since we have been able to solve the rate equations analytically within the approximation given in Eq. (9.217) for $W_f(\nu_f, v)$, the problem reduces to formula evaluation. The heart of these formulas lies in the accurate calculation of C_p.

SUBMMW assumes that the energy levels of the lasing molecule are those of a symmetric top which is pumped by an intense laser source. Given the parameters of the pump radiation, the constants associated with the lasing molecule, and the relaxation times, SUBMMW calculates the various performance characteristics. The program input for SUBMMW is presented in Table 25.

Table 25. Program Input for SUBMMW

Card	Variable	Description	Units	Format
1	IDEBUG (I)	Debug switches		10 I5
	ISW1	$\neq 0$ calculates absorption coefficient only		I5
	TARGET SYSTEM	16 characters to specify target's name		2A8

continued

Table 25 (continued)

Card	Variable	Description	Units	Format
2	K	Axial projection of J1		I5
	J1	Rotational quantum Number of lower state		I5
	J2	Number of upper state		I5
	J3	Number of lower FIR state		I5
	B0	Rotational constants of a symmetric top	CPS	E15.6
	RC0		CPS	E15.6
	B1		CPS	E15.6
	ABFREQ	Absorption frequency of pump transition, ν_P^0	sec^{-1}	E15.6
3	WAVEF	Submillimeter wavelength, λ_f	cm	E15.6
	EMASS	Molecular mass, M_0	a.u.	E15.6
	TAUF	Spontaneous lifetime of FIR transition, τ_{sp}^f	sec	E15.6
	TAUP	Spontaneous lifetime of pump transition, τ_{sp}^P	sec	E15.6
	RATEV	Velocity relaxation rate, W_{K0}	sec^{-1} $Torr^{-1}$	E15.6
4	GAMPE1	Diffusion coefficient constant, Γ_0		E15.6
	DELNUH	Homogeneous line width, $\Delta\nu_h$	MHz	E15.6
	T	Gas temperature	°K	E15.6
	PINIT	Initial gas pressure	Torr	E15.6
	DIAM	Tube diameter, d	cm	E15.6
5	ELENG	Length of tube, L	cm	E15.6
	PF	Initial pump intensity, usually = 0	watts cm^{-2}	E15.6
	OCF	Output coupling factor, t		E15.6
	OLOSS	Other losses, a		E15.6
	PUMP	Pump intensity, I_p	watts cm^{-2}	E15.6
6	DELNUI	Initial pump frequency offset, $\Delta\nu_p$	sec^{-1}	E15.6
	NSTEPS	The number of times the calculation is repeated for either Δp or $\Delta\nu_p^0$		I5
	PINC	$<10^4$, pressure increment $\geq 10^4$, ν_p increment	Torr MHz	E15.6
	DELTA	Convergence criterion for CP integral		E15.6
	STEP	Step length for Simpson's rule in CP integral evaluation		E15.6

10

Injection Locking

10.1. Steady-State Theory

10.1.1. Regenerative (Resonant) Ring Power Amplifiers[129]

In Fig. 61 we present a schematic diagram of a reflection regenerative (i.e., resonant) ring amplifier. The purpose of this positive-feedback technique is to design high-gain CO_2 power amplifiers characterized by the compactness and efficiency associated with oscillators, yet possessing phase characteristics that will not degrade the frequency stability of master oscillator sources.

Figure 61 shows a single-port device employing only one transmitting mirror, permitting all the power to be extracted from that mirror. The ring geometry prevents any power from being directed back to the master oscillator from the amplifier causing deleterious frequency changes.

Let G_0 be the small-signal power gain and R the output mirror *power* reflectivity; then for all wavelengths, if

$$G_0 R < 1 \tag{10.1}$$

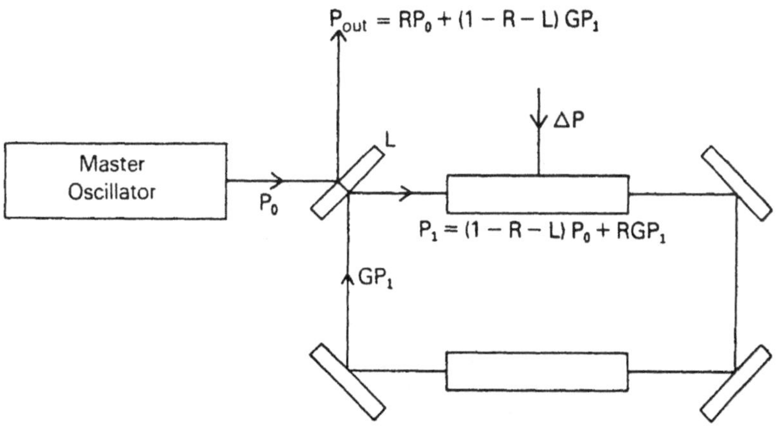

Fig. 61. Reflection regenerative ring amplifier, where L is the fractional power loss at the output mirror.

then the ring is *unconditionally stable*, i.e., it does *not* oscillate, and power is extracted from the amplifier only if the ring is driven by the master oscillator.

For

$$G_0R > 1$$

the ring can oscillate without drive (see below).

Let G be the amplifier saturated *power* gain. Under *conditions of drive* and proper tuning, G decreases such that $GR < 1$ and power is extracted from the amplifier at the desired frequency when the amplifier is tuned and locked to the master oscillator—this is called *conditionally stable* operation, which is also described as *classical injection locking* (see Tang and Statz[(130)]).

Let P_1 be the *circulating power* inside the ring amplifier and p_1 the field amplitude associated with this power:

$$P_1 \propto |p_1|^2 \tag{10.2a}$$

If we normalize p_1 such that $|p_1|^2$ is a photon number density, expressed in cm^{-3}, then

$$P_1 = ch\nu|p_1|^2 \times 10^{-7}(\text{W-cm}^{-2}) \tag{10.2b}$$

Let g be the saturated gain of the field amplitude, then

$$g^2 = G \tag{10.3}$$

and let q be the ring amplifier's perimeter.

If the *initial* amplitude of the electromagnetic wave entering the amplifier from the drive [see Eq. (10.10)] is

(A) $$a_1$$

assuming negligible random cavity photon noise, then the amplitude once around the ring will be

$$a_1 g\, e^{ikq}$$

and the *second* amplitude entering the amplifier will be

(B) $$r a_1 g\, e^{ikq}$$

where

$$r^2 \equiv R \tag{10.4}$$

We have taken the change in phase of the wave into account by introducing the factor, e^{ikq}, where the wavenumber is

$$k = \frac{2\pi}{\lambda} \tag{10.5}$$

We introduce the round-trip phase shift, also called the *cavity tuning angle*:

$$\theta \equiv kq = \frac{2\pi q}{\lambda} \tag{10.6}$$

From (B) we see that twice around the ring gives an amplitude

$$r a_1 g^2 e^{2ikq}$$

and the *third* amplitude entering the amplifier will be

$$\text{(C)} \qquad a_1 r^2 g^2 e^{2ikq}$$

Note the factor rg multiplied by the initial amplitude a_1.

The positive-feedback series in the regenerative amplifier will be obtained by summing (A), (B), (C), etc., to give in steady-state operation:

$$\begin{aligned} p_1 &= a_1 + a_1 rg\, e^{ikq} + a_1 r^2 g^2 e^{2ikq} + \cdots \\ &= a_1(1 + rg\, e^{ikq} + r^2 g^2 e^{2ikq} + \cdots) \\ &= \frac{a_1}{1 - rg\, e^{ikq}} \qquad rg < 1 \end{aligned} \tag{10.7}$$

having summed a geometric series. When Eq. (10.7) is substituted into Eq. (10.2a) we have the circulating power given by

$$P_1 \propto \frac{|a_1|^2}{|1 - rg\, e^{ikq}|^2} \tag{10.8}$$

Let $P_0 > 0$ be the drive power from the master oscillator, which is related to $P_1 \propto |a_1|^2$ by the definition of R, namely,

$$P_1 = (1 - R - L)P_0 \tag{10.9}$$

where L is the fractional power lost in the mirror, and as in Eq. (10.2a):

$$P_0 \propto |a_0|^2$$

Hence,

$$|a_1|^2 = (1 - R - L)|a_0|^2 \tag{10.10}$$

From Eqs. (10.8) and (10.10) we have

$$\frac{P_1}{P_0} = \frac{(1 - r^2 - L)}{|1 - rg\, e^{ikq}|^2} \tag{10.11}$$

We can rewrite the denominator in Eq. (10.11) as

$$\begin{aligned}(1-rg\,e^{ikq})(1-rg\,e^{-ikq}) &= 1-rg\,e^{ikq}-rg\,e^{-ikq}+r^2g^2 \\ &= (1-rg)^2+2rg-rg(e^{ikq}+e^{-ikq}) \\ &= (1-rg)^2+4rg\sin^2\frac{kq}{2} \qquad (10.12)\end{aligned}$$

and therefore the ratio of circulating to drive powers becomes

$$\frac{P_1}{P_0}=\frac{(1-r^2-L)}{(1-rg)^2+4rg\sin^2(\theta/2)} \qquad (10.13)$$

which is Eq. (4) of Buczek *et al.*,[129] when $L=0$.

The power extracted from the amplifier must be

$$\begin{aligned}\Delta P &\equiv GP_1-P_1 \\ &= (G-1)P_1 \qquad (10.14)\end{aligned}$$

Hence

$$\begin{aligned}\frac{\Delta P}{P_0} &= (g^2-1)\frac{P_1}{P_0} \\ &= \frac{(g^2-1)(1-r^2-L)}{(1-rg)^2+4rg\sin^2(\theta/2)} \qquad (10.15)\end{aligned}$$

At the output mirror we have the power balance equation (see Fig. 61),

$$P_{out}+P_1=(1-L)(P_0+GP_1) \qquad (10.16a)$$

or the *external* power gain of the resonant ring will be

$$P_{out}=(1-L)P_0+[(1-L)G-1]P_1 \qquad (10.16b)$$

or

$$\frac{P_{out}}{P_0}=(1-L)+\frac{[(1-L)g^2-1](1-r^2-L)}{(1-rg)^2+4rg\sin^2(\theta/2)}$$

having used Eq. (10.13).

The numerator will include the factor

$$\begin{aligned}&(1-L)(1-rg)^2+[(1-L)g^2-1](1-r^2-L) \\ &= 1-2rg+r^2g^2+g^2-1+r^2-r^2g^2 \\ &\quad -L+2Lrg-Lr^2g^2-Lg^2+Lg^2r^2-Lg^2+L+L^2g^2 \\ &= g^2(1+L^2-2L)+g(-2r+2rL)+r^2\end{aligned}$$

Hence

$$\frac{P_{out}}{P_0}=\frac{[g(1-L)-r]^2+4rg(1-L)\sin^2(\theta/2)}{(1-rg)^2+4rg\sin^2(\theta/2)} \tag{10.17a}$$

which is Eq. (2) of Buczek *et al.*[129], when $L=0$, for $P_0>0$.

To include the saturation characteristics of the active medium, it is assumed that the laser intensity being amplified through an incremental length is given by[131]

$$\frac{dI}{dx}=I\frac{\alpha_0}{1+I/I_s} \tag{10.18}$$

where I_s is the saturation intensity, in cm^{-3}, and α_0 is the small-signal gain coefficient. Analogously the conditions in the active medium can be described accurately in terms of the laser power, P, and the saturation power, P_s, by multiplying I factors in Eq. (10.18) by $ch\nu$ to obtain

$$\frac{dP}{dx}=P\frac{\alpha_0}{1+P/P_s}$$

which can be integrated, that is,

$$\int_{P_1}^{GP_1}\frac{1}{P}\left(1+\frac{P}{P_s}\right)dP=\alpha_0\int_0^d dx$$

where d is the discharge length; hence

$$\int_{P_1}^{GP_1}\left(\frac{dP}{P}+\frac{dP}{P_s}\right)=\alpha_0 d$$

$$\left(\log_e P+\frac{P}{P_s}\right)_{P_1}^{GP_1}=\alpha_0 d$$

$$\log_e\left(\frac{GP_1}{P_1}\right)+\frac{GP_1}{P_s}-\frac{P_1}{P_s}=\alpha_0 d$$

$$\log_e G+\frac{P_1}{P_s}(G-1)=\alpha_0 d$$

$$\log_e G=-(G-1)\frac{P_1}{P_s}+\alpha_0 d$$

$$G=\exp\left[-(G-1)\frac{P_1}{P_s}+\alpha_0 d\right] \tag{10.19}$$

Thus

$$G \exp\left[(G-1)\frac{P_1}{P_s}\right] = G_0 = g_0^2 = e^{\alpha_0 d} \tag{10.20a}$$

which is Eq. (7) of Buczek *et al.*(129)

When Eq. (10.14) is substituted for $(G-1)P_1$, Eq. (10.20a) becomes

$$\exp\left(\frac{\Delta P}{P_s}\right) = \frac{G_0}{G}$$

or, the power extracted from the amplifier is

$$\Delta P = P_s \ln\left(\frac{G_0}{G}\right) \tag{10.21a}$$

$$= 2P_s \ln\left(\frac{g_0}{g}\right) \tag{10.21b}$$

which can be regarded as an expression for the saturable gain, g, in terms of small-signal gain, g_0, saturation power P_s, and amplifier input power P_1.

Numerical Examples

Consider a reflective ring whose total discharge length d equals 1.5 m, and α_0 equals 0.8% cm^{-1}, from which we have

$$G_0 = e^{\alpha_0 d} = 3.32$$

If the saturation power P_s equals 80 W, then we can use Eqs. (10.15) and (10.21a) to obtain $G(P_0, R)$, which is substituted into Eq. (10.17a) to give P_{out}.

When the amplifier is tuned to the drive frequency, then q/λ is an integer, m, and $\sin 2\pi m = 0$; hence from Eqs. (10.15) and (10.21a):

$$\Delta P = P_0 \frac{(G-1)(1-R-L)}{(1-R^{1/2}G^{1/2})^2} = P_s \ln\left(\frac{3.32}{G}\right) \tag{10.22a}$$

which is an equation for determining g and is plotted in Fig. 62.

For a drive power $P_0 = 1$ W, $L = 0$ and a reflectivity $R = 49\%$ (hence $G_0 R = 1.62 > 1$, i.e., above threshold for self-oscillation) we have the on-resonance gain, modified as a result of the saturation characteristics of the amplifying medium given by

$$\frac{(G-1)0.51}{[1-(0.49G)^{1/2}]^2} = 80 \ln\left(\frac{3.32}{G}\right)$$

$$G \simeq 1.7156; \qquad \text{therefore } GR = 0.8406 < 1 \tag{10.22b}$$

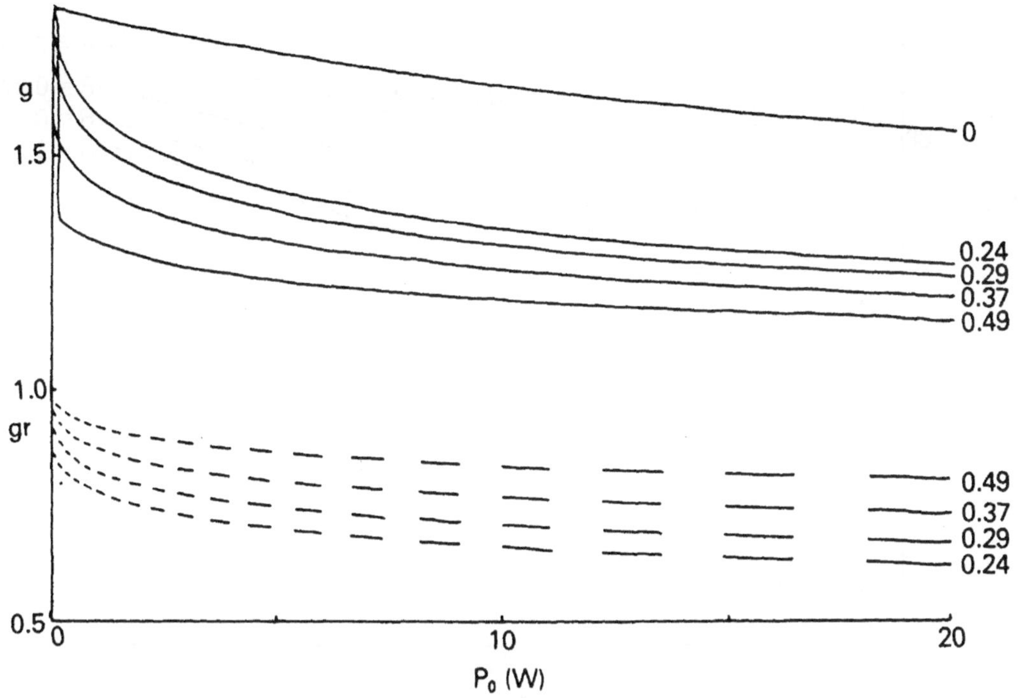

Fig. 62. Saturable gain, g, as a function of drive power, P_0 (solid curves) and $gr = gR^{1/2}$ (dashed curves). The numbers on the curves represent value of R.

and

$$P_{out} = P_0 \frac{(G^{1/2} - R^{1/2})^2}{(1 - R^{1/2}G^{1/2})^2}$$

$$= 53.806 \quad \text{W} \tag{10.23}$$

That is, 53.8 W of power can be extracted under conditionally stable conditions $GR < 1 < G_0R$ by using only 1 W of drive power, compared with 3 W of output power for a single-pass amplifier ($R = 0$).

For the condition where the ring is tuned to the input frequency, that is, $\theta = 0$, we have the power gain G, given by Eq. (10.22a). The value of G so calculated is substituted into Eq. (10.17a) to give P_{out}, i.e.,

$$P_{out} = P_0 \left[\frac{g(1-L) - r}{1 - rg} \right]^2, \qquad P_0 > 0, \, \theta = 0 \tag{10.17b}$$

In Fig. 63 we plot P_{out} vs. P_0 for various R. Buczek *et al.*[132] have plotted the power output, P_{out}, vs. mirror reflectivities, R, for different drive powers, as shown in Fig. 64. In this figure we have plotted P_{out} for $P_0 = 0$, the case of self-oscillation, which is treated next.

Figure 64 should show all the curves crossing the axis at $R = 1 - L$, since from Eq. (10.10) we have

$$a_1 = 0$$

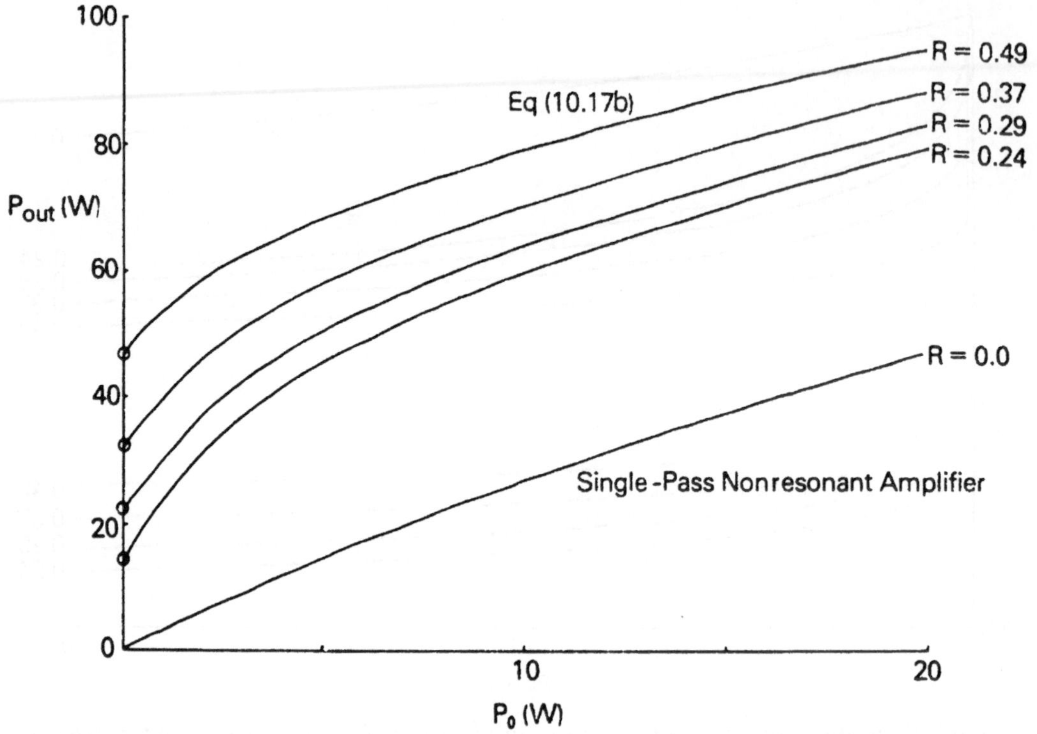

Fig. 63. Power output as a function of drive power for various mirror reflectivities. $P_s = 80$ W, $g_0 = 1.822$, $\theta = 0$, $L = 0$. The circles correspond to P_{out} $(P_0 = 0,\ L = 0) = P_s \ln G_0 R$ [Eq. (10.26)].

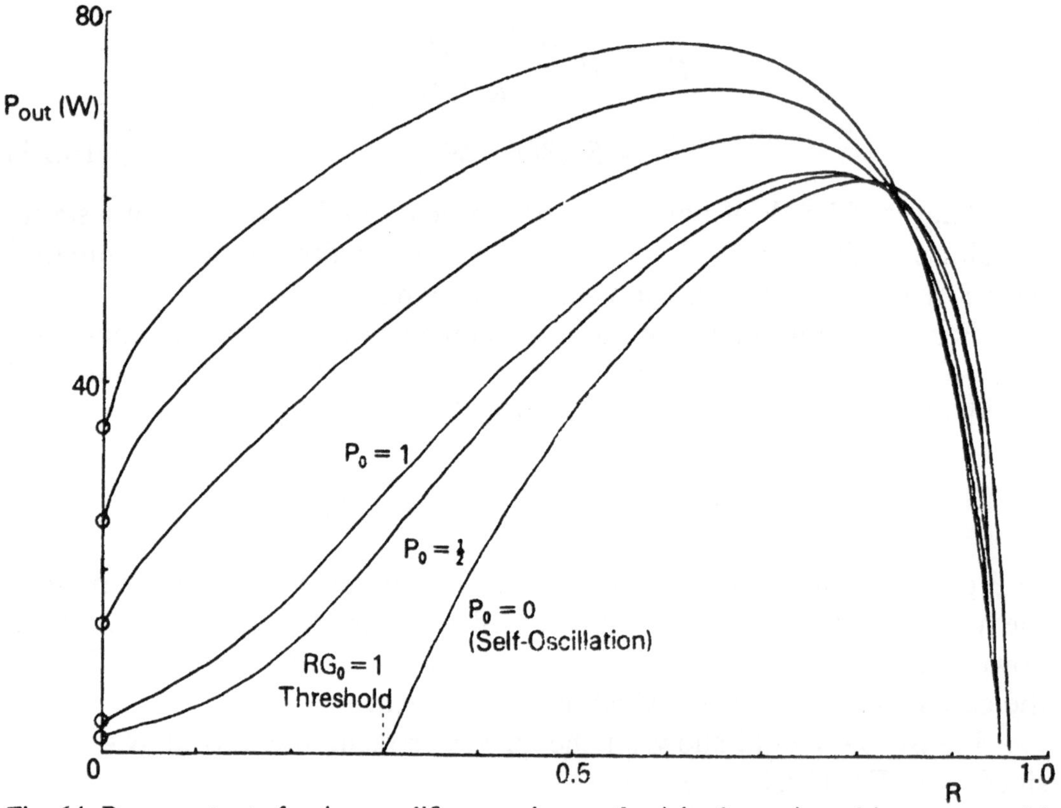

Fig. 64. Power output of a ring amplifier vs. mirror reflectivity for various drive powers with loss $L = 0.04$. The circles correspond to $P_0 g(1-L)$.

and so there will be no circulating power due to the drive. The discrepancy of this result with the actual plots in Fig. 64 is due to the coarseness of finding the roots of the g-equation.

For *self-oscillation*, i.e., $P_0 = 0$, we have the power balance condition shown in Fig. 65.

Let L be the fraction of power lost in the mirror. For the steady state,

$$\begin{aligned} P_1 &= R(GP_1) \\ &= (RG)P_1 \end{aligned} \tag{10.24}$$

Therefore

$$GR = 1 \quad \text{for self-oscillation} \tag{10.25}$$

The solution to the saturation equation was given earlier by

$$(G-1)\frac{P_1}{P_s} = \ln\left(\frac{G_0}{G}\right) \tag{10.20b}$$

Hence the circulating power is

$$P_1 = \frac{P_s}{G-1}\ln\left(\frac{G_0}{G}\right)$$

Power balance at the mirror gives

$$\begin{aligned} P_{out} &= (1-R-L)GP_1 \\ &= (1-R-L)P_s\frac{G}{G-1}\ln\left(\frac{G_0}{G}\right) \end{aligned}$$

When Eq. (10.25) is used, we have

$$\begin{aligned} P_{out} &= (1-R-L)P_s\frac{1}{1-R}\ln(G_0R), \qquad P_0 = 0 \\ &= 0 \qquad \text{when } R = 1-L \text{ or } G_0R = 1 \end{aligned} \tag{10.26}$$

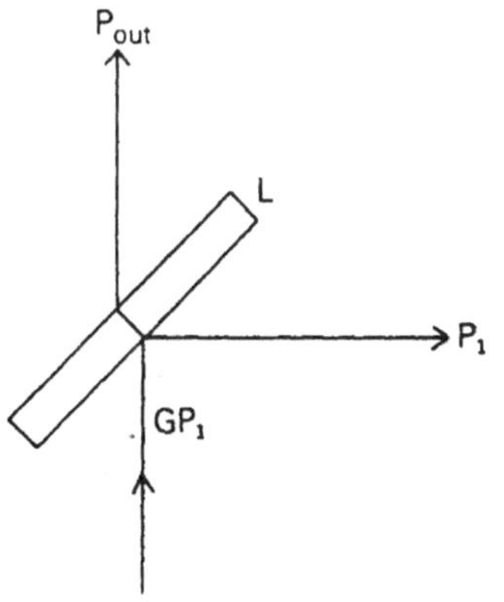

Fig. 65. Output mirror without drive.

Equation (10.26) is used to give the curve labeled $P_0 = 0$ in Fig. 64.

Hence self-oscillation is *not* possible for any R; the threshold is

$$R = 1/G_0 \tag{10.27}$$

By equating Eqs. (10.15) and (10.21b) we have an equation for determining the field amplitude gain, $g(\theta)$, as a function of the cavity tuning angle, θ:

$$P_0 \frac{(g^2-1)(1-r^2-L)}{(1-rg)^2+4rg\sin^2(\theta/2)} = 2P_s \ln\left(\frac{g_0}{g}\right), \qquad P_0>0 \tag{10.28}$$

In Fig. 66 we have plotted $g(\theta)$ for $L=0$, $P_s = 80$ W, $R = 0.24$, and a variety of P_0. From such plots the gain bandwidth can be seen to decrease

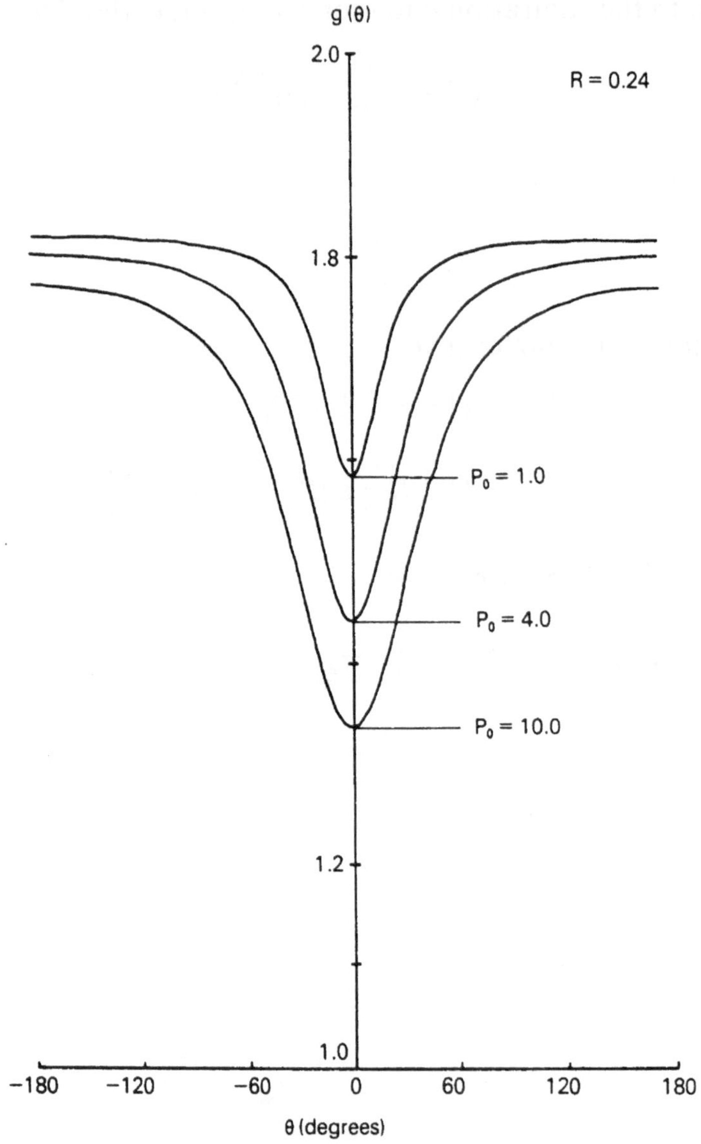

Fig. 66. Field amplitude gain $g(\theta)$ for reflectivity $R = 0.24$ and various drive powers P_0.

as R increases for fixed P_0, and it is observed that on resonance, $\theta = 0$, the gain decreases with increasing R.

Freiberg and Buczek[133] have plotted $P_{out}(\theta)$ for various P_0, for $L = 0$ and $P_s = 80$ W. That is, $g(\theta)$ is calculated from Eq. (10.28), which is then substituted into Eqs. (10.17a) and (10.17b) to obtain $P_{out}(\theta)$. In Fig. 67 we have plotted these output powers vs. cavity tuning angle.

For $R = 0.29$, when $P_0 = 1$ W, the on-resonance peak power input is about 30 W and the *half-power bandwidth* is about 30°, while for $P_0 =$ 10 W the output power is only about six times greater, while the half-power bandwidth is increased to about 100°.

From Eq. (10.6),

$$\theta = \frac{2\pi q}{\lambda} \tag{10.6a}$$

At resonance, for a given resonator, i.e., fixed q:

$$\frac{q}{\lambda} = m, \quad \text{an integer}$$

m characterizes the axial-mode resonance. Tuning the resonator through one axial mode involves changing λ such that

$$m \rightarrow m \pm 1$$

i.e., there is a 2π *change in* θ *between adjacent axial modes.* To determine

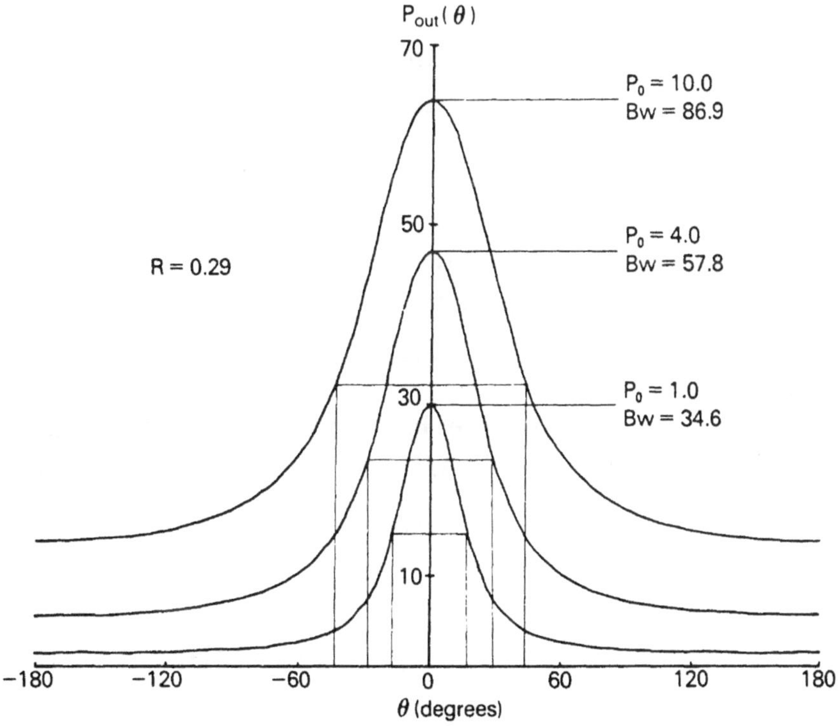

Fig. 67. Output power vs. cavity tuning angle for reflectivity $R = 0.29$ and various drive powers. Bw, half-power bandwidth.

the *intermode frequency spacing* we rewrite Eq. (10.6) as

$$\frac{\theta}{q/c}=\frac{2\pi c}{\lambda}=2\pi\nu=\omega \tag{10.29}$$

Hence for two frequencies ν_1 and ν_2:

$$\frac{1}{q/c}(\theta_1-\theta_2)=2\pi(\nu_1-\nu_2)$$

and for intermode spacing $(\theta_1-\theta_2)=2\pi$, so the *intermode frequency spacing* is

$$\nu_1-\nu_2\equiv\Delta\nu=\frac{c}{q} \tag{10.30}$$

Similarly, if ν_i is the injected frequency and ν_a is the frequency of the closest axial mode, then from Eq. (10.29):

$$\Delta\nu(\theta_i-\theta_a)=2\pi(\nu_i-\nu_a) \tag{10.31}$$

or

$$\begin{aligned} d\theta_{ia} &= \frac{2\pi d\nu}{\Delta\nu}=\frac{d\omega}{\Delta\nu} \\ &= \frac{2\pi\nu_i}{\Delta\nu}-\frac{2\pi q}{c}\cdot\frac{c}{\lambda_a} \\ &= \frac{2\pi\nu_i}{\Delta\nu}-2\pi m \end{aligned} \tag{10.32}$$

For a ring perimeter $q=2.2m$, the axial-mode frequency spacing is given by Eq. (10.30) to be

$$\begin{aligned} \Delta\nu &= \frac{3\times10^{10}}{2.2\times10^{2}}=1.36\times10^{8} \\ &= 136 \quad \text{MHz} \end{aligned} \tag{10.33}$$

If we define $\Delta\theta_{1/2}$ to be the half-power bandwidth, then the ring amplifier *frequency bandwidth, in MHz*, will be given by Eq. (10.31) to be

$$\begin{aligned} d\nu_{1/2} &= \frac{d\theta_{1/2}}{2\pi}\cdot\Delta\nu \\ &= \frac{d\theta_{1/2}}{2\pi}\cdot 136 \quad \text{MHz} \end{aligned} \tag{10.34}$$

Equation (10.34) can be used to convert the units on the abscissas of Fig. 67 from degrees to megahertz.

10.1.2. Hybrid Injection Locking[134]

The objectives of injection locking are to separate the optimization of frequency stability into the small low-power master oscillator, while the power and efficiency optimization is carried out in the larger high-power oscillator.

We recall that in conventional injection locking, the injected signal is tuned to a transition at which strong *self-oscillation* of the higher power laser occurs. In the case of two CO_2 lasers, the injected external signal and the self-oscillation are on the same P or R branch of the vibrational transition. This way of operating the system is also called "conditionally stable regenerative amplification."

In the hybrid way of operating, the higher-power laser is tuned to operate on a self-oscillation line different from that of the injection (or reference) laser.

Consider a ring laser capable of self-oscillation at some radian frequency ω_0, driven by a low-power oscillator at ω_1. The total power extracted from the amplifier will, in general, consist of a driven and an undriven contribution:

$$\Delta P = \Delta P(\omega_1) + \Delta P(\omega_0) \tag{10.35}$$

As before, the reflective ring laser uses only one partially transmitting mirror for both input and output coupling.

From Eq. (10.15) we have, for $L = 0$:

$$\Delta P(\omega_1) = \frac{P_0(g^2-1)(1-r^2)}{(1-rg)^2 + 4rg\sin^2(\theta/2)} \tag{10.36}$$

The second term in the denominator represents the off-resonance contribution. From Eq. (10.29),

$$\Delta\theta = \frac{q}{c}\Delta\omega$$

When $\Delta\theta$ measures the off-set from the mth axial mode, we have

$$\theta = 2m\pi + \frac{q}{c}\Delta\omega$$

$$= 2m\pi + \frac{q}{c}(\omega_1 - \omega_0) \tag{10.37}$$

We recall that the threshold condition for self-oscillation is

$$gr = 1 \tag{10.25a}$$

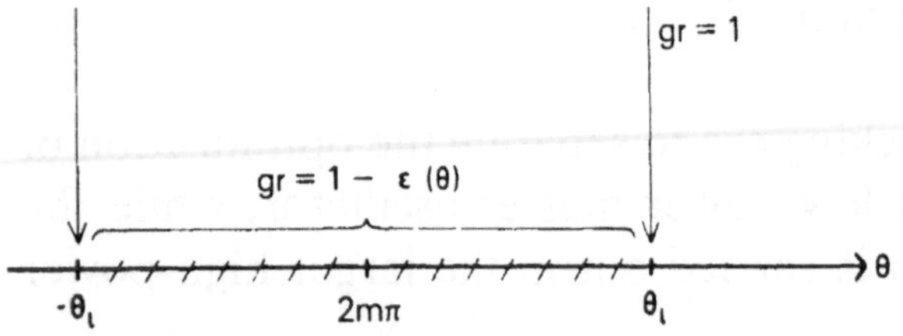

Fig. 68. Schematic of the locking range.

In a small region about $\theta = 2m\pi$, for $|\theta| < \theta_l$ say, the drive signal at ω_1 will decrease the gain below threshold:

$$gr = 1 - \varepsilon(\theta) \tag{10.38}$$

In other words, since we have $g(\theta)$ in this region the existence of the drive necessitates a smaller g (see Fig. 65). For example, g will be least at $\theta = 2m\pi$ [equivalent to $\theta = 0$ since θ always appears as $\sin(\theta/2)$] and greatest at θ_l, where $\varepsilon(\theta_l) = 0$. In this small region, since the gain is below threshold, the *self-oscillation is entirely quenched*, $\Delta P(\omega_0) = 0$, and the only power extracted is at the frequency of the injected signal. Hence, in this region, the higher-power laser is completely locked to the injected signal.

The locking angle θ_l, which defines the boundary of this small region, is the point at which $gr = 1$, and the total extracted power is

$$\Delta P_0 \equiv \Delta P|_{gr=1} = 2P_s \ln\left(\frac{g_0 r}{1}\right) \tag{10.39}$$

using Eq. (10.21b), where ΔP_0 is the maximum extracted power from the undriven laser.

At the boundary, $gr = 1$ and $\theta = \theta_l$, and $\Delta P(\omega_0) = 0$; the oscillation at ω_0 is quenched, so from Eq. (10.36):

$$\Delta P(\omega_1)|_{\theta_l} = \Delta P_0 = \frac{P_0(1-r^2)^2}{4r^2 \sin^2(\theta_l/2)} \tag{10.40}$$

In Fig. 68 we present a schematic of the locking range about the mth axial mode. We note again that within the locking range there is no self-oscillation.

When the small-angle approximation is used in Eq. (10.40):

$$\left(\frac{\theta_l}{2}\right)^2 \simeq \frac{P_0}{\Delta P_0}\frac{(1-r^2)^2}{4r^2}$$

or

$$\theta_l \simeq \pm\left(\frac{P_0}{\Delta P_0}\right)^{1/2}\frac{1-r^2}{r}$$

Hence, the *condition for quenching, also called the locking range, is*

$$\Delta\theta_l = 2\theta_l = 2\left(\frac{P_0}{\Delta P_0}\right)^{1/2}\frac{1-r^2}{r} \tag{10.41}$$

In terms of frequency, using Eq. (10.29):

$$\Delta\nu_l = \frac{c}{\pi q}\left(\frac{P_0}{\Delta P_0}\right)^{1/2}\frac{1-r^2}{r} \tag{10.42}$$

which is Eq. (5) of Buczek and Freiberg.[134]

To increase $\Delta\nu_l$, i.e., to broaden the locking range, we can either decrease r (*higher output coupling*) or increase P_0.

Outside the locking range, $|\theta| > \theta_l$, the extracted power contains both driven and undriven contributions as in Eq. (10.35) with the extracted power at the driven frequency, for small angles given by Eq. (10.36) with $gr = 1$:

$$\Delta P(\omega_1) = \frac{P_0(1-r^2)^2}{r^2\theta^2}, \qquad |\theta| > \theta_l \tag{10.43}$$

into which we substitute Eq. (10.41), which gives r as a function of θ_l:

$$\Delta P(\omega_1) = \theta_l^2 \cdot \frac{\Delta P_0}{P_0} \cdot \frac{P_0}{\theta^2}$$

$$= \Delta P_0\left(\frac{\theta_l}{\theta}\right)^2, \qquad |\theta| > \theta_l \tag{10.44}$$

Hence the contribution to the extracted power at the undriven frequency is

$$\Delta P(\omega_0) = \Delta P_0 - \Delta P(\omega_1)$$

$$= \Delta P_0\left[1 - \left(\frac{\theta_l}{\theta}\right)^2\right], \qquad |\theta| > \theta_l \tag{10.45}$$

Inside the locking range, $|\theta| < \theta_l$, we substitute Eq. (10.38) into Eq. (10.21b) to give

$$\Delta P = 2P_s\{\ln(g_0 r) - \ln[1 - \varepsilon(\theta)]\}$$

$$\simeq 2P_s \ln(g_0 r) + 2P_s\varepsilon(\theta)$$

or

$$\Delta P = \Delta P_0 + 2P_s\varepsilon(\theta), \qquad |\theta| < \theta_l \tag{10.46}$$

having used Eq. (10.39). Equation (10.46) is used to give the extracted power inside the locking range in Fig. 69 when we know $\varepsilon(\theta)$, and we would expect it to be an even function of θ.

When $P_s \simeq \Delta P_0$, we have the extracted power in the locking range, $\theta < \theta_l$, given by both Eqs. (10.36) and (10.46), so when we set them equal we have

$$\Delta P \simeq \Delta P_0(1+2\varepsilon) = \frac{P_0(g^2-1)(1-r^2)}{(1-rg)^2+4rg\sin^2(\theta/2)}$$

When Eq. (10.40) is substituted for ΔP_0 we have

$$\frac{P_0(1-r^2)^2(1+2\varepsilon)}{4r^2\sin^2(\theta_l/2)} = \frac{P_0(g^2-1)(1-r^2)}{(1-rg)^2+4rg\sin^2(\theta/2)}$$

$$\frac{(1-r^2)(1+2\varepsilon)}{4(\theta_l/2)^2} \simeq \frac{(r^2g^2-r^2)}{(1-1+\varepsilon)^2+4(1-\varepsilon)\theta^2/4} \tag{10.47}$$

$$(1+2\varepsilon)(1-r^2)(\varepsilon^2+\theta^2) \simeq (1-2\varepsilon-r^2)\theta_l^2$$

$$\varepsilon^2 \simeq \theta_l^2 - \theta^2$$

$$\varepsilon \simeq (\theta_l^2-\theta^2)^{1/2}, \qquad |\theta| \leqslant \theta_l$$

which is Eq. (12) of Buczek and Freiberg,[134] which gives Eq. (10.46) to be

$$\Delta P \simeq \Delta P_0 + 2P_s(\theta_l^2-\theta^2)^{1/2} \tag{10.48}$$

plotted as the dotted curve in Fig. 69.

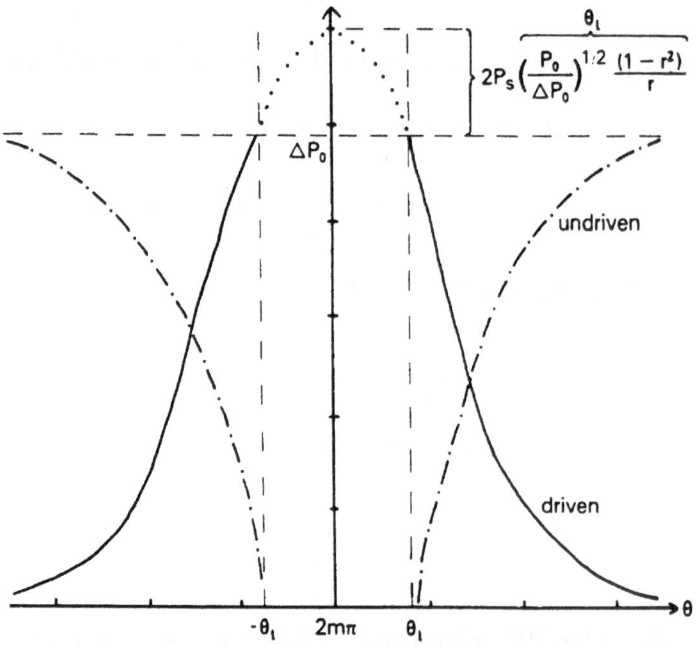

Fig. 69. Contributions to the total extracted power. (— · —) $\Delta P(\omega_0)$ [Eq. (10.45)]; (———) $\Delta P(\omega_1)$ [Eq. (10.44)]; (· · ·) ΔP [Eq. (10.48)].

To relate the phase of the output *amplitude* to that of the input amplitude, consider Eq. (10.7), the expression for the circulating amplitude, namely,

$$p_1 = \frac{a_1}{1 - rg\,e^{i\theta}} \tag{10.7a}$$

where

$$a_1 = \pm(1 - r^2 - L)^{1/2} a_0 \tag{10.10b}$$

Hence

$$\frac{p_1}{a_0} = \pm \frac{(1 - r^2 - L)^{1/2}}{1 - rg\,e^{i\theta}}$$

which is consistent with Eq. (10.13).

The value of the circulating amplitude incident on the output mirror is

$$g\,e^{i\theta} p_1$$

and so the fraction transmitted will be

$$\pm(1 - r^2 - L)^{1/2} g\,e^{i\theta} p_1$$

In other words, the contribution to the output power will be

$$p_{\text{out},1} = \frac{(1 - r^2 - L) g\,e^{i\theta} a_0}{1 - rg\,e^{i\theta}}$$

A fraction of the drive amplitude will be reflected at the output mirror. The choice of phase between the reflection and transmission coefficients must be such as to be consistent with the power balance equation, Eq. (10.16a), before we combine the amplitudes:

$$\begin{aligned} p_{\text{out}} &= p_{\text{out},1} + p_{\text{out},0} \\ &= \frac{(1 - r^2 - L) g\,e^{i\theta} a_0}{(1 - rg\,e^{i\theta})} + r\,e^{i\pi} a_0 \end{aligned} \tag{10.49a}$$

To see if our choice of phase, $e^{i\pi}$, is consistent, we combine terms in Eq. (10.49a):

$$\begin{aligned} \frac{p_{\text{out}}}{a_0} &= \frac{(1 - r^2 - L) g\,e^{i\theta} + re^{i\pi}(1 - rg\,e^{i\theta})}{1 - rg\,e^{i\theta}} \\ &= \frac{(1 - L) g\,e^{i\theta} - r}{1 - rg\,e^{i\theta}} \end{aligned} \tag{10.49b}$$

Hence

$$\frac{P_{out}}{P_0}=\frac{|(1-L)g\,e^{i\theta}-r|^2}{|1-rg\,e^{i\theta}|^2} \tag{10.17b}$$

where the denominator has the form required for consistency with Eq. (10.17a). To check the consistency of the numerator we write

$$\begin{aligned}
&[(1-L)g\,e^{i\theta}-r][(1-L)g\,e^{-i\theta}-r]\\
&\quad=(1-L)^2g^2+r^2-r(1-L)g[e^{i\theta}+e^{-i\theta}]\\
&\quad=(1-L)^2g^2+r^2-r(1-L)g\{2[1-2\sin^2(\theta/2)]\}\\
&\quad=[g(1-L)-r]^2+4r(1-L)g\sin^2(\theta/2)
\end{aligned}$$

which is consistent with the numerator of Eq. (10.17a). Therefore our choice of phase difference between reflection and transmission coefficients is consistent.

From Eq. (10.49b) we see that the ratio of the output to drive amplitudes is a complex quantity. In other words for real a_0, the output amplitude has suffered a change in phase; write Eq. (10.49b) as

$$\frac{p_{out}}{a_0}\equiv\rho\,e^{i\phi} \tag{10.49c}$$

where ρ is a real quantity.

Outside the injection-locked region, $rg=1$ and

$$\rho(\cos\phi+i\sin\phi)=\frac{[(1-L)g\,e^{i\theta}-r][1-e^{-i\theta}]}{2(1-\cos\theta)}$$

Equating the real and imaginary parts of this expression and taking their ratio gives

$$\frac{\sin\phi}{\cos\phi}=\frac{\sin\theta(1-L-r^2)}{(\cos\theta-1)(1-L+r^2)}$$

or

$$\phi=\tan^{-1}\frac{(1-L-r^2)\sin\theta}{(1-L+r^2)(\cos\theta-1)},\qquad |\theta|>\theta_l \tag{10.50}$$

Inside the injection-locked region, gr is given by Eq. (10.38), which upon substitution into Eq. (10.49b) gives

$$\rho\,e^{i\phi}=\frac{(1-L)g\,e^{i\theta}-r}{1-(1-\varepsilon)\,e^{i\theta}}$$

$$r\rho\,e^{i\phi}=\frac{(1-L)(1-\varepsilon)\,e^{i\theta}-r^2}{1-(1-\varepsilon)\,e^{i\theta}}$$

The numerator becomes

$$[(1-L)(1-\varepsilon)e^{i\theta}-r^2][1-(1-\varepsilon)e^{-i\theta}]$$

$$=(1-L)(1-\varepsilon)e^{i\theta}-r^2-(1-L)(1-\varepsilon)^2+r^2(1-\varepsilon)e^{-i\theta}$$

$$\text{real part}=(1-L)(1-\varepsilon)\cos\theta-r^2-(1-L)(1-\varepsilon)^2+r^2(1-\varepsilon)\cos\theta$$

$$\text{imaginary part}=(1-L)(1-\varepsilon)\sin\theta-r^2(1-\varepsilon)\sin\theta$$

Hence

$$\tan\phi=\frac{(1-\varepsilon)\sin\theta(1-L-r^2)}{(1-\varepsilon)\cos\theta(1-L+r^2)-r^2-(1-\varepsilon)^2(1-L)}$$

$$\tan\phi\simeq\frac{(1-\varepsilon)\theta(1-L-r^2)}{(1-\varepsilon)[1-(\theta^2/2)](1-L+r^2)-r^2-(1-\varepsilon)^2(1-L)}$$

$$\simeq\frac{(1-\varepsilon)(1-L-r^2)}{(1-\varepsilon)(1+r^2-L)-r^2-(1-\varepsilon)^2(1-L)-(\theta^2/2)(1-\varepsilon)(1+r^2-L)}$$

$$\simeq\frac{\theta(1-r^2-L)}{\varepsilon(1-r^2-L)-(\theta^2/2)(1+r^2-L)},\qquad |\theta|<\theta_l \tag{10.51}$$

Within the locking region, Eq. (10.51) shows that

$$\tan\phi\simeq\frac{\theta}{\varepsilon}\simeq\frac{\theta}{(\theta_l^2-\theta^2)^{1/2}}$$

or

$$\phi\simeq\frac{\theta}{\theta_l}\qquad\text{for small }\theta$$

and

$$\frac{d\phi}{d\theta}\simeq\frac{1}{\theta_l}$$

Since the injection-locked region is broadened by increasing the ratio $P_0/\Delta P_0$, or by reducing the reflectivity, the slope in Fig. 70 progressively decreases.

For a given $\theta\neq 0$, i.e., injected signal off-set from cavity-resonance frequency, ϕ indicates degree of pulling (algebraically) of injected signal toward the cavity-resonance frequency (Fig. 71).

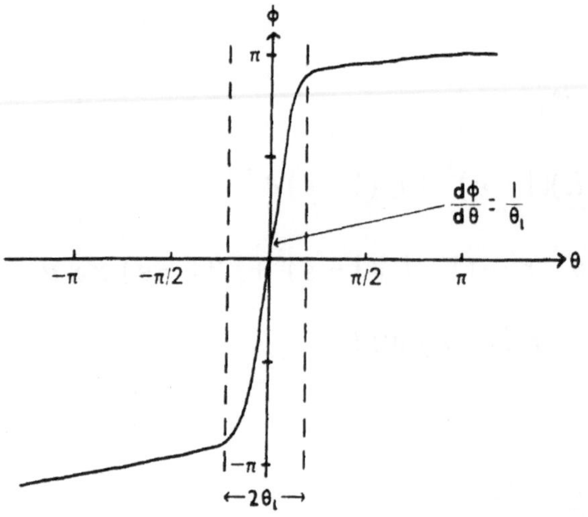

Fig. 70. External phase angle vs. cavity tuning angle.

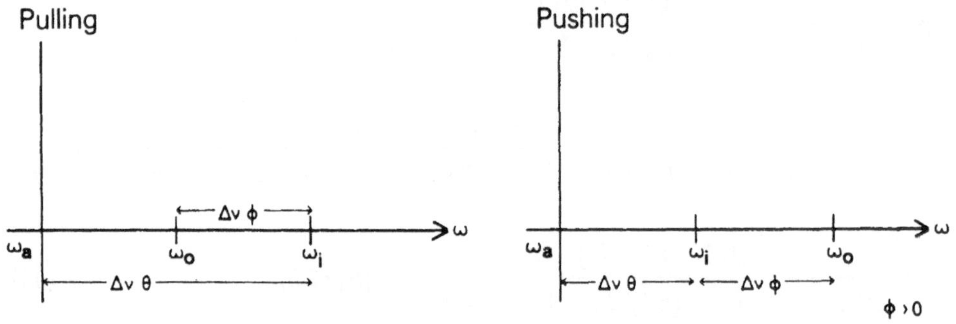

Fig. 71. Schematic of frequencies involved in pulling and pushing of injected signal.

10.2. *Time-Dependent Theory*[135]

When an external signal is applied to the laser cavity and we wish to study its amplification and phase shift, it is necessary to keep track of both the amplitude and the phase of the signal in the laser cavity rather than just its intensity. The intensity rate equation approach also breaks down when (a) several signals at slightly different frequencies are simultaneously present in the same cavity and interference effects between these signals is important, or (b) when we want to study the phase and frequency fluctuations of a laser signal.

Siegman (Reference 16, pages 449–454) has given the derivation of the equation necessary to describe both the amplitude and phase of the signal in a laser cavity in a single selected mode. This equation will turn out to be a first-order ODE for the *complex* phasor amplitude of the laser signal, rather than a real equation. This equation is called the *circuit equation*, because it is very similar in form to the ordinary differential equation for $V(t)$ or $I(t)$ in a lumped electric circuit.

The wave amplitude of a single transverse and axial mode of the laser cavity will be represented by a complex phasor amplitude

$$\tilde{E}_i(t)\, e^{i\omega_0 t}$$

The *complex* wave amplitude $\tilde{E}_i(t)$ may represent, for example, the positive-traveling wave at a *reference plane* just inside the input mirror of reflectivity R. Let us assume that there is an external signal

$$\tilde{E}_0(t)\, e^{i\omega_0 t}$$

i.e., mode matched to the selected cavity mode and incident on the input mirror from outside with amplitude

$$\frac{1}{(1-R)^{1/2}}\tilde{E}_0(t)\, e^{i\omega_0 t}$$

The total wave amplitude $\tilde{E}_i(t+2\tau)$ passing through the reference plane at $t+2\tau$ must (for an oscillator) equal the total wave amplitude $\tilde{E}_i(t)$ that passed through that plane one round-trip transit time 2τ earlier, multiplied by the net round-trip gain g_{rt} for a wave in the cavity, plus the added wave amplitude $\tilde{E}_0(t+2\tau)$ being transmitted through the input mirror at that same instant, that is,

$$\tilde{E}_i(t+2\tau) = g_{rt}\tilde{E}_i(t) + \tilde{E}_0(t+2\tau) \tag{10.52}$$

If k represents the wavenumber in the resonator of round-trip length $2L$, then the amplitude suffers a phase shift e^{ik2L}.

If T is the power gain of an *empty* cavity, while $\alpha(t)$ is the power-gain coefficient, then for an amplifier of active length l,

$$G = T e^{\alpha(t)l} \equiv g^2$$

In Section 10.1 we saw that

$$g_{rt} = gr\, e^{2ikL}$$

Hence

$$g_{rt} = (RT)^{1/2}\, e^{[\alpha(t)l]/2+2ikL} \tag{10.53}$$

We can multiply this expression by

$$+1 = e^{2in\pi}$$

where n is an integer chosen to bring the net cavity-tuning angle into the range $-\pi$ to π. We now have

$$g_{rt} = (RT)^{1/2}\, e^{[\alpha(t)l]/2+2i(kL-n\pi)}$$

We showed previously [Eq. (10.30)] that

$$2kL-2n\pi=2\pi\frac{(2L)}{\lambda}-2n\pi$$

$$=\frac{2\pi}{\Delta\nu}\frac{c}{\lambda}-2n\pi$$

$$=d\theta \tag{10.32a}$$

and so the factor in Eq. (10.53) becomes

$$\frac{\ln(g_{rt})}{2\tau}=\frac{-1}{2\tau/\{\ln[1/(RT)^{1/2}]\}}+\frac{1}{2\tau}\left[\frac{\alpha(t)l}{2}+id\theta\right]$$

$$\frac{\ln(g_{rt})}{2\tau}\equiv\frac{-1}{2T_0}+\frac{c\alpha(t)l}{4L}+i\frac{c\,d\theta}{2L}$$

$$\equiv m(t) \tag{10.54}$$

as defined in Lachambre *et al.*[135], since

$$(\Delta\nu)^{-1}=2\tau=\frac{2L}{c}\qquad\text{and}\qquad T_0\equiv\frac{2\tau}{\ln[1/(RT)^{1/2}]} \tag{10.55}$$

To derive a time-dependent differential equation describing the evolution of the wave amplitude, we expand the left-hand side of Eq. (10.52), using a Taylor series in the form

$$\tilde{E}_i(t+2\tau)=e^{2\tau D}\tilde{E}_i(t) \tag{10.56}$$

where

$$D\equiv d/dt$$

and rewrite Eq. (10.54) as

$$g_{rt}=e^{2\tau m(t)}$$

Hence Eq. (10.52) becomes

$$e^{2\tau D}\tilde{E}_i(t)=e^{2\tau m}\tilde{E}_i(t)+\tilde{E}_0(t+2\tau)$$

or

$$\tilde{E}_i(t)=\frac{\tilde{E}_0(t+2\tau)}{e^{2\tau D}-e^{2\tau m}}$$

$$=\frac{\tilde{E}_0(t+2\tau)}{e^{2\tau m}[e^{2\tau(D-m)}-1]}$$

To obtain Lachambre *et al.*'s Eq. (2), we rewrite this result as

$$\tilde{E}_i(t) = \frac{(1-e^{-2\tau m})\tilde{E}_0(t+2\tau)}{[e^{2\tau(D-m)}-1]\,e^{2\tau m}(1-e^{-2\tau m})}$$

$$\simeq \frac{2\tau m}{2\tau(D-m)}\,\frac{\tilde{E}_0(t+2\tau)}{e^{2\tau m}-1}$$

or

$$(D-m)\tilde{E}_i(t) \simeq \frac{m(t)\tilde{E}_0(t+2\tau)}{e^{2\tau m(t)}-1}$$

or

$$\frac{d\tilde{E}_i}{dt} \simeq m(t)\tilde{E}_i + \frac{m(t)\tilde{E}_0(t)}{e^{2\tau m(t)}-1} \tag{10.57}$$

having assumed that $\tilde{E}_0$ varies very slowly over 2τ.

Equation (10.57) is a first-order ODE, which represents the temporal evolution of the field, $\tilde{E}_i$, that builds up in the cavity under the direct influence of the externally injected field [see Eq. (10.54)] as manifest by [see Eq. (10.31)] the presence of $d\theta$:

$$\Delta\nu\, d\theta = d\omega = \omega_i - \omega_a \tag{10.58}$$

While the external beam is being injected, the noise present in the laser cavity at different mode frequencies gives rise to other radiation amplitudes. In other words, for the jth longitudinal mode amplitude E^j, with $d\theta = 0$ and therefore $m_j(t)$ given by the real part of Eq. (10.54), we have an analogous equation to Eq. (10.57), with $\tilde{E}_0$ replaced by the noise field corresponding to one photon per resonator mode, namely, $(\psi_N)^{1/2}$; hence,

$$\frac{dE^j}{dt} = m_j(t)E^j + \frac{(\psi_N)^{1/2}m_j(t)}{\exp[2\tau m_j(t)]-1} \tag{10.59}$$

To calculate ψ_N we note that if T_0 is the photon lifetime per mode, then cT_0 is the average distance traveled by a photon per mode, while $\sigma c T_0$ is the probability of making a collision per $CO_2(001)$, where σ is the laser radiative cross section. If N_3 is the number density of 0, 0, 1 molecules, then $N_3\sigma c T_0/V$ is the photon noise density or $N_3\sigma c T_0/2V$ is the one-way photon noise density per mode, ψ_N:

$$\psi_N = \frac{N_3\sigma c T_0}{2V} \tag{10.60}$$

The total photon density, in cm^{-3}, is defined by

$$\psi = \sum_{j=1}^{N} |E^j|^2 + |E_i|^2$$

$$\equiv \psi_s + \psi_i \tag{10.61}$$

The interaction of the active medium with the cavity photon density, as well as the pumping and relaxation of the different energy levels, are described by a set of differential equations of the form

$$\frac{dN_1}{dt} = \sigma c \psi \Delta N + \cdots \tag{10.62}$$

which can be converted to the more usual energy density equations by multiplying by $h\nu_1$; hence

$$\frac{dE_1}{dt} = \nu_1 \Delta N \sigma c h \psi + \cdots \tag{10.63}$$

which, upon comparison with Eq. (2.70b), gives

$$WI = \sigma c h \psi \tag{10.64}$$

To check the dimensions we note [see Eqs. (2.44) and (2.58)]

$$[WI] = [L^2T \cdot EL^{-2}T^{-1}] = [E]$$

while

$$[\sigma c h \psi] = [L^2 \cdot LT^{-1} \cdot ET \cdot L^{-3}] = [E]$$

The coupling of the population rate equations to the field equations is accomplished through the expression for the gain [see Eq. (2.64b)]

$$\alpha(t) = (\nu W)\,\Delta N = \sigma\,\Delta N \tag{10.65}$$

where σ is, strictly speaking, a function of ν!

The energy radiated per unit volume is obtained using Eq. (2.151) with $\bar{L} \approx 0$. We divide by $2L$ since radiation is moving both ways; hence

$$\frac{E_{\text{out}}}{A(2L)} \approx \int_{t=0}^{t=t'} \frac{I_\nu(t)}{2L}\frac{1}{2}\ln\left(\frac{1}{R}\right)dt$$

$$= \int \frac{ch\nu\psi}{4L}\ln\left(\frac{1}{R}\right)dt$$

$$= \int_{t=0}^{t=t'} \frac{h\nu\psi}{2\tau}\ln\left(\frac{1}{R}\right)dt \qquad (10^{-7}\ \text{J cm}^{-3}) \tag{10.66}$$

where ψ is either ψ_s or ψ_i. The peak power radiated per unit volume is

$$P = \frac{h\nu\psi}{2\tau}\ln\left(\frac{1}{R}\right) \qquad (10^{-7}\text{ W cm}^{-3}) \tag{10.67}$$

where $P_i(\psi_i)$ denotes the output power induced by the injection signal, while $P_s(\psi_s)$ denotes the output power from the cavity field. Equation (10.59) is the field analog of the intensity equation (2.57), and becomes, using Eqs. (10.54), (10.60), and (10.65):

$$\frac{dE^j}{dt} = \left(\frac{-1}{2T_0} + c\nu_j \frac{W\,\Delta Nl}{4L}\right)E^j + \frac{m_j}{\exp(2\tau m_j) - 1}\left[\frac{\nu_j W_c T_0 N_{0,0,1} P(J)}{2V}\right]^{1/2} \tag{10.68}$$

where the E^j are expressed in $\text{cm}^{-3/2}$. Lachambre *et al.* solve this equation for 12 values of ν_j, in addition to the complex equation (10.57) for the longitudinal mode in the vicinity of the injected signal, which becomes

$$\frac{d\tilde{E}_i}{dt} = \left(-\frac{1}{2T_0} + \frac{c\nu_0 W\Delta Nl}{4L} + \frac{icd\theta}{2L}\right)\tilde{E}_i + \frac{\tilde{E}_0(t)m(t)}{e^{2\tau m(t)} - 1} \tag{10.69}$$

The incident photon density is given by $|E_0|^2$ (cm^{-3}) and so the incident energy density will be $|E_0|^2 h\nu_i$ ($\text{erg}\cdot\text{cm}^{-3}$), where ν_i is the incident master oscillator frequency. Finally, we have the incident energy flux $c|E_0|^2 h\nu_i$ in $\text{erg}\cdot\text{cm}^{-2}\,\text{sec}^{-1}$, or in W-cm^{-2} if we use the factor 10^{-7}.

Equation (10.69) is complex, and if we write

$$\tilde{E}_i(t) = |\tilde{E}_i(t)|\,e^{i\theta(t)} \tag{10.70}$$

where $\theta(t)$ is the instantaneous phase, then

$$\frac{d\tilde{E}_i}{dt} = \frac{d|\tilde{E}_i|}{dt}e^{i\theta} + i|\tilde{E}_i|\frac{d\theta}{dt}e^{i\theta}$$

and if we write

$$\frac{m(t)}{e^{2\tau m(t)} - 1} = \frac{x\gamma + y\beta + i(y\gamma - x\beta)}{\gamma^2 + \beta^2}$$

where

$$x \equiv \left(\frac{-1}{2T_0} + \frac{c\alpha l}{4L}\right), \qquad y \equiv \frac{cd\theta}{2L}$$

$$\gamma \equiv e^{2\tau x}\cos 2\tau y - 1, \qquad \beta \equiv e^{2\tau x}\sin 2\tau y$$

then Eq. (10.69) becomes

$$e^{i\theta}\frac{d|\tilde{E}_i|}{dt} + i\tilde{E}_i\frac{d\theta}{dt} = (x + iy)\tilde{E}_i + [x\gamma + y\beta + i(y\gamma - x\beta)]\frac{\tilde{E}_0}{\gamma^2 + \beta^2}$$

If we take $\tilde{E}_0$ to be real, then the real and imaginary parts of the equation are given by

$$\text{(A)}\quad \cos\theta\frac{d|\tilde{E}_i|}{dt}-\sin\theta|\tilde{E}_i|\frac{d\theta}{dt}=x|\tilde{E}_i|\cos\theta-y|\tilde{E}_i|\sin\theta+\frac{(x\gamma+y\beta)\tilde{E}_0}{\gamma^2+\beta^2}$$

$$\text{(B)}\quad \sin\theta\frac{d|\tilde{E}_i|}{dt}+\cos\theta|\tilde{E}_i|\frac{d\theta}{dt}=x|\tilde{E}_i|\sin\theta+y|\tilde{E}_i|\cos\theta+\frac{(y\gamma+x\beta)\tilde{E}_0}{\gamma^2+\beta^2}$$

When (A) is multiplied by $\cos\theta$ and added to (B) multiplied by $\sin\theta$ we have

$$\frac{d|\tilde{E}_i|}{dt}=x|\tilde{E}_i|+[\cos\theta(x\gamma+y\beta)+\sin\theta(y\gamma-x\beta)]\frac{\tilde{E}_0}{\gamma^2+\beta^2}$$

When we define

$$\eta\equiv\frac{|\tilde{E}_i|}{\tilde{E}_0}$$

we have

$$\frac{d\eta}{dt}=x\eta+\frac{[\cos\theta(x\gamma+y\beta)+\sin\theta(y\gamma-x\beta)]}{(\gamma^2+\beta^2)}\tag{10.71a}$$

When (A) is multiplied by $-\sin\theta$ and added to (B) multiplied by $\cos\theta$, we have

$$\frac{d\theta}{dt}=y+\frac{[-\sin\theta(x\gamma+y\beta)+\cos\theta(y\gamma-x\beta)]}{\eta(\gamma^2+\beta^2)}\tag{10.71b}$$

Equations (10.71a) and (10.71b) are a pair of coupled first-order nonlinear ordinary differential equations for $|\tilde{E}_i|$ and θ. Consequently, two boundary conditions are required.

During the gain buildup, both the noise signal at the mode frequency and the injection signal at the master oscillator frequency circulate simultaneously in the laser cavity. Competition sets in between the two waves as to which will grow more rapidly and depopulate the active medium before the other enters into the saturation regime. In general, the injection field is stronger than the spontaneous emission; however, it experiences a smaller gain because it suffers a phase change through each cavity traversal and no longer adds in phase with the injection source amplitude. Depending on the detuning angle and on the injection level, either wave may reach the saturation level first. In addition, the growing field governed by the injection signal adds vectorially to the fixed-phase injection amplitude to produce a resultant that changes in amplitude and in phase with time. This time-varying phase is equivalent to a frequency modulation that pulls the frequency of the field toward the frequency of the nearest resonator mode. The amount of pulling varies with time and depends on the instantaneous gain, on the injection level, and on the detuning angle.

Davies *et al.*[136] have modified the computer model TLASER (see Section 2.12) to take injection locking into account.

References

1. K. F. Herzfeld and T. A. Litovitz, *Absorption and Dispersion of Ultrasonic Waves*, Academic Press, New York, p. 53 (1959).
2. W. J. Witteman, *J. Chem. Phys.* **35**, 1 (1961).
3. N. F. Mott and H. S. W. Massey, *Theory of Atomic Collisions*, Oxford University Press, New Jersey, 3rd ed., p. 24 (1965).
4. W. G. Vincent and C. H. Kruger, Jr., *Introduction to Physical Gas Dynamics*, John Wiley and Sons, New York, p. 46 (1965).
5. A. March, *Quantum Mechanics of Particles and Wave Fields*, John Wiley and Sons, New York, Sections 17, 20 (1951).
6. G. S. Rushbrooke, *Introduction to Statistical Mechanics*, Oxford University Press, New York, Chapter 2 (1949).
7. R. Marriott, Abstracts VII. *International Conference on the Physics of Electronic and Atomic Collisions*, North-Holland, Amsterdam, p. 22 (1971).
8. R. N. Schwartz, Z. I. Slawsky, and K. F. Herzfeld, *J. Chem. Phys.* **20**, 1591 (1952).
9. G. Herzberg, *Molecular Spectra and Molecular Structure, Vol. 2*, Van Nostrand, New York, p. 66 (1945).
10. L. Pauling and E. B. Wilson, *Introduction to Quantum Mechanics*, McGraw–Hill, New York (1935).
11. K. R. Manes and H. J. Seguin, *J. Appl. Phys.* **43**, 5073 (1972).
12. A. L. Hoffman and G. C. Vlases, *IEEE J. Quantum Electron.* **8**, 46 (1972).
13. A. C. G. Mitchell and M. W. Zemansky, *Resonance Radiation and Excited Atoms*, Cambridge University Press, New York, p. 94 (1971).
14. M. J. Beesley, *Lasers and Their Applications*, Barnes and Noble, New York, p. 39 (1971).
15. J. Gilbert, J. L. Lachambre, F. Rheault, and R. Fortin, *Canad. J. Phys.* **50**, 2523 (1972).
16. A. E. Siegman, *An Introduction to Lasers and Masers*, McGraw–Hill, New York, p. 415 (1971).
17. E. E. Stark Jr., *Appl. Phys. Lett.* **23**, 335 (1973).
18. D. B. Rensch, *Appl. Optics* **13**, 2546 (1974).
19. A. R. Davies, K. Smith, and R. M. Thomson, *Comp. Phys. Commun.* **10**, 117 (1975).
20. E. R. Fisher, private communication.
21. R. L. Taylor and S. Bitterman, *Rev. Mod. Phys.* **41**, 26 (1969).
22. B. F. Gordietz *et al.*, *IEEE J.* **QE-4**, 796 (1968).
23. B. A. Lengyel, *Lasers*, John Wiley, New York, p. 24 (1971).
24. R. J. Harrach and T. H. Einwohner, *Four-Temperature Kinetic Model for a CO_2 Laser Amplifier*, Lawrence Livermore Laboratory, University of California, Livermore UCRL-51399, May 1973.

25. C. A. Fenstermacher, M. I. Nutter, W. T. Leland, and K. Boyer, *Appl. Phys. Lett.* **20**, 56 (1972).
26. J. Tulip, *IEEE J. Quantum Electron.* **6**, 206 (1970).
27. D. H. Douglas-Hamilton, Avco-Everett, unpublished report (1974).
28. A. R. Davies, K. Smith, and R. M. Thomson, *J. Appl. Phys.* **47**, 2037 (1976).
29. W. L. Nighan and J. R. Bennett, *Appl. Phys. Lett.* **14**, 240 (1969).
30. O. P. Judd, *J. Appl. Phys.* **45**, 4572 (1974).
31. J. Reid, E. A. Ballik, and B. K. Garside, *Opt. Commun.* **12**, 354 (1974).
32. E. A. Ballik, B. K. Garside, J. Reid, and T. Tricker, *J. Appl. Phys.* **46**, 1322 (1975).
33. J. M. Hoffman, F. W. Bingham, and J. B. Moreno, *J. Appl. Phys.* **45**, 1798 (1974).
34. J. Reid and K. Siemsen, *Appl. Phys. Lett.* **29**, 250 (1976).
35. R. M. Thomson and H. M. Lamberton, *J. Phys. D.*
36. J. W. Bond, K. M. Watson, and J. A. Welch, *Atomic Theory of Gas Dynamics*, Addison–Wesley, Reading, Massachusetts, p. 241 (1965).
37. W. L. Nighan, *Phys. Rev. A* **2**, 1989 (1970).
38. W. F. Bailey, in: 25th Gaseous Electronics Conference, London–Ontario, Paper DD5 (1972).
39. H. S. W. Massey, *Electronic and Ionic Impact Phenomena*, Vol. 3, Oxford Press, New York, p. 1302 (1971).
40. T. Holstein, *Phys. Rev.* **70**, 367 (1946).
41. B. Sherman, *J. Math. Anal. Appl.*, **1**, 342 (1960).
42. W. H. Long, Jr., W. F. Bailey, and A. Garscadden, *Phys. Rev.* **A13**, 471 (1976).
43. S. Chapman and T. C. Cowling, *The Mathematical Theory of Non-Uniform Gases*, 3rd ed., Cambridge University Press, New York, p. 386 (1970).
44. L. S. Frost and A. V. Phelps, *Phys. Rev.* **127**, 1621 (1962).
45. E. R. Fisher, A. Lightman and R. Marriott, *Modeling of CO/N_2 Molecular Laser Systems*, Wayne State University, ARPA Order No. 675, Amendment 11 (1973).
46. C. J. Elliott, O. P. Judd, A. M. Lockett, and S. D. Rockwood, Los Alamos Report La-5562-MS, April 1974.
47. R. M. Thomson, Kenneth Smith, and A. R. Davies, *Comput. Phys. Commun.* **11**, 369 (1976).
48. S. D. Rockwood, *Phys. Rev.* **A8**, 2348 (1973).
49. L. J. Kieffer, Joint Institute for Laboratory Astrophysics Information Centre, Report 13, September 1973.
50. A. R. Hochstim (Ed.), *Kinetic Processes in Gases and Plasmas*, Academic Press, New York, p. 178 (1969).
51. P. K. Cheo and R. L. Abrams, *Appl. Phys. Lett.* **14**, 47 (1969); R. L. Abrams and P. K. Cheo, *Appl. Phys. Lett.* **15**, 177 (1969).
52. E. A. Ballik, B. K. Garside, and J. Reid, *Appl. Phys. Lett.* **26**, 380 (1975).
53. G. T. Schappert, *Appl. Phys. Lett.* **23**, 319 (1973).
54. F. A. Hopf and C. K. Rhodes, *Phys. Rev.* **A8**, 912 (1973).
55. R. R. Jacobs, K. J. Pettipiece, and S. J. Thomas, *Appl. Phys. Lett.* **24**, 375 (1974).
56. R. J. Harrach, *IEEE J. Quantum Electron.* **11**, 349 (1975).
57. L. M. Frantz and J. S. Nodvik, *J. Appl. Phys.* **34**, 2346 (1963).
58. E. Armandillo and I. J. Spalding, *J. Phys. D.* **8**, 2123 (1975).
59. G. T. Shappert and M. J. Herbert, *Appl. Phys. Lett.* **26**, 314 (1975).
60. B. J. Feldman, *IEEE J. Quantum Electron.* **9**, 1070 (1973).
61. R. H. Pantell and H. E. Puthoff, *Fundamentals of Quantum Electronics*, John Wiley & Sons, New York, Chapter 1 and Appendix (1969).
62. M. Sargent III, M. O. Scully, and W. E. Lamb, Jr., *Laser Physics*, Addison–Wesley, Reading, Massachusetts, Chapters 2, 8, and 13 (1974).

63. S. A. Roberts and K. Smith, *Comput. Phys. Commun.* **12**, 323 (1976).
64. J. R. Reitz and F. J. Milford, *Foundations of Electromagnetic Theory*, Addison–Wesley, Reading, Massachusetts, Chapter 2 (1966).
65. E. U. Condon and H. Obishaw (Eds.), *Handbook of Physics*, 2nd ed., McGraw–Hill, New York, Section 4–189 (1967).
66. J. O. Hirschfelder, C. F. Curtiss, and R. E. Bird, *Molecular Theory of Gases and Liquids*, John Wiley, New York, p. 444 (1954).
67. S. Chapman and F. W. Rootson, *Philos. Mag.* **33**, 248 (1917).
68. R. M. Thomson, Centre for Computer Studies, University of Leeds, Report No. 103, August 1977.
69. A. G. Engelhardt, A. V. Phelps, and C. G. Risk, *Phys. Rev. A* **135**, 1566 (1964); R. D. Hake and A. V. Phelps, *Phys. Rev. A* **158**, 70 (1967).
70. R. A. Haas, *Phys. Rev. A* **8**, 1017 (1973).
71. W. J. Wiegand and W. L. Nighan, *Appl. Phys. Lett.* **22**, 583 (1973).
72. W. L. Nighan and W. J. Wiegand, *Phys. Rev. A* **10**, 922 (1974).
73. A. L. S. Smith *et al.*, *Sealed CO_2 Lasers*, School of Physical Sciences, University of St. Andrews, CVD Research Project RU 16/3, Annual Report 1975.
74. J. F. Prince and A. Garscadden, *Appl. Phys. Lett.* **27**, p. 13 (1975).
75. P. Bletzinger, D. A. LaBorde, W. F. Bailey, W. Y. Long, P. D. Tanner, and A. Garscadden, *IEEE J. Quantum Electron.* **11**, 317 (1975).
76. D. S. Stark, P. H. Cross, and H. Foster, *IEEE J. Quantum Electron.* **11**, 774 (1975).
77. A. L. S. Smith and P. G. Browne, *J. Phys. D: Appl. Phys.* **7**, 1652 (1974).
78. F. E. Niles, *J. Chem. Phys.* **52**, 408 (1970).
79. S. C. Browne, *Basic Data of Plasma Physics, 1966* 2nd ed. M. I. T. Press, Cambridge, Massachusetts (1967), and references therein; see also Reference 46.
80. D. Rapp and D. D. Briglia, *J. Chem. Phys.* **43**, 1480 (1965).
81. E. W. McDaniel, V. Cermak, A. Dalgarno, E. G. Ferguson, and L. Friedman, *Ion–Molecule Reactions*, John Wiley, New York (1970).
82. F. Stuhl and J. J. Niki, *Chem. Phys.* **55**, 3943 (1971).
83. E. Ratajczak and A. F. Trotman-Dickenson, *Supplementary Tables of Bimolecular Gas Reactions*, University of Wales Institute of Science and Technology (1969).
84. D. L. Baulch, D. D. Drysdale, D. G. Horne, and A. R. Lloyd, *Evaluated Kinetic Data for High Temperature Reactions*, Vols. I and II, Butterworth, Reading, Massachusetts (1972).
85. G. S. Bahn, *Reaction Rate Compilation for the H–O–N System*, Gordon and Breach, New York (1968).
86. T. J. Keneshea, Air Force Cambridge Research Laboratories AFCRL-67-022 (1967); AFCRL-62-828 (1962).
87. C. E. Treanor, *Math. of Comp.* **20**, 39 (1966).
88. H. Nosrati, *Math. of Comp.* **27**, 267 (1973).
89. S. A. Roberts, K. Smith, and R. M. Thomson, Centre for Computer Studies, University of Leeds, Report No. 92, November 1976.
90. Nottingham Algorithms Group, ICL 1900 System, N.A.G. Library Manual, Document No. 504 (1973).
91. L. R. Peterson, *Phys. Rev.* **187**, 105 (1969).
92. L. R. Peterson and A. E. S. Green, *J. Phys. B.* (*Proc. Phys. Soc.*), **1**, 1131 (1968); see also L. R. Peterson, T. Sawada, J. N. Bass, and A. E. S. Green, *Comput. Phys. Commun.* **5**, 239–262 (1973) for a description of a computer code used to calculate electron energy loss in the CSDA approximation.
93. R. W. Schunk and R. B. Hays, *Planet Space Sci.* **19**, 113 (1971).

94. A. Dalgarno and G. Lejeune, *Planet Space Sci.* **19**, 1653 (1971).
95. D. H. Douglas-Hamilton and S. A. Mani, *Appl. Phys. Lett.* **23**, 508 (1973).
96. D. H. Douglas-Hamilton, S. A. Mani, and R. M. Patrick, *Investigation of the Production of High Density Uniform Plasmas.* Avco-Everett Research Laboratory Inc., Research Report 399, January 1974; *J. Appl. Phys.* **45**, 4406 (1974).
97. P. A. Sturrock, *Phys. Rev.* **112**, 1488 (1958).
98. E. T. Whittaker and G. N. Watson, *A Course of Modern Analysis*, Cambridge University Press, New York, p. 172 (1962).
99. E. C. Titchmash, *Introduction to the Theory of Fourier Integrals*, Clarendon Press, Oxford, p. 11*ff* (1937).
100. H. Margenau and G. N. Murphy, *The Mathematics of Physics and Chemistry*, Van Nostrand, New York, 2nd ed., p. 90 (1967).
101. P. E. Dyer and D. J. James, *J. Appl. Phys.*, **46**, 1679 (1975).
102. R. J. Briggs, *Electron Stream Interaction with Plasmas*, M.I.T. Press, Cambridge, Massachusetts, p. 11*ff* (1964).
103. A. L. Bloom, *Gas Lasers*, John Wiley and Sons, New York, pp. 23, 69 (1968).
104. G. D. Smith, *Numerical Solution of Partial Differential Equations*, Oxford University Press, New York, p. 58 (1965).
105. E. C. Du Fort and S. P. Frankel, *Math. Tables and Aids to Comput.* **7**, 135 (1953).
106. A. T. Jones, Rolls-Royce Advanced Research Laboratory, Report RR(OH)586 (1974).
107. M. Born and E. Wolf, *Principles of Optics*, Pergamon, New York, 5th ed., p. 92 (1975).
108. A. T. Jones, Rolls-Royce Advanced Research Laboratory, Report RR(OH)620 (1975).
109. J. J. Degnan, *Appl. Phys.* **11**, 1 (1976).
110. J. J. Degnan and D. R. Hall, *IEEE Quantum Electron.* **9**, 901 (1973).
111. J. H. McElroy, N. McAvoy, E. H. Johnson, J. J. Degnan, F. E. Goodwin, D. M. Henderson, T. A. Nussmeier, L. S. Stokes, B. J. Peyton, and T. Flatteau, *Proc. IEEE* **65**, 221 (1977).
112. M. W. Zemansky, *Heat and Thermodynamics*, McGraw–Hill, New York, p. 81 (1957).
113. D. J. Day, private communication (1975).
114. A. J. Laderman and S. R. Byron, *J. Appl. Phys.* **42**, 3138 (1971).
115. J. C. AuYeung and H. A. Haus, *Theoretical Prediction of Capillary Tube Amplifier Gain*, R. L. E. Progress Report No. 115, M. I. T. Press, Cambridge, Massachusetts (1974).
116. S. C. Cohen, *IEEE Quantum Electron.* **12**, 237 (1976).
117. B. F. Gordietz, N. N. Sobolov, and L. A. Shelepic, *Sov. Phys. JETP* **26**, 1039 (1968).
118. T. Y. Chang and T. J. Bridges, *Opt. Commun.* **1**, 423 (1970).
119. T. Y. Chang, *IEEE Trans. Microwave Theory and Tech.* **22**, 983 (1974).
120. H. J. A. Bluyssen, R. E. McIntosh, A. F. van Etteger, and P. Wyder, *IEEE J. Quantum Electron.* **11**, 341 (1975).
121. C. H. Townes and A. L. Schawlow, *Microwave Spectroscopy*, McGraw–Hill, New York, p. 78 (1955).
122. J. O. Henningsen and H. G. Jensen, *IEEE Quantum Electron.* **11**, 248 (1975).
123. D. T. Hodges, J. R. Tucker, and T. S. Hartwick, *Infrared Physics*, **16**, 175 (1976).
124. J. R. Tucker, *Conference Digest, International Conference on Submillimeter Waves and Their Applications*, Atlanta, Georgia (*IEEE* Cat. No. 74 CHO 856-5MTT), p. 17 (1974); also *Aerospace Tech. Memo.* 74 (8170)-1.
125. V. I. Krylov, *Approximate Calculation of Integrals* (translated by A. H. Stroud), Macmillan, New York, p. 129 (1962).
126. K. Smith, *Comput. Phys. Commun.* **14**, (1978).
127. S. M. Freund, G. Duxburg, M. Römheld, J. T. Tiedge, and T. Oka, *J. Molec. Spectrosc.* **52**, 38 (1974).
128. F. A. Andersen, B. Bak, and S. Brodersen, *J. Chem. Phys.* **24**, 989 (1956).

129. C. J. Buczek, R. J. Freiberg, and M. L. Skolnick, *J. Appl. Phys.* **42**, 3133 (1971).
130. C. L. Tang and H. Statz, *Proc. IEEE* **51**, 120 (1963).
131. O. Svelto, *Principles of Lasers*, Plenum Press, New York, Eqs. (2.61) and (2.129) (1976).
132. C. J. Buczek, R. J. Freiberg, and M. L. Skolnick, *Proc. IEEE* **61**, 1411 (1973).
133. R. J. Freiberg and C. J. Buczek, *Opt. Commun.* **4**, 139 (1971).
134. C. J. Buczek and R. J. Freiberg, *IEEE J. Quantum Electron.* **8**, 641 (1972).
135. J. L. Lachambre, P. Lavigne, G. Otis, and M. Noel, *IEEE J. Quantum Electron.* **12**, 756 (1976).
136. A. R. Davies, S. A. Roberts, K. Smith, and R. M. Thomson, Centre for Computer Studies, University of Leeds, Report No. 105, November 1977.
137. I. P. Shkarofsky, T. W. Johnston, and M. P. Bachynski, *The Particle Kinetics of Plasmas*, Addison-Wesley, Reading, Massachusetts, p. 283 (1966).
138. W. H. Long, Jr., *Appl. Phys. Lett.* **31**, 391 (1977).

Index